MATHEMATICS

MATHEMATICS

An Informal Approach

SECOND EDITION

Albert B. Bennett, Jr.
University of New Hampshire

Leonard T. Nelson
Portland State University

ALLYN AND BACON INC. Boston London Sydney Toronto

Library of Congress Cataloging in Publication Data

Bennett, Albert B.
    Mathematics, an informal approach.

    Includes bibliographical references and index.
    1. Mathematics—1961–   . I. Nelson, Leonard T.
II. Title.
QA39.2.B475 1985   510   84-18540
ISBN 0-205-08305-6

Printed in the United States of America.

10 9 8 7 6 5 4 3 2 1   88 87 86 85

ACKNOWLEDGEMENTS

Chapter 1

pg 2 (top) Reprinted by courtesy of Hale Observatories. **pg. 3** (top left) Courtesy of Carl Zeiss, Inc., New York. (top right) Reproduced from *Art Forms in Nature,* by Ernst Haeckel (New York: Dover Publications, Inc., 1959). (bottom right) Courtesy of Garry Bennett. Photo by Ron Bergeron. **pg. 4** (bottom right) Photo by Ron Bergeron. **pg. 7** © Lorenz 1970– Reprinted from *Look* Magazine. **pg. 8** (bottom) Courtesy Italian Government Travel Office. **pg. 14** (exercise 16) Reproduced from *An Introduction to Color,* by Ralph M. Evans, © 1948 John Wiley & Sons, Inc. Reprinted by permission of John Wiley & Sons, Inc. **pg. 21** and **pg 25** Reproduced from *Logic,* Nuffield Mathematics Project, © 1972 The Nuffield Mathematics Project. Reprinted by permission of John Wiley & Sons, Inc. **pg. 30** B.C. by permission of Johnny Hart and Field Enterprises, Inc. **pg. 34** (middle) Reprinted from *How To Take a Chance,* by Darrell Huff. Illustrated by Irving Geis., by permission of W.W. Norton & Company, Inc. Copyright © 1959 by W.W. Norton & Company, Inc. **pg. 39** B.C. by permission of Johnny Hart and Field Enterprises, Inc.

Chapter 2

pg. 48 (top) Courtesy of Dr. Jean de Heinzelin.
pg. 49 (bottom right) courtesy of Musée de l'Homme.

pg. 51 Copyright © 1975. Reprinted by permission of *Saturday Review* and Robert D. Ross. **pg. 56** Reprinted by courtesy of Hale Observatories **pg. 65** (top) Courtesy of the Trustees of the British Museum.
pg. 67 (top) Drawing by Mal; © 1973 The New Yorker Magazine, Inc. **pg. 69** (top right) Drawing by Kovarsky; © 1960 The New Yorker Magazine, Inc. **pg. 73** (middle) Photo by Ron Bergeron. **pg. 74** (top) *Chip Trading Activities,* © 1975 Davidson, Galton, Fair. Courtesy of Scott Resources, Inc., P.O. Box 2121, Fort Collins, Colorado. **pg. 75** Reprinted from *Margarita Philosophica* of Gregor Reisch, 1503. **pg. 76** (bottom) Reprinted from *Summa de Arithmetica,* by Luca Pacioli. **pg. 82** Photo Courtesy of Radio Shack, A Division of Tandy Corporation. **pg. 83** (top) Photos by Ron Bergeron. (bottom) Courtesy of M.I.T. News Office. **pg. 89** (bottom) Reprinted from *Saturday Review* Courtesy of James Estes. **pg. 92** Reprinted with permission from Sidney Harris. **pg. 94** Courtesy of Leo D. Geoffrion, Department of Education, University of New Hampshire. **pg. 96** (top right) Courtesy of Leonard Todd. **pg. 100** Courtesy of John Demchuck.

Acknowledgements continue on page 709

Contents

Preface

> "I write so that the learner may always see the inner ground of the things he learns, even so that the source of the invention may appear, and therefore in such a way that the learner may understand everything as if he had invented it by himself."
>
> GOTTFRIED W. LEIBNITZ

TO STUDENTS

If you are among the many people who feel they dislike mathematics, this book may change your attitude. It was written to help you relate mathematical ideas to your life. You will see many illustrations of the ways in which mathematics occurs in nature and in a wide variety of applications. Just as Leibnitz stated in the above quote, we have attempted to arrange the ideas so that everything may be understood as easily as possible.

Calculators and computers are important aids in studying mathematics. As you read this book, keep a calculator handy to work out the computations. There are instructions for programming in Logo and BASIC, and access to a microcomputer will make this material more interesting and enjoyable.

Mathematics is not a spectator sport. To help you become more actively involved there is a wide variety of exercises which are classified under three headings: Applications and Skills; Problem Solving; and Computers. The starred (★)

exercises are either completely or partially answered for your convenience. The exercises which require a computer are marked with the symbol ☞. The exercises contain puzzles, tricks, and other recreational mathematics which we think you will enjoy.

You will discover at least three facts by using this book. First, mathematics is more than a collection of computational skills—it is a way of thinking. Second, mathematics occurs around you and in your life in many surprising and interesting ways. Third, mathematics can be enjoyable. You may also discover, as have many students who used this book, that you like mathematics.

TO THE INSTRUCTOR

This text together with the activities in *Mathematics An Activity Approach* represent the content and style of the introductory mathematics courses we teach. These texts were written because of the growing opinion that inductive thinking should be given a more prominent role in mathematics courses for teachers. The content and methods in *Mathematics An Informal Approach,* 2nd edition, closely parallels courses I and II in the 1983 *Recommendations On The Mathematical Preparation of Teachers* by the Committee on the Undergraduate Program in Mathematics (CUPM) and contains most of the topics in courses III and IV of this report. The recommendations in both CUPM's report and NCTM's 1980 *Agenda For Action,* have influenced the expansion of problem solving and the use of computers throughout the second edition of our text.

SPECIAL FEATURES

Problem Solving   *An Agenda For Action* contains recommendations that "problem solving be the focus of school mathematics in the 1980's." CUPM recommends that "courses for teachers emphasize the central role of problem solving in mathematics." The section on problem solving and its exercise set in Chapter 1 have been rewritten to provide more detail and practice in the use of Polya's four-step problem solving process. The problem solving strategies from this section are used in the problem solving exercises in Chapters 2 through 10.

Computer Applications   The NCTM and CUPM reports recommend that computers be integrated into the core mathematics curriculum for teachers. Elementary programming commands in Logo and BASIC are introduced in Chapter 2. These commands are then used in subsections and exercise sets in Chapters 3 through 10. BASIC is the computer language in Chapters 3, 5, 6, 7, 8, and 10 and Logo is the language for Chapters 4 and 9. The commands which are used for BASIC and Logo are the common ones which are found in all versions of these languages.

Calculators   There are instructions for using a calculator in the sections on numeration, whole numbers, fractions, integers, and decimals. There are calculator exercises for each of these topics.

Mathematical Reasoning   *An Agenda for Action* contains recommendations for the use of deductive and inductive reasoning to draw conclusions. It states that "mathematics curricula and teachers should set as objectives the development of logical processes, concepts, and language." Chapter 1 contains sections on inductive and deductive reasoning. The deductive reasoning includes Venn diagrams, and valid and invalid forms of reasoning.

Pedagogy   "If one reviews the major changes of direction of school mathematics throughout history, it will be found that nearly all of them have been changed in content. . . . It seems quite apparent that content is more easily changes than the manner of presentation."* The 1983 recommendations by CUPM, unlike those of the past two decades, contain many suggestions involving pedagogy. There are frequent references to the use of models for the basic operations, whole numbers, integers, fractions, and geometric shapes and properties. There are recommendations that properties of two dimensional space be illustrated by models such as geoboards and paper folding, and that properties of three dimensional space be illustrated by geometric solids in nature, such as honeycombs and crystals. These recommendations describe the types of pedagogy in *Mathematics An Informal Approach,* 2nd edition.

History   Many of the topics in this book are introduced by historical highlights. As Aristotle has said, "Here and elsewhere we shall not obtain the best insight into things until we actually see them growing from the beginning . . .".

Exercises   "I hear and I forget. I see and I remember. I do and I understand." This ancient Chinese proverb shows that getting students actively involved in learning is not a new idea. The exercise sets in this text are extensive, occupying over 40 percent of the book. They are divided into three parts: Applications and Skills; Problem Solving; and Computer Applications. A star (★) beside an exercise indicates that the question is either completely or partially answered in the back of the text. Many of the computer exercises do not require a computer. Those which do are marked with the symbol ☞.

Approximation and Mental Calculation   The sections on the four basic operations for whole numbers and the sections on fractions, integers, and decimals contain methods of approximation and mental calculation.

Diskettes   There are two diskettes for Apple II Plus, IIe, IIc computers which contain all the computer programs in the text and exercise sets. *BASIC Diskette for Mathematics An Informal Approach* is programmed in Apple-Soft BASIC for Chapters 2, 3, 5, 6, 7, 8, and 10; and *Logo Diskette for Mathematics An Informal Approach* is for Chapters 2, 4, and 9. The Logo diskette is programmed in Apple logo on one side and Terrapin/Krell Logo on the other side.

---

*Lloyd F. Scott, "Increasing Mathematics Learning through Improving Instructional Organization," *Learning & the Nature of Mathematics* (Chicago: Science Research Associates, 1972), pp. 19–32.

ACTIVITY BOOK

> "The laboratory method proposes that the experimental origin of mathematics be fully recognized; that the pupil be led to feel the need of the mathematical tool through some material experiment. . . . Experiment, estimation, approximation, observation, induction, intuition, and common sense are to have honored places in every mathematical classroom in which the laboratory method has sway."*

Mathematics An Activity Approach is an activity book which can be used to supplement the text. It contains 36 activity sets, one for each section of the text. Each activity set is a sequence of inductive activities and experiments which enables the student to build up mathematical ideas through the use of models and the discovery of patterns. The activity sets extend the ideas presented in the corresponding sections of the text. As examples: attribute pieces are used to illustrate properties of sets; operations in various bases are carried out with multibase pieces, chip trading, and the abacus; area formulas are developed from geoboard activities; properties of polyhedra are illustrated by models; operations with integers, fractions, and decimals are carried out with various models; measurements are approximated by devices for indirect measurement; and empirical probabilities are obtained from probability experiments. There is an appendix of 42 material cards for use with the activities. Many of the activity sets are followed by *Just For Fun* enrichment activities.

ORGANIZATION AND FORMAT

The ten chapters of this book, together with the activity book, provide sufficient material for a two semester course. Each chapter contains from three to six sections, and each section is followed by one or more exercise sets. Complete answer guides are available for the activity book as well as the text. The text and activity book can be used together in several ways:

Lecture course with assignments from the exercise sets. The activity book can be used as a supplement to the text.

Combination lecture and lab course in which lectures are followed by a lab class on the related activity set, either during the same class period or in two consecutive class meetings.

Lab course based on the activity sets. The text readings and exercise sets can be assigned as a supplement.

---

*J. W. A. Young, "The Teaching of Mathematics," *The Mathematics Teacher,* 61, No. 3 (March 1968), 287–95.

SEQUENCE OF TOPICS

This edition contains new sections on deductive reasoning (Section 1.2), problem solving (Section 1.3), programming in Logo and BASIC (Section 2.3), solving equations (Section 8.1), and functions and sequences (Section 8.3). If time does not permit teaching the entire section on deductive reasoning then conditional statements and their contrapositives are recommended for minimal coverage. Programming in Logo and BASIC are presented in the same section but either one may be taught separately. Since variables occur throughout the text in formulas, number properties, and computer programs, one possible sequence is to follow Chapter 1 with Section 8.1 on variables and equations, and then return to earlier chapters. The beginning of Section 8.3 contains the topic of functions which is needed for the geometric mappings in Chapter 9.

Here are four possibilities for selections of topics for one-semester courses. Recommendations III and IV contain combinations of topics from number systems, geometry, algebra, probability, and statistics

I. Number Systems

    Chapters: 1 Mathematical Reasoning and Problem Solving
               2 Sets, Numeration, and Computers
               3 Whole Numbers and Their Operations
               6 Fractions and Integers
               7 Decimals: Rational and Irrational Numbers

II. Geometry and Algebra

    Chapters: 1 Mathematical Reasoning and Problem Solving
               2 Sets, Numeration, and Computers (Section 2.3)
               4 Geometric Figures
               5 Measurement
               8 Algebra and Functions
               9 Motions in Geometry

III. Chapters: 1 Mathematical Reasoning and Problem Solving
               2 Sets, Numeration, and Computers (Section 2.3)
               3 Whole Numbers and Their Operations
                   (Sections 3.1, 3.2, 3.3, 3.4)
               4 Geometric Figures
               5 Measurement (Sections 5.1 and 5.2)
               6 Algebra and Functions (Section 8.1)
             10 Probability and Statistics

IV. Chapters: 1 Mathematical Reasoning and Problem Solving
               2 Sets, Numeration, and Computers (Section 2.3)
               3 Whole Numbers and Their Operations
                   (Sections 3.1, 3.2, 3.3, and 3.4)
               4 Geometric Figures (Sections 4.1 and 4.2)

MATHEMATICS

Mathematical Reasoning and Problem Solving

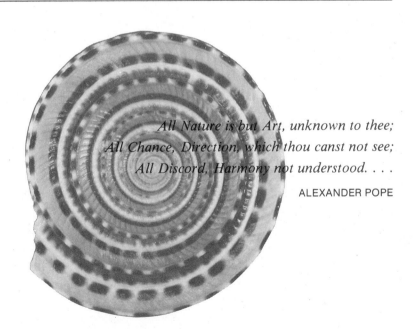

All Nature is but Art, unknown to thee;
All Chance, Direction, which thou canst not see;
All Discord, Harmony not understood. . . .

ALEXANDER POPE

The great spiral galaxy in Andromeda

When Plato was asked what God does, he is said to have answered, "God eternally geometrizes."

1.1 PATTERNS AND INDUCTIVE REASONING

PATTERNS IN NATURE

From the smallest microscopic plants and organisms to the great galaxies of the universe, Nature has shown no lack of imagination in creating mathematical patterns. The variety of shapes shown on the top left of the following page are sea plants, so small that a thimbleful contains millions of them. Which shapes have mathematical names you are familiar with?

Diatoms, minute unicellular algae

Microscopic sea urchin

The spiral is a common pattern in nature. It is found in spider webs, seashells, plants, animals, weather patterns, and the shapes of galaxies.

Web of an Epurid spider
*Courtesy of The American Museum
of Natural History.*

Sundial seashell

The frequent occurrence of spirals in living things can be explained by different growth rates. Living forms curl so that the faster growing or longer surface lies outside and the slower growing or shorter surface lies inside. Consider, for example, the chambered nautilus, shown on the next page. As it grows it lives in successively larger compartments. The outer surfaces of these compartments have faster growing rates than the inner surfaces. Similarly, in the ram's horns the leading edges grow more than the trailing edges, and so the horns curl back.

Chambered nautilus
Courtesy of the American Museum of Natural History.

Dall sheep
Courtesy of the American Museum of Natural History.

FIBONACCI NUMBERS

Patterns occur in plants and trees in a variety of ways. Many of these patterns are related to a famous sequence of numbers called *Fibonacci numbers*. After the first two numbers of this sequence, which are 1 and 1, each successive number may be obtained by adding the two previous numbers.

$$1 \quad 1 \quad 2 \quad 3 \quad 5 \quad 8 \quad 13 \quad 21 \quad 34 \quad 55 \quad \ldots$$

The center of a daisy has two intersecting sets of spirals, one turning clockwise and one turning counterclockwise. The number of spirals in each set is a Fibonacci number. Also, the number of petals will often be a Fibonacci number. The daisy in this photo has 21 petals.

If you select a particular bud or leaf from a branch and then move in a spiral motion around the branch until reaching the bud or leaf directly above your selection, the number of leaves will be a Fibonacci number. This Fibonacci number is governed by the type of plant. For the cherry tree this number is 5. After leaf 0, there are five leaves, up to and including leaf 5, which is directly above leaf 0. Notice that the beginning leaf is not counted. The number of complete turns in passing around the stem from one leaf to a leaf that is directly above is also a Fibonacci number. Following the dotted path around this stem, you will see that the number of complete turns for the cherry tree is two. (Further examples of this type can be found on page 339, Exercise 3.)

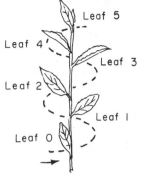

Cherry tree branch

Fibonacci numbers were discovered by the Italian mathematician Leonardo Fibonacci (ca. 1175–1250). He used these numbers in studying the birth patterns of rabbits. Suppose that in the first month a pair of baby rabbits are too young to produce more rabbits, but in the second month they become adult rabbits and adult rabbits produce a pair of baby rabbits every month. Each new pair of baby rabbits and adult rabbits will follow the same pattern. Let's look at the numbers of pairs of rabbits for each of the first few months. In the first month there is 1 pair of baby rabbits. In the second month the baby rabbits mature and there is 1 pair of adult rabbits. In the third month there are 2 pairs of rabbits, because the adult rabbits produce a pair of baby rabbits. See if you can account for the 3 pairs of rabbits during the fourth month and the 5 pairs during the fifth month. The sequence formed by the number of pairs of rabbits for each month is the Fibonacci sequence.

	Month
	1st
	2nd
	3rd
	4th
	5th

The realization that Fibonacci numbers could be applied to the science of plants and trees occurred several hundred years after they had been developed for studying the birth patterns of rabbits.

NUMBER PATTERNS

Number patterns have fascinated people since the beginning of recorded history. One of the earliest patterns was the distinction between even numbers (0, 2, 4, 6, 8, . . .) and odd numbers (1, 3, 5, 7, 9, . . .). The game of "Even and Odd" has been played for generations. To play this game, one person picks up some stones and the other person guesses "odd" or "even." If the guess is correct, that person wins.

Pascal's Triangle   One of the most familiar number patterns is Pascal's triangle. It has been of interest to mathematicians for hundreds of years, appearing in China as early as 1303. This triangle is named after the French mathematician Blaise Pascal (1623–1662) who wrote a book on some of its uses. There are many patterns in the rows and diagonals. One

pattern is used to obtain each row of the triangle from the previous row. In the sixth row, for example, each of the numbers 5, 10, 10, and 5 can be obtained by adding the two numbers in the row above it. What numbers are in the seventh row of Pascal's triangle?

Arithmetic Sequence   Sequences of numbers are often generated by patterns. In the sequences shown to the right, each number is obtained from the previous number in the sequence by adding the

7, 11, 15, 19, 23, . . .
10, 20, 30, 40, 50, . . .
1, 3, 5, 7, 9, . . .

same number throughout. This number is called the *common difference*. Such a sequence is called an *arithmetic sequence* or *arithmetic progression*. The first arithmetic sequence shown here has a common difference of 4. The common differences for the second and third sequences are 10 and 2.

Geometric Sequence   In this type of sequence each number is obtained by multiplying the previous number by some common number. This number is called the *common ratio*, and the resulting sequence

3, 6, 12, 24, 46, . . .
1, 5, 25, 125, 625, . . .
2, 6, 18, 54, 162, . . .

is called a *geometric sequence*. The common ratio in the first sequence shown here is 2, in the second sequence it is 5, and in the third sequence it is 3.

Triangular and Square Numbers   The following sequence of numbers is neither arithmetic nor geometric. These numbers are called *triangular numbers* because of the arrangement of dots that is associated with each number. What is the next triangular number?

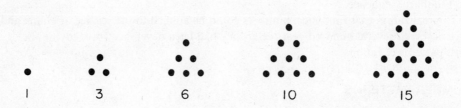

The numbers of dots in square arrays are called *square numbers* or *perfect squares*. The first few of these numbers are shown on the right.

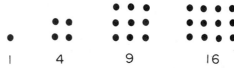

1   4   9   16

There are other types of numbers that receive their names from the numbers of dots in geometric figures (page 11, Exercise 8). Such numbers are called *polygonal* or *geometric numbers* and represent one kind of link between geometry and arithmetic.

INDUCTIVE REASONING

The process of forming conclusions on the basis of observations or experiments is called *inductive reasoning*. Scientists use inductive reasoning when they perform experiments to learn about the laws of nature. Isaac Newton (1642–1727), for example, conducted

experiments (but probably not with apples) involving gravity and then used inductive reasoning to arrive at general conclusions. His results are known today as Newton's laws. Mathematicians and scientists aren't the only ones who use inductive reasoning; everybody uses this type of reasoning. When our conclusions are based on too few examples, we hear the phrase, "don't jump to conclusions."

Let's look at an example of inductive reasoning. Each of the following sums of three consecutive whole numbers is divisible by 3:*

$$4 + 5 + 6 = 15 \qquad 2 + 3 + 4 = 9 \qquad 7 + 8 + 9 = 24$$

If we conclude, on the basis of these examples, that "the sum of any three consecutive whole numbers is divisible by 3," we are using inductive reasoning. Inductive reasoning is the process of making an "informed guess," based on examples.

Although inductive reasoning is important in mathematics, it sometimes leads to incorrect results. Consider, for example, the number of regions that can be obtained in a circle by connecting points on the circumference of the

*The numbers 0, 1, 2, 3, 4, and so on, are called *whole numbers*.

circle. Connecting 2 points gives 2 regions; connecting 3 points gives 4 regions; and so on. Each time a new point on the circle is used, the number of regions appears to double.

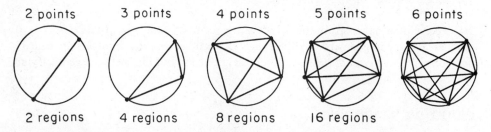

| 2 points | 3 points | 4 points | 5 points | 6 points |
| 2 regions | 4 regions | 8 regions | 16 regions | |

The numbers of regions in the above circles are the beginning of the geometric sequence 2, 4, 8, 16, . . . , and it is tempting to conclude that 6 points will produce 32 regions. However, no matter how the 6 points are located on the circle, there will not be more than 31 regions.

In spite of the uncertainty of inductive reasoning, its contributions to mathematics and science have been enormous. Galileo Galilei (1564–1643), Italian astronomer and physicist, made a discovery concerning swinging pendulums, and it became one of the more famous applications of inductive reasoning. While he was attending a church service, his mind was distracted by a large bronze lamp that oscillated back and forth. Using his pulse to keep time he discovered that the time required for the lamp to complete one swing through an arc was the same, whether the arc was small or large!

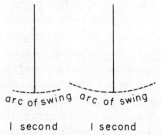

arc of swing   arc of swing

I second   I second

Counterexample Conclusions that are based on inductive reasoning are sometimes later shown to be false. An example which shows that a statement is false is called a *counterexample*. One famous counterexample by Galileo disproved a statement that had been accepted as true for 2000 years. Aristotle (384–322 B.C.), ancient Greek philosopher and scientist, had reasoned that heavy objects fall faster than lighter ones. Galileo demonstrated this to be false by dropping two pieces of metal from the Leaning Tower of Pisa. In spite of the fact that one was 10 times heavier than the other, both hit the ground at the same moment.

Leaning Tower of Pisa, Pisa, Italy

If you have a general statement, test it to see if it is true for a few special cases. You may be able to find a counterexample to show that the statement is not true. The statements in the following two examples are both false. Obtain counterexamples by finding whole numbers for which these statements are false.

Example 1   The sum of any two whole numbers is divisible by 2. (False)

This statement is true for the following pairs of numbers:

$$3 + 7 = 10 \qquad 4 + 8 = 12 \qquad 5 + 9 = 14$$

However, it is not true for 7 and 4, since $7 + 4 = 11$ and 11 is not divisible by 2. This counterexample shows that the statement in Example 1 is not true for all pairs of whole numbers.

Example 2   The sum of any four consecutive whole numbers is divisible by 3. (False)

This statement is true for the following sums:

$$3 + 4 + 5 + 6 = 18 \qquad 6 + 7 + 8 + 9 = 30 \qquad 0 + 1 + 2 + 3 = 6$$

However, $1 + 2 + 3 + 4 = 10$ is not divisible by 3. This counterexample shows that the statement in Example 2 is false.

SUPPLEMENT (Activity Book)

Activity Set 1.1 Geometric Number Patterns (Geometric patterns on arrays of dots and their corresponding number patterns)
Just for Fun: Fibonacci Numbers in Nature

Exercise Set 1.1

1. There are many patterns and number relationships on the calendar that can be easily discovered. Here are a few.

   ★   a. The sum of the three circled dates on this calendar is 45. For any sum of three consecutive numbers (from the rows), there is a quick method for determining the numbers. Explain how this can be done. Try your method to find the three consecutive numbers whose sum is 54.

<table>
<tr><td colspan="7">Nov. 1985</td></tr>
<tr><td>Sun</td><td>Mon</td><td>Tue</td><td>Wed</td><td>Thu</td><td>Fri</td><td>Sat</td></tr>
<tr><td></td><td></td><td></td><td></td><td></td><td>1</td><td>2</td></tr>
<tr><td>3</td><td>4</td><td>5</td><td>6</td><td>7</td><td>8</td><td>9</td></tr>
<tr><td>10</td><td>11</td><td>12</td><td>13</td><td>(14)</td><td>(15)</td><td>(16)</td></tr>
<tr><td>17</td><td>18</td><td>19</td><td>20</td><td>21</td><td>22</td><td>23</td></tr>
<tr><td>24</td><td>25</td><td>26</td><td>27</td><td>28</td><td>29</td><td>30</td></tr>
</table>

   b. If you are told the sum of any three adjacent dates from a column, it is possible to determine the three numbers. Explain how this can be done and use your method to find the numbers whose sum is 48.

   ★   c. The sum of the 3 by 3 array of numbers outlined on the calendar is 99. There is a shortcut method for using this sum to find the 3 by 3 array of numbers.

Explain how this can be done. Try your method to find the 3 by 3 array whose sum is 198.

★ 2. Here are the first few Fibonacci numbers: 1, 1, 2, 3, 5, 8, 13, 21, 34, 55. Compute the sums shown here and compare the answers with the Fibonacci numbers. Find a pattern and explain how this pattern can be used to find the sums of consecutive Fibonacci numbers.

$1 + 1 + 2 =$
$1 + 1 + 2 + 3 =$
$1 + 1 + 2 + 3 + 5 =$
$1 + 1 + 2 + 3 + 5 + 8 =$
$1 + 1 + 2 + 3 + 5 + 8 + 13 =$
$1 + 1 + 2 + 3 + 5 + 8 + 13 + 21 =$

3. The products of 1089 and the first few digits produce some interesting number patterns. Describe one of these patterns. Will this pattern continue if 1089 is multiplied by 5, 6, 7, 8, and 9?

$1 \times 1089 = 1089$
$2 \times 1089 = 2178$
$3 \times 1089 = 3267$
$4 \times 1089 = 4356$
$5 \times 1089 =$

4. There are many patterns in Pascal's triangle. Add the first few numbers in the 2nd diagonal, starting from the top. This sum will be another number from the triangle. Will this be true for the sums of the first few numbers in the other diagonals?

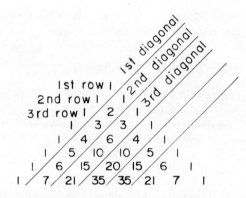

★ 5. Compute the sums of the numbers in the first few rows of Pascal's triangle.

★   a. What kind of sequence (arithmetic or geometric) do these sums form?

    b. What will be the sum of the numbers in the 12th row of this triangle?

6. In the familiar song "The Twelve Days of Christmas," the number of gifts that were received each day are triangle numbers. On the 1st day there was one gift, on the 2nd day there were three gifts, on the 3rd day, six gifts, etc., until the 12th day of Christmas. How many gifts were received on the 12th day? What is the total number of gifts received for all 12 days?

7. Identify each of the following sequences as being arithmetic or geometric. State a rule for obtaining each number from the preceding number. Write the next three numbers in each sequence.

★   a. 4.5, 9, 13.5, 18, . . .
    b. 15, 30, 60, 120, . . .
★   c. 24, 20, 16, 12, . . .
    d. 729, 243, 81, 27, . . .

8. The Greeks (500–200 B.C.) were deeply interested in numbers associated with patterns of dots in the shapes of geometric figures. Write the next three numbers in each sequence in parts a, b, and c.

a. Triangular numbers

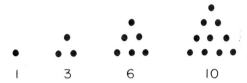

b. Square numbers

★   c. Pentagonal numbers

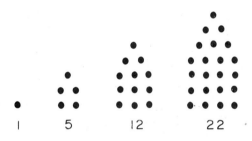

9. Suppose the interior of a circle is to be partitioned into the maximum number of regions by line segments. One line will divide it into 2 regions; 2 lines will divide it into 4 regions; 3 lines will produce 7 regions; and for 4 lines there are 11 regions.

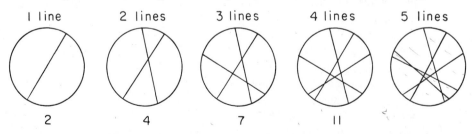

★   a. Find a pattern for the numbers 2, 4, 7, and 11, and use inductive reasoning to predict the maximum number of regions for 5 lines. Check your conclusion by counting the regions.

b. Use inductive reasoning to predict the maximum number of regions for 10 lines.

10. Find a pattern in each of the following sets of equations and use inductive reasoning to predict the next equation. Evaluate both sides of your equation.

★   a. $1^2 + 2^2 + 2^2 = 3^2$ b. $1^3 + 2^3 = 3^2$
       $2^2 + 3^2 + 6^2 = 7^2$ $1^3 + 2^3 + 3^3 = 6^2$
       $3^2 + 4^2 + 12^2 = 13^2$ $1^3 + 2^3 + 3^3 + 4^3 = 10^2$

    c. $1 + 2 = 3$
       $4 + 5 + 6 = 7 + 8$
       $9 + 10 + 11 + 12 = 13 + 14 + 15$

★ 11. If we begin with the number 6, then double it to get 12, and then place the 12 and 6 side by side, the result is 126. This number is divisible by 7. Try this procedure for some other numbers. Find a counterexample which shows that the result is not always divisible by 7.

★ 12. Using whole numbers there is just one combination of two numbers whose sum is 2 and one combination whose sum is 3. For the numbers 4 and 5 there are two different ways of writing the sum using two numbers. Predict the number of different sums for the numbers from 6 to 10 and then show that your conjecture is either true or false. In general, how many sums of different pairs of numbers are there for a given number?

$2 = 1 + 1$
$3 = 1 + 2$
$4 = 1 + 3 = 2 + 2$
$5 = 1 + 4 = 2 + 3$
$6 =$
$7 =$
$8 =$
$9 =$
$10 =$

13. Find a counterexample for each of the following statements:

    a. The product of any two whole numbers is divisible by 2.

    b. Every whole number greater than 5 is the sum of either two or three consecutive whole numbers. For example, 11 is equal to $5 + 6$, and 18 is equal to $5 + 6 + 7$.

★ 14. After the first term the top sequence shown here is a geometric sequence. Write the next two numbers in this sequence. Add 4 to each number in the top sequence and divide the results by 10 to complete the lower sequence.

0   3   6   12   24   __   __
.4   .7 __ __ __ __ __

★   a. This famous sequence of numbers is known as Bode's law and gives an amazingly close approximation of the distances from the first seven planets to the sun in astronomical units. (An astronomical unit of 1 is the earth's distance from the

Planet	Distance from Sun in Astronomical Units	Bode's Law
Mercury	0.4	0.4
Venus	0.7	0.7
Earth	1.0	1.0
Mars	1.52	1.6

sun.) Using this sequence of numbers and inductive reasoning, astronomers predicted that there would be a planet between Mars and Jupiter. What was the predicted distance of this planet from the sun? (Asteroids were eventually found between Mars and Jupiter, the biggest of which is Ceres, about 800 kilometers in diameter.)

b. In the 1770s when Bode's law was discovered, only the first five planets in the table had been discovered. Using Bode's law astronomers found Uranus. What should its distance have been to satisfy Bode's law?

Ceres	2.77	
Jupiter	5.2	5.2
Saturn	9.5	10.0
Uranus	19.2	
Neptune	30.1	38.8
Pluto	39.5	77.2

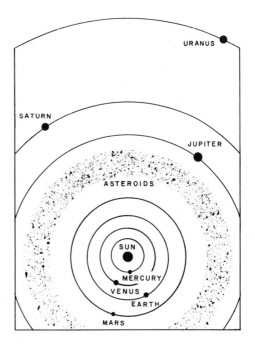

15. *Calendar Trick:* Here is a trick you can have some fun with. Give a friend a calendar and ask him or her to secretly circle one number in each row and add these numbers. Without seeing the results, you then ask how many Sundays were circled. For the circled dates shown here the answer is 2. Then ask how many Mondays were circled, etc., repeating the question for each day of the week. After receiving this information you will be able to quickly tell your friend the sum of the circled numbers. The following questions will help you understand how this trick works:

★   a. What would be the sum if each of the five dates under Wednesday were circled? (This number is the key to the trick.)

★   b. What is the sum of any three dates under Wednesday and one date under both Tuesday and Thursday? (Remember, select only one date from each row.) Does it matter which of the three Wednesday dates are chosen?

       c. What is the sum of any three dates under Wednesday and one each under the Thursday and Friday columns?

       d. Explain how this trick works.

16. *Fraser Spiral:* The background in this photo produces an illusion called the Fraser spiral. Can you explain what is wrong with this "spiral"? (*Hint:* Try tracing it.)

The only complete safeguard against reasoning ill, is the habit of reasoning well; familiarity with the principles of correct reasoning and practice in applying those principles.   JOHN STUART MILL

1.2
LOGIC AND DEDUCTIVE REASONING

Lewis Carroll, well-known author of *Alice's Adventures in Wonderland,* also wrote books on logic. At the beginning of his *Symbolic Logic** he states that logic will give you

> The power to detect fallacies, and to tear to pieces flimsy illogical arguments which you will so continually encounter in books, in newspapers, in speeches, and even in sermons, and which so easily delude those who have never taken the trouble to master this fascinating Art. Try it. That is all I ask of you!

As Lewis Carroll noted, we do not have to go far for examples of illogical conclusions. Consider the following statement:

If you don't eat your spinach, then you won't get dessert.

We all know what this statement means if the spinach isn't eaten. The person won't get dessert! However, suppose the spinach is eaten. Does this imply that the person will get dessert? No! This conclusion does not follow from the given statements, as we shall see in this section.

DEDUCTIVE REASONING

There are two main types of reasoning: inductive and deductive. In Section 1.1 we obtained conclusions by inductive reasoning. With this type of reasoning, conclusions

*Lewis Carroll, *Symbolic Logic* and *Game of Logic* (New York: Dover Publications, Inc., 1958).

are based on observations. A conclusion from inductive reasoning might be called an "informed guess." Deductive reasoning, on the other hand, is the process of obtaining a *conclusion* from one or more given statements, called *premises*. Here is an example of deductive reasoning in which the conclusion is obtained from two premises.

> ### *Premises*
>
> 1. All whales are mammals.
> 2. All dolphins are whales.
>
> ### *Conclusion*   All dolphins are mammals.

Sometimes more than one conclusion is possible. What other conclusions can be obtained from the three premises in the following example?

> ### *Premises*
>
> 1. There are six children in a family.
> 2. Five of the children have blond hair.
> 3. There are four boys in the family.
>
> ### *Conclusion*   At least three of the boys have blond hair.

If we accept the premises in the preceding examples, then we are forced to accept the conclusions. This is the nature of deductive reasoning, deriving conclusions from given statements.

VENN DIAGRAMS

Circle diagrams like those in the next two examples are a common means of visualizing information and drawing conclusions. They are called *Venn diagrams* after John Venn (1834–1923), one of the pioneers in the early development of logic.

> ### *Premises*
>
> 1. All salamanders are amphibians.
> 2. All animals that develop an amnion are not amphibians.
>
> ### *Conclusion*   Salamanders do not develop an amnion.

Statement 1 is represented by drawing a circle for all salamanders inside the circle for all amphibians. Statement 2 is represented by drawing a circle for all animals with an amnion outside the circle for the amphibians. The two separate circles, for salamanders and for animals with an amnion, show that the conclusion follows from statements 1 and 2.

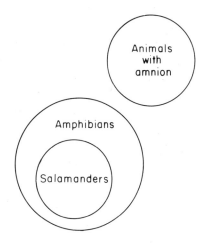

When a conclusion follows from the given information, it is called *valid*; if the conclusion does not follow, it is called *invalid*. Venn diagrams are used in the next example to show that the conclusion does not follow from the given information. It is possible to represent the information in the premises so that the circles for the Appropriations Committee and the Welfare Committee have nothing in common. So, we are not forced to accept the conclusion and therefore it is invalid.

Premises

1. Some members of the Appropriations Committee are Republicans.
2. Some Republicans are on the Welfare Committee.

Conclusion Some members of the Appropriations Committee are members of the Welfare Committee. (Invalid)

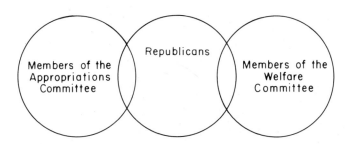

CONDITIONAL STATEMENT

One of the most common types of statements used in reasoning is a statement of the form, if . . . then . . . .

Hypothesis   *Conclusion*
If <u>a number is divisible by 4</u> then <u>it is an even number.</u>

A statement of this form is called a *conditional statement.* A conditional statement has two parts: the "if" part, called the *hypothesis,* and the "then" part, called the *conclusion.* In this example the hypothesis is, "a number is divisible by 4," and the conclusion is, "it is an even number."

Conditional statements can be represented by Venn diagrams. The diagram shown here is for the preceding conditional statement. It shows that the even numbers (large circle) contain the numbers that are divisible by 4 (small circle). This conditional statement can also be worded in the following way:

All numbers that are divisible by 4 are even numbers.

A conditional statement is *true* if whenever the hypothesis is true, the conclusion is true. The conditional statement that is illustrated above is true. It is impossible to have a number that is divisible by 4 (making the hypothesis true) but that is not an even number (making the conclusion false).

Here are two more examples of conditional statements. Both of these examples show that the word "then" does not have to occur, and the second example shows that the hypothesis does not have to occur first in a conditional statement.

Hypothesis   *Conclusion*
If <u>I can keep this up,</u> <u>I will never get old.</u>

Conclusion   *Hypothesis*
<u>Gasoline will not be rationed</u> if <u>it is used sparingly.</u>

Sometimes a statement can be rewritten in if-then form to clarify its meaning. In this form it is easier to see which condition is the hypothesis and which is the conclusion. Here is an example:

You will stay in good condition by jogging every day.

Hypothesis   *Conclusion*
If <u>you jog every day</u> then <u>you will stay in good condition.</u>

CONVERSE OF A STATEMENT

For every conditional statement there is another conditional statement called the *converse,* which can be obtained by interchanging the hypothesis and conclusion.

> *Statement* If the product of two numbers is zero then at least one of the numbers is zero. (True)

> *Converse* If at least one of the numbers is zero then the product of two numbers is zero. (True)

In this example both the statement and its converse are true. However, just because a conditional statement is true, it does not follow that the converse will also be true. One of the most common errors in everyday reasoning is to assume that the converse is true just because a conditional statement is true. Here is the conditional statement from page 18 and its converse.

> *Statement* If a number is divisible by 4, then it is an even number. (True)

> *Converse* If a number is even, then it is divisible by 4. (False)

To show a statement is false we need only to find one counterexample. The diagram on page 18 shows that the converse is false because there are even numbers (2, 6, 10, 14, . . .) that are not divisible by 4.
Here is another example of a true conditional statement whose converse is false.

> *Statement* If a person is on the Boston School Board then the person lives in Boston. (True)

> *Converse* If a person lives in Boston then the person is on the Boston School Board. (False)

Here is a diagram for these two statements. The points in the large circle represent the people who live in Boston and the points in the small circle represent the people on the school board. The converse is not true because there are people who live in Boston who are not on the school board.

CONTRAPOSITIVE OF A STATEMENT

The following two conditional statements have the same meaning, that is, they are different ways of saying the same thing:

Statement 1

Hypothesis   *Conclusion*

If a number is divisible by 6 then it is divisible by 3. (True)

Statement 2

Hypothesis   *Conclusion*

If a number is not divisible by 3 then it is not divisible by 6. (True)

Compare the underlined parts in Statement 1 with the underlined parts in Statement 2. The hypothesis and conclusion in Statement 1 were negated and interchanged to obtain the hypothesis and conclusion in Statement 2. When two conditional statements are related like this, each is called the *contrapositive* of the other.

The following diagram shows that Statements 1 and 2 in this example are different ways of saying the same thing. The numbers that are not divisible by 3 are outside the large circle. Therefore, if a number is not divisible by 3, then it can't be divisible by 6 because all the numbers that are divisible by 6 are inside the small circle.

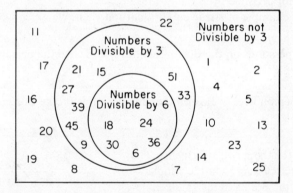

Since a conditional statement and its contrapositive have the same meaning, whenever a conditional statement is true, its contrapositive is also true; if the conditional statement is false, its contrapositive is false.

The diagram in the next example was done by an elementary school student to show that the statement and its contrapositive have the same meaning.* The region outside the large circle represents the times when the dog does not wear a lead.

*Nuffield Mathematics Project, *Logic* (New York: John Wiley & Sons, Inc., 1972).

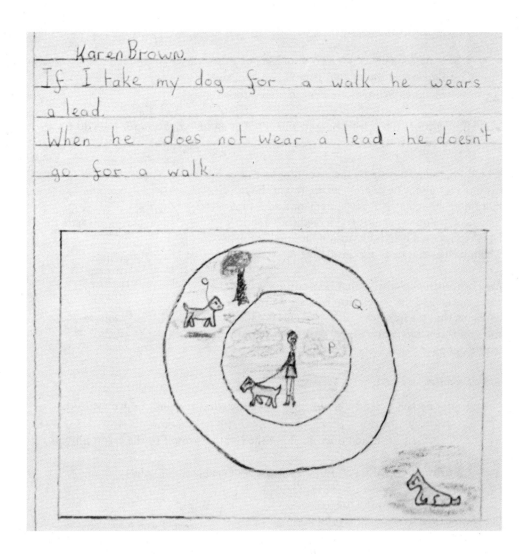

Karen Brown.
If I take my dog for a walk he wears a lead.
When he does not wear a lead he doesn't go for a walk.

VALID REASONING

Here are two common rules or laws of logic for deriving conclusions from given statements.

Law of Detachment   This is a familiar reasoning technique that we use frequently. Whenever we know a conditional statement is true and also that the hypothesis is true, we may conclude that the conclusion is true. This is called the *law of detachment*. Here is an example.

Premises

1. If a person challenges a creditor's report, then the credit bureau will conduct an investigation for that person.
2. Ronald C. Whitney challenged a creditor's report.

Conclusion   The credit bureau will conduct an investigation for Ronald C. Whitney. (Valid)

This diagram shows why the reasoning in this example is valid. The two circles represent the information in statement 1. Since statement 2 says that Ronald C. Whitney challenged a creditor's report, he is represented by a point inside the small circle, which means that he is also inside the large circle.

Law of Contraposition   When a conditional statement is true and the conclusion is false, we may conclude that the hypothesis is false. This is called the *law of contraposition* and is illustrated in the next example.

Premises

1. If a person is denied credit, then the person has the right to protest to the credit bureau.
2. Cynthia F. Johnson does not have the right to protest to the credit bureau.

Conclusion   Cynthia F. Johnson was not denied credit. (Valid)

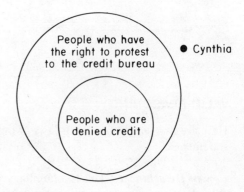

The diagram shows why this reasoning is valid. The two circles represent the information in statement 1. The people who do not have the right to protest to the credit bureau are outside the large circle. According to statement 2, Cynthia F. Johnson is outside the large circle, and therefore she cannot be inside the small circle.

In summary, for any true conditional statement,

If _(hypothesis)_ then _(conclusion)_ .

there are two situations that lead to valid reasoning.

1. When the hypothesis is true, the conclusion is true.

2. When the conclusion is false, the hypothesis is false.

In other words, for a true conditional statement only a true hypothesis can lead to a true conclusion. If the conclusion is false, the hypothesis must also be false.

INVALID REASONING

One common abuse of logic is called *reasoning from the converse*. Just because a conditional statement is true and the conclusion is true, we cannot conclude that the hypothesis is true. Here is an example of this type of invalid reasoning.

Premises

1. If a company fails to have an annual inspection, then its license will be terminated.
2. The Samson Company's license was terminated.

Conclusion   The Samson Company failed to have an annual inspection. (Invalid)

This conclusion is not necessarily true because the converse or statement 1 may not be true. There may be other reasons for the termination of the company's license. For example, the company may not have paid its license fee. The diagram shows that it is possible for the Samson Company to be inside the large circle (satisfying statement 2) but outside the small circle.

We are now prepared to analyze the example from the beginning of this section.

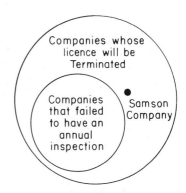

Premises

1. If you don't eat your spinach, then you won't get dessert.
2. You eat your spinach.

Conclusion   You get dessert. (Invalid)

First, let's replace Statement 1 by its contrapositive.

1. If you get dessert, then you eat your spinach.
2. You eat your spinach.

Conclusion   You get dessert. (Invalid)

With statement 1 written in this form, it is obvious that the conclusion is obtained by "reasoning from the converse" and therefore is invalid.

Here is another example of invalid reasoning. Just because a conditional statement is true and the hypothesis is false, it does not necessarily mean that the conclusion is false.

Premises

1. If this year's tests are successful, then the United States will be using laser communications by 1992.
2. This year's tests were not successful.

Conclusion The United States will not be using laser communications by 1992. (Invalid)

Statement 1 in this example tells us what will happen if "this year's tests are successful." However, it does not say what will happen if the tests are not successful. In the following diagram the region outside the small circle (representing "unsuccessful tests"), but inside the large circle, shows that it is possible that the United States may be using laser communications by 1992. Therefore, the conclusion is invalid.

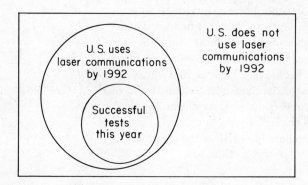

In summary, for any true conditional statement,

If (hypothesis) then (conclusion) ,

there are two situations that lead to invalid reasoning.

1. When the hypothesis is false, it does not follow that the conclusion is false.

2. When the conclusion is true, it does not follow that the hypothesis is true.

Exercise Set 1.2

1. Here is a diagram of two statements by an elementary school student.*
   a. If the first statement is true, is the second statement necessarily true?

★ b. Do these two statements have the same meaning?

If it is windy, Susan wears her hat.
If Susan does not wear her hat, it is not windy.

Susan wears her hat

It is windy

It is not windy

Susan does not wear her hat

2. Write the converse of each statement.
   a. If you take a deduction for your home office, then you must itemize your deductions.

★ b. The knight will land on a black square, if the knight begins on a white square.

   c. If there are peace talks, the prisoners will be set free.

★ d. The Democrats will win the election, if they take California.

*Nuffield Mathematics Project, *Logic* (New York: John Wiley & Sons, Inc., 1972).

3. Rewrite each of the following statements in "if . . . then . . . " form:

   a. Taking a hard line with a bill collector may lead to a lawsuit.

★ b. All employees in Tripak Company must retire by age 65.

   c. A person who files a written application within 31 days of a termination notification will be issued a new policy.

★ d. Every pilot must have a physical examination every 6 months.

   e. People under 16 years of age cannot get a driver's license.

4. Write the contrapositive for each of these conditional statements.

   a. If you subtract $750 for each dependent, then the computer will reject your return.

★ b. The cards should be dealt again, if there is no opening bid.

   c. If not delighted, return the books at the end of the week's free sing-a-long.

5. When a conditional statement and its converse are both true, they can be combined into one statement called a *biconditional statement* by using the words "if and only if." Here is how this can be done by using the statement and its converse from page 19.

       The product of two numbers is zero if and only if at least one of the numbers is zero.

Combine the statement and its converse into a biconditional statement.

   a. If you pay the Durham poll tax, then you are 18 or older. If you are 18 or older, then you pay the Durham poll tax.

★ b. If Smith is guilty, then Jones is innocent. If Jones is innocent, then Smith is guilty.

Write the biconditional statement as a conditional statement and its converse.

   c. Robinson will be hired if and only if she meets the conditions set by the board.

★ d. There will be negotiations if and only if the damaged equipment is repaired.

6. Which of the following examples have valid conclusions? State the law (detachment or contraposition) if the reasoning is valid.

   a. (1) If an illegal move is made, the game pieces should be set up as they were before the move.

     (2) An illegal move was made.

   Conclusion: The game pieces should be set up as they were before the move.

★ b. (1) If the game pieces cannot be set up as they were before the illegal move, then the game is annulled.

     (2) The game pieces were set up as they were before the illegal move.

   Conclusion: The game was not annulled.

c. (1) If a person is healthy, then he or she has about 10 times as much lung tissue as necessary.

(2) Frank has less lung tissue than he needs.

Conclusion: Frank is not a healthy person.

★ d. (1) You may keep the books if you like everything about them.

(2) John kept the books.

Conclusion: John liked everything about the books.

7. Form a valid conclusion for each set of premises. Use all the premises in each example. Name the law (detachment or contraposition) for obtaining each conclusion.

a. (1) If anemia occurs, then something has interfered with the production of red blood cells.

(2) The production of red blood cells in this patient is normal.

Conclusion: _____

★ b. (1) If poison is present in the bone marrow, then production of red blood cells will be slowed down.

(2) This patient has poison in her bone marrow.

Conclusion: _____

c. (1) If there is insufficient vitamin K in the body, there will be a prothrombin deficiency.

(2) Mr. Keene does not have a prothrombin deficiency.

Conclusion: _____

8. A common practice in presenting information is to support a particular point of view with a conditional statement and then to state the contrapositive for emphasis. This situation is similar to what is done in the following quoted passage:

> If government is to improve, we must pay attention to it. If we don't [pay attention to it], we'll get sluggish and unresponsible government, concerned principally with its own self preservation.*

a. Write the first statement in this passage in if . . . then form.

★ b. Write the contrapositive of the statement in part a.

c. Write the second statement in the passage in if . . . then form.

★ d. Compare the statements in parts b and c. Which is the stronger statement?

9. Advertisements are often misleading and tempt people to draw conclusions that are favorable to a certain product. Here are some examples of valid and invalid reasoning that have been adapted from ads. Determine which examples have valid

*Mark Frazier and Jim Lewis, "New Ways to Cut Your Taxes," *Reader's Digest* (March 1979), p. 162

conclusions, based on the given statements, and state the law (detachment or contraposition) if the reasoning is valid.

   a. Great tennis players use Hexrackets.
     Therefore, if you use a Hexracket, you are a great tennis player.

★ b. People who use our aluminum siding are satisfied.
     Therefore, if you don't use our aluminum siding, you won't be satisfied.

   c. If you take Sleepwell, you will have extra energy.
     Therefore, if you don't have extra energy, you are not taking Sleepwell.

10. Use Venn diagrams to show the conclusion in part a is valid and the conclusion in part b is invalid.

   a. (1) John and some other boys were absent today, but all the girls were present.
      (2) All the children who were absent yesterday were absent again today.
     Conclusion: The girls were not absent yesterday. (Valid)

★ b. (1) If people are happy, then they have enough to eat.
      (2) All rich people have enough to eat.
     Conclusion: Some rich people are happy. (Invalid)

11. There is a common type of problem that can be analyzed by recording the given information in a table. For example, using the information from the problem below we can write "No" in three of the boxes in this table. Why? Each row and column of the table should have exactly one "Yes." Continue filling out the table to solve this problem.

	Clerk	Farmer	Chef
Lang	No	No	
Murphy			
Ramsay	No		

   Problem: Determine each person's occupation.

   (1) Lang, Murphy, and Ramsay are a clerk, farmer, and chef.

   (2) Lang is not the clerk or the farmer.

   (3) Ramsay is not the clerk.

12. Use tables (as suggested in Exercise 11) to solve these problems.

   a. Janet Davis, Sally Adams, Collette Eaton, and Jeff Clark have the following occupations: architect, carpenter, diver, and engineer. Determine each person's occupation.

   (1) The first letters of a person's last name and occupation are different.

   (2) Jeff and the engineer go sailing together.

   (3) Janet lives in the same neighborhood as the carpenter and the engineer.

★   b. Dow, Eliot, Finley, Grant, and Hanley have the following occupations: appraiser, broker, cook, painter, and singer. If three of these people are men, determine each person's sex and occupation.

     (1) The broker and the appraiser attended a father-and-son banquet.

     (2) The singer, the appraiser, and Grant all belong to the same club.

     (3) Dow and Hanley are married to two waiters.

     (4) The singer told Finley that he liked science fiction.

     (5) The cook owes Hanley $25.

13. *Recreational Logic:* Lewis Carroll popularized logic by writing comically worded statements and conclusions. Use Venn diagrams to determine which of the following conclusions are valid. (*Hint:* The statement "No professors are ignorant" can be written as "All professors are not ignorant.")

    a. No professors are ignorant. All ignorant people are vain.
      Conclusion: No professors are vain.

★   b. All who are anxious to learn, work hard. Some of these boys work hard.
      Conclusion: Some of these boys are anxious to learn.

    c. Babies are illogical. Nobody is despised who can manage a crocodile. Illogical persons are despised.
      Conclusion: Babies cannot manage crocodiles.

14. *Logic Puzzle: What Is the Name of This Book,* by Raymond M. Smullyan, has many original and challenging problems in recreational logic. Here is a puzzle from this book.*

An enormous amount of loot has been stolen from a store. The criminal (or criminals) took the heist away in a car. Three well-known criminals, A, B, and C, were brought to Scotland Yard for questioning. The following facts were ascertained.

1. No one other than A, B, or C was involved in the robbery.
2. C never pulls a job without using A (and possibly others) as an accomplice.
3. B does not know how to drive.

Is A innocent or guilty?

*Raymond M. Smullyan, *What Is the Name of This Book* (Englewood Cliffs, N.J.: Prentice-Hall, Inc., 1978), p. 67.

Students should obtain, as a result of their education, the ability to reason carefully and to make intelligent and efficient use of their resources when confronted with a problem. * ALAN H. SCHOENFELD

1.3
PROBLEM SOLVING

"Learning to solve problems is the principal reason for studying mathematics. In solving problems, students need to be able to apply the rules of logic necessary to arrive at valid conclusions. They should be unfearful of arriving at tentative conclusions, and they must be willing to subject these conclusions to scrutiny."† This quote by the National Council of Supervisors of Mathematics stresses the importance of learning to solve problems and the role of deductive ("valid conclusions") and inductive ("tentative conclusions") reasoning in problem solving.

George Polya is one of the foremost authorities on problem solving. In his book, *How To Solve It,* he outlines a four-step process for solving problems:

Understanding the problem

Devising a plan

Carrying out the plan

Looking back

The purpose of this section is to help you become familiar with Polya's four-step process and to acquaint you with some of the common strategies for solving problems: drawing pictures, simplifying, guessing and checking, listing possibilities, using models, and working backward.

---

*Alan H. Schoenfeld, "Heuristics in the Classroom," *Problem Solving in School Mathematics, 1980 Yearbook* (Reston, Virginia: The National Council of Teachers of Mathematics, 1980), p. 15.
†National Council of Supervisors of Mathematics, "Position Paper on Basic Mathematical Skills," Washington, DC: National Institute of Education, 1977.

DRAWING PICTURES

One of the most helpful suggestions for understanding a problem and obtaining ideas for a solution is to draw pictures and diagrams. There is a widely known story about a mathematician who got stuck in the middle of a proof while lecturing before a class. He went over to the side of the board, drew a small diagram that the class could not see, and then was able to continue his lecture.* In the following problem the drawings will help you to think through the solution.

PROBLEM   Mr. Jones is planning a dinner party for his wife's birthday. There will be 22 people, and he needs to borrow card tables, the size that seat one person on each side. He wants to arrange them in a rectangular shape so that they look like one large table. What is the smallest number of tables Mr. Jones needs to borrow?

Understanding the Problem   The tables must be placed next to each other, edge to edge, so they form one large rectangular table. Therefore, it will not be possible to put 4 people at each card table.

One large table

Devising a Plan   Drawing pictures of the different possible arrangements of card tables is a natural approach to solving this problem. There are only a few possibilities. The tables can be placed in one long row; they can be placed side by side with two abreast; etc.

Carrying Out the Plan   Here are drawings for two of the five possible arrangements that will seat 22 people. The x's show that 22 people can be seated at each arrangement. Since the remaining arrangements (3 by 8, 4 by 7, and 5 by 6) require 24, 28, and 30 card tables, the smallest number of card tables is 10.

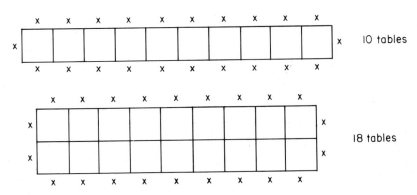

*Related by M. Kline, *Why Johnny Can't Add* (New York: St. Martin's, 1973), p. 166.

Looking Back   The drawings show that the single row of tables requires the fewest tables because each end table has places for 3 people and the remaining tables each have places for 2 people. In all of the other arrangements the corner tables seat only 2 people and the remaining tables seat only 1 person. Therefore, regardless of the number of people, the single row is the arrangement with the smallest number of card tables. This solution can be extended for a dinner party with any number of people. For example, for 38 people, there will be 6 people at the end tables and 32 people in between. Therefore, there will be 2 end tables and 16 tables in between for a total of 18 tables. There are many variations to this problem. For example, what if the length of the room enables only 6 tables to be placed end to end? What if the arrangement of the tables can be L-shaped? or U-shaped?

SIMPLIFYING

Simplifying a problem or solving a related but easier problem is another important strategy for understanding the given information and devising a plan for the solution. Sometimes the numbers in a problem are large or inconvenient and finding a solution for smaller numbers can lead to a plan for solving the original problem.

PROBLEM   There are 15 people in a room, and each person shakes hands exactly once with everybody else. How many handshakes have taken place?

Understanding the Problem   For each pair of people there will be one handshake. For example, if Paul and Sue shake hands, this is counted as one handshake. The problem is to determine the total number of different ways that 15 people can be paired up.

Paul   Sue

Devising a Plan   Fifteen people is a lot of people to work with at one time. Let's simplify the problem and count the number of handshakes for 3 people and the number for 4 people. Solving these special cases may give us an idea for solving the original problem.

Carrying Out the Plan   We have already noted that there is just one handshake for 2 people. If there are 3 people, there will be 3 handshakes, as shown on the next page, and if there are 4 people there will be 6 handshakes.

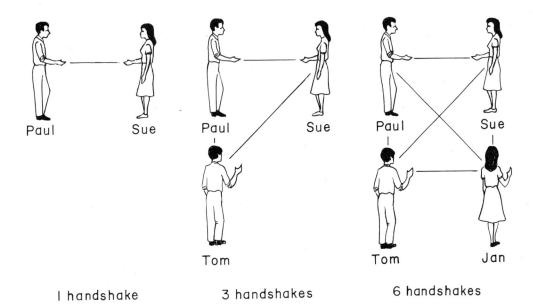

Paul   Sue   Paul   Sue   Paul   Sue

Tom   Tom   Jan

1 handshake   3 handshakes   6 handshakes

These numbers of handshakes, 1, 3, 6, are the beginning of a familiar pattern of numbers, the triangular numbers (page 6).

1, 3, 6, 10, 15, 21, 28, 36, 45, 55, 66, 78, 91, 105

At this point, we could use inductive reasoning to conclude that the total number of handshakes for 5 people is 10; for 6 people is 15; etc., until finally, the total number of handshakes for 15 people would be 105. Remember, however, that inductive reasoning is only an "educated guess" and we cannot be sure we have found the correct pattern.

Let's count the number of handshakes for 5 people. This diagram shows there are 6 handshakes for 4 people. If we consider a fifth person, this person will shake hands with the first 4 people, and so there will be 4 more handshakes. This will bring the total to 10 handshakes. Similarly, if we bring in a sixth person, this person will shake hands with the first 5 people, and so there will be 5 more handshakes. Suddenly, we can see why the triangular numbers are the solution to the problem. The fifth person adds 4 new handshakes, the sixth person adds 5 new handshakes, etc. The total number of handshakes for 15 people is 105.

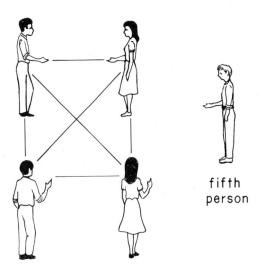

fifth person

Looking Back   By looking at special cases with numbers smaller than 15, we obtained a better understanding of the problem and an insight for solving it. Another solution to this problem is to notice that person number 1 will shake hands with exactly 14 other people, resulting in 14 handshakes. Since person number 1 cannot produce any more handshakes, we now consider the remaining 14 people. Person number 2 will shake hands with exactly 13 people, resulting in 13 handshakes. Continuing in this manner, the number of handshakes is the sum of whole numbers from 14 to 1.

$$14 + 13 + 12 + 11 + 10 + 9 + 8 + 7 + 6 + 5 + 4 + 3 + 2 + 1 = 105$$

GUESSING AND CHECKING

Sometimes it doesn't pay to guess, as illustrated by this cartoon. On the other hand, many problems can be solved by trial and error procedures. As Polya once said, "mathematics in the making consists of guesses." If your first guess is off, it may lead to a better guess. Even if guessing doesn't produce the correct answer, you may increase your understanding of the problem and strike upon an idea for solving it.

PROBLEM   How far is it from town A to town B in this cartoon?

© 1968 United Features Syndicate, Inc.

Understanding the Problem   There are several bits of information. Let's see how Peppermint Patty could have gotten a better understanding of the problem with a diagram. First, the towns A, B, C, and D are one after the other, so they can be represented by 4 points on a line, as shown in the top figure. Next, it is 10 miles farther from A to B than B to C, so we can move point B closer to point C. It is also 10 miles farther from B to C than from C to D, so point C can be moved closer to point D. Finally, the distance from A to D is given as 390 miles. The problem is to find the distance from A to B.

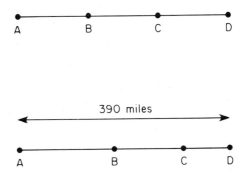

Devising a Plan   One method of solving this problem is to let the unknown distance be x and to use algebra (page 35, Exercise 7a). Another method is to guess and then use the result to make a better guess. The guessing approach puts this problem within the reach of elementary school students.

Carrying Out the Plan   If the 4 towns were equally spaced, as in the top diagram, the distance between each town would be 130 miles (390 ÷ 3). However, the distance from town A to town B is the greatest, so let's begin with a guess of 150 miles.

Guess 1   If the distance from A to B is 150 miles (AB = 150), then BC = 140 miles and CD = 130 miles. But these distances are too great because their total is 420 miles.

Guess 2   If the distance from A to B is 145 miles, then BC = 135 miles and CD = 125 miles. The sum of these distances is 405, which is still too great.

Guess 3   If the distance from A to B is 140 miles, then BC = 130 miles and CD = 120 miles. The sum of these distances is 390 miles. Therefore, the distance from A to B is 140 miles.

Looking Back   One of the reasons for looking back at a problem is to allow students a chance to explain their reasoning. For example, you might have noticed that the first guess gave a total distance of 420 miles, which is 30 miles too much. This means that if the distance between each of the three towns in guess 1 is decreased by 10 miles, we will have the right distances. Therefore, the distance between town A and town B is 140 miles. This is an example of how *guessing and checking* is sometimes a very appropriate method of problem solving that can quickly lead to the correct answer.

LISTING POSSIBILITIES

A problem can sometimes be solved by listing some or all of the possibilities. A table is often convenient for organizing such a list, as in the next example.

PROBLEM   John and Harry earned the same amount of money, although one worked 6 days more than the other. If John earns $12 a day and Harry earns $20 a day, how many days did each work?

Understanding the Problem   Answer a few simpler questions to get a feeling for the problem. How much did John earn in 3 days? Did John earn as much in 3 days as Harry did in 2 days?

Devising a Plan   One method of solving this problem is to let x equal the number of days Harry worked and use algebra (page 472). Another method is to list each day and the total earnings for each person through that day. This list will eventually reach points at which both people have earned the same amount and one has worked 6 days more than the other.

Number of Days	John's Pay	Harry's Pay
1	12	20
2	24	40
3	36	60
4	48	80
5	60	100
6	72	120
7	84	140
8	96	160
9	108	180
10	120	200
11	132	220
12	144	240
13	156	260
14	168	280
15	180	300

Carrying Out the Plan   The complete table is shown above. There are 3 amounts under John's column that equal amounts in Harry's column. It took John 15 days to earn $180, and Harry required only 9 days (6 days less) to earn $180.

Looking Back   You may have noticed that for every 5 days John earns $60 and for every 3 days Harry earns $60. This observation suggests a different way to answer the original question. When John and Harry have worked 10 days and 6 days, respectively, they have each earned $120; when each has worked 15 and 9 days, respectively, they have each earned $180.

USING MODELS

Models are important aids for visualizing a problem and suggesting a solution. The new recommendations by the Committee on the Undergraduate Program in Mathematics (CUPM) contain frequent references to the use of models for illustrating number relationships and geometric properties.*

*Committee on the Undergraduate Program in Mathematics, *Recommendations On the Mathematical Preparation of Teachers,* 1983.

PROBLEM   Find an easy method for computing the sum of whole numbers,
          $1 + 2 + 3 + \ldots$, to any specified number.

Understanding the Problem   If the last whole number in the sum is 8, then the sum
    is $1 + 2 + 3 + 4 + 5 + 6 + 7 + 8$. If the last number in the sum is 100, then the
    sum is $1 + 2 + 3 + \ldots + 99 + 100$.

Devising a Plan   One method of solving this
    problem is to cut "staircases" out of graph
    paper. The one shown here is a *1-through-8
    staircase*. There is 1 square at the top; 2
    squares at the next level; and so forth, down
    to the bottom level, which has 8 squares. The
    total number of squares is the sum $1 + 2 +
    3 + 4 + 5 + 6 + 7 + 8$. By using 2 copies of
    a staircase and placing them together, we can
    obtain a rectangle whose total number of
    squares can be easily found by multiplying
    "length by width."

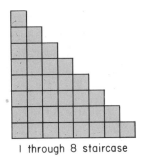

I through 8 staircase

Carrying Out the Plan   Cut out 2 copies of the
    1-through-8 staircase and place them together
    to form a rectangle. Since the total number of
    squares is 8×9, the number of squares in one
    of these staircases is $(8 \times 9) \div 2 = 36$. Check
    this answer by computing the sum

$$1 + 2 + 3 + 4 + 5 + 6 + 7 + 8$$

Predict the sum of whole numbers from 1
through 13 by using 2 staircases. Check your
answer by computing the sum.

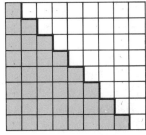

Two, I through 8 staircases

Looking Back   By placing 2 staircases together to form a rectangle, we see that the
    number of squares in 1 staircase is just half the number of squares in the rec-
    tangle. This geometric approach to the problem shows that the sum of whole
    numbers from 1 to a specific number is the product of the last number times the
    next number, divided by 2. Using this method the sum of whole numbers from 1
    to 100 is $(100 \times 101) \div 2 = 5050$; and in general, the sum of whole numbers from
    1 to n is $n(n + 1)/2$.

WORKING BACKWARD

When something of value has been lost or misplaced, it is natural to backtrack your
steps in hope of discovering where the loss might have occurred. This is similar to a

common strategy for solving problems in mathematics. The solution to the following problem illustrates the *working backward* strategy.

PROBLEM   A certain gambler took his week's paycheck to a casino. Aside from a $2 entrance fee and a $1 tip to the hatcheck person upon leaving, there were no other expenses. Bad luck plagued the gambler. The first day he lost one-half of the money he had left after paying the entrance fee, and the same thing happened on the second, third, and fourth days. At the end of the fourth day he had $5 remaining. What was his weekly paycheck?

Understanding the Problem   Let's try a number to get a better feeling for the problem. If the gambler entered the casino with $40, how much money would he have at the end of the day, after leaving the casino?

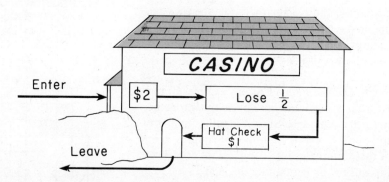

Devising a Plan   Guessing the amount of the paycheck requires too many computations. Since we know the gambler has $5 at the end of the fourth day, a more appropriate strategy for solving the problem is to retrace his steps back through the casino (see the diagram below). First, he receives back $1 for the hatcheck, so that he has $6. Doubling this, because of his loss, gives $12. Receiving the entrance fee brings his total to $14, which is the amount of money he had at the beginning of the fourth day.

Carrying Out the Plan   Continuing to work backward through each day shows that the gambler's paycheck was $140.

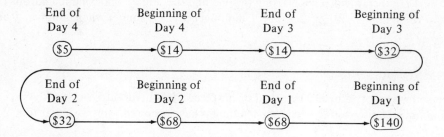

Looking Back   The solution should now be checked by beginning with $140 and going through the expenditures for the 4 days to see if $5 is the remaining amount. This problem can be varied by replacing the $5 at the end of the fourth day by any amount and working backward to the beginning paycheck.

SUPPLEMENT (Activity Book)

Activity Set 1.3 Tower of Brahma
Just For Fun: Instant Insanity

Sometimes the main difficulty in solving a problem is knowing what question is to be answered.

Exercise Set 1.3

Exercises 1 through 7 use the problem-solving strategies presented in this section and are analyzed by Polya's four-step process. Answer each question in parts a, b, c, and d.

1. *Drawing Pictures:* A well is 20 feet deep. A snail at the bottom climbs up 4 feet each day and slips back 2 feet each night. How many days will it take the snail to reach the top of the well?

   ★ a. *Understanding the Problem:* What is the greatest height the snail reaches during the first 24 hours? How far up the well will the snail be at the end of the first 24 hours?

   b. *Devising a Plan:* One plan that is commonly chosen is to compute 20 ÷ 2, since it appears that the snail gains 2 feet each day. However, 10 days is not the correct answer. A second plan is to *draw a diagram* and plot out the snail's daily progress. What is the snail's greatest height during the second day, and how far up the well is the snail at the end of the day?

   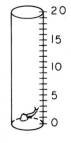

   c. *Carrying Out the Plan:* Trace out the snail's daily progress and mark its position at the end of each day. On which day does the snail get out of the well?

d. *Looking Back:* There is a surprise ending at the top of the
   well because the snail does not slip back on the ninth day. Make up a new snail
   problem by changing the numbers, so that there will be a similar surprise ending
   at the top of the well.

2. *Guessing and Checking:* Assume that a person can carry a 4-day supply of food
   and water for a trip across a desert that takes 6 days to cross. One person cannot
   make the trip alone because the food and water would be gone after 4 days. How
   many persons would have to start out in order for one person to get across and for
   the others to get back to the starting point?

   a. *Understanding the Problem:*
      Draw a line segment and divide
      it into 6 equal parts to represent
      the 6 days of travel across the
      desert. How far can one person
      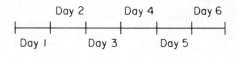
      go alone? Mark this point on the drawing. The problem requires that one
      person, whom we will called the *leader,* get across the desert and that any other
      people, whom we will call *helpers,* get back to the starting point. The helpers do
      not have to return to the starting point together.

   b. *Devising a Plan:* Guessing and checking is a natural strategy to use for solving
      this problem. Can the leader complete the trip with help from the food and
      water supplies of just one helper?

   c. *Carrying Out the Plan:* Use guessing and checking until you find the number of
      helpers that are needed to get the leader across the desert. Explain your answer.

   d. *Looking Back:* This problem can be extended in many ways. For example, if
      the trip across the desert takes 7 days, can the leader (with the aid of helpers)
      complete the trip? Or, if a person can carry a 5-day supply of food and water
      and the trip across the desert takes 7 days, how many helpers must the leader
      have?

3. *Listing Possibilities:* There are two 2-digit numbers that satisfy the following conditions: (1) each number has the same digits; (2) the sum of the digits in each number
   is 10; and (3) the difference between the two numbers is 54. What are the two
   numbers?

   a. *Understanding the Problem:* The numbers 58 and 85 are 2-digit numbers which
      have the same digits. However, the sum of the digits in each number is 13
      (5 + 8 = 13). Find a 2-digit number such that the sum of the digits is 10.

   b. *Devising a Plan:* Since there are only nine 2-digit numbers whose digits add up
      to 10, the problem can be solved by *listing all the possibilities* and computing
      the differences between pairs *that have the same digits.* List these pairs of
      numbers.

   c. *Carrying Out the Plan:* Compute the differences between the pairs of numbers
      in part b. Which pair of numbers has a difference of 54?

★   d. *Looking Back:* This problem can be extended by changing the requirement that the sum of the two digits equal 10 or that the difference between the two numbers is 54. Solve the original problem with "10" replaced by "12."

4. *Simplifying the Problem:* There are 8 coins and a balance scale. The coins are alike in appearance, but one of them is counterfeit and lighter than the other 7. Find the counterfeit coin using two weighings on the balance scale.

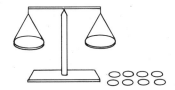

   a. *Understanding the Problem:* If there were only 2 coins and 1 was counterfeit and lighter, the bad coin could be determined in just one weighing. The balance scale at the right shows this situation. Is the counterfeit coin on the left or right side of this balance beam?

   b. *Devising a Plan:* One method of solving this problem is guessing and checking. It is natural to begin with 4 coins on each side of the balance beam. Explain why this method will not lead to the correct solution. Another method is to *simplify* the problem and try solving it for a fewer number of coins.

★   c. *Carrying Out the Plan:* Explain how the counterfeit coin can be found with: 1 weighing if there are only 3 coins, and 2 weighings if there are 6 coins. By now you may have an idea for solving this type of problem. Explain the solution to the original problem.

   d. *Looking Back:* Eight coins are not the maximum number for which this problem can be solved. Explain how the problem can be solved with just 2 weighings if there are 9 coins.

5. *Listing Possibilities:* A bank that has been charging a monthly service fee of $2 for checking accounts plus 15 cents for each check announces that it will change its monthly fee to $3 and each check will cost 8 cents. The bank claims the new plan is cheaper than the old plan and that it will save the customer money. How many checks must a customer write per month before the new plan is cheaper than the old plan?

   a. *Understanding the Problem:* Try some numbers to get a feeling for the problem. Compute the cost of 10 checks under the old plan and the new plan. Which plan is cheaper for 10 checks?

b. *Devising a Plan:* One method of solving this problem is to let the number of checks be x and use algebra (page 477, Exercise 7b). Another method is to list systematically the costs of 1 check, 2 checks, etc., as shown in this table. How much more does the new plan cost than the old plan for 6 checks?

c. *Carrying Out the Plan:* Extend the table until you reach a point at which the new plan is cheaper than the old plan. How many checks does this require?

Checks	Cost for Old Plan	Cost for New Plan
1	2.15	3.08
2	2.30	3.16
3	2.45	3.24
4	2.60	3.32
5	2.75	3.40
6		
7		
8		

★ d. *Looking Back:* The difference in costs for one check between the old plan and the new plan is 93 cents. What happens to these differences as the number of checks increases? How many checks must a customer write per month before the new plan is 33 cents cheaper?

6. *Using Models:* The following patterns can be used to form a cube. A cube has 6 faces: there are the top and bottom faces, the left and right faces, and the front and back faces. Two faces have been labeled on each of the following patterns. Label the remaining 4 faces on each pattern so that when the cube is assembled with the labels on the outside, each face will be in the right place.

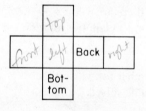 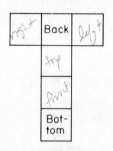 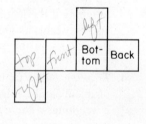

a. *Understanding the Problem:* A baby's block or a die is a cube. It has 6 faces that are squares. There are 11 patterns that can be folded into a cube. Find a pattern that is different from the 3 above. (*Hint:* Start with a cube and cut it apart.)

Cube

b. *Devising a Plan:* One method of solution for labeling the faces of the patterns is to think through the folding of each pattern and try to predict the names of each face. Then fold the patterns into cubes and check your answers. Describe another plan.

c. *Carrying Out the Plan:* Use the plan suggested in part b or your own plan to label the faces of the three patterns.

★   d. *Looking Back:* Another variation of this problem is to label 4 of the faces of a pattern with the letters M A T H, so that when the pattern is assembled to form a cube, it spells MATH around its 4 vertical faces. Solve the problem by writing the missing letters on this pattern.

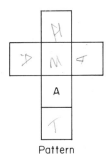

Pattern

7. *Working Backward:* Three girls play three rounds of a game. On each round there are two winners and one loser. The girl who loses on a round has to double the number of chips that each of the other two girls has by giving up some of her own chips. Each girl loses one round. At the end of three rounds each girl has 40 chips. How many chips did each player have at the beginning of the game?

   a. *Understanding the Problem:* Let's select some numbers to get a feeling for this game. Suppose girl A, girl B, and girl C have 20, 10 and 5 chips, respectively, and girl A loses the first round. Girl B and girl C will have their numbers of chips doubled by taking chips from girl A. How many chips will each girl have after this round?

   b. *Devising a Plan:* Since we know the *end result* (each girl finished with 40 chips), a natural strategy is to *work backward* through the three rounds to the beginning. Assume that girl C loses the third round. How many chips did each girl have at the end of the second round?

	A	B	C
Beginning			
End of first round			
End of second round			
End of third round	40	40	40

   c. *Carrying Out the Plan:* Assume that girl B loses the second round and girl A loses the first round. Continue working backward through the three rounds to determine the number of chips each girl had at the beginning.

★   d. *Looking Back:* Check your answer by working forward from the beginning. Girl A had the most chips at the beginning of this game, and she lost the first round. Could the girl with the fewest number of chips at the beginning of the game have lost the first round? Try it.

Try using the suggested strategies in Exercises 8 through 12 to solve the problems.

8. *Drawing Pictures* and *Guessing and Checking:* In driving from town A to town D, you pass first through town B and then through town C. It is 10 times farther from

A to B than from B to C, and 10 times farther from B to C than from C to D. If it is 1332 miles from A to D, how far is it from A to B?

★ 9. *Listing Possibilities* or *Guessing and Checking:* It costs 13 cents to mail a postcard and 20 cents to mail a letter. Harold wrote to 15 people and it cost him $2.37. How many postcards did he write?

10. *Simplifying the Problem:* There are 5 identical-looking coins and a balance scale. One of these coins is counterfeit and either heavier or lighter than the other 4. With only 3 weighings on the balance scale, explain how the counterfeit coin can be identified and whether it is lighter or heavier than the others.

★ 11. *Working Backward:* Suellen and Angela both have $510 in their savings accounts. Each week Suellen adds $10 to her account and Angela adds $20 to her account. They both started their accounts on the same day, and at that time Suellen's account had twice as much money as Angela's. How much money did Suellen open her account with?

12. *Using Models* and *Guessing and Checking:* The figure at the right is a domino donut for which the number of dots on each side is 11. Arrange the 4 dominoes below into a domino donut so that all 4 sides equal the same sum (not necessarily the sum in this example).

 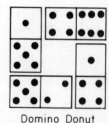

Domino Donut

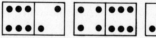

13. *Pail Puzzle:* How can you bring exactly 6 liters of water from the river when you have only two containers, a 4-liter pail and a 9-liter pail? (Guess and Check)*

★ 14. *Chain Puzzle:* Given 4 pieces of chain that are 3 links each, explain how all 12 links can be joined into a single circular chain by cutting and rejoining only 3 links. (Work backward by beginning with a circular chain.)

*It is possible to obtain any whole number of liters from 1 to 13 using these two containers.

15. *Tricky-Fun Problems:* Trick questions such as the following can help improve problem-solving ability because they require that a person listen and think carefully about the given information and the question. They also can be lots of fun!

    a. If you went to bed at 8 o'clock and set the alarm to get up at 9 o'clock in the morning, how many hours of sleep would you get?

    b. Take 2 apples from 3 apples and what do you have?

    c. A farmer has 17 sheep. All but 9 died. How many did he have left?

★   d. I have two U.S. coins that total 30 cents. One is not a nickel. What are the two coins?

    e. A bottle of cider costs 86 cents. The cider costs 60 cents more than the bottle. How much does the bottle cost?

★   f. How much dirt is in a hole 3 meters long, 2 meters wide, and 2 meters deep?

    g. A woman weighs 70 pounds plus half her weight. How much does she weigh?

    h. The score at the end of the baseball game was 7 to 2 and not a man crossed home plate. Explain how this could happen.

★   i. Which is correct to say: (1) "the whites of the egg *are* yellow"; or (2) "the whites of the egg *is* yellow"?

2

Sets,
Numeration,
and Computers

Our minds are finite, and even in these circumstances of finitude we are surrounded by possibilities that are infinite; and the purpose of human life is to grasp as much as we can out of that infinitude.

ALFRED NORTH WHITEHEAD

2.1
COUNTING AND SETS

Long before numbers were invented
numerical records were kept by means
of tallies. These were maintained by
making collections of pebbles, cutting
notches in wood, or making marks in
dirt or stone. The oldest example of the
use of a tally stick dates back 30,000
years and was found in 1937 in
Czechoslovakia. It is a 7-inch wolf
bone engraved with 55 notches
occurring in 11 sets of 5.

A more recent example of tallies is
found on the 8000-year-old Ishango
bone, which was discovered on the
shore of Lake Edward in the Congo.
The marks on this bone occur in
several groups that are arranged in
three distinct columns (see pages 57–58,
Exercise 2). The tip of this bone is
provided with a quartz flake, which,
according to geologist-archaeologist
Dr. de Heinzelin, is very likely a
writing tool.*

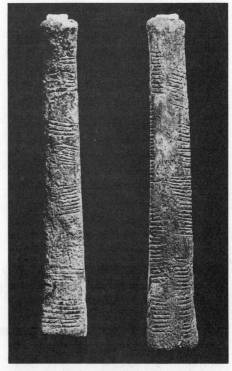

Two views of the Ishango bone, found on
the shores of Lake Edward in the Congo

At one time, tally sticks were common for keeping records of debts and
payments. The tally stick would be split
lengthwise across the notches, with the debtor receiving one piece and the creditor the
other. Such tally sticks were used by the British Royal Treasury in the thirteenth century
and continued in use until the nineteenth century.

ONE-TO-ONE
CORRESPONDENCE AND
COUNTING

It is not difficult to imagine our ancient
ancestors keeping track of their flocks
by matching each animal with a pebble,
a mark in the dirt, or a notch on a
stick. Such methods of tally keeping
are examples of one-to-one
correspondence.

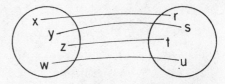

Two sets in one-to-one correspondence

*Jean de Heinzelin, "Ishango," *Scientific American,* **206,** June 1962, pp. 113–14.

Definition: Two sets or collections of objects are said to be in *one-to-one correspondence* if for every element in the first set there is just one element in the second set and for every element in the second set there is just one element in the first set.

Counting is an extension of the idea of one-to-one correspndence. The numbers 0, 1, 2, 3, 4, etc., are called *whole numbers*. To *count* the elements of a set we match these elements with the whole numbers that are greater than zero: 1, 2, 3, 4, and so forth. To count the number of letters in the word "straight" (one of the longest one-syllable words) we match these letters with the numbers 1 through 8.

$$s \leftrightarrow 1$$
$$t \leftrightarrow 2$$
$$r \leftrightarrow 3$$
$$a \leftrightarrow 4$$
$$i \leftrightarrow 5$$
$$g \leftrightarrow 6$$
$$h \leftrightarrow 7$$
$$t \leftrightarrow 8$$

WHOLE NUMBERS AND THEIR USES

Fifty thousand years ago, people were living in caves and using fire. Implements from the remains of this period suggest the existence of barter and the need for numbers. During the Old Stone Age (10,000–15,000 B.C.) figures of people, animals, and abstract symbols were painted in caves in Spain and France. The symbols were composed of many geometric forms: straight lines, spirals, circles, ovals, and dots. The rows of dots and rectangular figures in this photograph are from El Castillo caves, Spain (12,000 B.C.). It is conjectured by some scholars that this was a system for recording the days of the year. It seems likely that numbers were in existence by this time. The earliest dated event in history is the Egyptian calendar, 4236 B.C. This calendar predates the earliest known writing by over 500 years.

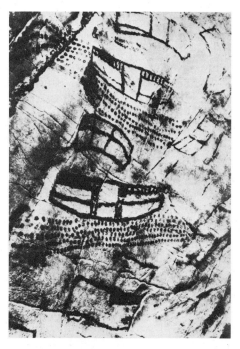

Art symbols (1200 B.C.)
El Castillo caves, Spain

At first it may only have been necessary to distinguish between one object, two objects, and many objects. The first words for numbers were probably associated with specific things. This influence can be seen in the expressions we have for two, such as a "couple" of people, a "brace" of hens, and a "pair" of shoes. Eventually the concepts of twoness, threeness, etc., were separated from physical objects, and the abstract notion of "number" developed.

Definition: Two sets of objects are said to have the *same number* of elements if they can be put into one-to-one correspondence.

Look around you and find a set of objects that can be put into one-to-one correspondence with the fingers of your hand. We express this relationship by saying that the set has "5" elements. The number 5 is an abstract idea expressing a common property of sets. Similarly, the numbers 0, 1, 2, 3, etc., are each abstract ideas associated with sets.

Some of the earliest symbols for numbers were sets of marks that could be put into one-to-one correspondence with the set of objects being counted. For example, 2 was often represented by two marks, //; 3 by three marks, /// ; etc. Such symbols are easy to remember because they suggest the meanings of the numbers.

It is usually assumed that numbers arose in answer to the practical needs of people, such as counting possessions, days, years, and so on. However, there is some evidence that numbers developed in connection with religious rituals and that they may have been used first to denote the rank or order of people in a group. We now recognize three different uses of whole numbers: cardinal, ordinal, and naming.

Cardinal Use   If a number gives information about quantity and answers questions such as, "How many?" or "How much?", this is the *cardinal use* of numbers. The following are examples of cardinal uses of whole numbers: There are 50 states; he is a 21-year-old lieutenant; and the House voted by 162 to 52 to shrink itself by one-third.

Ordinal Use   If a number gives information about the order or location of objects, this is an *ordinal use*. Here are a few examples: She is in row 27; the fifth-ranked player beat the second-ranked; and we meet on the third Wednesday of every month.

Naming Use   Social Security numbers, license plate numbers, and checking account numbers are tags to name and identify objects. This is a *naming use* of numbers. For example: Flight 972 is bound for Kennedy Airport; she flies a Sikorsky S-61 helicopter; and number 82 is a guard.

In the following sentence you will recognize all three of the uses of whole numbers:

Nineteen cars have arrived for the 8th annual stock car derby, but car 27 has withdrawn.

HISTORICAL BACKGROUND OF SETS

The notion of sets or collections of objects is not only fundamental to counting and numbers, but it permeates every branch of mathematics. The importance of sets was first recognized during the nineteenth-century development of logic. George Boole (1815–1864) was one of the first mathematicians to systematically use sets in logic. He developed an algebra of sets which is now called *Boolean algebra*. The person most often associated with sets is Georg Cantor (1845–1918), the Russian-born mathematician who moved to Germany at the age of 11. He is referred to as "the father of set theory" because of his extensive development of sets. In 1874, he published a controversial paper on set theory, and the resulting criticism by other mathematicians led to a series of mental breakdowns, which continued to his death. In recent years Cantor has won widespread recognition for his theory of sets, which has contributed to every area of mathematics.

SETS AND THEIR ELEMENTS

There are many words for sets of objects: a *flock* of birds, a *herd* of cattle, a *collection* of paintings, a *bunch* of grapes, a *group* of people, and so forth. The word "set" is so basic in mathematics it is one of the undefined terms. Intuitively, it is described as a collection of objects called *elements,* which can be either concrete, such as the set of seats in a theater, or abstract, as a set of numbers.

There are two conditions that a set of objects must satisfy. For the first condition it must be possible to determine if a given object is in the set. We refer to this condition by saying that the set is *well-defined.*
This just means that sets must be described carefully enough to avoid confusion about which elements belong to the set. For example, "the set of all armed forces personnel in a particular country" is a well-defined set. Given any person, it can be determined whether or not he or she is in this set. On the other hand, "the set of all beautiful paintings" is not well-defined because what is beautiful is a matter of opinion. The second condition is that a set must not contain an element more than once. A set of numbers, for example, should not have the same number listed twice.

"We understand you tore
the little tag off your mattress."

There are two common methods of specifying a set. One is by *describing* the elements of the set, such as, "the capitals of the six New England states." The second method is by *listing* each element in the set. When this is done the members of the set are written between braces. Here is the set of capitals in New England:

{Augusta, Concord, Boston, Hartford, Providence, Montpelier}

If the set of elements is large, we sometimes begin the list and then use three dots to mean "and so forth." The following set contains the multiples of 10 from 10 to 500:

$$T = \{10, 20, 30, 40, 50, 60, 70, \ldots, 490, 500\}$$

Sometimes a set containing no elements will be described. The set of all prime numbers greater than 31 and less than 37 is a set with no elements. This is called the *empty set* or *null set* and is denoted by the set braces with no elements between them, { }, or by the Greek letter phi, ϕ. This is a very useful set and occurs quite frequently.

It is customary to denote sets by uppercase letters and the elements of sets by lowercase letters. If k is an *element of* a set S we write $k \in S$ and if it is *not an element of S,* this is written $k \notin S$. Using the previous set T as an example, $60 \in T$ and $55 \notin T$.

Attribute Pieces Attribute pieces are geometric models of various shapes, sizes, and colors for illustrating sets. Those shown here have three different shapes [triangular (*t*), rectangular (*r*), and hexagonal (*h*)], two sizes [large (*l*) and small (*s*)], and two colors [black (*b*) and white (*w*)].

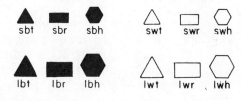

These objects can be classified into *sets* in many different ways. Here are a few: S is the set of all small attribute pieces; L is the set of all large attribute pieces; T is the set of all triangles; H is the set of all hexagons; W is the set of all white attribute pieces; LB is the set of all large black attribute pieces; etc. These attribute pieces will be used in the following paragraphs to provide examples of relationships between sets and operations on sets.

RELATIONSHIPS BETWEEN SETS

There are relationships between sets, just as there are relationships between numbers. For any two numbers, either one is less than the other or they are equal. Two sets may have no elements in common, some elements in common, or all elements in common. The following sets of attribute pieces illustrate two of these situations. On the left, set H (hexagons) has some attribute pieces in common with set S (small pieces). On the right, every attribute piece in BT (black triangles) is also in T (triangles).

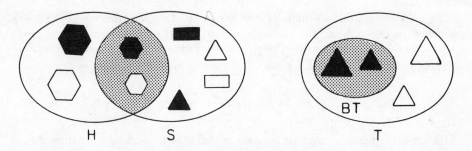

Subsets The relationship between BT and T is described by saying BT is a *subset* of T, because every attribute piece in BT is also in T.

Definition: If every element of set *A* is also an element of set *B*, then *A* is a *subset* of *B*, written $A \subset B$.

In the preceding example, $BT \subset T$. The sets *H* and *S* have some elements in common but neither is a subset of the other. This can be indicated by writing $H \not\subset S$ and $S \not\subset H$. According to the definition of subset, every set is a subset of itself. For example $BT \subset BT$, $T \subset T$, $H \subset H$, etc., because every element in the first set is also an element of the second set.

If we know $A \subset B$ and that one or more elements in *B* are not in *A*, then *A* is sometimes called *a proper subset* of *B*. As examples, *BT* is a proper subset of *T*, but *H* is not a proper subset of itself. The null set, { }, is a proper subset of every nonempty set. To see this we must show that Statement 1 is true for any set *B*.

(1) If an element is in { } then it is in *B*. However, we know that Statement 1 has the same meaning as Statement 2, its contrapositive (page 20), and it is easy to see that Statement 2 is true.

(2) If an element is not in *B* then it is not in { }. So, Statement 1 is also true.

Equal Sets   Sets that contain the same elements are *equal*. Sometimes two equal sets may look different or have different descriptions. The set of all solutions to the equation $x^2 - 5x + 6 = 0$ and the set {2, 3} are examples of equal sets.

Definition: If *A* is a subset of *B*, and *B* is a subset of *A*, then both sets have exactly the same elements and they are *equal*, written $A = B$. In this case, *A* and *B* are just different letters for the same set.

The set of attribute pieces with less than four sides is equal to the set of triangular attribute pieces, as shown by the following two sets. Notice that equality of sets does not depend on the order of the elements.

$$\{lbt, sbt, lwt, swt\} = \{lwt, lbt, swt, sbt\}$$

Equivalent Sets   The set of small black attribute pieces can be put into one-to-one correspondence with the set of large white attribute pieces. We refer to this fact by saying the two sets are *equivalent*. In other words, they have the same number of elements.

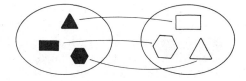

Definition: Two sets are *equivalent* if their elements can be placed in one-to-one correspondence.

Note that two equal sets are also equivalent, but it is possible for two sets to be equivalent without being equal. The sets in the previous diagram are equivalent but they are not equal.

Disjoint Sets The two sets of attribute pieces shown here have no elements in common. We describe this fact by saying the two sets are *disjoint*.

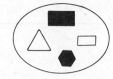 

Definition: Two sets are *disjoint* if they have no elements in common.

OPERATIONS ON SETS

There are operations that replace two sets by a third set just as there are operations on numbers which replace two numbers by a third number. An operation that assigns each pair of elements to another element is called a *binary operation*. Addition and multiplication are two examples of binary operations on whole numbers. Intersection and union are binary operations on sets.

Intersection of Sets The set of small attribute pieces and the set of black attribute pieces have three elements in common. If we form a third set containing these common elements, it is called the *intersection* of the two sets.

Definition: The *intersection* of two sets A and B is the set of all elements that are in both A and B. This operation is written as $A \cap B$.

The intersection of the above two sets is the set of attribute pieces that are small and black. This is illustrated by shading the common region inside the two curves.

$$\{swr, swh, swt, sbt, sbr, sbh\} \cap \{sbt, sbr, sbh, lbt, lbr, lbh\} = \{sbt, sbr, sbh\}$$

If two sets are disjoint, such as the set of large attribute pieces (L) and the set of small attribute pieces (S), their intersection is the null set. This can be briefly written as $L \cap S = \{\ \}$, or $L \cap S = \phi$.

Union of Sets The set of small attribute pieces and the set of black attribute pieces have nine different elements. A third set containing all of these elements is called the *union* of the two sets.

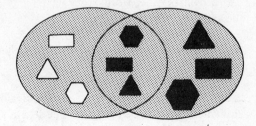

Definition: The *union* of two sets *A* and *B* is the set of all elements that are either in *A*, or in *B*, or in both *A* and *B*. This operation is written as *A* ∪ B.

The union of the two sets pictured above is the set of all attribute pieces that are small or black. This is illustrated by shading the total region inside the two curves.

$$\{swr, swh, swt, sbt, sbr, sbh\} \cup \{sbt, sbr, sbh, lbt, lbr, lbh\} =$$
$$\{swr, swh, swt, sbt, sbr, sbh, lbt, lbr, lbh\}$$

Notice that in the preceding example, the union of the set of small attribute pieces and the set of black attribute pieces contains each attribute piece only once, even though three of these elements are contained in both sets. The union of the two disjoint sets shown here contains the four attribute 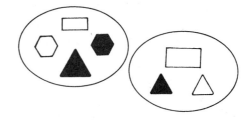 pieces from one set and three from the other. The union of disjoint sets can be written by listing the elements in the first set followed by those in the second.

$$\{lbt, swr, swh, sbh\} \cup \{lwr, sbt, swt\} = \{lbt, swr, swh, sbh, lwr, sbt, swt\}$$

Complement of a Set Frequently, we wish to compare a set that has a certain property with another set that does not have this property. Consider the set of small black attribute pieces and the remaining ones without this property. In the following diagram the small black pieces are inside the circle and the others are outside. These two sets are called *complements* of each other.

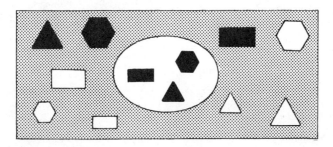

Definition: For any given set, if two subsets *A* and *B* are disjoint and their union is the whole set, then *A* is the *complement* of *B*, written *A* = *B′*, *and B* is the *complement* of *A*, written *B* = *A′*.

The complement of *SB*, the set of small black attribute pieces, is the set of attribute pieces not contained in *SB*.

$$SB' = \{lbt, lbh, lbr, lwt, lwh, lwr, swt, swr, swh\}$$

The "given set" in the previous definition is sometimes called the *universal set*. We have been using a universal set of 12 attribute pieces. In solving problems with whole numbers, the universal set is often the set of all whole numbers. If the universal set is all whole numbers, the set of odd numbers and the set of even numbers are complements of each other. To give another example, the complement of the set of whole numbers less than 10 is the set of whole numbers greater than or equal to 10.

The universal set can be any set, but once it is established, each subset has a unique (one and only one) complement. In other words, *complement* is an operation that assigns each set to another set.

INFINITE SETS

Are time and space infinite quantities? What do we mean by "infinitely large" and "infinitely small"? The problems of the infinite have challenged the human mind as no other single problem in the history of thought.

First of all, we must realize that "very big" and "infinite" are entirely different concepts. There are about 100 billion stars in our galaxy, The Milky Way, and about 1 billion other galaxies. The number of electrons and protons in all of these stars and in interspace has been estimated to be less than 10^{80} (1 followed by 80 zeros). While this number of objects is very large, it is still not infinite.

In the sense that "infinite" means "without end" or "without bound," we have an intuitive notion of its meaning. These vague terms, however, were not precise enough to be useful to mathematicians, and in 1874 Cantor developed the following definition of infinite.

Galaxy Messier 81,
10 million light-years away

Definition: A set is *infinite* if it has a proper subset with which it can be put into one-to-one correspondence.

Let's apply this definition to the set of whole numbers to show it is infinite. The set of even numbers is a proper subset of the set of whole numbers, because every even number is a whole number and the set of whole numbers contains numbers that are not even, namely, the odd numbers. By matching every whole number with the even number that is twice as big, the numbers of both sets can be placed in one-to-one correspondence (see the following diagram). Intuitively, it may seem that we should "run out" of even numbers before whole numbers, but this does not happen. For every whole number n there corresponds an even number $2n$, and for every even number there is a whole number that is half as big.

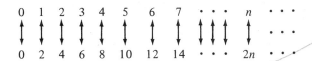

Notice that this definition distinguishes between finite and infinite sets. Try as we may, there is no way to form a one-to-one correspondence between the finite set {1, 2, 3, 4} and any of its proper subsets.

We have shown that the set of whole numbers is infinite because it can be put into one-to-one correspondence with the set of even numbers. Remember that earlier we defined two sets to have the *same number* of elements if they can be put into one-to-one correspondence. This means that the set of whole numbers and the set of even numbers have the same number of elements.

Georg Cantor was especially interested in infinite sets and was the first to recognize that there are many different types of these sets. That is, there is a hierarchy of infinite sets, in which each set is so much bigger than the previous set they cannot be put into one-to-one correspondence. The first or smallest of the infinite sets is the type that can be put into one-to-one correspondence with the set of whole numbers. These sets are said to be *countably infinite*. (For another example of a countably infinite set, see page 61, Exercise 12.)

SUPPLEMENT (Activity Book)

Activity Set 2.1 Sorting and Classifying (Operations on sets of attribute pieces)
Just for Fun: Games with Attribute Pieces

Exercise Set 2.1: Applications and Skills

1. The notches in the 30,000-year-old Czechoslovakian wolf bone are arranged in two groups. There are 25 notches in one group and 30 in the other. Within each series the notches are in groups of 5. Could this recording system have been accomplished without: number names? number symbols?

2. Both sides of the 8000-year-old Ishango bone are shown next. There is one row of marks on one side of the bone and two rows of marks on the other side. Anthropologists have questioned the significance of the numbers of these marks: Are they

arbitrary records of game killed or belongings, or are they intended to show relationships between numbers?* For each group of marks in these rows, write the number of marks.

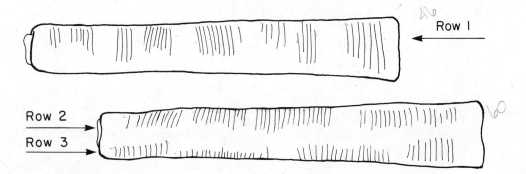

★ a. These sets of numbers form definite number patterns. Which of the three rows of notches represent prime numbers?

  b. Which of these rows suggests a knowledge of multiplication by 2?

★ c. Explain how the numbers in the remaining row are related.

  d. Do you think these marks were intended to show relationships between numbers?

3. For each number in the following sentences determine whether it is a cardinal, ordinal, or naming use of numbers.

★ a. About 25 percent of the output is harvested by hand.

★ b. This report continues on page 742.

  c. Loch Ness, more than 900-feet deep in places, is part of the Great Glen, a fault line bisecting Scotland.

  d. The Datsun 810 has a tilt steering wheel and maintenance warning system.

  e. He is seeded number 5 in the tennis tournament.

4. Which of these sets are well-defined?

★ a. All the letters on this page

  b. All difficult mathematics problems

★ c. All useful books

  d. All Ford trucks built in 1980

*For a discussion of these marks, by de Heinzelin, see A. Marshack, *The Roots of Civilization* (New York: McGraw-Hill, 1972), pp. 21–26.

5. Given the universal set $U = \{0, 1, 2, 3, 4, 5, 6, 7, 8\}$ and sets $A = \{0, 2, 4, 6, 8\}$, $B = \{1, 3, 5, 7\}$, and $C = \{3, 4, 5, 6\}$, list the elements in the following sets.

a. $A \cap C$ ★ b. $C' \cup B$ c. $C' \cap A$ ★ d. $(A \cap C) \cup B$

e. Answer parts **b** and **c** if the universal set is all whole numbers. 4+6

6. The 12 attribute pieces shown here have 3 shapes, 2 sizes, and 2 colors. Use the following sets to answer parts a through g. *W:* white attribute pieces; *H:* hexagonal attribute pieces; *SW:* small white pieces; *SB:* small black pieces; *L:* large pieces.

lbt   lbr   lbh   sbt   sbr   sbh

lwt   lwr   lwh   swt   swr   swh

★ a. Which pairs of sets, if any, are equivalent?

b. Which pairs of sets, if any, are equal?

★ c. Which pairs of sets, if any, are disjoint?

d. Which set is a subset of another?

★ e. Which attribute pieces are in the intersection of W and L?

f. Which attribute pieces are in the union of W and L?

g. The union of which two sets is equal to the complement of SB?

7. Sometimes the best way to find the elements which have a given property is to find those elements which do not have this property. That is, we find one set and then take its complement. There are nine attribute pieces that are not(large and black), and these can be found by taking the complement of the set of large black attribute pieces. Use complements to find the elements in the following sets.

★ a. All attribute pieces that are not(hexagonal and small).

b. All attribute pieces that are not(white and triangular).

8. The attribute pieces shown here are hexagonal or small (some pieces have both properties). The elements in the complement of this set are the remaining four attribute pieces that are not(hexagonal or small). Use complements to find the elements in the following sets.

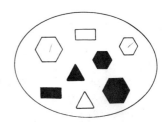

★ a. All attribute pieces that are not(small or white).

b. All attribute pieces that are not(triangular or large).

9. Illustrate the set under each figure by shading the figure.

a.

★ b.

c.

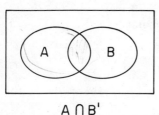

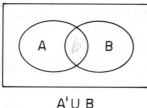

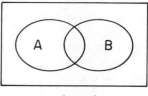

$A \cap B'$

$A' \cup B$

$A' \cup B'$

Use set operations on A and B to name the shaded region.

★ d.

e.

f.

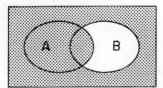

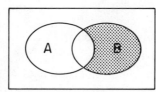

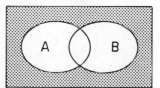

10. The shaded region in the accompanying diagram represents $(A \cap B)'$, the complement of $A \cap B$. Shade in the regions below for $(A \cup B)'$, $A' \cup B'$, and $A' \cap B'$. (*Hint:* Two of the four diagrams should be shaded exactly like the remaining two.)

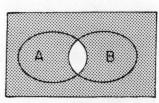

$(A \cap B)'$

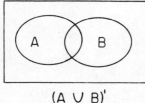

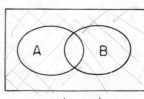

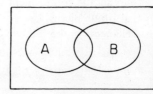

$(A \cup B)'$

$A' \cup B'$

$A' \cap B'$

Only two of the following equations are true for all sets A and B. These two equations are called De Morgan's laws. Use the diagrams to determine which two equations hold and which two do not.

★   a. $(A \cap B)' = A' \cap B'$ b. $(A \cap B)' = A' \cup B'$

★   c. $(A \cup B)' = A' \cap B'$ d. $(A \cup B)' = A' \cup B'$

11. This Venn diagram of human populations was used in investigations correlating the presence or absence of B26+ (a human antigen), RF+ (an antibody protein), spondylitis (an inflammation of the vertebrae), and arthritis (an inflammation of the joints), with the incidence of various rheumatic diseases.

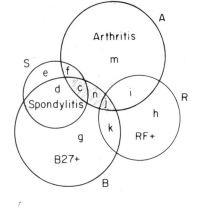

   ★ a. What is the letter of the region that corresponds to $(S \cap B) \cap A'$?

   b. What is the letter of the region that corresponds to $(A \cap B) \cap (R \cup S)'$

12. Show that the set of counting numbers and the set of multiples of 10, {10, 20, 30, 40, . . . }, have the same number of elements, by putting them into one-to-one correspondence.

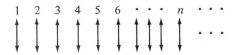

13. *Hotel Puzzle:* There once was a certain hotel in never-never land with an infinite number of rooms. On one very busy holiday in which every room was taken, the hotel manager found that a room had not been saved for a very important person who was due to arrive at any minute. Luckily, there was a mathematician staying at the hotel and he proposed the following solution. Have the people in room 1 move to room 2; those in room 2 move to room 4; those in room 3 move to room 6; and in general whatever the number of a room, its occupants moved to a room whose number was twice as big. The important and late-arriving guest was then placed in room 1, and in the end, after some grumbling about moving around, everyone was happy, especially the hotel manager.*

   a. Explain why this solution was possible.

   ★ b. How many new rooms were made available by this solution?

   c. Would the mathematician's plan have worked if each person had been moved to a room whose number was 10 times greater than their first room number?

---

*For more problems and solutions involving a hotel with an infinite number of rooms, see N. Y. Vilenkin, *Stories about Sets* (New York: Academic Press, 1968), pp. 4–15.

Exercise Set 2.1: Problem Solving

1. *Network Survey:* Here are the results of a survey of 120 people which show the television networks they watched in a given evening. How many people did not watch any of these three networks?

Networks	Numbers of People
ABC	55
NBC	30
CBS	40
ABC and CBS	10
ABC and NBC	12
NBC and CBS	8
NBC and CBS and ABC	5

   a. *Understanding the Problem:* Let's use a Venn diagram to visualize the given information. Each region inside the circles represents the number of people who watched these networks. For example, those people in section y watch NBC and CBS but not ABC. Note that some of the 55 people who watched ABC also watched NBC, and some watched CBS. What region of the diagram represents the people who did not watch any of these networks?

   ★ b. *Devising a Plan:* We can find the number of people who did not watch any of these networks by first finding the missing numbers in the regions of the circles. For example, v is in the intersection of all three circles and the table shows that $v = 5$. Use this number and the fact that there are eight people in the intersection of NBC and CBS to find the value of y.

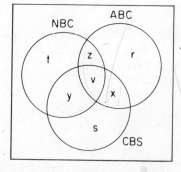

   c. *Carrying Out the Plan:* Continue the process in part b to find the missing numbers in the circles. What are the values of z, x, r, s, and t? How many people did not watch any of the three networks?

   d. *Looking Back:* Another approach to this problem is to add the numbers of people who watched ABC, NBC, and CBS $(55 + 30 + 40)$. Then, since this sum contains the numbers 10, 12, and 8, which are in the intersections of pairs of ABC, NBC, and CBS, they must be subtracted. Shade these three pairs of intersections in the diagram. Finally, by subtracting 10, 12, and 8, we remove the number of people in the intersection of all three sets, and so 5 must be added. Use this method to find the number of people who watched at least one of these networks.

2. *Rainy Day Puzzle:* During a vacation it rained on 13 days, but when it rained in the morning, the afternoon was fine and every rainy afternoon was preceded by a fine morning. There were 11 fine mornings and 12 fine afternoons. How long was the vacation?

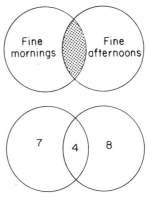

   a. *Understanding the Problem:* Let's use a diagram to visualize the given information. The sets shown here represent the days of the vacation. What region represents the days with both fine mornings and fine afternoons? Rainy afternoons? Rainy mornings?

   ★ b. *Devising a Plan:* One approach is to guess and try numbers in the three regions of these circles. The numbers shown here satisfy the conditions that there were 11 fine mornings and 12 fine afternoons, but the number of rainy days is not satisfied. Explain why. Try some different numbers.

   c. *Carrying Out the Plan:* Find the numbers for the regions of the diagram which satisfy all the given conditions. How long was the vacation?

   ★ d. *Looking Back:* Let's vary the conditions of this problem by changing the number of days when it rained. What is the maximum number of days that there could have been rain and still have 11 fine mornings and 12 fine afternoons?

3. *Hundred-Men Puzzle:* In a certain town there live 100 men: 85 are married; 70 have a telephone; and 75 own a car. What is the least number of men who are married, have a telephone, and own a car?

   a. *Understanding the Problem:* A Venn diagram can be used to visualize the problem. We would like to find the least number of men in the intersection of the three sets. Explain why the intersection of the three sets is not empty, that is, why there may be some men satisfying all three conditions.

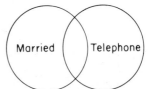

   ★ b. *Devising a Plan:* One approach is to solve the original problem for two of the conditions and then bring in the third condition. If 85 men are married and 70 have a telephone, what is the least number that satisfies both conditions? Fill in the three regions of the diagram so that the smallest possible number is in the intersection and the sum of the numbers is 100. (*Hint:* How many men would be required if there were no men in the intersection of the two sets?)

c. *Carrying Out the Plan:* By answering the question in part **b** the problem has been reduced to two conditions: Those men who are married and have a telephone (the intersection in part **b**) and the 75 men who own a car. What is the least number of men who satisfy both of these conditions? Fill in the three regions so that the sum of the numbers is 100.

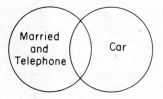

★   d. *Looking Back:* What if there is a fourth condition? Suppose that 80 of the 100 men own a home. Continue the approach in parts **b** and **c** to find the least number of men who are married, have a telephone, own a car, and own a home.

Egyptian stone giving an account of the expedition of Amenhotep III in 1450 B.C.

2.2
NUMERATION
SYSTEMS

The first uses of numbers, their names, and their symbols have been lost in early history. Number symbols or numerals probably preceded number words, since it is easier to cut notches in a stick than to establish phrases to identify a number.

Early numeration systems appear to have grown from such tallying. In many of these systems one, two, and three were represented by $|$, $||$, and $|||$. By 3400 B.C. the Egyptians had an advanced system of numeration for numbers up to and exceeding one million. Their first few number symbols show the influence of the simple tally strokes.

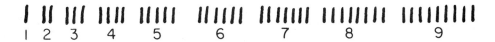

Their symbol for "3" can be seen in the third row from the bottom of the stone inscriptions shown above. What other symbols for single-digit numerals can you see on this stone?

GROUPING AND NUMBER BASES

As soon as it became necessary to count larger numbers of objects, the counting process was extended by grouping. Since the fingers furnished a convenient counting device, grouping by fives was used in some of the oldest methods of counting. Even today, people keep tallies by four straight marks and one drawn across, ⦀⦀. In certain parts of South America and Africa, it is still customary to "count by hands":

one, two, three, four, hand, hand and one, hand and two, hand and three, etc.

The left hand was generally used to keep a record of the number of objects being counted, while the right index finger pointed to the objects. When all five fingers had been used, the same hand would be used again to continue counting.

The number of objects used in the grouping process is called the *base*. Here are some examples of different bases.

Base Five   In the example of counting by hands, the base is five. By using names for the first four numbers and "hand" for the name of the base, the numbers up to and including 24 (4 hands and 4) can be named. Here are two examples of names for the numbers of dots.

one hand and three   two hands and four

To continue this system, we need a name for 25 or 5^2. This could be called "hand of hands." With this new power of the base, numbers up to 124 (4 hands of hands, 4 hands, and 4) can be named. The next step is to name 125 or 5^3. In this manner, number names can be generated indefinitely by creating new names for powers of the base five.

Base Ten   As soon as people became accustomed to counting by the fingers on one hand, it was natural to use the fingers on both hands to group by tens. In most numeration systems today, grouping is done by tens. The names of our numbers reflect this grouping process. "Eleven" derives from "ein lifon," meaning *one left over,* and "twelve" is from "twe lif," meaning *two over ten.* There are similar derivations for the number names from 13 to 19. "Twenty" is from "twe-tig," meaning *two tens,* and hundred means *ten times ten.** When the grouping is done by tens, the system is called a *base ten numeration system.*

---

*H. W. Eves, *An Introduction to the History of Mathematics,* 3rd ed. (New York: Holt, Rinehart and Winston, 1969), pp. 8–9.

Base Twenty   Grouping by twenty originates from the barefoot days when the toes as well as the fingers were accessible aids for counting. A system which uses grouping by twenty is called a *base twenty numeration system*. There are many traces of base twenty. The Mayas of Yucatan and Aztecs of Mexico had elaborate number systems based on twenty. In Greenland they used the expression "one man" for twenty, "two men" for forty, and so on. A similar system was used in New Guinea. For example, their word for 90 was "4 men and 2 hands." Evidence of grouping by twenty among the ancient Celtics can be seen in the French use of "quatre-vingt" (four-twenty) for 80. In our language the use of "score" suggests past tendencies to count by twenties. Lincoln's familiar Gettysburg Address begins with "Four score and seven years ago." Another example occurs in the childhood nursery rhyme "Four and twenty blackbirds baked in a pie."

"You're probably all wondering why I called you here today."

Base Two   Base two is the smallest base for which there can be any grouping. If we had been born with two fingers rather than ten, base two would probably be used today by most countries rather than base ten. Very likely, base two was the first base to be used by primitive people. In recent years, this base has been found among the African pygmies. Their first few number names show the influence of grouping by twos for numbers greater than three.

a	oa	ua	oa-oa	oa-oa-a	oa-oa-oa
1	2	3	4	5	6

ANCIENT NUMERATION SYSTEMS

The ancient Egyptian numeration system had a base of 10. The numerals for the first few powers of 10 were picture symbols called *hieroglyphics*. Here are the Egyptian hieroglyphics and the corresponding powers of 10.

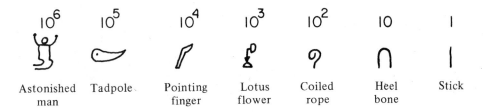

10^6	10^5	10^4	10^3	10^2	10	I
Astonished man	Tadpole	Pointing finger	Lotus flower	Coiled rope	Heel bone	Stick

2342

14,026

These symbols were repeated the required number of times to represent a number. Two examples are shown here. When a power of 10 occurred 0 times, as in 14,026, it was expressed by omitting the symbol for that power of 10. In this case, there are 0 hundreds and so the Egyptian numeral for 100 is omitted.

Some examples of the Egyptian numeration system can be seen in the stone inscriptions on the first page of this section. Notice the numeral for 743 in the third row from the bottom. The symbols for 3 ones, 4 tens, and 7 hundreds are written from left to right. It was the Egyptian custom to write increasing powers of 10 from left to right, rather than from right to left as we do today.

Roman Numeration   Roman numerals can be found on clock faces, buildings, gravestones, and the preface pages of books. Like the Egyptians, the Romans used a base ten numeration system. In addition to the symbol for one and powers of 10, there are symbols for five, fifty, and five hundred. In all, there are seven symbols.

I	V	X	L	C	D	M
1	5	10	50	100	500	1000

There is historical evidence that C is from "centum," meaning hundred, and M is from "milli," meaning thousand. The origin of the other symbols is uncertain. The Romans wrote their numerals with the highest powers of 10 on the left and the units on the right.

1000 900 70 9
MDCCCCLXXVIIII   1979

When a Roman numeral is placed to the left of a numeral for a larger number, such as IX for nine, it indicates subtraction. The subtractive principle was recognized by the Romans but they did not make much use of it.* Compare the preceding Roman numeral for 1979 with the following numeral written with the subtractive principle.

1000 900 70 9
MCMLXXIX   1979

The Romans had relatively little need for large numbers, and so they developed no general system for writing them. One early example of this is the inscription on a

*D. E. Smith, *History of Mathematics,* **2** (Lexington, Mass.: Ginn, 1925), p. 60.

monument commemorating the victory over the Carthaginians in 260 B.C. The symbol , for 100,000 is repeated 23 times to represent 2,300,000.

Mayan Numeration There is archaeological evidence that the Mayas were in Central America before 1000 B.C. During the Classical Period (A.D. 300–900) they had a highly developed knowledge of astronomy and a 365-day calendar with a cycle going back to 3114 B.C. The Mayas used a base twenty system and had a symbol for zero. The Mayan numerals for zero through 19 are shown below. Notice that there is grouping by fives within the first twenty numbers.

"No, no, no! *Thirty* days hath September!"

0	⬯	5	—	10	⚌	15	☰
1	•	6	•—	11	•⚌	16	•☰
2	••	7	••—	12	••⚌	17	••☰
3	•••	8	•••—	13	•••⚌	18	•••☰
4	••••	9	••••—	14	••••⚌	19	••••☰

The Mayas wrote their numbers vertically. In the example shown here the position of the top numeral, ••⚌, represents 12 times 20, and the lower numeral, •☰, represents 16. This "two-place numeral" represents

$$12 \times 20 + 16 \text{ or } 256$$

••⚌ 12 × 20

•☰ 16

Whenever the Mayas wrote one number over the other, as in this example, the top numeral indicated the number of twenties and the bottom numeral the number of ones. This type of system is called a *place value* or *positional numeration system* because the position of the numerals indicates the number of times the power of the base are to occur. Here are three examples of Mayan numerals.*

For further details on two Mayan numeration systems, see James K. Bidwell, "Mayan Arithmetic," *The Mathematics Teacher,* 60 No. 7 (November 1967), 762–68.

• • • (3 twenties)   —— (5 twenties)   • • (7 twenties)

◯ (0)   •—— (6)   • • • (18)

60   106   158

HINDU-ARABIC NUMERATION

Our system for writing numbers is another example of positional numeration. It is named for the Hindus, who invented it, and the Arabs, who transmitted it to Europe. Since we have a base ten numeration system, the only number symbols that are needed are the digits 0, 1, 2, 3, 4, 5, 6, 7, 8, and 9.

There are various theories about the origin of these digits. However, it is widely accepted that they originated in India. Notice the resemblance of the Brahmi numerals for 6, 7, 8, and 9 to our numerals. The Brahmi numerals for 1, 2, 4, 6, 7, and 9 were found on stone columns in a cave in Bombay dating from the second or third century B.C.* It seems likely that our numerals 2 and 3 are merely cursive forms of these early symbols for two and three.

Our method of numeration is remarkably efficient. Each digit represents a power of 10 according to its position (place value), and these positions are referred to by name. In 75,063 the 3, 6, 0, 5, and 7 are called the *units digit, tens digit, hundreds digit, thousands digit,* and *ten thousands digit,* respectively. When we write a number as the sum of numbers represented by each digit in the numeral, the number is said to be written in *expanded form.* Here are three ways of writing a number in expanded form.

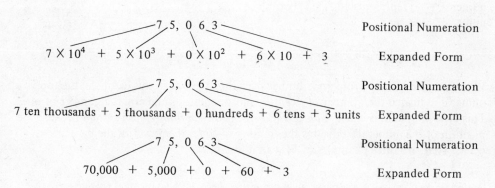

$$7,5,0\ 6\ 3 \quad \text{Positional Numeration}$$
$$7 \times 10^4 + 5 \times 10^3 + 0 \times 10^2 + 6 \times 10 + 3 \quad \text{Expanded Form}$$

$$7,5,0\ 6\ 3 \quad \text{Positional Numeration}$$
$$7\ \text{ten thousands} + 5\ \text{thousands} + 0\ \text{hundreds} + 6\ \text{tens} + 3\ \text{units} \quad \text{Expanded Form}$$

$$7,5,0\ 6\ 3 \quad \text{Positional Numeration}$$
$$70,000 + 5,000 + 0 + 60 + 3 \quad \text{Expanded Form}$$

*J. R. Newman, *The World of Mathematics,* **1** (New York: Simon and Schuster, 1956), pp. 452–54.

Many years of development and struggle lie behind the Hindu-Arabic numeration system. The oldest dated European manuscript that contains our numerals was written in Spain in A.D. 976. In 1299, merchants in Florence were forbidden to use these numerals. Gradually, over a period of centuries, the Hindu-Arabic numeration system replaced the more cumbersome Roman numeration.

Reading and Writing Numerals   The number names for the whole numbers from one to twenty are all single words. The names from twenty-one to ninety-nine, with the exceptions of thirty, forty, fifty, etc., are compounded number names that are hyphenated. These names are also hyphenated when they occur as parts of other names. For example, we write, "three hundred forty-seven," for 347. Some people write or read this number as "three hundred and forty-seven," but the "and" is not necessary.

Numerals with more than three digits are read by naming periods for each group of three digits. Within each period the digits are read as we would read any number from 1 to 999, and then the name of the period is recited. The names of several periods are listed at the right. Here is an example of a number in the trillions.

thousand
million
billion
trillion
quadrillion
quintillion
sextillion

2 3, 4 7 8, 5 0 6, 0 4 2, 3 1 9

trillion   billion   million   thousand

This number is read as "twenty-three trillion, four hundred seventy-eight billion, five hundred six million, forty-two thousand, three hundred nineteen."

ROUNDING OFF

If you were to ask a question, such as, "How many people voted in the 1980 presidential election?" you might be told that in "round numbers" it was about 85 million. Approximations such as this are often as helpful as knowing the exact number, which in this example was 85,100,119.

To *round off* a number to a given place value, locate the place value and then check the digit to its right. If the digit to the right is 5 or greater, then all the digits to the right are replaced by 0's and the given place value is increased by 1. If the digit to the right is 4 or less, then all digits to the right of the given place value are replaced by 0's. Rounded off to the nearest million, the number of votes in the 1980 election is 85 million, because the 1 to the right of the millions place is less than 5.

millions place

85,100,119

Here are some more examples of rounding off to different place values in 85,100,119.

Rounding to	Hundred thousands place	Rounded off
1. Hundred thousands	85,100,119	85,100,000
2. Ten thousands	Ten thousands place 85,100,119	85,100,000
3. Thousands	Thousands place 85,100,119	85,100,000
4. Hundreds	Hundreds place 85,100,119	85,100,100

MODELS FOR NUMERATION

Of all the arithmetical concepts, positional numeration and place value lend themselves most readily to the use of physical materials. Here are several models which will be described: bundles of sticks, bean sticks, multibase pieces, the abacus, and chip trading.

Bundles of Sticks (or Straws)   In this model units and tens are represented by single sticks and bundles of 10 sticks. One hundred is represented by a bundle of 10 bundles.

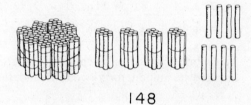

148

Bean Sticks   In the *bean stick model*, single beans represent units; a stick (or strip of cardboard) with 10 beans glued to it represents 10; and 100 is a collection of 10 bean sticks.

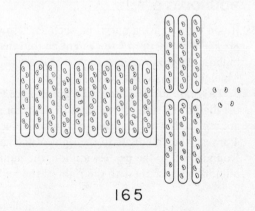

165

Multibase Pieces   The powers of the base in this model are represented by objects called *units, longs, flats,* and *blocks.* In base 10, a *long* equals 10 units, a *flat* equals 10 longs, and a *block* equals 10 flats. These pieces represent the increasing powers of 10, namely, 1, 10, 100, and 1000. This model can be extended to higher powers of the base by using groups of blocks. For example, 10,000 is represented by a column of 10 blocks, and is called a *long-block*. The following set of multibase pieces represents 1347.

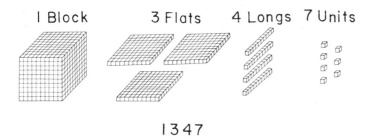

1 Block   3 Flats   4 Longs   7 Units

1 3 4 7

Abacus The abacus is considered to be the oldest of all computing devices. Some form of it was used in Egypt at least as early as 500 B.C. There are several types in use in the world today. They are found in the Middle East, Russia, and the Orient. The basic idea of the abacus is to represent numbers by beads on rods. A Japanese abacus, called a *soroban,* is pictured below. Each bead below the horizontal bar (from right to left) represents a power of 10 ($1, 10, 10^2, 10^3, \ldots$), and the corresponding bean above the bar represents five times as much ($5, 50, 500, 5000, \ldots$). The two beads in the first column, which have been pushed to the horizontal bar, represent 6 ones. The one bead in the second column represents 10, and the four beads in the third column represent 400. All together, the beads represent 416.

4   1   6

This Japanese soroban is the generous gift of Seiichi Kaida, Tokyo, Japan.

In Western countries today the abacus is used mainly as a model for teaching place value and the basic operations. In the simplified abacus shown here, any number of markers may be placed on a column. For base ten, each column represents increasing powers of 10 from right to left (from 10^0 to 10^6). The three markers on the 10^4 column of this abacus represent 3×10^4 or 30,000. The seven markers on the 10^3 column represent 7000, etc. If there are no markers on a column, this means that particular power of the base occurs zero times. The markers shown represent 37,204.

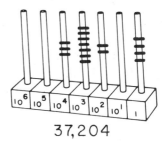

37,204

Chip Trading This model uses colored chips and a mat and is similar to the abacus. There are columns for the chips, and in base ten the columns represent powers of 10. One yellow chip represents a unit, a blue chip stands for 10, green corresponds to 100, and each red chip is worth 1000. Viewed in another way, 10 yellow chips can be traded for 1 blue chip, 10 blue chips can be traded for 1 green chip, etc.

4 3 7 5

The mat keeps the chips in columns as a model for the place-value concept of positional numeration. The 5 yellow chips on the mat illustrated represent 5 ones; the 7 blue chips represent 7 tens; etc. All together, these chips represent 4375.

PLACE VALUE ON CALCULATORS

On most calculators the digits that are entered first appear on the right in the display, then they move left as successive digits are entered. For example, to enter 428 on the calculator, we press 4, then 2, then 8.

4   2   8

First a "4" appears in the display, then the "4" moves over and "42" appears, and finally the "42" moves and "428" appears. The three stages of the display are thus:

| 4. | 42. | 428. |

On some calculators the numbers appear in the display from left to right. When 428 is entered, first a "4" appears at the left end of the display, then "42," and finally "428."

| 4. | 42. | 428. |

The calculator can be used to increase understanding of positional numeration and place value by writing a number in expanded form and computing the sum. If we enter the value indicated by the position of each digit in 48,023 and *add* these numbers, then "48023" will appear in the display. Here is the sum to enter:

40000   +   8000   +   0   +   20   +   3

This process can be reversed by entering a number into the calculator and then removing one digit at a time from the left in the display. If 93706 is entered, it can be changed to 3706 by subtracting 90000. Next, the 3706 can be changed to 706 by subtracting 3000. Finally, subtracting 700 from 706 leaves the units digit in the display.

93706
3706
706
6

SUPPLEMENT (Activity Book)

Activity Set 2.2 Models for Numeration (Base five multibase pieces and the variable-base abacus)

Exercise Set 2.2 Applications and Skills

1. In the fifteenth and sixteenth centuries there were two opposing opinions on the best numeration system and methods of computing. There were the "abacists" who used Roman numerals and computed on the abacus and the "algorists" who used the Hindu-Arabic numerals and place value. The accompanying sixteenth-century picture shows an abacist competing against an algorist. The abacist has a reckoning table with four horizontal lines and a vertical line down the middle. Counters or chips placed on lines represented powers of 10, as shown in the next diagram. The thousand's line was marked with a cross to aid the eye in reading numbers. If more lines were needed, every third line was marked with a cross. This was the origin of separating groups of three digits in a numeral by a comma.*

An algorist computing with numerals and an abacist computing with counters

*D. E. Smith, *History of Mathematics,* 2 (Lexington, Mass: Ginn, 1925), pp. 183–85.

★ a. What number is represented on the left side of this reckoning table?

b. Each counter in a space between the horizontal lines represents half as much as it would on the line above. What number is represented on the right side of this reckoning table?

★ c. Each line and the spaces between them can be labeled by Roman numerals similar to the labeling of lines and spaces on sheet music. Use the seven Roman numerals (I, V, X, L, C, D, M) to label the lines and spaces on the reckoning table.

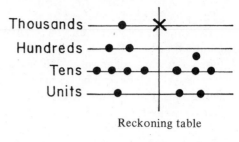

Reckoning table

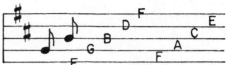

2. Most ancient civilizations developed ways to represent numbers with various positions of their hands and fingers. By the Middle Ages this method had become international for business transactions. In the system shown here the numbers from 1 to 10,000 can be represented.

★ a. Represent 9090 on your hands with this system. On this chart circle the hand positions that you used.

b. Circle the hand positions for representing 404.

★ c. Which pairs of these columns are similar?

d. How can this chart be used to explain the following phrase found in Juvenal's tenth satire: "Happy is he indeed who has postponed the hour of his death so long and finally numbers his years upon his right hand."*

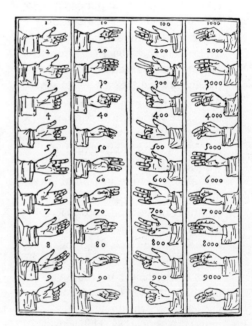

Finger counting for numbers
from 1 to 10,000

*H. W. Eves, *Mathematical Circles* (Boston: Prindle, Weber and Schmidt, 1969), p. 29.

e. In medieval times the finger symbol for 30 was called "the tender embrace" and was a symbol for marriage.* Have you ever seen this symbol used as a gesture of something else?

3. Here is the complete counting system of a twentieth-century Australian tribe that uses only four numbers.

Neecha	Boola	Boolla Neecha	Boolla Boolla
1	2	3	4

a. If this system were continued, what would be the names for 5 and 6?

b. How would even numbers differ from odd numbers in a continuation of this system?

4. Using the base five system of counting by fingers and grouping by hands, the name for 7 is "one hand and two."

★ a. What is the name for 22 in this system?

★ b. If 25 is called a "hand of hands," what is the name for 37 in this system?

★ 5. The names of the following numbers are some of the literal meanings of number words taken from primitive languages in various parts of the world.† Write in the missing names that follow this pattern.

5, whole hand; 6, one on the other hand; 8, _____ ;
10, _____ ; 11, one on the foot; 15, _____ ;
16, _____ ; 20, man; 21, one to the hands of the next man;
25, _____ ; 30, _____ ;
40, _____ .

6. The following numeration systems were used at widely varying times in different geographical locations. Compare these sets of numerals for the numbers 1 through 10. What similarities can you find? What evidence is there of grouping by fives?

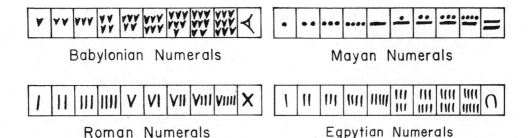

Babylonian Numerals   Mayan Numerals

Roman Numerals   Egpytian Numerals

*H. W. Eves, *Mathematical Circles Squared* (Boston: Prindle, Weber and Schmidt, 1972), p. 4.
†D. Smeltzer. *Man and Number* (London: A. and C. Black, 1970), pp. 14–15.

7. Write each number in the Egyptian, Roman, and Mayan numeration systems.
   a. 15 ★ b. 48 c. 172

8. Perhaps the greatest achievement in the development of numeration systems was the invention of a symbol for zero. The following ancient numeration systems had no zero symbol. How is 603 written in each of these systems?
   a. Egyptian ★ b. Roman

★ 9. These Greek numerals have been dated about 1200 B.C. Use these symbols in the same manner the Egyptians used their symbols to write 2483.

| | 10 100 1000

10. The Attic-Greek numerals were developed sometime prior to the third century B.C. and came from the first letters of the Greek names for numbers. Use the clues in the following table to find the missing numerals. What base is used in this system?

Hindu–Arabic	I	4	8		26	32	52	57	206	
Attic–Greek			ΓΙΙΙ	ΔΓΙ		ΔΔΔΙΙ		ᗰΓΙΙ	HHΓΙ	ᗰΔΙ

11. Write each number in three different ways using expanded form.
    a. 256,049 b. 7088

12. Write the names of the numbers in parts **a** through **d**.
★ a. 5,438,146 b. 31,409
★ c. 816,447,210,361 d. 62,340,782,000,000

13. Round off 43,668,926 to the nearest:
    a. Hundred thousand (hundred thousands place)
★ b. Ten thousand (ten thousands place)
    c. Thousand (thousands place)
    d. Hundred (hundreds place)

14. Use a sketch to represent the following numbers as they would be illustrated by the given models.
★ a. 136 by multibase pieces b. 47 by the bundle-of-sticks model
    c. 64 by the bean-stick model

15. Draw markers or chips on the following diagrams to represent the numbers.

★   a. 4908   b.   7052

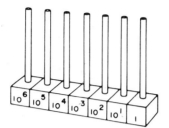

Red	Green	Blue	Yellow

16. The Japanese abacus (soroban) on page 73 has 23 columns.

★   a. What is the largest number that can be represented on this abacus?

     b. Describe the positions of the beads for the number in part **a**.

     c. Write the name of this number.

17. *Calculator Exercise:* In parts **a** through **c**, explain what you would do to the calculator to change the top display to the one under it without changing the digits that are the same in both displays.

★   a.   b.   c.

1034692.	938647.	40000.
1834692.	908047.	400000.

18. *Binary Numbers:* Powers of 2 are used in base two numeration systems. These powers are called *binary numbers*. Here are the first few.

1	2	4	8	16	32	64	128	. . .
2^0	2^1	2^2	2^3	2^4	2^5	2^6	2^7	. . .

Every positive whole number can be written as a sum of binary numbers so that each binary number is used only once or not at all. For example, $25 = 16 + 8 + 1$. Write each of the following numbers as a sum of binary numbers, using no binary number more than once.

a. 35   ★   b. 42   c. 64

Exercise Set 2.2: Problem Solving

1. *Chess Legend:* There is a legend that chess was invented for the Indian King Shirham by the Grand Vizier Sissa Ben Dahir. As a reward, Sissa requested that he be given 1 grain of wheat to place on the first square of the chess board, 2

grains for the second square, 4 grains for the third square, then 8 grains, 16 grains, etc. until each square of the board was accounted for. The King was surprised at such a meager request until Sissa informed him that this was more wheat than existed in the entire kingdom. What is the sum of all the grains for the 64 squares of the chess board?

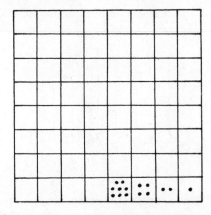

a. *Understanding the Problem:* The number of grains for each square is a binary number. It is sometimes convenient to express these numbers as powers of 2.

$$1 \qquad 2 \qquad 2^2 \qquad 2^3 \qquad 2^4 \qquad \ldots$$

Write the number of grains for the sixty-fourth square as a power of 2.

b. *Devising a Plan:* It would be a difficult task to compute the sum of all 64 binary numbers. Let's form a table for the first few sums and look for a pattern. Compute the next three totals in this table. How is each total related to a power of 2?

No. of Squares	Number of Grains	Total
1	1	1
2	$1 + 2$	3
3	$1 + 2 + 2^2$	7
4	$1 + 2 + 2^2 + 2^3$	
5	$1 + 2 + 2^2 + 2^3 + 2^4$	
6	$1 + 2 + 2^2 + 2^3 + 2^4 + 2^5$	

c. *Carrying Out the Plan:* Use the pattern in part **b** to express the sum of the grains for all 64 squares. Write your answer using a power of 2.

★ d. *Looking Back:* King Shirham was surprised at the total amount of grain because the number of grains for the first few squares is so small. There is more grain for each additional square than the total number of grains for all the preceding squares. Show that the number of grains for the sixty-fourth square is greater than the total number of grains for the first 63 squares.

2. *Single-Elimination Tournaments:* A single-elimination basketball tournament has 247 teams competing for the championship. If the tournament sponsors must pay $20 to have each game refereed, what is the total cost of referees for the tournament?

   a. *Understanding the Problem:* For every two teams that play each other, there is a winner and a loser. The loser is eliminated from the tournament and the winner plays another team. The following brackets for a four-team tournament show that three games are needed to determine a champion.

   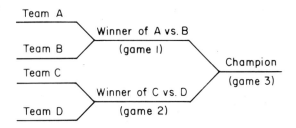

   If the number of teams entered in the tournament is not a power of 2, byes are necessary. That is, some teams will not have to play the first round so that the number of teams for the second round will be a power of 2. Draw a set of brackets for a seven-team tournament. How many teams will not play the first round?

   b. *Devising a Plan:* One approach to solving this problem is to use small numbers and look for a pattern. Complete the following table.

Number of teams	2	3	4	5	6	7	8
Total number of games	1		3				

★ c. *Carrying Out the Plan:* Use the approach suggested in part **b** and inductive reasoning, or use your own plan, to solve this problem. What is the total cost of referees for the tournament?

   d. *Looking Back:* The brackets in part **a** have directed our attention to the winning teams. The total number of games played can be more easily determined by thinking about the losing teams. Each game that is played determines one loser. How many losing teams were there in the tournament?

2.3
COMPUTER
PROGRAMMING

The computer revolution is here. Computers have invaded every aspect of life from space research to elementary school instruction. The first electronic computer, the ENIAC, was built in 1946 at the University of Pennsylvania. It consisted of 18,000 vacuum tubes, weighed 30 tons, and tended to overheat and break down. Since that time the replacement of tubes by miniaturized circuits has resulted in smaller, more reliable computers.

MICROCOMPUTERS

The most widely available computer in homes and precollege education is the micro-computer. The heart of a microcomputer contains miniaturized circuits called *chips*. A chip contains thousands of electronic circuits. The following photo shows a chip inside its housing. Microcomputers first appeared in the 1970s and are now common in businesses, schools, and homes.

The typical microcomputer uses a typewriter-style keyboard for input, a television screen for displaying input and output, and some type of magnetic system for recording programs and data. The cheapest magnetic system for storage of programs and data is a cassette tape recorder. An alternative to the cassette tape recorder is the floppy disk (or diskette) and disk drive. A floppy disk is a thin circular piece of magnetically coated plastic that is inserted into the disk drive.

Before a computer can perform a task it must be given step-by-step instructions using words it understands. A sequence of instructions and data for carrying out a task is called a *program* or a *procedure*. Each program is written in one of many existing computer languages.

Two of the easiest computer languages to learn are BASIC and Logo. BASIC was developed in 1964 at Dartmouth College by John Kemeny and Thomas Kurtz. Logo was developed in the 1970's at MIT under the direction of Seymour Papert. The remainder of this chapter contains instructions for programming in these two languages. We have used only those commands in Logo and BASIC which are common in most versions of these languages. However, it is possible that some adjustments in computer programs will be needed. In particular, specific computer manuals should be consulted for such tasks as editing programs and starting and turning off the computer.

PROGRAMMING IN LOGO

Logo was written for children. In experiments at MIT's Artificial Intelligence Laboratory, children drew geometric figures by giving instructions to a mechanical robot called a *turtle*. As the turtle moved across the floor on large sheets of paper, a pen at its center traced a path. The turtle could be commanded to move from one point to another by giving it an angle to turn through and a distance to move.

Now geometric figures are drawn on the computer

A robot at M.I.T.'s Artificial Intelligence Lab

screen by moving a small triangular pointer called a *turtle*. The turtle has both a *position* and a *heading*. Its *start position* is at the center of the screen, heading north. This is called *home*. Regardless of where the turtle is on the screen, if you type HOME and then press RETURN (or ENTER), the turtle will go to the center of the screen and head in the north direction. For each position on the

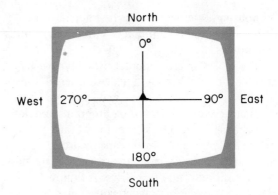

screen the turtle points in some direction, called its *heading*. The heading is some number of degrees from 0 to 360.

The turtle can be moved about by giving it commands. Type FORWARD 30 and press RETURN and the turtle will move forward 30 turtle steps, tracing its path. Type RIGHT 90 and press RETURN, and the turtle will turn 90° to the right. Type FORWARD 50 and press RETURN, and the turtle will move forward 50 steps, tracing its path. These moves are shown on the screen.

```
FORWARD 30
RIGHT 90
FORWARD 50
```

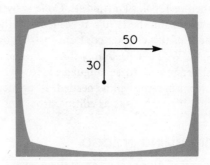

To return the turtle to its home position and clear the screen of all turtle tracks type HOME CLEARSCREEN and press RETURN.* Sometimes it is helpful to hide the turtle to get a better view of a geometric figure. You can make the turtle invisible by typing HIDETURTLE and you can make it appear again by typing SHOW-TURTLE. The turtle can be turned right or left any number of degrees and it can be moved backward as well as forward. Here are seven commands and their abbreviations. The RETURN

Command	Abbreviation
FORWARD	FD
BACKWARD	BK
RIGHT	RT
LEFT	LT
CLEARSCREEN	CS
HIDETURTLE	HT
SHOWTURTLE	ST

*In Apple Logo this can be accomplished by typing CLEARSCREEN.

key must be pressed before the
computer will carry out any
command.*

Creating Commands   A line can be
drawn by moving the turtle foward and
backward. The following commands
will produce the line shown here and
leave the turtle in its start position at
the center of the screen.

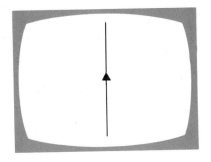

```
FD 80
BK 160
FD 80
```

One of the advantages of Logo is that we can define new commands called
procedures. For example, we can give a list of commands a name, and whenever we
type the name the turtle will carry out these commands. We can think of a procedure as
teaching the turtle a new word. Here is a procedure for drawing the line in the
preceding example. We need to make up a name for this procedure, so let's call it
LINE.

```
TO LINE
 FD 80
 BK 160
 FD 80
END
```

The TO tells the computer you are defining a new command. The END tells the
computer you have finished the definition. After you type END and press RETURN,
the computer will respond with, LINE DEFINED. Now, if you type LINE, the turtle
will draw a line in the direction it is heading and finish in the position it started in.

LINE is used three times in the following set of commands. The turtle's initial
position is the center of the screen and its heading is north. The figure shows three lines
which are at 60° angles to each other. The turtle's final heading is 180 (south) because it
has turned through three 60° angles.

```
LINE
RT 60
LINE
RT 60
LINE
RT 60
```

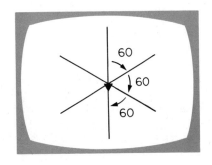

*In the remaining development, the need to press RETURN (or ENTER) will usually not be stated.

There are times when we want the turtle to move to a new location but not draw a path. This can be accomplished by using the command PENUP. When we want the turtle to draw again, we use the command PENDOWN. These commands are used in the next example together with the command LINE to draw two parallel lines.

```
LINE
RT 90
PENUP
FD 20
PENDOWN
LT 90
LINE
```

Two parallel lines

Procedures in Logo are built up by using simpler procedures. Here is an example that begins with a procedure called STEP for drawing a right angle. The 90° left turn leaves the turtle heading north. Then STEP is used four times to define STAIRS. Finally, STAIRS is used four times to define MANYSTAIRS.

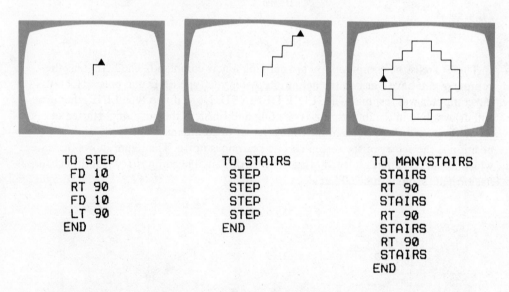

```
TO STEP          TO STAIRS        TO MANYSTAIRS
   FD 10            STEP             STAIRS
   RT 90            STEP             RT 90
   FD 10            STEP             STAIRS
   LT 90            STEP             RT 90
END              END                STAIRS
                                    RT 90
                                    STAIRS
                                    RT 90
                                    STAIRS
                                 END
```

Repeat Command STEP was used four times to define STAIRS. Rather than type STEP four times, the same result can be accomplished by using a command called REPEAT. This command requires a number to tell the turtle the number of times the instructions are to be repeated and a list of instructions which are typed inside square brackets.* Notice that the final "S" is missing from STAIRS in the procedure below. Since this is a new procedure, we need a new word.

```
TO STAIR
  REPEAT 4 [STEP]
END
```

The commands LINE and RT 60 which were used three times to produce three lines at 60° angles (page 86) can now be replaced by one command.

```
REPEAT 3 [LINE RT 60]
```

The following commands produce a square whose sides have length 60. Notice that two commands are written on each line. Several commands may be typed on a line before RETURN is pressed.

```
FD 60 RT 90
FD 60 RT 90
FD 60 RT 90
FD 60 RT 90
```
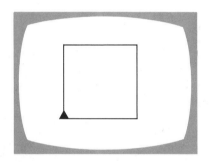

Since FD 60 RT 90 is repeated four times, we can accomplish the same result by using the REPEAT command.

```
REPEAT 4 [FD 60 RT 90]
```

Now let's use the REPEAT command to define a procedure for drawing this square.

```
TO SQ
  REPEAT 4 [FD 60 RT 90]
END
```

By typing SQ the turtle will draw a square whose sides have length 60.

Variables   With variables we can define procedures for drawing figures with variable dimensions. Words as well as letters can be used for variables but they must be preceded by a colon (:). Here is a procedure for drawing various-size squares which uses

*On some computers the left and right square brackets are obtained by holding down the shift key and pressing N and M.

SIDE as a variable. Once the procedure is defined and SQUARE 80 is typed, the turtle will draw a square whose sides have length 80.

```
TO SQUARE :SIDE
  REPEAT 4 [FD :SIDE RT 90]
END
```

This figure was produced by using the procedure SQUARE four times.

```
SQUARE 80
SQUARE 60
SQUARE 40
SQUARE 20
```

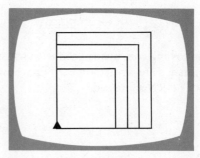

Nested squares

The procedure for drawing a square is carried out six times in the next example. After each square is drawn the turtle is turned 60° before drawing the next square. The first command hides the turtle so the intersection of the squares at the center of the screen will be more visible.

```
TO SPINSQUARE
  HIDETURTLE
  REPEAT 6 [SQUARE 50 RT 60]
END
```

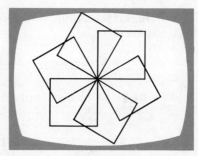

Six squares

The next procedure has two variables, one for width and one for length. It instructs the turtle to draw rectangles of various sizes.

```
TO RECTANGLE :WIDTH :LENGTH
 REPEAT 2 [FD :WIDTH RT 90 FD :LENGTH RT 90]
END
```

Once this procedure is defined and RECTANGLE 40 60 is typed, the turtle will draw a rectangle with a width of 40 and a length of 60.

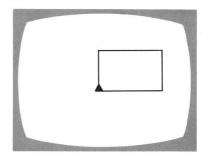

Rectangle 40 60

PROGRAMMING IN BASIC

BASIC (*B*eginner's *A*ll-Purpose *S*ymbolic *I*nstruction *C*ode) was designed as a simple and easy-to-learn language which allows beginners to write elementary computer programs. While Logo is a relatively new computer language, BASIC, on the other hand, has been widely used for over 20 years and is the "built-in" language of most microcomputers. In the next few pages you will learn some BASIC commands and how to organize them to write computer programs.

PRINT and END Commands Program 2.3A has two statements on lines numbered 10 and 20. PRINT is a command that instructs the computer to do something and to print the result. END is a command which tells the computer there are no more statements. In BASIC the lines of a program must be numbered. It is a good programming habit to use multiples of 10 in case additional lines need to be inserted later.

"Nope, no chess problems until you've finished your quadratic equations!"

Program 2.3A

```
10   PRINT 36 + 17
20   END
```

After you have typed each line of a program, the RETURN (or ENTER) key must be pressed to enter the instructions into the computer. When the complete program has been typed, the command RUN is typed and the RETURN (or ENTER) key is pressed to order the computer to carry out its instructions.* If RUN is typed for the preceding program, the computer will print out 53. When you are finished with a program and wish to type a new program, the command NEW (or SCRATCH) must be typed to erase the old program. The commands RUN and NEW do not require line numbers; the computer carries them out as soon as they are typed.

Program 2.3B demonstrates how the PRINT command and quotation marks can be used for printing statements. The printout below the program shows that the computer has printed exactly what is contained inside the quotation marks and has done the computations that are outside and printed the results. The asterisk (*) which is used in line 20 indicates multiplication.

Program 2.3B

```
10   PRINT "THE SUM 14 + 6 IS EQUAL TO "14 + 6"."
20   PRINT "5 AND 9 HAVE A SUM OF "5 + 9" AND A PRODUCT OF
        "5 * 9"."
30   PRINT "THIS IS AS EASY AS A, B, C."
40   END
```

```
RUN

THE SUM 14 + 6 IS EQUAL TO 20.
5 AND 9 HAVE A SUM OF 14 AND A PRODUCT OF 45.
THIS IS AS EASY AS A, B, C.
```

INPUT Command   Sometimes it is necessary to compute a formula for different values of a variable. The INPUT command allows the operator to type in numbers for the variables when the program is run. Program 2.3C shows how this is done. This program computes the sum of whole numbers from 1 to N. The slash (/) which is used in line 30 indicates division. When the program is run, the computer will print out the quoted message in line 10 and then line 20 causes a question mark to be printed. At this point the operator should type a number and press the RETURN (or ENTER) key. The printout shows that 6 was typed in when this program was run.

Program 2.3C

```
10   PRINT "PLEASE TYPE A WHOLE NUMBER."
20   INPUT N
30   PRINT "THE SUM OF WHOLE NUMBERS FROM 1 TO "N" IS "N * (N
        + 1) / 2"."
40   END
```

---

*From now on the need to press the RETURN (or ENTER) key will usually not be mentioned.

```
RUN

PLEASE TYPE A WHOLE NUMBER.
?6
THE SUM OF WHOLE NUMBERS FROM 1 TO 6 IS 21.
```

Several variables may be used with an INPUT command. When a program is run and the computer comes to an INPUT command with two or more variables, it still types only one question mark. The numbers which are typed in by the operator should be separated by commas.

Program 2.3D has three variables. This program computes the monthly profit for the sale of X Type I heaters, Y Type II heaters, and Z Type III heaters. The profits for Type I, Type II, and Type III heaters are $14.80, $21.65, and $25.90, respectively. When the operator types RUN and presses the RETURN key, the computer will print the quoted message in line 10. Then line 20 causes a question mark to be printed, and the operator should type in three numbers. The printout for this program shows that 4280, 3561, and 1577 were typed in when the program was run.

Program 2.3D

```
10   PRINT "THIS PROGRAM COMPUTES THE MONTHLY PROFIT FOR THE
       SALE OF THREE TYPES OF HEATERS. TYPE THE NUMBER OF TY
       PE I, TYPE II, AND TYPE III HEATERS."
20   INPUT X,Y,Z
30   PRINT "THE MONTHLY PROFIT IS $"14.80 * X + 21.65 * Y +
       25.90 * Z"."
40   END
```

```
RUN

THIS PROGRAM COMPUTES THE MONTHLY PROFIT FOR THE SALE OF
THREE TYPES OF HEATERS. TYPE THE NUMBER OF TYPE I, TYPE II,
AND TYPE III HEATERS.
? 4280, 3561, 1577
THE MONTHLY PROFIT IS $181283.95.
```

LET Command LET is a command for assigning a variable to a number or formula. Lines 10 and 20 of Program 2.3E set the variables A and B equal to numbers and line 30 sets C equal to a formula. The printout is the computed value of C.

Program 2.3E

```
10   LET A = 5
20   LET B = 7
30   LET C = 2 * A + B
40   PRINT C
50   END
```

```
RUN

17
```

The INPUT command can be combined with the LET command to allow the operator to type in different numbers for a formula each time the program is run. Program 2.3F computes the total cost for H students buying hot-lunch tickets and C students buying cold-lunch tickets. Each hot-lunch ticket costs $1.65 and a cold-lunch ticket costs $1.20. When this program was run, 174 and 118 were typed in for H and C, respectively.

Program 2.3F

```
10    INPUT H,C
20    LET T = 1.65 * H + 1.20 * C
30    PRINT "THE TOTAL COST FOR "H" STUDENTS BUYING HOT LUNC
      H TICKETS AND "C" STUDENTS BUYING COLD LUNCH TICKETS
      IS $"T"."
40    END


RUN

? 174, 118
THE TOTAL COST FOR 174 STUDENTS BUYING HOT LUNCH TICKETS
AND 118 STUDENTS BUYING COLD LUNCH TICKETS IS $428.7.
```

GO TO and IF-THEN Commands In the programs so far the computer has carried out instructions by proceeding from one line to the next-higher-numbered line. Sometimes it is convenient to "jump" to some other part of the program. The GO TO command can be used to send the computer to any line of the program.

Program 2.3G revises Program 2.3C (page 90) for computing the sum of whole numbers from 1 to N, by inserting line 36. Now the RUN command needs to be typed only once. When the computer gets to line 36 it is sent back to line 10 for another value of N. The computer continues going through lines 10, 20, 30, and 36 as long as the operator types a number for the INPUT command. The computer will never reach the END command in this

"What I especially like about this baby is this little drawer where I can keep my lunch."

program. In order to get out of the program there are certain keys which the operator can press. Two examples are ⌐BREAK⌐, and ⌐CONTROL⌐ ⌐RESET⌐. The printout below RUN in Program 2.3G shows that sums were computed for N = 4 and N = 5.

Program 2.3G

```
10    PRINT "PLEASE TYPE A WHOLE NUMBER."
20    INPUT N
30    PRINT "THE SUM OF WHOLE NUMBERS FROM 1 TO "N" IS "N * (N
        + 1) / 2"."
36    GOTO 10
40    END
```

```
RUN

PLEASE TYPE A WHOLE NUMBER.
?4
THE SUM OF WHOLE NUMBERS FROM 1 TO 4 IS 10.
PLEASE TYPE A WHOLE NUMBER.
?5
THE SUM OF WHOLE NUMBERS FROM 1 TO 5 IS 15.
PLEASE TYPE A WHOLE NUMBER.
```

The program you have just seen will continue to run as long as the operator types numbers for N. What we need is some method of sending the computer to line 40 to end the program. This is accomplished by the next command.

IF-THEN Command An IF-THEN command enables the computer to make a test of a condition and then to take action depending on the result of this test. Let's revise the preceding program by inserting an IF-THEN command (see line 22) in Program 2.3H. When the computer comes to line 22 it examines the value of N. If N is equal to 0, the computer is sent to line 40 to end the program. If N is not equal to 0, the computer will continue past line 22 to line 30.* The printout shows that the operator typed two nonzero values for N before typing 0.

Program 2.3H

```
10    PRINT "PLEASE TYPE A WHOLE NUMBER. TO END THE PROGRAM TY
        PE 0."
20    INPUT N
22    IF N = 0 THEN  GOTO 40
30    PRINT "THE SUM OF WHOLE NUMBERS FROM 1 TO "N" IS "N * (N
        + 1) / 2"."
36    GOTO 10
40    END    .
```

```
RUN

PLEASE TYPE A WHOLE NUMBER. TO END THE PROGRAM TYPE 0.
?7
THE SUM OF WHOLE NUMBERS FROM 1 TO 7 IS 28.
```

*Note: The IF-THEN command has been combined with the GO TO command in line 22. This could be written as, IF X = 0 THEN 40, by omitting the words GO TO.

```
PLEASE TYPE A WHOLE NUMBER. TO END THE PROGRAM TYPE 0.
?8
THE SUM OF WHOLE NUMBERS FROM 1 TO 8 IS 36.
PLEASE TYPE A WHOLE NUMBER. TO END THE PROGRAM TYPE 0.
?0
```

Program 2.3I tests the product of two numbers to determine if it is positive. If the product is positive then the hypothesis in line 20 is true and the computer is sent to line 50. If the product is not greater than zero, the computer continues on to line 30.

Program 2.3I

```
10   INPUT A, B
20   IF A * B ) 0 THEN   GOTO 50
30   PRINT "THE PRODUCT IS NOT POSITIVE."
40   GOTO 60
50   PRINT "THE PRODUCT IS POSITIVE."
60   END
```

Program 2.3J determines if the total cost of N items is greater than $500. Each item costs $14.80. If the cost is greater than 500, line 30 sends the computer to line 60. If the cost is not greater than 500, the computer continues past line 30 to line 40.

Program 2.3J

```
10   INPUT N
20   LET C = 14.80 * N
30   IF C ) 500 THEN   GOTO 60
40   PRINT "THE COST IS NOT ABOVE $500."
50   GOTO 70
60   PRINT "THE COST IS ABOVE $500."
70   END
```

Loops and FOR-NEXT Commands

A *loop* is a sequence of statements in a program which is carried out several times. We have already seen examples of a loop in the programs which computed the sums of whole numbers from 1 to N (Programs 2.3G and 2.3H). These programs were inconvenient in that we had to type in the next value of N each time the computer went through the program. This situation can be avoided by a FOR-NEXT command. This command assigns the values which the variable is to take on.

Program 2.3K revises Program 2.3H by using a FOR-NEXT command. The FOR in line 10 stipulates that N is to take on the whole numbers from 2 through 12. The NEXT in line 30 assigns the next value of N, as long as N is less than 12, and returns the computer to line 10. When N equals 12 the computer makes its last pass through the program and since there are no more values for N it goes to line 40 to end the program.

Program 2.3K

```
10   FOR N = 2 TO 12
20   PRINT "THE SUM OF WHOLE NUMBERS FROM 1 TO "N" IS "N * (N
       + 1) / 2"."
30   NEXT N
40   END
```

Any letter can be used for the variable in a FOR-NEXT command. In Program 2.3L the variable is X. The first line of this program shows that X will take on the whole numbers from 1 to 5, beginning with 1.

Line 30 keeps sending the computer back to line 10 as long as X is less than or equal to 5. When the computer comes to line 30 for the fifth time, it continues on to line 40 to end the program.

Program 2.3L

```
10   FOR X = 1 TO 5
20   PRINT "PRIVATE PROPERTY"
30   NEXT X
40   END
```

Here is another program with a FOR-NEXT command. This program prints the first N even numbers. The operator types in a number for N in line 10 and X takes on the values from 0 to N − 1 in line 20. Notice the semicolon used in the PRINT statement in line 30. This causes the printout to be printed on one line until that line is full. The printout below this program was obtained when the operator ran the program and typed 56 for N.

Program 2.3M

```
10   INPUT N
20   FOR X = 0 TO N - 1
30   PRINT 2 * X" ";
40   NEXT X
50   END


RUN

? 56
0 2 4 6 8 10 12 14 16 18 20 22 24 26 28 30 32 34 36 38 40 42
44 46 48 50 52 54 56 58 60 62 64 66 68 70 72 74 76 78 80 82
84 86 88 90 92 94 96 98 100 102 104 106 108 110
```

SUPPLEMENT (Activity Book)

Activity Set 2.3 Computer Games (Target Practice, Hi-Lo, and Pica-Centro)

Exercise Set 2.3: Logo

1. Sketch the figure the turtle will
   draw by carrying out the following
   commands.

```
FD 80 BK 160 FD 80
RT 90 FD 50 LT 90
FD 80 BK 160 FD 80
PENUP RT 90 FD 50 LT 90 PENDOWN
FD 80 BK 160 FD 80
```

"Looks like it might be a nice day
tomorrow!"

2. The procedure ONELINE draws a
   line and then moves the turtle 15
   steps to the right.

```
TO ONELINE
 FD 60 BK 120 FD 60
 PENUP RT 90 FD 15 LT 90 PENDOWN
END
```

ONELINE

a. Sketch the figure which is drawn by PARALLELS.

```
TO PARALLELS
 PENUP LT 90 FD 60 RT 90 PENDOWN
 REPEAT 9 [ONELINE]
END
```

★   b. Sketch the figure which is obtained from these commands.

RT 90 PARALLELS

★   3. Define a procedure called **GRID** for instructing the turtle to draw a grid which is similar to the one shown here. (*Hint:* Use PARALLELS from Exercise 2.)

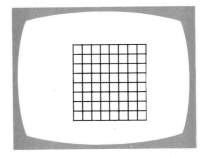

4. Procedures can be defined for drawing letters, and then letters can be combined to form procedures for drawing words. Here are procedures for the letters L and O. Notice that the last four commands in each procedure lift the turtle's pen and move the turtle to the right for the next letter.

```
TO O
 FD 30 RT 90
 FD 15 RT 90
 FD 30 RT 90
 FD 15 RT 90
 PENUP RT 90
 FD 25 LT 90
 PENDOWN
END
```

```
TO L
 FD 30
 BK 30
 RT 90
 FD 15
 PENUP
 FD 10
 LT 90
 PENDOWN
END
```

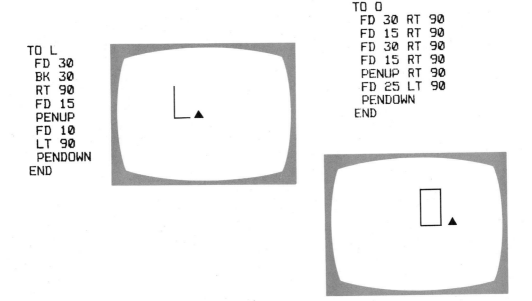

★   a. Write procedures for the letters V and E.

b. Define a procedure for printing LOVE.

c. Define procedures for the letters in your name and use them to define a procedure for printing your name.

5. Six spokes of a wheel were drawn on page 85 by using LINE and right turns of 60°. Use this approach to write the commands for drawing 10 spokes. (*Hint:* Use REPEAT.)

★ 6. Here is the beginning of a procedure for drawing a variable number of spokes (odd or even). The word NUMBER is a variable and the procedure is called SPOKE. Finish the definition so this command will draw any number of spokes.

```
TO SPOKE :NUMBER
   HIDETURTLE
```

7. The command RECTANGLE that was defined on page 88 is repeated here four times with turns of 90°. Sketch the resulting figure.

```
REPEAT 4 [RECTANGLE 30 60 RT 90]
```

★ 8. Define a procedure called MANY-KITE that will produce the figure shown here. Use the procedure KITE.

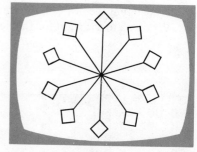

```
TO KITE
   FD 60 LT 45
   SQUARE 10
   RT 45 BK 60
END
```

MANYKITE

9. Revise the procedure in Exercise 8 to draw an arbitrary number of kites. Use the variable :NUMBER. Here is the beginning. The "S" in MANYKITES provides a new word for this new procedure.

TO MANYKITES :NUMBER

10. Define a procedure called THREE-SQUARES for drawing three concentric squares as shown here.

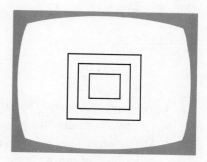

THREESQUARES

Exercise Set 2.3: BASIC

1. Determine the printout for each program.

    a.
    ```
    10   PRINT "THE PRODUCT OF 7 AND 8 IS EQUAL TO "7 * 8"."
    20   END
    ```

★   b.
    ```
    10   PRINT "3 + 9 IS EQUAL TO "3 + 9"."
    20   PRINT "THE AVERAGE OF 3 AND 9 IS "(3 + 9) / 2"."
    30   END
    ```

    c.
    ```
    10   PRINT "41 + 35 AND 35 + 41 ARE BOTH EQUAL TO "41 + 35"."
    20   PRINT "BASIC IS AN EASY COMPUTER LANGUAGE TO LEARN."
    30   PRINT "I AM READY FOR MORE CHALLENGING EXAMPLES."
    40   END
    ```

2. Use the following programs to determine the printouts for the given values.

    a. Program 2.3C (page 90) for N = 35
    ★   b. Program 2.3D (page 91) for X = 250, Y = 135, and Z = 190
    ★   c. Program 2.3F (page 92) for H = 200 and C = 150
    d. Program 2.3I (page 94) for A = 3417 and B = 3608
    ★   e. Program 2.3J (page 94) for N = 30
    f. Program 2.3M (page 95) for N = 20

3. Each of these programs produces a sequence of numbers. Determine the printout for N = 10.

    a.
    ```
    10   INPUT N
    20   FOR X = 1 TO N
    30   PRINT 2 ^ X - 1" ";
    40   NEXT X
    50   END
    ```

    ★   b.
    ```
    10   INPUT N
    20   FOR X = 1 TO N
    30   PRINT 2 * X - 1" ";
    40   NEXT X
    50   END
    ```

★   4. An exercise and weight loss specialist says that if a person walks briskly for 30 minutes a day, the person will use 600 calories and will lose 1 pound in 4 days. Assuming these conditions, complete the following program to determine the weight W lost in D days. Then show the printout for the program with D = 14. The printout should read: THE NUMBER OF POUNDS LOST IN 14 DAYS IS 3.5.

    ```
    10   INPUT D
    20   LET W = [            ]
    30   PRINT [                              ]
    40   END
    ```

5. Write a program that prints out the cost of N decals, if each decal costs $1.30.

★ 6. Write a program that prints out the total amount of money received from the sale of three types of theater tickets: X tickets which cost $8 each; Y tickets which cost $11 each; and Z tickets which cost $15 each.

7. Write a program which prints out your name 200 times.

8. Write a program that determines if the product of three numbers is greater than 1000.

★ 9. A copying business charges these prices: 10 cents per sheet if the number of sheets is less than or equal to 25; 8 cents per sheet if the number of sheets is greater than 25 and less than or equal to 50; and 6 cents per sheet if the number of sheets is greater than 50. The following program computes the cost C for N sheets. Fill in the missing parts and determine the printouts for N = 20, N = 30, and N = 60.

"When do you want it?"

```
10    INPUT N
20    IF N ( = 25 THEN  GOTO □
30    IF N ( = 50 THEN  GOTO 80
40    LET C = .06 * N
50    GOTO □
60    LET C = .10 * N
70    GOTO 90
80    LET C = □
90    PRINT "THE COST OF COPYING "□" SHEETS IS $"□"."
100   END
```

10. Write a program which computes the cost C of renting a car and driving it N miles at the following rates: 50 cents per mile if the number of miles is less or equal to 100; 40 cents per mile if the number of miles is greater than 100 and less than or equal to 1000; and 30 cents per mile if the number of miles is greater than 1000. (*Hint:* See Exercise 9.)

Exercise Set 2.3: Problem Solving

1. *Programming in Logo:* Write a procedure called GRID.SQUARES which instructs the turtle to draw a 3 by 5 array of nonintersecting squares of the same size.

   a. *Understanding the Problem:* Here is a 2 by 3 array of squares. Draw a 3 by 5 array. Notice that the size of the squares and the distance between them is not given.

   ★ b. *Devising a Plan:* First we need a command to draw a square (see page 88). Then, since the same size square is needed 15 times the REPEAT command will be helpful. (*Hint:* Do one row or one column at a time.)

   c. *Carrying Out the Plan:* Write the commands to draw the array.

   d. *Looking Back:* Revise your program to increase the space between the squares.

2. *Programming in BASIC:* Write a program in BASIC to print out the first 11 numbers of an arithmetic sequence.

   a. *Understanding the Problem:* Each number in an arithmetic sequence (after the first number) is obtained by adding the common difference to the preceding number.

   $$2 \quad 5 \quad 8 \quad 11 \quad 14 \quad 17 \quad 20 \quad 23 \quad 26 \quad \ldots$$

   We want a program that will print out the first 11 numbers of any arithmetic sequence. Such a program will require only two numbers for input. What are these two numbers for the above sequence?

   b. *Devising a Plan:* Let's use D to represent the common difference. One approach is to obtain each number X of the sequence by adding D to the preceding number. This can be accomplished by repeatedly using the LET statement:

   $$\text{LET } X = X + D$$

   Another approach is to note that each number X can be obtained by adding some whole number times D to the first number of the sequence. For the sequence shown above this can be accomplished by repeatedly using the following LET statement:

   $$\text{LET } X = 2 + ND$$

   What values of N and D will produce the sixth number of the above sequence?

★   c. *Carrying Out the Plan:* The repeated use of the LET statement for the plans in part **b** can be accomplished by a FOR-NEXT command. Use one of these plans or one of your own to write a program to print out the first 11 numbers of any arithmetic sequence.

   d. *Looking Back:* In a geometric sequence each number (after the first) is obtained by multiplying the preceding number by the common ratio. What changes need to be made to your program in part **c** so that it will print out the first 11 numbers of any geometric sequence?

Whole Numbers and Their Operations

In learning any kind of subject matter it is necessary to concentrate on the high spots, the most significant things. A good memory should be like a fisherman's net—that is, it should retain all the big fishes but let the little ones escape.

ERASMUS

3.1
ADDITION

Addition is first encountered with concrete objects.
If 4 clams are *put together* with 3 clams, the total
number of clams is the *sum* 4 + 3. The close association between the operation of *addition* and the set
operation of *union* can be seen in the words that
have been used over the years. In the thirteenth
through fifteenth centuries the words *aggregation,
composition, collection, assemble,* and *join* were
all used at various times to mean addition.

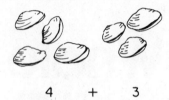

DEFINITION OF ADDITION

The idea of *putting sets together* or *taking their union* is often used as a definition for
addition. If set R has r elements and set S has s elements, and if these two sets have no
elements in common, then r added to s is defined as the number of elements in the
union of sets R and S. This is written symbolically as

$$r + s = n(R \cup S)$$

In the definition of addition the requirement that R and S be disjoint sets is necessary. Suppose, for example, that there are 8 people in a group who play a guitar and 6
who play a piano. We cannot assume that the total
number of people in the group who play a guitar or
a piano is 14, because there may be some who play
both. The case in which there are 2 people who play
both instruments is illustrated in the accompanying
diagram. The total number of people is 12, as
shown by the total number of different points inside
the circles. In this example, $n(G) = 8$, $n(P) = 6$,
but $n(G \cup P) \neq 8 + 6$.

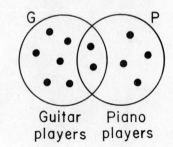

Guitar   Piano
players   players

MODELS FOR ADDITION
ALGORITHMS

An *algorithm* is a step-by-step routine for computing. The algorithm for addition involves two separate procedures: (1) adding digits, and (2) "regrouping" or "carryir (when necessary) so that the sum is written in positional numeration. Each of the models used on page 72 through 74 for numeration is also appropriate for illustrating the four basic operations on whole numbers. Several of these models are presented in the following paragraphs and in Exercise Set 3.1: Applications and Skills to illustrate the addition algorithm.

Bundles of Sticks To illustrate the sum of two numbers, the sticks representing these numbers can be placed below each other, just as the numerals are in the addition algorithm. The sum is the total number of sticks in the bundles plus the total number of individual sticks. To compute $46 + 33$ there are 7 bundles of sticks (7 tens) and 9 sticks (9 ones), so the sum is $7 \times 10 + 9$ or 79.

Whenever there is a total of 10 or more individual sticks, they can be regrouped into a bundle of 10. In the following example the total number of individual sticks is 14, which groups into 1 bundle of 10 sticks and 4 more. Thus, there is a total of 6 bundles and 4 sticks. In the algorithm a "4" is recorded in the units column and the extra 10 is recorded by writing a "1" in the tens column.

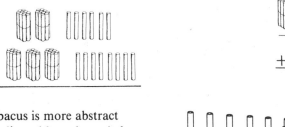

Abacus The abacus is more abstract (place value is indicated by columns) than the bundles of sticks model and in this respect is closer to the pencil-and-paper algorithm for addition. Addition is accomplished by pushing the markers on each column together and regrouping or carrying if necessary. The expression "to carry"

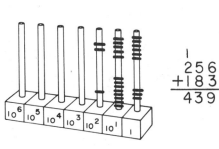

is an old one and probably originates from the time when a counter or chip was actually carried to the next column on an abacus. When the markers for 256 and 183 are pushed together, column by column, on the abacus, regrouping will be needed on the tens column. This is done by removing 10 markers from the tens column and placing another marker on the hundreds column. There will be three markers left in the tens column. In the algorithm this step is indicated by recording a "3" for the sum in the tens column and writing a "1" above the hundreds column.

ADDITION ALGORITHM

The primary purpose of an algorithm is to compute in the quickest and most efficient manner. The familiar pencil-and-paper algorithm for addition does just this. The digits of the numbers to be added are placed under one another, units under units, tens under tens, etc., and regrouping from one column is indicated by writing small numerals above the next column, as shown in the accompanying example.

$$\begin{array}{r} 121 \\ 4375 \\ 893 \\ + \ 1464 \\ \hline 6732 \end{array}$$

Expanded Forms   There are several ways of providing intermediate steps between the physical models for computing sums (bundles of sticks, abacus, etc.) and the pencil-and-paper algorithm for addition. One of these is to represent numbers in their *expanded forms*. Here are two examples of this method. The regrouping in Example 2 has been done in two stages.

EXAMPLE 1

$$\begin{array}{l} 452 = 4 \text{ hundreds} + 5 \text{ tens} + 2 \\ + \ 243 = 2 \text{ hundreds} + 4 \text{ tens} + 3 \\ \hline \quad\quad\; 6 \text{ hundreds} + 9 \text{ tens} + 5 \\ \quad\quad\; = 695 \end{array}$$

EXAMPLE 2

$$\begin{array}{l} 345 = 3 \text{ hundreds} + \ 4 \text{ tens} + \ 5 \\ + \ 278 = 2 \text{ hundreds} + \ 7 \text{ tens} + \ 8 \\ \hline \quad\quad\; 5 \text{ hundreds} + 11 \text{ tens} + 13 \end{array}$$

Regrouping $\begin{cases} 5 \text{ hundreds} + 12 \text{ tens} + \ 3 \\ 6 \text{ hundreds} + \ 2 \text{ tens} + \ 3 \end{cases}$

$$= 623$$

Partial Sums   Another intermediate step for providing understanding of the addition algorithm is computing *partial sums*. In this method the digits in each column are added, and these partial sums are placed on separate lines. This eliminates carrying (in most examples), because the digits of the sums are placed in the appropriate columns. The "18" in the first line of the partial sums shown here represents 18; the "12" in the second line of the sums represents 120; and the "8" in the third line represents 800. Some people prefer to write in the missing 0's.

$$\begin{array}{r} 238 \\ 521 \\ + \ 179 \\ \hline \end{array}$$

Partial Sums $\begin{cases} 18 \quad \text{1st line} \\ 12 \quad \text{2nd line} \\ \underline{\ \ 8} \quad \text{3rd line} \\ 938 \end{cases}$

Scratch Method Sometimes it is instructive to examine algorithms from the past. While it is now customary to add from right to left, beginning with the units digits, this was not always the practice. The early Hindus, and later the Europeans, added from left to right. For instance, to compute 897 + 537 from left to right, the first step is to add 8 and 5 in the hundreds column. In the second step the 9 and 3 are added in the tens column, and because carrying is necessary, the "3" in the hundreds column is scratched out and replaced by a "4." In the third step we add the units digits, and again, carrying is necessary. So the "2" in the tens column is scratched out and replaced by a "3." The Europeans called this the *scratch method*.

First step
```
  897
+ 537
 1 3
```

Second step
```
  897
+ 537
 1 3 2
  4
```

Third step
```
  897
+ 537
 1 3 4
  4 3
```

Lattice Method In another early algorithm for addition, called the *lattice method*, the numbers in the columns were added and their sums placed in cells under the columns. This is similar to the method of partial sums. The final sum is obtained by adding the numbers along the diagonal cells.

Calculator Method We have been examining pencil-and-paper algorithms, that is, sequences of steps for computing sums on paper. Similarly, there are sequences of steps for computing sums on calculators, referred to as *calculator algorithms*. Most calculators are designed for algebraic (equation-type) logic, such as the equation format 475 + 381 + 906 = ☐. On these calculators this sum is computed by the sequence of steps shown next. In Step 4, pushing ⊞ gives the total of the first two numbers. The final sum can be obtained by pushing ⊞ or ⊟ in Step 6.

Calculator Algorithm	Displays at the End of Each Step
Step 1 Enter 475	475.
Step 2 ⊞	475.
Step 3 Enter 381	381.
Step 4 ⊞	856.
Step 5 Enter 906	906.
Step 6 ⊟ or ⊞	1762.

NUMBER PROPERTIES

To illustrate some of the abuses of what has been popularly called "modern mathematics" or "new math," Morris Kline relates the following conversation in his book *Why Johnny Can't Add.*

    Teacher: Why is $2 + 3 = 3 + 2$?
    Student: Because both equal 5.
    Teacher: No, the correct answer is because the commutative law of addition holds.

At the beginning of the new math era one of the major criticisms of traditional mathematics was the amount of rote learning and unthinking repetition. One attempt to promote understanding in the new math programs was the use of number properties, such as the commutative and associative properties shown in the paragraphs that follow. While it is doubtful that the student in the previous example has acquired a better understanding of mathematics, these number properties, nevertheless, can be helpful in computing and in providing a better understanding of the steps in computing.

Associative Property for Addition Sometimes we will want to add two numbers by breaking one of the numbers into a convenient sum of numbers. The sum $147 + 26$ can be easily added in your head by computing $147 + 20$ and then adding 6. The change in the second of the following two equations is permitted by the associative property for addition.

$$147 + 26 = 147 + (20 + 6) = (147 + 20) + 6$$

Associative Property
for Addition

In a sum of three numbers the middle number can be added to (associated with) either of the two end numbers. This is called the *associative property of addition.* In general, for any three numbers that are used for the placeholders, $\square$, $\triangle$, and $\diagup\!\!\!\square$,

$$(\square + \triangle) + \diagup\!\!\!\square \;=\; \square + (\triangle + \diagup\!\!\!\square)$$

The associative property of addition also plays a roll in arranging numbers to produce sums of 10, called "making tens." As an example,

$$8 + 7 = 8 + (2 + 5) = (8 + 2) + 5 = 10 + 5 = 15$$

Associative Property
for Addition

Commutative Property for Addition This property cuts in half the number of basic addition facts that must be memorized. It also allows us to select convenient combinations of numbers when adding. The numbers 26, 37, and 4 are arranged more con-

veniently in the right side of the following equation than on the left because
$26 + 4 = 30$ and it is easy to compute $30 + 37$. This rearrangement of numbers is
permitted by the commutative property for addition.

$$26 + (37 + 4) = 26 + (4 + 37)$$

Commutative Property
for Addition

In any summand the positions of two numbers can be interchanged or commuted
without changing the sum. This is called the *commutative property for addition.* In
general, for any two numbers that are used for the placeholders $\square$ and $\triangle$,

$$\square + \triangle = \triangle + \square$$

APPROXIMATION AND
MENTAL CALCULATION

Looking for convenient combinations of numbers makes mental calculations of sums
easier. The associative and commutative properties for addition permit us to compute
a sum of numbers in any order we wish. For example, in the following sum we might
begin by adding 17 and 23 to get 40 and then count up by tens and fives. Compute this
sum in your head.

$$5 + 17 + 10 + 5 + 10 + 23 + 10 + 5 + 10$$

Rounding Off   Often we need to make quick calculations to check a bill we received;
to estimate a quantity of paint, fertilizer, or lumber needed to cover certain areas; and
so on. This calculation does not have to be exact but good enough for the purpose at
hand. Suppose, for example, you have written checks for $417, $683, and $228, and
need to determine quickly (before writing another check) an approximate balance. You
may round to the nearest leading digit and think to yourself: $4 + 7 + 2$ is 13 (hundred).
Here is a written record of this thinking. (The symbol $\approx$ means "approximately equal.")

$$417 + 683 + 228 \approx 400 + 700 + 200 = 1300$$

An even better mental approximation could be obtained by rounding to the nearest
tens digit, using number properties and thinking: $4 + 6 + 2$ is 12 (hundred) and
$20 + 80 + 30$ is 130 so my sum is 1330. A written record of this thought might look
like the following:

$$417 + 683 + 228 \approx (400 + 600 + 200) + (20 + 80 + 30) = 1200 + 130 = 1330$$

Did you notice how much faster the mental version is than the written version?

INEQUALITY OF WHOLE NUMBERS

Inequality of whole numbers can be explained intuitively in terms of the locations of these numbers as they occur in the counting process. For example, 3 is less than 5 because it is named before 5 in the counting sequence. This ordering of numbers can be pictured by a number line with the numbers located at equally spaced intervals. For any two numbers, the one that occurs on the left is less than the other.

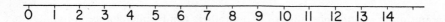

Inequality of whole numbers is defined in terms of addition.

Definition: For any two whole numbers, m and n, m is *less than n* if there is a nonzero whole number k such than $m + k = n$. This is written as $m < n$. We also say that n is *greater than m* and write $n > m$.

The symbols $<$ and $>$ were first used by the English surveyor Thomas Harriot in 1631. There is no record of why Harriot chose these symbols, but the following conjecture is logical and will help you to remember their use. The distances between the ends of the bars in the equality symbol (e.g., $3 = 12/4$) are equal, and the number on the left of the sign equals the number on the right. Similarly, $3 < 4$ would indicate that 3 is less than 4, because the distance on the left between the bars is less than the distance on the right between the bars. The reasoning is the same whether we write $3 < 4$ or $4 > 3$. These symbols could easily have evolved to our present notation, $<$ and $>$, with the bars completely converging to prevent any misjudgment of these distances.*

COMPUTER APPLICATIONS

Computers can be a helpful aid to learning and applying mathematics. One student was looking at the first few triangular numbers (1, 3, 6, 10, 15, 21) and noticed that the sequence of units digits in those numbers (1, 3, 6, 0, 5, 1) started with 1 and returned to 1 at the sixth triangular number. Would the same units digits now repeat? After calculating the first 18 triangular numbers it was found that the units digits in those numbers did not repeat as expected but revealed the following symmetrical pattern.

1, 3, 6, 0, 5, 1, 8, 6, 5, 5, 6, 8, 1, 5, 0, 6, 3, 1

Knowing that the nth triangular number is $n(n + 1)/2$ (page 37) the student wrote Program 3.1A to generate triangular numbers for studying patterns in the units digits.

---

*This is one of two conjectures on the origin of the inequality symbols, described by H. W. Eves, *Mathematical Circles* (Boston: Prindle, Weber and Schmidt, 1969), pp. 111–13.

Program 3.1A

```
10   INPUT X
20   FOR N = 1 TO X
30   PRINT N * (N + 1) / 2" ";
40   NEXT N
50   END
```

What patterns can you find in the first 40 digits?

1, 3, 6, 0, 5, 1, 8, 6, 5, 5, 6, 8, 1, 5, 0, 6, 3, 1, 0, 0, 1, 3, 6, 0, 5, 1, 8, 6, 5, 5, 6, 8, 1, 5, 0, 6, 3, 1, 0, 0

Palindromic Numbers A number is called a *palindromic number* if it is equal to its *reverse,* that is, it reads the same both forward and backward. For example, 31413 is a palindromic number. In the example at the right a number (87) is added to its reverse (78), then the sum (165) is added to its reverse (561), and so forth. After repeating this process four times a palindromic number is obtained. Will this process always result in a palindromic number?

$$
\begin{array}{r}
87 \\
+\,78 \\
\hline
165 \\
+\,561 \\
\hline
726 \\
+\,627 \\
\hline
1353 \\
+\,3531 \\
\hline
4884
\end{array}
$$

Program 3.1B is for experimenting with the sums of numbers and their reverses. When a number equals its reverse the hypothesis in line 50 is true and the computer is sent to line 90.

Program 3.1B

```
10   PRINT "TYPE A NUMBER. "
20   INPUT N
30   PRINT "TYPE THE REVERSE OF THIS NUMBER. "
40   INPUT M
50   IF M = N THEN   GOTO 90
60   LET N = N + M
70   PRINT "THE SUM IS "N"."
80   GOTO 30
90   PRINT "YOU HAVE A PALINDROMIC NUMBER. "
100  END
```

SUPPLEMENT (Activity Book)

Activity Set 3.1 Adding with Multibase Pieces and Chip Trading

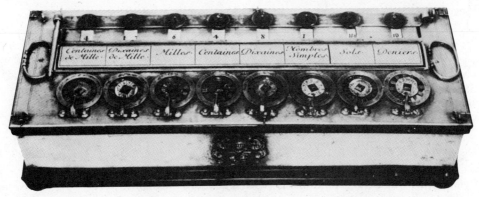

Blaise Pascal's arithmetic machine, 1642

Exercise Set 3.1: Applications and Skills

1. This calculating machine was developed by Blaise Pascal for computing sums. The machine is operated by dialing a series of wheels bearing numbers from 0 to 9 around their circumferences. To carry a number to the next column, when a sum is greater than 9, Pascal devised a ratchet mechanism that would advance a wheel one digit when the wheel to the right made a complete revolution. The wheels from right to left represent units, tens, hundreds, etc. To compute 854 + 629, first turn the hundreds, tens, and units wheels 8, 5, and 4 turns, respectively. These same wheels are then dialed 6, 2, and 9 more turns. The sum will appear on indicators at the top of the machine.

   ★   a. Which of these wheels will make more than one revolution for this sum?

        b. Which two wheels will be advanced one digit due to carrying?

   ★   c. Can this sum be computed by turning the hundreds wheel for both hundreds digits, 8, and 6; then turning the tens wheel for 5 and 2; and then turning the units wheel for 4 and 9?

2. *Error Analysis:* Some types of errors and misuses of addition are very common, and others such as the one in this sign, are fairly rare. Describe the type of error occurring in each of the following examples.

   ★   a.   47
            + 86
            ——
            123

        b.   16
            + 48
            ——
            91

   ★   c.   56
            + 78
            ——
            1214

        d.   35
            + 46
            ——
            171

3. Pairs of numbers are represented by the multibase pieces in part **a** and the bean stick model in part **b**. Do the necessary regrouping to obtain the sum that is represented by each model, and sketch the answer.

★ a.

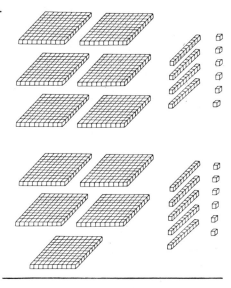

b.

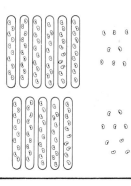

4. Pairs of numbers are represented on the top and bottom halves of the abacus in part **a** and the Chip Trading mat in part **b**. Add these numbers and do the regrouping for each sum. Sketch the answers on the abacus and mat to the right.

★ a.

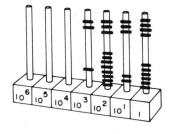

Regroup

b.

Red	Green	Blue	Yellow

Regroup

Red	Green	Blue	Yellow

5. Addition is illustrated on the number line by a series of arrows as shown here.

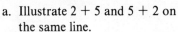

3 + 4

a. Illustrate 2 + 5 and 5 + 2 on the same line.

★ b. Devise a way of illustrating 0 plus any number on the number line. Then illustrate 0 + 6 and 6 + 0 on the same line.

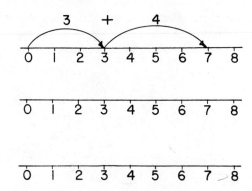

c. Use the number line to show that (3 + 4) + 1 and (4 + 1) + 3 are equal.

6. Find the number for each of the following expressions in parts **a** through **d**. In which cases is the number of people in the union of the two sets equal to the sum of the numbers of people in each set?

$A = \{$John, Mary$\}$
$B = \{$Sue, Joan$\}$
$C = \{$John, Sue, Al$\}$
$D = \{$Bill, Paul, Jane$\}$

★ a. $n(A \cup B)$ ★ b. $n(A \cup C)$ ★ c. $n(B \cup C)$ ★ d. $n(B \cup D)$

7. Compute the sums for the given methods. Name an advantage and disadvantage of each method. Does the lattice method eliminate the need for carrying?

★ a. Scratch Method

```
   482
  6731
+ 2064
```
9277

8177
92

b. Lattice Method

```
4 7 6 2
+ 8 7 6
```

5638

8. Which property, commutativity for addition or associativity for addition, is being used in equalities **a** through **d**?

★ a. (38 + 13) + 17 = 38 + (13 + 17)   assoc.

b. (47 + 62) + 12 = (62 + 47) + 12

★ c. 2 × (341 + 19) = 2 × (19 + 341)   comm

d. $\dfrac{13 + (107 + 42)}{3 \times (481 + 97)} = \dfrac{(13 + 107) + 42}{3 \times (481 + 97)}$

9. *Approximation and Mental Calculation:* Round off each number to its nearest leading digit and compute the sum.

		Think 80	Think 50	Think 100	Think 100	Think 10	Approximate Sum
		83	47	112	98	12	340
a.		102	38	21	75	42	
★ b.		26	43	59	18	66	
c.		25	212	81	39	44	
★ d.		27	68	18	91	198	

10. *Approximation:* A homeowner has the following bills to pay for the month of March, and a monthly salary of $1800. To estimate whether or not these bills can be paid, round off each bill to its nearest leading digit and compute the sum.

   Electricity $86, Heat $128, Water and sewage $94, Property taxes $163, Life insurance $230, Car insurance $65, House insurance $58, Food $541, Doctor's bills $477, Gas and oil $73, Car payments $148, Home mortgage $225, Dentist's bills $109, and Pleasure $14.

   a. Can the homeowner pay these bills with $1800?

★ b. What is the actual sum of these bills?

11. *Calculator Exercise:* Continue the pattern of numbers on the left side of the equations. Does the pattern continue to hold on the right side of the equations?

$$1 \times 9 + 2 = 11$$
$$12 \times 9 + 3 = 111$$
$$123 \times 9 + 4 = 1111$$

★ 12. *Missing Dollar Puzzle:* One day three men went to a hotel and were charged $30 for their room. The desk clerk later realized that he had overcharged them by $5 and sent the refund up with the bellboy. The bellboy knew that it would be difficult to split the $5 three ways. Therefore, he kept a $2 "tip" and gave the men only $3. Each man had originally paid $10 and was given back $1. Thus, it cost each man $9 for the room. This means that they spent $27 for the room plus the $2 tip. What happened to the other dollar?

Exercise Set 3.1: Problem Solving

1. *Dart Board Scores:* Karen and Angela are playing darts. Each player throws three darts on her
   turn and adds the numbers on the regions that
   are hit. They always hit the dart board and when
   a dart lands on a line the score is the larger of
   the two numbers. After four turns they noticed
   their eight sums were all different. How many
   different sums are possible? 19

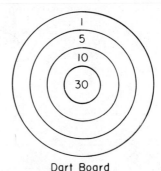

Dart Board

a. *Understanding the Problem:* Find two more
   sums for their fifth turns which are different
   from the first eight sums. What is the smallest
   possible sum for one turn? What is the largest
   possible sum for one turn?

Karen	Angela
21	40
15	7
61	32
12	16

b. *Devising a Plan:* Here are two approaches to finding all the different sums. Since
   the lowest sum is 3 and the highest sum is 90, we could list the numbers from 3
   through 90 and determine which can be obtained. Or, we can make a systematic list
   showing the different regions that the three darts can strike in. Which approach
   would you use?

c. *Carrying Out the Plan:* Use one of the above approaches, or one of your own, to
   find the number of different sums. Show how each sum can be obtained from the
   dart board numbers.

★ d. *Looking Back:* Let's solve this problem for different numbers of regions. Complete the following table for a dart board with two regions having the numbers
   1 and 5 and a dart board with three regions having the numbers 1, 5, and 10.
   Find a pattern and predict the number of different sums for a dart board with
   five regions having the numbers 1, 5, 10, 30, and 100. Verify your answer by
   listing the sums.

Number of Regions	1	2	3	4	5
Number of Different Sums	1				

2. *Sums of Neighbor Numbers:* This circle has whole numbers from 1 to 7. Adding *neighbor numbers,* numbers which are next to each other, we can get every number from 8 to 28 by using two or more numbers. A few sums for numbers greater than 7 are shown here. Find the remaining sums up to 28.

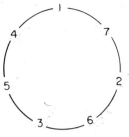

$$8 = 1 + 7$$
$$9 = 7 + 2$$
$$10 = 1 + 7 + 2$$
$$11 = 2 + 6 + 3$$

a. Arrange the numbers 1, 2, 3, 4, 5, 6, 7, and 8 in a circle so that all sums from 9 to 36 can be obtained. (*Hint:* See the above circle. The arrangement follows a pattern.) Show how each sum can be obtained.

★ b. Is there a solution to the neighbor number problem for the numbers from 1 to 6?

c. How can the neighbor number problem be extended?

3. *Faded Documents:* Computations in which some of the digits are missing are called *faded document puzzles.* The idea underlying these puzzles is that ancient documents have been discovered, but so decayed that some of the digits are not recognizable. Supply the missing digits.

★ a.
```
    7 □ 2 □
 + □ 4 □ 2
 ─────────
  1 3 1 9 1
```

b.
```
    □ 7 □
 +  9 □ 2
 ────────
  1 9 7 0
```

c. Use every digit from 1 to 9 exactly once to get the correct sum.

4. *Cryptarithms:* A *cryptarithm* is a puzzle in which a different letter is substituted for each digit (0, 1, 2, 3, 4, 5, 6, 7, 8, 9) in a simple arithmetic problem. In part **a**, a test of each digit shows that B must be 5 and, consequently, A must be 2. Knowing that K is 2, deduce the values of C, G, and F in part **b**.

a.
```
   B
   B
   B
   B
 + B
 ───
  A B
```

b.
```
   C C C
 +     K
 ──────
  G F F G
```

Here are two more cryptharithms. The first one is easy, but the second one is difficult and intended for the ambitious. Supposedly, it was a message written by a college student to his parents.

★ c.
```
  T H I S
      I S
 + V E R Y
 ────────
  E A S Y
```

d.
```
  S E N D
 + M O R E
 ────────
 M O N E Y
```

Exercise Set 3.1: Computers

1. Change the computer program for generating triangular numbers (see page 111) so that it will print all triangular numbers from the xth triangular number to the yth triangular number. (*Hint:* Begin by changing line 10 to INPUT X,Y.)

2. Increase the list of units digits for the triangular numbers from 40 digits to 80 digits.

   a. Describe the patterns you find.

   ★ b. Using your patterns, predict the 121st units digit. Check your conjecture with the computer.

   c. Describe a way to predict the units digit in any triangular number.

3. The following computer program revises the program for adding numbers and their reverses (see page 111) by inserting lines 25 and 75. These lines provide a counter to count the number of sums needed to obtain a palindromic number. Line 25 begins with X=0 and line 75 increases X by 1 each time through the program.* Notice that the hypothesis of the conditional command in line 50 has two conditions which must both be satisfied before the computer is sent to line 90.

```
10   PRINT "TYPE A NUMBER."
20   INPUT N
25   LET X = 0
30   PRINT "TYPE THE REVERSE OF THIS NUMBER."
40   INPUT M
50   IF M = N AND X > 0 THEN  GOTO 90
60   LET N = N + M
70   PRINT "THE SUM IS "N"."
75   LET X = X + 1
80   GOTO 30
90   PRINT "YOU HAVE A PALINDROMIC NUMBER. THE NUMBER OF SUMS
        IT TOOK IS "X"."
100  END
```

   a. What is the printout for this program if N=96 in line 20?

   ★ b. Find a two-digit number for which the process of adding numbers and their reverses requires six sums to obtain a palindromic number.

   c. The process in part **b** will produce a palindromic number for all two-digit numbers in six or fewer steps, except for one number and its reverse. Find this number.

---

*The X=X+1 in line 75 is not an ordinary equation since it would not make any sense. What we are doing is replacing X by X+1, that is, increasing the value of X by 1.

3.2
SUBTRACTION

Subtraction is usually explained by *taking away* a subset of objects from a given set. In the preceding cartoon, Peter has had 2 clams taken away from his 5 clams by a series of trades. The accompanying figure shows 2 clams being taken away from 5 clams and illustrates $5 - 2 = 3$. The word "subtract" literally means to *draw away from under,* and the

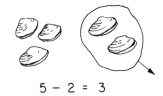

$$5 - 2 = 3$$

words which have been used for this operation are variations of this meaning. In 1202 the Italian mathematician Fibonacci used words meaning *I take* for the operation of subtraction. Over the years the terms *detract, subduce, extract, diminish, deduct,* and *rebate* have all been used as synonyms for "subtract." We still hear "deduct" being used in place of "subtract": "deduct the tax"; and "deduct what I owe."

DEFINITION OF SUBTRACTION

The process of *take away* for subtraction may be thought of as the opposite of the process of *put together* for addition. Because of this dual relationship, subtraction and addition are called *inverse operations.* This relationship is used to define subtraction in terms of addition. For any two whole numbers, *r* and *s,*

$$r - s = \square \quad \text{if and only if} \quad r = s + \square$$

The preceding definition says we can compute the difference $17 - 5$ by determining what number must be added to 5 to give 17. We use this approach when making change. Rather than subtract 83 cents from \$1.00 to determine the difference, we pay back the change by counting up from 83 to 100.

In the definition of subtraction there is no need of restricting *r* to be greater than or equal to *s.* However, in the early school grades, before negative integers are introduced, all of the examples involve subtracting a smaller number from a larger one.

Historically, subtraction was limited to this case, as shown in the following explanation by the English author Digges (1572):

> To subduce or subtray any sūme, is wittily to pull a lesse frō a bigger nūber.*

MODELS FOR SUBTRACTION ALGORITHMS

There are two types of examples to consider in explaining the steps for finding the difference between two multidigit numbers: those examples where borrowing or regrouping is not needed and those where borrowing or regrouping is needed. Each of these cases is illustrated by models in the following paragraphs. The amounts to be taken away in each example will be encircled.

Bundles of Sticks To find the difference between 48 and 23, 4 bundles of sticks and 8 sticks are used to represent 48. Then, 2 bundles of sticks and 3 single sticks are taken away so that the remaining bundles and sticks represent the difference between 48 and 23.

Sometimes regrouping is necessary before subtracting. The usual method of computing 53 − 29 is to borrow a 10 from the tens column so that there will be 13 ones in the units column. This is accomplished with the sticks by regrouping 1 bundle of sticks so that there are 13 single sticks. Once this is done, the subtraction can be completed by removing 2 bundles of sticks and 9 single sticks.

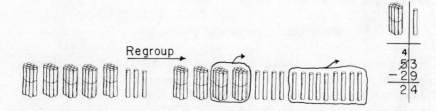

Abacus The abacuses shown next illustrate subtraction with regrouping. The markers on abacus (a) represent 721. In order to compute 721 − 384, the markers on abacus (a) have been regrouped on abacus (b). One marker on the tens column has been replaced

*See D. E. Smith, *History of Mathematics,* **2** (Lexington, Mass.: Ginn, 1925), p. 95.

by 10 markers on the units column so that 4 markers can be removed from the units column. One marker on the hundreds column has been replaced by 10 markers on the tens column so that 8 markers can be removed from the tens column. The difference, 721 − 384, is computed by removing 4 markers, 8 markers, and 3 markers, from the units, tens, and hundreds columns, respectively.

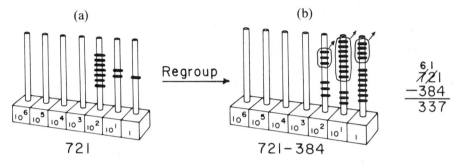

ALGORITHMS FOR SUBTRACTION

There are several algorithms for subtraction and each one has its advantages. Of the three methods described in the following paragraphs, the decomposition method is the most common. •

Decomposition Method This is the method of borrowing or regrouping, which was illustrated by the bundles of sticks model and the abacus in the previous examples. In the accompanying example one of the tens is regrouped from the tens column so that 8 can be subtracted from 14. In the next step, one of the hundreds is regrouped from the hundreds column so that 6 can be subtracted from 11.

$$\begin{array}{r} {\scriptstyle 4\ 1}\ \ {\scriptstyle 14} \\ \not{5}\not{2}\not{4} \\ -\ 3\ 6\ 8 \\ \hline 1\ 5\ 6 \end{array}$$

Equal Additions Method The equal additions method is one of the most rapid ways of subtracting. It appeared in Europe in the thirteenth century, and by the fifteenth and sixteenth centuries it had become popular. It is based on the principle that the difference between two numbers remains the same when both numbers are increased by the same amount.

As in the decomposition method in the preceding example, 10 is added to the 4 in the units column so that 8 can be subtracted from 14. However, to compensate for this change, the 6 in the tens column is changed to a 7, as shown in the first of the accompanying 3 steps. In the second step, 10 is added to the 2

$$\begin{array}{r} 5\ 2\ 4 \\ -\not{3}\not{6}\ 8 \\ \hline 1\ 5\ 6 \end{array}$$

First Step	Second Step	Third Step
$5^{1}2\,{}^{1}4$	$5^{1}2\ 4$	$5\,{}^{1}2\,{}^{1}4$
${}_{7}$	$4\ \ 7$	$4\ 7$
$-\,3\,\not{6}\,8$	$-\,\not{3}\,\not{6}\,8$	$-\,\not{3}\,\not{6}\,8$
6	$5\ 6$	$1\ 5\ 6$

in the tens column and this is paid for by changing the 3 in the hundreds column to a 4. Then 7 is subtracted from 12. Finally, in the third step, 4 is subtracted from 5.

/ **Complementary Method** The complementary method of subtraction was taught in many schools in the United States in the nineteenth century. As with the equal additions method, this approach has been used since medieval times.

This clever approach to subtraction requires only adding and is perhaps the fastest of the three methods. Beginning with the units column in Example 1, the complement of 7 (for the number 10) is 3. So 3 is added to 5. For each of the remaining lower digits we determine the complement of 9 and add this complement to the above digit. For example, the complement of 6 (for the number 9) is 3, so 3 is added to 1. Next the complement of 9 (for the number 9) is 0, so 0 is added to 4. Finally, the complement of 2 (for the number 9) is 7, so 7 is added to 8. To compensate for the complements that have been added, subtract 1 from the first digit on the left. This last step in Example 1 was accomplished by crossing out the 1 in the ten thousands column. The complementary method is also used in Example 2. In the units column the complement of 1 is 9 and $9 + 2 = 11$. Therefore, a "1" is carried from the units to the tens column. In the final step of the algorithm, 1 is subtracted from 7.

EXAMPLE 1

$$\begin{array}{r} 8\ 4\ 1\ 5 \\ -\ 2\ 9\ 6\ 7 \\ \hline \not{1}\ 5\ 4\ 4\ 8 \end{array}$$

EXAMPLE 2

$$\begin{array}{r} 1 \\ 7\ 4\ 3\ 2 \\ -\ \ 8\ 5\ 1 \\ \hline \not{7}\ 5\ 8\ 1 \\ 6 \end{array}$$

Calculator Method To find the difference of two numbers on a calculator the numbers and operation are entered in the order in which they appear (from left to right) when written horizontally. The steps for computing $247 - 189$ are shown on the right.

Steps	Displays
1. Enter 247	247.
2. $\boxed{-}$	247.
3. Enter 189	189.
4. $\boxed{=}$	58.

Some calculators will subtract a constant number from any number that is entered into the calculator. Suppose, for example, that you wanted to subtract $26 from each of the weekly paychecks listed. The first step in the next sequence subtracts 26 from 561. To subtract 26

Weekly Paychecks

$561
$452
$608
$513
.
.
.

from each of the remaining paychecks it is only necessary to enter the amount of pay and press [=].

Steps	Displays
1. 561 [−] 26 [=]	535.
2. 452 [=]	426.
3. 608 [=]	582.
4. 513 [=]	487.

APPROXIMATION AND MENTAL CALCULATION

There are many ways to manipulate numbers mentally to subtract.

Increase or Decrease Both Numbers Because the difference between two numbers remains constant when both numbers are increased or decreased by the same amount, a difference like $47 - 18$ can be thought of in many ways.

$49 - 20$	(increase both by 2)
$50 - 21$	(increase both by 3)
$30 - 1$	(decrease both by 17)
$40 - 11$	(decrease both by 7)

The difference, 29, is easy to compute in any of these forms.

Add-up Method Another convenient mental method is to "add-up" from the smaller to the larger number. Here are two examples.

$53 - 17 = ?$	$135 - 86 = ?$
$17 + \textcircled{3} = 20$	$86 + \textcircled{14} = 100$
$20 + \textcircled{33} = 53$	$100 + \textcircled{35} = 135$
So, $17 + \textcircled{36} = 53$	So, $86 + \textcircled{49} = 135$

Rounding Off If an approximate difference is all that is needed, the two numbers can be rounded off. Here is an example in which the numbers are rounded off to the highest leading digit.

$$624 - 289 \approx 600 - 300 = 300$$

Both numbers do not have to be rounded off to the same place value. In the next example, 812 is rounded off to 800 (nearest hundred) and 245 is rounded off to 250 (nearest ten). We can then use the add-up method to obtain a difference of 550.

$$812 - 245 \approx 800 - 250 = 550$$

COMPUTER APPLICATIONS

Here is a number curiosity involving the difference between a number and its reverse. The example at the right begins with 542 and its reverse, 245. The process of forming the reverse of a number and subtracting the two numbers has been continued until the difference is zero. If we begin with any three-digit number, will the process always end with zero?

$$
\begin{array}{r}
542 \\
- 245 \\
\hline
297
\end{array}
\qquad
\begin{array}{r}
594 \\
- 495 \\
\hline
99
\end{array}
$$

$$
\begin{array}{r}
792 \\
- 297 \\
\hline
495
\end{array}
\qquad
\begin{array}{r}
99 \\
- 99 \\
\hline
0
\end{array}
$$

Program 3.2A helps us to study the differences between numbers and their reverses. Lines 50 and 60 ensure that the smaller number is subtracted from the larger. In both of these lines the LET command is combined with the IF . . . THEN command. Notice the use of the colon in line 50. This allows us to put another command on this line, which in this case sends the computer to line 70 to print the difference between the number and its reverse. Follow through this program beginning with N = 542 and M = 245.

Program 3.2A

```
10   PRINT "TYPE A THREE-DIGIT NUMBER."
20   INPUT N
30   PRINT "TYPE THE REVERSE OF THE NUMBER."
40   INPUT M
50   IF N > = M THEN  LET N = N - M: GOTO 70
60   IF N < M THEN  LET N = M - N
70   PRINT "THE DIFFERENCE IS "N"."
80   IF N = 0 THEN  GOTO 100
90   GOTO 30
100  PRINT "THE PROGRAM IS OVER."
110  END
```

SUPPLEMENT (Activity Book)

Activity Set 3.2 Subtracting with Multibase Pieces

Exercise Set 3.2: Applications and Skills

1. *Error Analysis:* One common source of errors
   in subtraction occurs when elementary school
   students *add* rather than subtract. For example,
   when addition is taught *first,* the student learns
   to respond to the pair of numbers 5 and 3 by writing 8. This response becomes
   so automatic that later on we find the student writing 8 for the difference 5 − 3.
   Try to detect the reason for the error in each of the following computations.

$$\begin{array}{r} 5 \\ + 3 \\ \hline 8 \end{array} \qquad \begin{array}{r} 5 \\ - 3 \\ \hline 8 \end{array}$$

   ★   a. $\begin{array}{r} 84 \\ -36 \\ \hline 52 \end{array}$ ★   b. $\begin{array}{r} 52 \\ -38 \\ \hline 24 \end{array}$ ★   c. $\begin{array}{r} 46 \\ -27 \\ \hline 73 \end{array}$ ★   d. $\begin{array}{r} 94 \\ -37 \\ \hline 12 \end{array}$

2. In parts **a** and **b** do the necessary regrouping so that you can remove (or cross out)
   the required number of objects. Use your results to fill in the missing numbers in
   the given boxes.

   ★   a.   Compute 625 − 63 by removing
   6 longs and 3 units.

$$\begin{array}{r} 625 \\ -63 \\ \hline \boxed{} \end{array}$$

   b.   Compute 68 − 29 by removing
   2 bean sticks and 9 beans.

$$\begin{array}{r} 68 \\ -29 \\ \hline \boxed{} \end{array}$$

3. In parts **a** and **b** do the necessary regrouping by drawing new sets of markers and
   chips so that you can remove the required numbers of objects. Use your results to
   fill in the missing numbers in the given boxes.

   a.   Compute 1358 − 472 by regrouping and then removing 2 markers from the
   units column, 7 markers from the tens column, and 4 markers from the
   hundreds column.

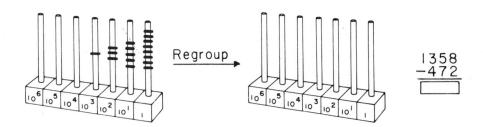

$$\begin{array}{r} 1358 \\ -472 \\ \hline \boxed{} \end{array}$$

★   b. Compute 6245 − 2873 by removing 3 yellow chips, 7 blue chips, 8 green chips, and 2 red chips.

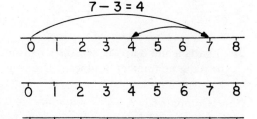

Red	Green	Blue	Yellow
○ ○ ○ ○ ○ ○	○ ○	○ ○ ○ ○ ○ ○	○ ○ ○ ○ ○

Regroup →

Red	Green	Blue	Yellow

6245
−2873
[ ]

4. Subtraction is illustrated on the number line by arrows that represent numbers. The number being subtracted is represented by an arrow from right to left.

7 − 3 = 4

0 1 2 3 4 5 6 7 8

★   a. Illustrate (6 − 3) − 2 on the given number line.

0 1 2 3 4 5 6 7 8

    b. Use this number line to illustrate 6 − 6.

0 1 2 3 4 5 6 7 8

5. Addition is often used to check subtraction by adding the two "lower numbers." In this example the sum of the two lower numbers is the top number, 903. In the units column 6 + 7 = 13, so "1" is carried to the tens column. In the tens column, 1 + 2 + 7 = 10, so "1" is carried to the hundreds column. Find out which of the following differences is wrong by adding the two lower numbers in each example.

$$\begin{array}{r} 903 \\ -\ 576 \\ \hline 327 \\ 11 \end{array}$$

a.   436
     − 197
     239

b.   1702
     − 486
     1316

c.   500
     − 164
     336

6. Try some numbers in parts **a, b** and **c** to determine if these properties hold.

    a. Is subtraction commutative?

$$\square - \triangle \overset{?}{=} \triangle - \square$$

★   b. Is subtraction associative?

$$(\square - \triangle) - \square \overset{?}{=} \square - (\triangle - \square)$$

    c. Is multiplication distributive over subtraction?

$$\square(\triangle - \square) \overset{?}{=} (\square \times \triangle) - (\square \times \square)$$

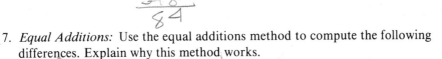

7. *Equal Additions:* Use the equal additions method to compute the following differences. Explain why this method works.

    a. 732
       − 348
       ‾‾‾‾

    ★ b. 1746
         − 382
         ‾‾‾‾

    c. 910
       − 462
       ‾‾‾‾

8. *Complementary Method:* Use the complementary method to compute the following differences.

    a. 347
       − 169
       ‾‾‾‾

    ★ b. 8023
         − 476
         ‾‾‾‾

    c. 6023
       − 2184
       ‾‾‾‾

9. *Calculator Exercise:* Slate and brick are sold by weight. At one company the slate or brick is placed on a loading platform and a forklift moves the total load, platform and all, onto the scales. The weight of the platform is then subtracted from the total weight. Assuming the weight of the platform is 83 kilograms (83 kg), find the weight of the slate for each of these total weights. (If you use a calculator as described on page 122, the 83 needs only to be entered onto the calculator one.)

Total Load Weight (kg)	Weight of Slate (kg)
748	665
807	
1226	
914	
1372	
655	

10. *Mental Calculation:* Compute these differences in your head by using the suggested methods from pages 123 and 124. Explain your steps for each exercise.

    a. *Increase or Decrease Both Numbers*

    $435 - 198 =$ $775 - 260 =$ $245 - 85 =$

    b. *Add-up Method*

    $400 - 185 =$ $535 - 250 =$ $135 - 47 =$

11. *Approximation:* Approximate each difference by rounding off one or both numbers to more convenient numbers. Explain your steps for obtaining each approximation.

    a. $359 - 192$ ★ b. $712 - 293$

    c. $800 - 245$ d. $1522 - 486$

★ 12. *Number Curiosity:* Select any three-digit number whose first and third digits are different, and reverse the digits. Find the difference between these two numbers. By knowing only the units digit in this difference, it is possible to determine the other two digits.

834
− 438
‾‾‾‾

   Try this trick with some three-digit numbers. Look for a pattern in the differences and explain how the trick works.

Exercise Set 3.2: Problem Solving

1. *Force Out:* This is a two-person game. An arbitrary number is selected and from it the players take turns subtracting any single digit number greater than zero. The player who is forced to obtain zero loses the game. Describe a strategy for winning this game.

   a. *Understanding the Problem:* On each player's turn only one single-digit number (1, 2, 3, 4, 5, 6, 7, 8, or 9) may be subtracted. If you can get the remaining number to be 1, then you will win. Select a number and play the game to become familiar with the rules. If the number is 15 and it is your turn to play, what number should you subtract?

   ★ b. *Devising a Plan:* One approach to solving this problem is to play the game for small numbers to see if you can hit upon an idea for a winning strategy. Another approach is to work *backward* to find the numbers that will guarantee you a win. Explain how you can win if the remaining number is greater than 1 and less than 11 and it is your turn.

   c. *Carrying Out the Plan:* Select an approach to look for a solution to this problem. Explain why you can win if you can get the remaining number to be 11 and it's your opponent's turn to play. Describe a strategy for winning the game if it's your turn to play and the final digit in the number is not a 1 (1, 11, 21, 31, 41, etc.)

   d. *Looking Back:* Let's revise the game so that the players subtract any number from 1 through 19. Suppose you are to take the first turn and the starting number is 76. Describe a strategy for winning this game.

★ 2. *Differences of Neighbor Numbers:* Put the whole numbers from 1 to 50 on the circumference of a circle in such a way that the difference between neighbor numbers is at most 2. (*Hint:* Simplify the problem).

★ 3. *Cryptarithm:* Use each of the digits 0 through 9 exactly once to obtain the smallest positive difference.

4. Use the digits 2, 5, 7, and 8 to form 2 two-digit numbers with no repeated digit; that is, a different digit in each box. What is the maximum number of positive differences? (*Hint:* Make a systematic list.)

Exercise Set 3.2: Computers

1. Use the following numbers in Program 3.2A (page 124). Does the process of repeatedly subtracting a number and its reverse result in zero?

   a. 472   ★ b. 581   c. 127   ★ d. 917

2. Revise Program 3.2A by making the following additions and changes.

```
25  LET X = 1
70  PRINT "THE DIFFERENCE AT STEP "X" IS "N"."
85  LET X = X + 1
```

Follow through the revised program by beginning with N = 742. What is the printout for line 70?

★ 3. The number of differences, before the process of repeatedly subtracting a three-digit number and its reverse results in zero, depends on the difference between the units digit and the hundreds digit. Use the program in Exercise 2 to help find this relationship. Record your results in the following table.

Difference between Units and Hundreds Digits	0	1	2	3	4	5	6	7	8
Number of Steps	1								

4. *The Number 6174:* There is something very special about 6174. Select any four-digit number whose digits are not all equal and arrange the digits to form the largest number possible, that is, put the digits in decreasing order. Then form the reverse and subtract the two numbers. Continue this process of forming the largest possible number from the difference and sub-

Step	Maximum Number	Reverse	Difference
1	8421	1248	7173
2	7731	1377	6354
3	6543	3456	3087
4	8730	0378	0352
5	8532	2358	(6174)

tracting its reverse. The original number in the example above is 4218. The process ended in five steps with 6174. Can you find a four-digit number whose digits are not all equal for which this process requires more than seven steps to produce 6174?

a. Write a computer program to help you investigate this question. (*Hint:* It may help you to look at the program in Exercise 2. The tests in lines 50 and 60 will not be needed. The N = 0 in line 80 should be replaced by N = 6174.)

☞ ★ b. Use your program (in part **a**) to see if there are "special numbers" when this process is carried out for three-digit or five-digit numbers.

State office buildings at the Empire State Plaza, Albany, New York

3.3
MULTIPLICATION

The skyscraper in the center of this picture is called the Tower Building. There is an innovative window washing machine mounted on top of this building. The machine lowers a cage on a vertical track so that each column of 40 windows can be washed. After one column of windows is washed, the machine moves to the next column. The rectangular face of the building which can be seen has 36 columns of windows. The total number of windows is $40 + 40 + 40 + \cdots + 40$, a sum in which 40 occurs 36 times. Using multiplication this sum equals the product 36×40, or 1440. We are led to different expressions for the sum and product by considering the rows of windows across the floors. There are 36 windows in each floor on this face of the building and 40 floors. Therefore, the number of windows is $36 + 36 + 36 + \cdots + 36$, a sum in which 36 occurs 40 times. This sum is equal to 40×36, which is also 1440. In sums such as these, where the addition of one number is repeated, multiplication is a short-cut for doing addition.

DEFINITION OF MULTIPLICATION *is repeated addition.*

Historically, multiplication was developed to replace certain special cases of addition, namely, the cases of *several equal addends*. This meaning is expressed by the word for "multiply" in Latin, and in several other languages.

For this reason we usually see multiplication of whole numbers explained and defined as "repeated addition." For whole numbers r and s, $r \times s$ is the sum with s occurring r times.

$$r \times s = s + s + s + \cdots + s$$

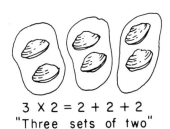

$3 \times 2 = 2 + 2 + 2$
"Three sets of two"

Another method of viewing multiplication of whole numbers is through rectangular arrays of objects, as suggested by the rows and columns of windows at the beginning of this section. The rectangular array shown here has four rows of seven squares each. The number of squares is 4 times 7. In general, $r \times s$ is the number of objects in a rectangular array having r rows with s objects in each row.

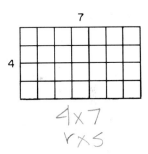

4×7
$r \times s$

MODELS FOR
MULTIPLICATION ALGORITHMS

Physical models for the basic operations can provide an understanding of these operations and suggest or motivate the rules and steps for computing. Bean sticks, bundles of sticks, multibase pieces, Chip Trading, and the abacus are all suitable models for illustrating the multiplication algorithm. Multibase pieces and Chip Trading are used in the following paragraphs.

Multibase Pieces   To multiply 3 times 234, using multibase pieces, the numbers of each type of multibase piece for 234 are tripled. The result is 6 flats, 9 longs, and 12 units. Regrouping is then needed: 10 units are replaced by 1 long, leaving 2 units; and 10 longs are replaced by 1 flat, leaving 0 longs. The result will be 7 flats, 0 longs, and 2 units.

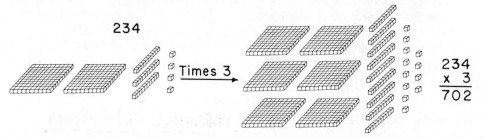

The next example illustrates how multiplication by 10 can be carried out with multibase pieces. If 10 units are placed together they form 1 long, 10 longs form 1 flat, and 10 flats form 1 block. Therefore, to multiply 234 by 10, each multibase piece for 234 is replaced by the multibase piece for the next higher power of ten. We begin with 2 flats, 3 longs, and 4 units; and we end with 2 blocks, 3 flats, 4 longs, and 0 units.

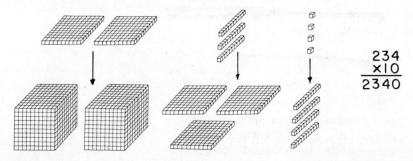

Chip Trading Multiplication by single-digit numbers with the Chip Trading model is done by multiplying the number of chips in each column of the mat and then regrouping when necessary. To compute 4×2063, the chips in each column of mat (a) have been replaced by four times as many chips on the corresponding columns of mat (b). After regrouping there will be 8 red chips, 2 green chips, 5 blue chips, and 2 yellow chips.

Red	Green	Blue	Yellow
⊙ ⊙		⊙⊙⊙ ⊙⊙⊙	⊙⊙⊙

(a)

Times 4 →

Red	Green	Blue	Yellow
⊙ ⊙ ⊙ ⊙ ⊙ ⊙ ⊙ ⊙		⊙⊙⊙ ⊙⊙⊙ ⊙⊙⊙ ⊙⊙⊙ ⊙⊙⊙ ⊙⊙⊙ ⊙⊙⊙ ⊙⊙⊙	⊙⊙⊙ ⊙⊙⊙ ⊙⊙⊙ ⊙⊙⊙

(b)

Multiplying by 10 is easy when we think in terms of regrouping. Referring to the next diagram, there are 3 yellow chips on mat (c). Multiplying by 10 replaces each yellow chip by 10 yellow chips. Regrouping, each group of 10 yellow chips is traded for 1 blue chip. Therefore, multiplying by 10 is accomplished by trading 3 yellow chips for 3 blue chips. Similarly, there are 7 blue chips on mat (c), and multiplying by 10 replaces each blue chip by 10 blue chips. Each group of 10 blue chips can be traded for 1 green chip. Therefore, the effect of multiplying by 10 is to trade the 7 blue chips for 7 green chips. The final product is represented by the chips on mat (d).

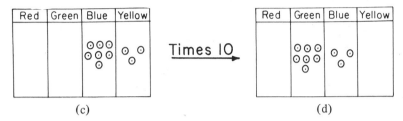

Red	Green	Blue	Yellow
		ooo oo oo o	o o o

(c)

Times 10 →

Red	Green	Blue	Yellow
		ooo ooo o	o o o

(d)

Multiplication by multiples of 10 can be illustrated by combinations of products with single-digit numbers and powers of 10. To multiply 30 times 47, we can first multiply 47 by 3 and then multiply this result by 10.

Times 3 → **Times 10** →

Red	Green	Blue	Yellow
		o o o o o o	o o o o o o

Red	Green	Blue	Yellow
	o	o o o o	o

Red	Green	Blue	Yellow
	o	o o o o	o

ALGORITHMS FOR MULTIPLICATION

One of the earliest methods of multiplication is found in the Rhind Papyrus. This ancient scroll (ca. 1650 B.C.) is more than 5 meters in length and was written to instruct the Egyptian scribes in computing with whole numbers and fractions. It begins with the words "Complete and thorough study of all things, insights into all that exists, knowledge of all secrets . . . ," and indicates the Egyptians' respect and awe of mathematics. Although most of its 85 problems have a practical origin, there are some of a theoretical nature. Their algorithm for multiplication was a succession of doubling operations, followed by addition. We call this algorithm the *duplation method*.

Duplation Method This method depends on the fact that any number can be written as the sum of binary numbers (1, 2, 4, 8, 16 . . .). To compute 37 $\times$ 52, the 52 is repeatedly doubled as shown here. This process stops when the next binary number in the list is greater than the number you are multiplying by.

$$\rightarrow 1 \times 52 = 52$$
$$2 \times 52 = 104$$
$$\rightarrow 4 \times 52 = 208$$
$$8 \times 52 = 416$$
$$16 \times 52 = 832$$
$$\rightarrow 32 \times 52 = 1664$$

In this example, we want 37 of the 52s, and since 37 is equal to the sum of the binary numbers 32, 4, and 1, the product of 37 $\times$ 52 is equal to $(32 + 4 + 1) \times 52$. This is $1664 + 208 + 52$, which equals 1924.

Russian Peasant Multiplication   There is a variation of the Egyptian method, called *Russian peasant
multiplication,* which was used in medieval Europe.
It involves repeatedly doubling and halving the two
numbers to be multiplied. In the example shown
here, 54 × 17 is computed by repeatedly doubling
the 17 and halving the 54. In the halving process,
fractions are disregarded. When 1 is reached in the
halving column, this process stops. Next, each row with even numbers in the halving
column is crossed out. Now, the sum of the remaining numbers in the doubling column
is equal to the product of the original two numbers. In this example, 54 × 17 is equal
to 34 + 68 + 272 + 544, which is 918.

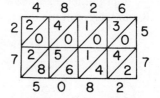

Halving		Doubling
~~54~~	X	~~17~~
27		34
13		68
~~6~~		~~136~~
3		272
1		+ 544
		918

Gelosia Method   *lattice*   One of the popular schemes used
for multiplying in the fifteenth century was called
the *gelosia* or *grating method.* This algorithm was
performed in a framework resembling a window
grating or jalousie, as shown here. The two numbers
to be multiplied, 4826 and 57, are written at the top
and right side of the grating. The partial products of
these numbers are written in the squares of the grating. The sums of the numbers in the
diagonal cells, from upper right to lower left, equal the product, 275082, of the original
two numbers.

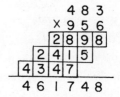

Pencil-and-Paper Method   In the algorithm we
now use for computing products, the two numbers
being multiplied are placed under each other with
units under units, tens under tens, etc. This method
first appeared in an Italian text dated 1470. Because
of the use of squares resembling a chessboard, the
Venetians called this method "per scachiere." By the
sixteenth century the squares had disappeared but
the "chessboard method" was generally adopted by
most writers.

```
      4 8 3
    x 9 5 6
    2 8 9 8
  2 4 1 5
4 3 4 7
4 6 1 7 4 8
```

   The pencil-and-paper method requires computing partial products of single-digit
numbers. When multiplying a one-digit number times a two-digit number there are two
partial products. In this example, the product of 3 times 17 is represented by three rows
of 17 squares. The two regions in the diagram correspond to the two partial products.

```
   17
 x  3
   21    (3 × 7)
   30    (3 × 10)
   51
```

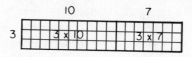

When multiplying a two-digit number by a two-digit number there are four partial products. Here we visualize 13×17 as 13 rows of 17 squares. Each region in the diagram represents one of the partial products.

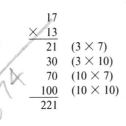

$$
\begin{array}{r}
17 \\
\times\ 13 \\
\hline
21 \quad (3 \times 7) \\
30 \quad (3 \times 10) \\
70 \quad (10 \times 7) \\
100 \quad (10 \times 10) \\
\hline
221 \\
\end{array}
$$

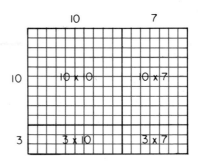

Calculator Method The steps for multiplying numbers with a calculator are similar to those for adding and subtracting with a calculator. The numbers and operations are entered in the order they occur from left to right when the product is written horizontally. The steps for computing $114 \times 238 \times 71$ are listed here.

1. Enter 114
2. ☒
3. Enter 238
4. ☒
5. Enter 71
6. ⊟

Special care must be taken on some calculators when multiplication is combined with addition or subtraction. The numbers and operations will not always produce the correct answer if they are entered onto the calculator in the order they appear. Try the following computation by entering the numbers onto your calculator as they appear from left to right.

$$3 + 4 \times 5$$

The correct answer is 23 because multiplication should be performed before additon. If a calculator produces the correct answer in this example, it is designed so that when ⊞ 4 is entered, 4 is not added to 3, but it is multiplied by 5 when ☒ 5 is entered. Then the product 4×5 is added to 3. On this type of calculator any combination of products with sums and differences can be computed by entering the numbers and operations in the order they occur from left to right and then pressing ⊟. For example, the answer to the following computation can be easily obtained on such a calculator.

$$114 \times 238 + 19 \times 605 - 32 \times 180 = 32867$$

If a calculator does not perform products before sums and differences, the products can be computed separately and recorded by hand or saved in the calculator by using the storage key.

3.3 MULTIPLICATION

135

NUMBER PROPERTIES

Distributive Property   When multiplying a sum of two numbers by a third number, we can add the two numbers and then multiply by the third number; or, we can multiply each number of the sum by the third number and then add the two products. For example, to compute $35 \times (10 + 2)$, we can compute 35×12, or add 35 times 10 to 35 times 2. This fact is described by saying that *multiplication is distributive (or distributes) over addition.*

$$35 \times 12 = \underbrace{35 \times (10 + 2)} = \underbrace{35 \times 10} + \underbrace{35 \times 2}$$

Distributive Property

In general, for any three numbers that are used for the placeholders $\square$, $\triangle$, and $\diamondsuit$

$$\square(\triangle + \diamondsuit) = (\square \times \triangle) + (\square \times \diamondsuit)$$

The distributive property can be extended for an arbitrary number of numbers in a sum. For example,

$$3 \times (17 + 26 + 32) = 3 \times 17 + 3 \times 26 + 3 \times 32$$

For sums of more than two numbers this is called the *generalized distributive property.* Multiplication also distributes over subtraction.

$$5 \times (20 - 3) = 5 \times 20 - 5 \times 3$$

Associative Property for Multiplication   Here is an easy way to multiply any number by 5. First divide the number by 2, or multiply by $\frac{1}{2}$, and then multiply by 10. Try this method to compute 5×124. The fact that dividing by 2 and multiplying by 10 produces the same result as multiplying by 5 is a consequence of the *associative property for multiplication.*

$$5 \times 124 = \underbrace{(10 \times \tfrac{1}{2}) \times 124} = \underbrace{10 \times (\tfrac{1}{2} \times 124)}$$

Associative Property
for Multiplication

In general, for any three numbers that are used for the place holders $\square$, $\triangle$, and $\diamondsuit$ the middle number may be grouped or "associated with" either the first number or the third number.

$$(\square \times \triangle) \times \diamondsuit = \square \times (\triangle \times \diamondsuit)$$

Commutative Property for Multiplication   Before reading beyond this sentence, try computing the product $25 \times 46 \times 4$ in your head. The easy way to do this is by rearranging the order of the numbers so that 25×4 is computed first and then 46×100. Changing the order of two numbers in a product is permitted by the *commutative property for multiplication.* In the following two equations both the commutative and associative properties for multiplication are used to change the order of the numbers in the product.

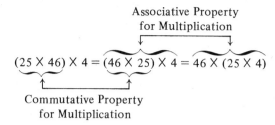

$$(25 \times 46) \times 4 = (46 \times 25) \times 4 = 46 \times (25 \times 4)$$

Commutative Property
for Multiplication

In general, for any two number replacements for $\triangle$ and $\square$ the order of multiplication does not change the resulting product.

$$\triangle \times \square = \square \times \triangle$$

APPROXIMATION AND MENTAL CALCULATION

The number properties for multiplication, especially the distributive property, are very useful for mental calculations. Those people who occasionally forget a multiplication fact, like $7 \times 8 = 56$, might find it helpful to think of 7×8 as:

$$7 \times 8 = 7 \times (7 + 1) = 49 + 7 = 56$$

Distributive Property

The distributive property is sometimes helpful in computing larger products. In the next example, 103 has been replaced by $100 + 3$.

$$21 \times 103 = 21 \times (100 + 3) = 2100 + 63 = 2163$$

Distributive Property

Occasionally, it is convenient to replace a number by the difference of two numbers and use the fact that multiplication distributes over subtraction. Rather than compute 45×98, we can compute 45×100 and subtract 45×2.

$$45 \times 98 = 45 \times (100 - 2) = 4500 - 90 = 4410$$

Distributive Property

Rounding Off The products of large numbers can be approximated by rounding off one number or both numbers. Here are some examples.

$$48 \times 98 \approx 48 \times 100 = 4800$$
$$62 \times 38 \approx 60 \times 40 = 2400$$
$$194 \times 26 \approx 200 \times 26 = 5200$$

NAPIER'S RODS

Even as late as the seventeenth century, multiplication of large numbers was a difficult task for all but professional clerks. In order to help people "do away with the difficulty and tediousness of calculations," the Scottish mathematician John Napier (1550–1617) invented a method of using rods for performing multiplication. Napier's rods, or "bones" as they are sometimes called, consist of the 10 rods shown here.

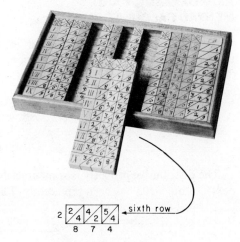

Napier's method is similar to the gelosia or grid method of multiplying. To compute 6 times 479, the rods for the multiples of 4, 7, and 9 are placed side by side, as shown in this set of wooden rods from the Andechs monastery in Bavaria. The sums of the numbers in the diagonal cells of the sixth row (indicated by VI), adding from right to left, give the product 2874.

Napier's rods, Andechs monastery
in Bavaria

To compute 73 times 61, the rods for the multiples of 6 and 1 are placed side by side. In this case, there are two partial products: one from the third row of the rods for computing 3 × 61; and one from the seventh row for computing 7 × 61. These products are 183 and 427. Why must 427 be changed to 4270? The product 73 × 61 equals 183 + 4270, or 4453.

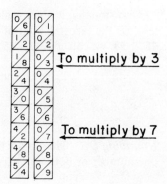

To multiply by 3

To multiply by 7

FINGER MULTIPLICATION

By the Middle Ages, the use of finger numbers and systems of finger computations were widespread. One such system for multiplication, which in recent years was still common in some parts of Europe, uses the finger positions shown here for computing the products of numbers from 6 through 10.

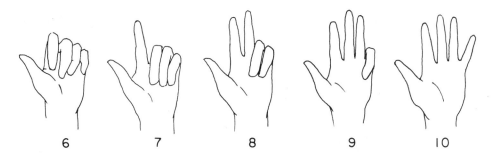

6   7   8   9   10

The two numbers to be multiplied are each represented on a different hand. The hands shown on the right represent 7 and 8. To compute 7×8, the sum of the raised fingers is the number of tens, and the product of the closed fingers is the number of ones. In this case, the 50 from the $2 + 3$ raised fingers and the 6 from the 3×2 closed fingers give a product of 56. Try this method to compute some other products of numbers from 6 through 10.

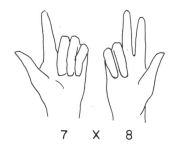

7   X   8

COMPUTER APPLICATIONS

This investigation was given to an elementary school class for practice in multiplication. Start with a whole number and then compute another number by doubling its units digit and adding its tens digit. Continue this process with each new number. The following *number chain* starts with 15.

$$\overbrace{(2 \times 5 + 1)} \quad \overbrace{(2 \times 3 + 0)} \quad \overbrace{(2 \times 2 + 1)}$$

$$15 \to 11 \to 3 \to 6 \to 12 \to 5 \to 10 \to 1 \to 2$$
$$\downarrow$$
$$15 \leftarrow 17 \leftarrow 18 \leftarrow 9 \leftarrow 14 \leftarrow 7 \leftarrow 13 \leftarrow 16 \leftarrow 8 \leftarrow 4$$

We ended with the number we started with and all positive whole numbers less than 19 are in this number chain. Notice that 19 gives itself back under our rules. Can we start with any whole number and always obtain a number chain?

Program 3.3A prints the numbers in a number chain. The X in line 20 is the first number in the chain. Lines 30 through 70 form a loop in which the units digit is doubled and added to the tens digit. Line 60 tests this number to see it if equals the first number in the chain. For each single digit number that occurs in the number chain, a 0 must be entered for the value of T in line 40.

Program 3.3A

```
10  PRINT "TYPE A WHOLE NUMBER. "
20  INPUT X
30  PRINT "TYPE THE TENS AND UNITS DIGITS, SEPARATED BY A
    COMMA. "
40  INPUT T,U
50  LET Y = 2 * U + T
55  PRINT Y
60  IF Y = X THEN   GOTO 80
70  GOTO 30
80  PRINT "THE NUMBER CHAIN HAS BEEN COMPLETED. YOU OBTAINED
    THE NUMBER YOU STARTED WITH. "
90  END
```

Cryptarithms   Here is a cryptarithm with two unknown digits. The computer can be programmed to find these digits by trying all possibilities. This is accomplished in the following program by a FOR–NEXT command (page 94). This program uses two loops to find digits for A and B. First, the computer sets A = 1 and then tries all values of B from 0 through 9 (lines 20 to 50). If these values do not work, it sets A = 2 and tries all values of B from 0 through 9. This process continues until A = 9 or a solution is found.

$$\begin{array}{r} ABA \\ \times \ \underline{AB} \\ 28BAA \end{array}$$

Program 3.3B

```
10  FOR A = 1 TO 9
20  FOR B = 0 TO 9
30  LET P = (A * 100 + B * 10 + A) * (A * 10 + B)
40  IF P = (28000 + B * 100 + A * 10 + A) THEN   GOTO 80
50  NEXT B
60  NEXT A
70  GOTO 90
80  PRINT A * 100 + B * 10 + A" * "A * 10 + B" = "P"."
90  END
```

SUPPLEMENT (Activity Book)

Activity Set 3.3 Multiplying with the Abacus and Chip Trading
Just for Fun: Crossnumbers for Calculators

An exhibit for illustrating multiplication, at the California Museum of
Science and Industry

Exercise Set 3.3: Applications and Skills

1. The children in the picture on the preceding page are computing products of numbers from 1 through 8. Each time three buttons are pressed on the switch box, the product is illustrated by lighted bulbs in the 8 by 8 by 8 cube of bulbs. Buttons 3, 4, and 1 are for the product 3 × 4 × 1. The 12 bulbs in the upper left-hand corner of part **a** will be lighted for this product. Whenever the third number of the product is 1, only those bulbs on the front face of the cube (facing children) will be lighted. Indicate the bulbs in parts **b** and **c** which will be lighted for the given products.

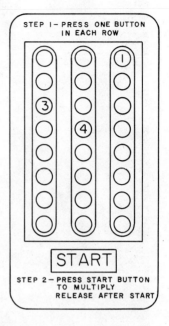

a. 3 × 4 × 1 ★   b. 7 × 3 × 1   c. 2 × 8 × 1

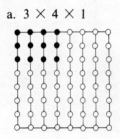 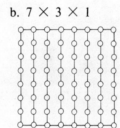 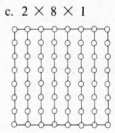

The first two numbers in the product of three numbers determine the rectangular array on the front face of the cube. The third number in the product determines the number of times the array on the front face is repeated in the cube. The 24 bulbs that are boxed in on the cube in part **d** will be lighted for 3 × 4 × 2. Box in the bulbs in parts **e** and **f** which will be lighted for the given products.

d. 3 × 4 × 2   e. 6 × 4 × 3 ★   f. 1 × 8 × 8

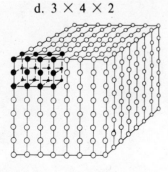 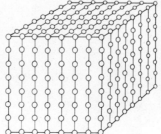 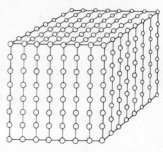

142   CHAPTER 3   WHOLE NUMBERS AND THEIR OPERATIONS

2. In parts **a** (multibase pieces) and **b** (bean sticks) sketch the new sets of objects for the given products. Then do the regrouping and write the numbers represented by the final collection of objects.

   a. Multiply by 4:

   b. Multiply by 3:

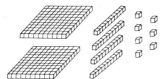

3. The products in parts **a** and **b** are each done in two steps. Sketch the new sets of chips and markers for each mat and abacus. After regrouping, what numbers will be represented by the final sets of chips and markers?

   ★ a. Multiply by 40:

Times 10										Times 4			

Red	Green	Blue	Yellow		Red	Green	Blue	Yellow		Red	Green	Blue	Yellow

   b. Multiply by 20:

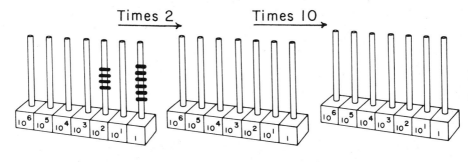

4. Multiplication of whole numbers can be illustrated on the number line by a series of arrows. This top number line shows 4×2. Draw arrow diagrams for the products in parts **a** and **b**.

   a. 3×4 b. 2×5

   c. Use the number line to show that $3 \times 4 = 4 \times 3$.

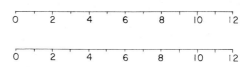

5. *Error Analysis:* Students who know their basic multiplication facts may still have trouble with the steps of the multiplication algorithm. Try to detect the type of error in each of the following computations.

a. $\overset{2}{2}7$ ★   b. $\overset{2}{1}8$ c. $\overset{4}{5}4$ ★   d. $\overset{1}{3}4$

$$\begin{array}{r} \overset{2}{2}7 \\ \times\ 4 \\ \hline 48 \end{array} \qquad \begin{array}{r} \overset{2}{1}8 \\ \times\ 3 \\ \hline 34 \end{array} \qquad \begin{array}{r} \overset{4}{5}4 \\ \times\ 6 \\ \hline 342 \end{array} \qquad \begin{array}{r} \overset{1}{3}4 \\ \times 24 \\ \hline 76 \end{array}$$

6. Use the given method to compute the product.

★   a. Gelosia   b. Egyptian duplation   c. Russian peasant

347×605

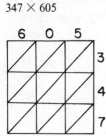

b. 27×35

$1 \times 35 = \ \ 35$
$2 \times 35 = \ \ 70$
$4 \times 35 = 140$
$8 \times 35 = 280$
$16 \times 35 = 560$

c. 42×30

Halving		Doubling
42	X	30
21	X	60
10	X	120
5	X	240
2	X	480
1	X	960

7. Use a grid (pages 134 and 135) to illustrate the partial products that occur in computing these products by pencil and paper. (*Suggestion:* Use graph paper.)

a. $\begin{array}{r} 24 \\ \times\ 7 \end{array}$ ★   b. $\begin{array}{r} 56 \\ \times 73 \end{array}$ c. $\begin{array}{r} 84 \\ \times 26 \end{array}$

8. *Estimation and Calculators:* Estimate the second factor so that the product will fall within the given range. Check your answer with a calculator. Count the number of tries to land in the range.

                           Range

Example $22 \times$ ____   (900, 1000)

              $22 \times 40 = 800$ Too small

              $22 \times 43 = 946$ In the range in two tries.

Product	Range	Number of Tries
a. $32 \times$ ____	(800, 850)	_____
b. $95 \times$ ____	(1650, 1750)	_____
c. $103 \times$ ____	(2800, 2900)	_____
d. $76 \times$ ____	(3500, 3600)	_____

9. Which number property is being used in each of the following equalities?

★   a. $3 \times (2 \times 7 + 1) = 3 \times (7 \times 2 + 1)$

       b. $18 + (43 \times 7) \times 9 = 18 + 43 \times (7 \times 9)$

★   c. $(12 + 17) \times (16 + 5) = (12 + 17) \times 16 + (12 + 17) \times 5$

       d. $4 \times (13 + 22)/(7 + 5) = 4 \times (13 + 22)/(5 + 7)$

       e. $(15 \times 2 + 9) + 3 = 15 \times 2 + (9 + 3)$

10. *Approximation and Mental Calculation:* Compute each product in your head. Explain your steps.

     a. Use the fact that multiplication distributes over addition to compute the product mentally.

         18×11 25×12 14×102

★   b. Use the fact that multiplication distributes over subtraction to compute the product mentally.

         35×19 51×9 30×99

     c. Round off one or both numbers to approximate the product mentally.

         18×20 97×34 49×52

11. *Finger Multiplication*

     a. Use your hands to compute 9×6, using the finger multiplication system on page 139. Will this method produce the correct results for the special cases of 9 times 10, or 10 times 10?

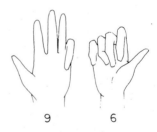

9   6

★   b. There is a similar method of finger multiplication for computing the products of numbers from 11 through 15. For example, the product of 12 and 14 is 168. Beginning with 100, how can the finger positions for 12 and 14 be used to get 68?*

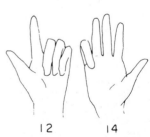

12   14

*For a description of how this method can be extended to numbers from 10 to 50, see L. P. Alger, "Finger multiplication," *The Arithmetic Teacher,* **15** No. 4 (April 1968), 341–43.

12. *Napier's Rods:* Use the rods given for multiples of 7, 8, and 4 to compute 6 × 784.

★   a. To compute 8 times 903, the eighth rows of the multiples of 9, 0, and 3 are needed. Write the numbers for these eighth rows and use them to compute 8 × 903.

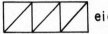 eighth row

b. One way to compute 94 times 288 is to use two rods with multiples of 8. Explain how this product can be computed without using any rods with multiples of 8. (*Hint:* Use the commutative property for multiplication.)

Exercise Set 3.3: Problem Solving

1. *The Lucas Problem:* Every day at noon a ship leaves Le Havre, France for New York and another ship leaves New York for Le Havre. The trip lasts 7 days and 7 nights. How many New York–Le Havre ships will the ship leaving Le Havre today meet during its journey to New York.*

   a. *Understanding the Problem:* This problem assumes that there are ships already en route. As each ship leaves the dock another arrives. So a given ship will meet one ship when leaving the dock in Le Havre, several ships en route, and one more when docking in New York. If the trip could be completed in one day, a ship leaving Le Havre would meet three ships in completing the trip. Explain why.

★   b. *Devising a Plan:* A diagram will be helpful. In addition, you may want to simplify the problem. If the trip could be completed in two days, how many ships would it meet? Draw a diagram to show the places where these ships are encountered.

   c. *Carrying Out the Plan:* Extend your plan to solve the problem for a seven-day trip. Mark each point of the diagram at which the ships will meet.

---

*This problem was stated by Edouard Lucas, a nineteenth-century French mathematician. See Boris A. Kordemsky, *The Moscow Puzzles* (New York: Charles Scribner's Sons, 1972), p. 108.

d. *Looking Back:* How many ships must a company have to provide continual service from Le Havre to New York and back under the conditions of the original problem? Complete this table which relates the number of ships needed to the number of days for the ocean crossing.

Number of Days for Trip	1	2	3	4	5	6	7
Number of Ships Needed	4						

2. *Product Patterns:* There is an easy method for computing the following products in your head. Find this method. (*Hint:* Look for patterns.) Try your method to compute the last three products.

$25 \times 25 = 625$ $22 \times 28 = 616$ $71 \times 79 = 5609$

$37 \times 33 = 1221$ $35 \times 35 = 1225$ $75 \times 75 = 5625$

$56 \times 54 =$ $62 \times 68 =$ $42 \times 48 =$

★ a. Explain this method for computing products. What types of numbers can be used?

b. We can see why this method works by representing the product on a grid. Here is a grid which illustrates 24×26. Rearrange these regions to show that $24 \times 26 = 20 \times 30 + 4 \times 6$.

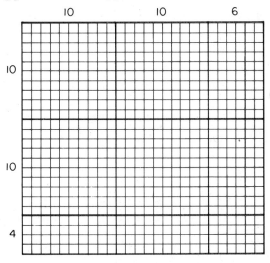

c. Draw a grid to show why $37 \times 33 = 30 \times 40 + 7 \times 3$.

d. Can the above method for computing products be used on three-digit numbers? For example, 124×126.

3. *Calculator Exercise:* The 1983–84 Manhattan telephone directory has approximately 1600 pages, each with an 8 inch by 10 inch printed surface. If every square inch of printed surface on these pages contained a 50 by 50 row of dots, as shown here, how many dots would there be in this directory? The world's population is approxi-

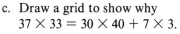

mately four billion. If each person is represented by one dot, how many of these telephone directories would be required?

4. *Faded Document:* Supply the missing digits for the boxes in this faded-document problem.

$$
\begin{array}{r}
4\ \square\ \square \\
\times\ \square\ \square\ 7 \\
\hline
\square\ \square\ 8\ 2 \\
1\ 2\ \square\ \square \\
\hline
\square\ \square\ \square\ \square\ \square
\end{array}
$$

5. *Cryptarithms:* Find the five digits represented by the letters in part **a**. The cryptarithm in part **b** is unusual, in that E and O do not represent only one digit. Each E stands for an even digit (0, 2, 4, 6, or 8) and each O for an odd digit (1, 3, 5, 7, or 9).

★   a.
$$
\begin{array}{r}
S\ T \\
\times\ R\ T \\
\hline
S\ T \\
P\ Q\ R \\
\hline
P\ T\ T\ T
\end{array}
$$

          b.
$$
\begin{array}{r}
E\ E\ O \\
\times\ O\ O \\
\hline
E\ O\ E\ O \\
E\ O\ O \\
\hline
O\ O\ O\ O\ O
\end{array}
$$

Exercise Set 3.3: Computers

1. What change is needed in the number chain computer program (Program 3.3A on page 140) if the rule is: Compute each new number by multiplying the units digit by 3 and adding the tens digit.

   ★   a. What number chain will be obtained if we begin the program with X = 17?

        b. What happens if 29 is used for the first number of this chain?

2. Will multiplying the units digit by 4 and adding the tens digit produce a number chain with numbers less than 39?

        a. Revise Program 3.3A on page 140 to investigate this question.

   ★   b. What happens when 39 is used as the first number of this chain?

3. What number chain will be obtained if we multiply the units digit by 5 and add the tens digit? Revise Program 3.3A on page 140 to investigate this question.

4. Write computer programs to solve the following cryptarithms. (*Hint:* Use three FOR–NEXT commands for parts **a** and **b**. Begin with R = 1 in part **a** and C = 1 in part **b** in the FOR–NEXT commands for R and C.)

   ☞   a.
$$
\begin{array}{r}
R\ S\ T \\
\times\ \ \ T\ R \\
\hline
3\ 8\ T\ T\ R
\end{array}
$$

   ☞ ★   b.
$$
\begin{array}{r}
C\ D\ E \\
\times\ \ \ E\ D \\
\hline
C\ C\ E\ E\ D
\end{array}
$$

General Motors Terex Titan and Chevrolet Luv pickup

3.4
DIVISION AND
EXPONENTS

Division, like multiplication, is a shortcut for computing. One way that division occurs is in comparing two quantities. Consider the relative size of the Terex Titan as compared to the Luv pickup on the Titan's dump body. The Terex Titan can carry 317250 kilograms while the Luv pickup has a limit of 450 kilograms. We can determine how many times greater the Titan's capacity is than the Luv's by dividing 317250 by 450. The answer is 705, which means that the Luv pickup will have to haul 705 loads to fill the Titan just once! If the boy's truck in the photo holds 3 kilograms of sand, how many of its loads will be required to fill the Titan?

DEFINITION OF DIVISION

The division example, which compares the size of the Terex Titan to the Luv pickup, can be checked by multiplication. The load weight of the smaller truck times 705 should equal the load weight of the larger truck. The close relationship between division and multiplication can be used to define division in terms of multiplication. This is the most popular method of defining division and perhaps the oldest. J. P. A. Erman, in *Life in Ancient Egypt,* has attributed this definition to the Egyptians. Symbolically, it is stated as follows. For any whole numbers r and s, with $s \neq 0$,

$$r \div s = \square \quad \text{if and only if} \quad r = s \times \square$$

This definition says that division can be done by multiplication. If I asked you to compute $54 \div 9$, you would immediately say 6, because you know that $54 = 9 \times 6$.

There are three common terms in the division process: dividend, divisor, and quotient. In the previous example, 54 is the *dividend,* 9 is the *divisor,* and 6 is the *quotient.*

Over the centuries, division has acquired two meanings or uses. David Eugene Smith, in *History of Mathematics,* speaks of the *twofold nature of division* and gives references to the sixteenth-century authors who first clarified the differences.* These two meanings of division, called partitive and measurement, are illustrated in the following two paragraphs.

Partitive Concept   Suppose you had 24 tennis balls, which you wanted to distribute equally among 3 people. How many tennis balls would each person receive? This can be determined by separating (partitioning) the tennis balls into 3 equivalent sets. The following arrangement of tennis balls shows the 24 balls partitioned into 3 groups and illustrates $24 \div 3$. The divisor, 3, indicates the number of groups, and the quotient, 8, is the number of balls in each group. This is an example of the *partitive* use of division.

Measurement Concept   Suppose you had 24 tennis balls and wanted to give 3 tennis balls to as many people as possible. How many people would receive tennis balls? This can be determined by subtracting away, or measuring off, as many sets of 3 as possible. The following arrangement of tennis balls shows the result of this measuring process and illustrates $24 \div 3$. The divisor, 3, is the number of balls in each group, and the quotient, 8, is the number of groups. This is an example of the *measurement* use of division.

*D. E. Smith, *History of Mathematics,* **2** (Lexington, Mass: Ginn, 1925), p. 130.

In both examples the answer to the questions is the same, 8. However, the process of grouping the tennis balls in each case is different. It is important to be aware of these differences, especially when introducing and explaining what is meant by division.

MODELS FOR
DIVISION ALGORITHMS

Each of the models used for addition, subtraction, and multiplication are also suitable for illustrating the long division algorithm. The multibase pieces and the abacus are used in the next examples, but similar demonstrations could be given with other models.

partitive

Multibase Pieces   To compute $378 \div 3$ using the multibase model and the partitive concept of division, we will perform the following steps.

(1) Begin with 3 flats, 7 longs, and 8 units.

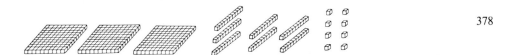

378

(2) Partition the flats by placing 1 flat in each of three groups. This leaves 7 longs and 8 units.

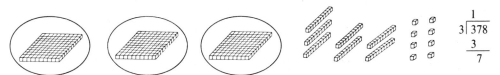

$$\begin{array}{r} 1 \\ 3\overline{)378} \\ 3 \\ \hline 7 \end{array}$$

(3) Partition the longs by placing 2 longs in each of the three groups, leaving 1 long and 8 units.

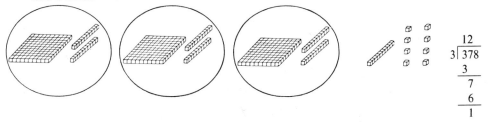

$$\begin{array}{r} 12 \\ 3\overline{)378} \\ 3 \\ \hline 7 \\ 6 \\ \hline 1 \end{array}$$

(4) Exchange the long for 10 units so that there are 18 units.

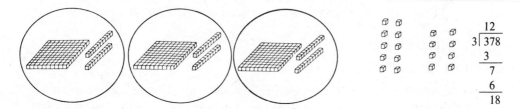

(5) Partition the units by placing 6 units in each of the three groups.

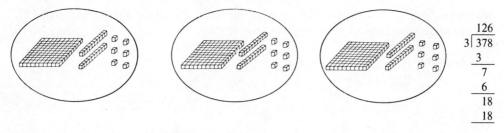

Notice that each of the final groups of multibase pieces represents the quotient, 126.

Abacus The <u>measurement concept</u> is used in the following examples to illustrate division on the abacus. Consider the example of dividing 639 by 3. Beginning with the hundreds column, measure off and remove 3 markers at a time. Each group of 3 markers represents 3 hundreds, which equals 100 threes. Why? So, to record this activity, 1 marker should be placed above the hundreds column for each group of 3 markers which is removed. In all, 2 markers should be placed above the hundreds column to represent a quotient of 200. We can see the reason for this procedure in the following equations.

$$600 \div 3 = (100 \times 6) \div 3 = 100 \times (6 \div 3) = 100 \times 2$$

We continue dividing by removing 3 markers at a time from the columns and placing 1 marker above a column for each group of 3. The markers that are placed above the columns represent the quotient.

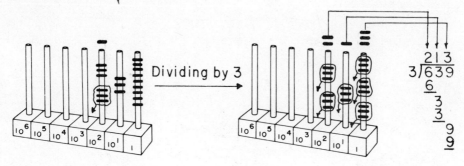

Dividing by 3

CHAPTER 3   WHOLE NUMBERS AND THEIR OPERATIONS

Division on the abacus can be done one column at a time when dividing by a single-digit number. If the number of markers on a column is less than the divisor, the markers can be regrouped. For example, to illustrate 432 ÷ 6, there are only 4 markers on the hundreds column. These markers can be regrouped to the tens column to give 43 markers, or we can simply think of the 4 markers on the hundreds column and the 3 markers on the tens column as 43 tens. Since there are 7 sixes in 42, 7 markers are placed above the tens column, and the remaining marker in the tens column is regrouped with the 2 markers on the units column. For the final step in this example, 12 ÷ 6 = 2, so 2 markers are placed above the units column.

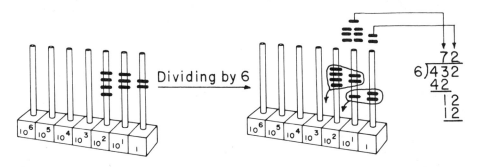

ALGORITHMS FOR DIVISION

Algorithms for division have historically been the most difficult of the basic operations of arithmetic. The Italian author Pacioli (1494) said, "If a man can divide well, everything is easy, for all the rest is involved therein."

Duplation Method The ancient Egyptians used an algorithm for division which is similar to their duplation method of multiplying and shows the close relationship between these two operations. For example, to divide 1710 by 90 the algorithm begins by repeatedly doubling the 90's as shown here. This process is stopped when the next number is greater than 1710. Next, we find the numbers in the right-hand column, whose sum is 1710. Since these numbers correspond to the binary numbers 1, 2, and 16, the answer is 19. This method can also be used when the quotient is not a whole number (see page 162, Exercise 6).

1	90 ←
2	180 ←
4	360
8	720
16	1440 ←

$$
\begin{array}{r}
90 \\
180 \\
+ \ 1440 \\
\hline
1710
\end{array}
$$

Long Division Algorithm Today's pencil-and-paper method, called the *long division algorithm,* has developed

error analysis - find error

over hundreds of years into an efficient method for computing. It is usually viewed as the process of subtracting off multiples of the divisor, until the remaining number is less than the divisor. One of the forerunners of our long division algorithm is similar to the form shown here, in which each digit of the divisor is multiplied separately by a digit in the quotient. This method has the advantage that the products involve only single-digit numbers.

Forerunner to Long Division Algorithm	Long Division Algorithm
25 38⟌952 $\underline{6}$ (20 × 30) 35 $\underline{16}$ (20 × 8) 19 $\underline{15}$ (5 × 30) 42 $\underline{40}$ (5 × 8) 2	25 38⟌952 $\underline{76}$ (20 × 38) 192 $\underline{190}$ (5 × 38) 2

Calculator Method The quotient of two numbers can be found by entering the numbers and the division operation as they appear from left to right. Here are the steps and displays for computing 27094 ÷ 7.

Steps	Displays
1. 27094	27094.
2. ÷	27094.
3. 7	7.
4. =	3870.5714

The sum or product of two whole numbers is always another whole number. This is called the *closure property,* and we say that each of these operations is *closed* in the set of whole numbers because they always produce another whole number. Subtraction and division of whole numbers, on the other hand, are not closed. That is, the difference or quotient of two whole numbers is not always another whole number. In the previous example, 27094 ÷ 7 is approximately equal to 3870, but the decimal part of the answer means that there is a remainder. To find this remainder, multiply 7 times .5714 and round off to the nearest whole numbers, which is 4.

$$7 \times .5714 = 3.9998$$
$$\approx 4$$

To check the quotient, 3870, and the remainder, 4, multiply 3870 by 7 and add 4.

APPROXIMATION AND MENTAL CALCULATION

To obtain the quotient of one number divided by another, it is sometimes helpful to halve both numbers.

$$144 \div 18 = 72 \div 9 = 8$$

This halving process can be carried out several times.

$$168 \div 28 = 84 \div 14 = 42 \div 7 = 6$$

It is also permissible to divide each number by 3.

$$180 \div 12 = 60 \div 4 = 15$$
$$900 \div 36 = 300 \div 12 = 100 \div 4 = 25$$

These methods of simplifying division by dividing by 2 or 3 can be combined, as in the next example.

Divide by 3   Divide by 2
$$336 \div 48 = 112 \div 16 = 56 \div 8 = 7$$

$$a \div b = ak \div bk \quad \text{for} \quad k \neq 0$$

Rounding Off   The quotient of two numbers may be approximated by rounding off one or both numbers to more convenient numbers.

$$472 \div 46 \approx 460 \div 46 = 10$$
$$145 \div 23 \approx 150 \div 25 = 6$$
$$8145 \div 195 \approx 8000 \div 200 = 40$$

Rounding off to obtain an approximate quotient can be combined with the process of dividing the divisor and dividend by the same number.

$$427 \div 72 \approx 430 \div 70 = 43 \div 7 \approx 6$$
$$139 \div 18 \approx 140 \div 18 = 70 \div 9 \approx 8$$

DIVISION WITH ZERO

The special cases of division involving zero are worth looking at carefully. Perhaps it is the association of zero with nothingness which causes confusion, but whatever the reason, misconceptions involving division and zero are common.

Dividing by a Nonzero Number   The definition of division shows that 0 divided by any nonzero number is 0. In particular, consider dividing 0 by 4. We will use the fact that any number times 0 is equal to 0.

$$0 \div 4 = 0 \quad \text{because} \quad 0 = 4 \times 0$$

KNow exponents

3.4   DIVISION AND EXPONENTS

ding by Zero The temptation in this case is to think that a number divided by 0, $4 \div 0$, is an infinite number. Actually, no such number exists. Using the definition ~~vision~~ again,

$$4 \div 0 = \square \qquad \text{if} \qquad 4 = 0 \times \square$$

However, since any number times 0 equals 0, there is no number that can be written into the placeholder, $\square$, to make $4 = 0 \times \square$. Therefore, division by zero is not possible.

Dividing Zero by Zero The confusion over this case may result from the familiar phrase that "any number divided by itself is 1," which is not true for zero divided by zero. In fact, $0 \div 0$ is never permitted because it leads to an absurd result. For example, we know that $0 \times 6 = 0$ and $0 \times 3 = 0$. Using the fact that division is the inverse of multiplication, these equations could be written as $6 = 0 \div 0$ and $3 = 0 \div 0$, if $0 \div 0$ were permitted. But this would mean that $3 = 6$. We avoid this contradiction by not allowing, or defining, $0 \div 0$ to be a number.

 EXPONENTS

The large numbers used today were unnecessary a few centuries ago. The word "billion" which is now commonplace was not adopted until the seventeenth century. Even now "a billion" means different things to different people. In the United States it represents 1,000,000,000 (a thousand million) and in England it is 1,000,000,000,000 (a million million).

Our numbers are named according to powers of 10. The first, second, and third powers of 10 are the familiar *ten, hundred,* and *thousand.* After this, only every third power of 10 has a new or special name: million, billion, trillion, etc. (page 71).

$10^0 = 1$	one
$10^1 = 10$	*ten*
$10^2 = 100$	one *hundred*
$10^3 = 1,000$	one *thousand*
$10^4 = 10,000$	ten thousand
$10^5 = 100,000$	one hundred thousand
$10^6 = 1,000,000$	one *million*
$10^7 = 10,000,000$	ten million
$10^8 = 100,000,000$	one hundred million
$10^9 = 1,000,000,000$	one *billion*
$10^{10} = 10,000,000,000$	ten billion
$10^{11} = 100,000,000,000$	one hundred billion
$10^{12} = 1,000,000,000,000$	one *trillion*

An exponent of 10 indicates the number of zeros which follow the 1. As examples, $10^1 = 10$ (1 followed by 1 zero), and $10^0 = 1$ (1 followed by 0 zeros). A number such as 10^{21}, which is the frequency of gamma rays per second, is a 1 followed by 21 zeros.

Raising a number to a power is the operation called *exponentiation.* Numbers written as b^n are said to be in *exponential form.* We call b^n the *nth power of b* or *b to the nth power,* with the exception that b^2 is called *b squared* and b^3 is *b cubed.* This terminology has been inherited from the early Greeks who pictured the products of numbers as geometric arrays of dots.

In general, for any number b and whole number n, b^n is the product with b occurring n times: $b^n = b \times b \times \cdots \times b$. In case $n = 1$ or 0, $b^1 = b$ and $b^0 = 1$. ($10^0 = 1$, $2^0 = 1$, etc.) The case in which $n = 0$ and $b = 0$, that is, 0^0, is not defined. The number b is called the *base.*

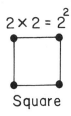

$$2 \times 2 = 2^2$$

Square

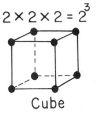

$$2 \times 2 \times 2 = 2^3$$

Cube

Laws of Exponents Multiplication and division can be performed easily for numbers that are written as powers of the same base. To multiply we add the exponents, and to divide, the exponents are subtracted. Here are two examples.

$$2^4 \times 2^3 = (2 \times 2 \times 2 \times 2) \times (2 \times 2 \times 2) = 2 \times 2 \times 2 \times 2 \times 2 \times 2 \times 2 = 2^7$$

$$\frac{2^8}{2^3} = \frac{2 \times 2 \times 2 \times 2 \times 2 \times 2 \times 2 \times 2}{2 \times 2 \times 2} = 2 \times 2 \times 2 \times 2 \times 2 = 2^5$$

These equations are special cases of the following theorems, which hold for all whole numbers a, n, and m ($a \neq 0$ in the rule for subtracting exponents).

Rule for Adding Exponents	$a^n \times a^m = a^{n+m}$
Rule for Subtracting Exponents	$a^n \div a^m = a^{n-m}$, for $a \neq 0$

The primary advantages of exponents are their compactness and convenience for computing, as shown by the following examples.

In our galaxy there are 10^{11} (one hundred billion) stars, and in the observable universe there are 10^9 (one billion) galaxies. If every galaxy had as many stars as ours, there would be $10^9 \times 10^{11}$ stars.

$$10^9 \times 10^{11} = 10^{9+11} = 10^{20}$$

A galaxy similar to the Milky Way, about 100,000 light-years across

If 1 out of every 1000 stars had a planetary system, there would be $10^{20} \div 10^3$ stars with planetary systems.

$$10^{20} \div 10^3 = 10^{20-3} = 10^{17}$$

To continue this example, if 1 out of every 1000 stars with a planetary system had a planet with conditions suitable for life, there would be $10^{17} \div 10^3$ such stars.

$$10^{17} \div 10^3 = 10^{17-3} = 10^{14}$$

You may wish to continue this line of reasoning to consider the numbers of planets which might have life or perhaps intelligent life. The odds in favor of life in other parts of the universe appear to be overwhelming.

Calculators Numbers raised to a power can be computed on a calculator, provided they do not exceed the capacity of the calculator's display. On some calculators, the steps on the right will produce the number for 4^{10}, if carried out to Step 10.

Steps	Displays
1. Enter 4 $\boxed{\times}$	4.
2. $\boxed{=}$	16.
3. $\boxed{=}$	64.
4. $\boxed{=}$	256.

The number of steps in the previous sequence can be decreased by applying the rule for adding exponents: $a^n \times a^m = a^{n+m}$. To compute 4^{10}, first compute 4^5 on the calculator and then multiply 1024 times 1024 ($4^5 \times 4^5 = 4^{10}$).

Some calculators have exponential buttons for evaluating numbers raised to a power. To compute a number y to some exponential power x, the base y is entered into the calculator first; then the exponential button $\boxed{y^x}$ is pressed; and then the exponent x is entered. The steps for evaluating 4^{10} are shown to the right.

Steps	Displays
1. Enter 4	4.
2. $\boxed{y^x}$	4.
3. Enter 10	10.
4. $\boxed{=}$	1048576.

Numbers that are raised to powers will frequently exceed the calculator display. If you try to compute 4^{15} on a calculator with only eight places in its display, there will not be room for the answer in positional numeration. Some calculators will automatically convert to scientific notation when numbers in positional numeration are too large for the display (page 431).

COMPUTER APPLICATIONS

The powers of the digits 1 through 9 involve some interesting patterns. Here are the fourth powers of these numbers.

1^4	2^4	3^4	4^4	5^4	6^4	7^4	8^4	9^4
1	16	81	256	625	1296	2401	4096	6561

Notice how the units digits of these numbers form a symmetric pattern with "5" in the center.

$$1 \quad 6 \quad 1 \quad 6 \quad 5 \quad 6 \quad 1 \quad 6 \quad 1$$

Will the second powers, third powers, fifth powers, etc. of the digits 1 through 9 have patterns? Program 3.4A helps answer these questions. The circumflex, ^, is the notation for the operation of exponentiation.*

Program 3.4A

```
10   PRINT "THIS PROGRAM PRINTS THE NTH POWERS OF THE DIGITS
        FROM 1 THROUGH 9. TYPE A WHOLE NUMBER FOR N. "
20   INPUT N
30   FOR X = 1 TO 9
40   PRINT X ^ N" ";
50   NEXT X
60   END
```

Perfect Cubes   The number 1331 is a perfect cube: $11^3 = 1331$. The sum of the digits of this number is 8, which is also a perfect cube. Are there any other such numbers? Are there perfect cubes such that the sum of the digits is a perfect square? Program 3.4B computes the cubes of whole numbers from 1 to N. The value of N is typed in by the operator when the computer prints out a question mark in line 20. The printout below is for N = 18.

Program 3.4B

```
10   PRINT "TYPE A WHOLE NUMBER. "
20   INPUT N
30   FOR X = 1 TO N
40   PRINT X ^ 3" ";
50   NEXT X
60   END

TYPE A WHOLE NUMBER.
? 18
1 8 27 64 125 216 343 512 729 1000 1331 1728 2197 2744
3375 4096 4913 5832
```

*The up arrow ↑ is also a computer symbol for exponentiation.

Is the sum of the digits in any of these numbers, other than 1331, a perfect cube? Is the sum of the digits in any of these numbers a perfect square?

SUPPLEMENT (Activity Book)

Activity Set 3.4 Dividing on the Abacus
Just for Fun: Calculator Games and Number Tricks

Exercise Set 3.4: Applications and Skills

1. Encircle groups of bean sticks and beans to illustrate the given quotients. Cross out and sketch new beans if regrouping is needed.

    a. 86 ÷ 2

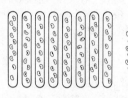

    ★   b. 96 ÷ 4

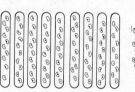

2. Use the abacus in parts **a** and **b** to compute the given quotients. Use the long division algorithm and compare the steps used on the abacus with those used in the algorithm.

    ★   a. Illustrate 8064 ÷ 2 by drawing markers on the second abacus and then encircling groups of 2 markers. How is the quotient determined by the groups of markers?

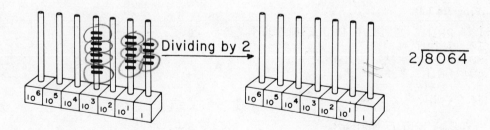

b. Illustrate 4272 ÷ 3 by drawing markers on the second abacus and then encircling groups of 3 markers. How is the quotient determined by the groups of markers? (Note: Regrouping is needed.)

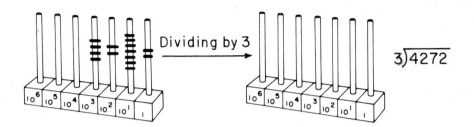

★ 3. a. What division fact is illustrated by the arrows on this number line?

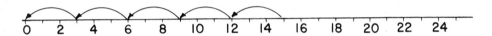

b. Draw arrow diagrams for 24 ÷ 4 and 18 ÷ 9.

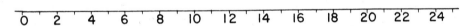

4. *Error Analysis:* These examples of long division have four different types of errors. Locate each type of error.

a.
```
    56 R4
8) 4052
   40
   ──
   52
   48
   ──
    4
```

★ b.
```
    68
3) 258
   24
   ──
   18
   18
   ──
```

c.
```
    370
7) 2149
   21
   ──
   49
   49
   ──
```

★ d.
```
    29 R20
4) 136
   8
   ──
   56
   36
   ──
   20
```

5. a. Compute each side of the following equation. Does the right side equal the left? Try some other numbers in the square, rhombus, and triangular placeholders. Can you find a case in which division is not distributive over addition?

$$(\boxed{6} + \triangle{15}) \div \triangle{3} = (\boxed{6} \div \triangle{3}) + (\triangle{15} \div \triangle{3})$$

b. Is division commutative or associative? Try some numbers in the following placeholders. It takes only one counterexample to show that a property does not hold.

★ $\square \div \triangle \overset{?}{=} \triangle \div \square$ NO

$\square \div (\triangle \div \varnothing) \overset{?}{=} (\square \div \triangle) \div \varnothing$ NO

2

4 =

6. Use the duplation method to compute the following quotients. This method requires a little trial and error to find the numbers for the correct sum. The exact sum cannot be found for one of these examples. In this case, a fraction will be needed in the answer.

★   a. $1232 \div 112$

1	112
2	224
4	448
8	896

b. $348 \div 24$

1	24
2	48
4	96
8	192

7. a. Compute these products and quotients. Leave your answers in exponential form.

add   ★ $5^{14} \times 5^{20}$ $10^{12} \times 10^{10}$ ★ $10^{32} \div 10^{15}$ $3^{22} \div 3^8$ ← *subtract*

b. Find numbers for m and n to show that
$2^n - 2^m \neq 2^{n-m}$ ★ $2^n + 2^m \neq 2^{n+m}$

8. The chart below shows the approximate frequencies of some common types of waves. Visible light waves, for example, are between 10^{14} and 10^{15} waves or cycles per second.

★   a. The frequency of television is 10^8 cycles per second. If a type of x-ray has a frequency that is 10^{11} times greater, what is the x-ray frequency?

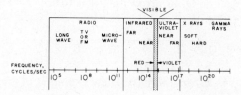

b. If the frequency for infrared is 10^{13} and it is 1000 times greater than the frequency for micro-waves, what is the microwave frequency?

★   c. If a radio frequency is 10^8 and gamma rays have a frequency of 10^{21}, how many times greater is the gamma-ray frequency?

★ 9. *Calculator Exercise:* Beneath each of the following equations there is a sequence of calculator steps. Which sequences produce the correct answers? If a sequence does not produce the correct answer, write the equation which corresponds to the sequence of steps.

a. $8 \times (12 \div 3) = 32$
1. Enter 8
2. $\boxed{\times}$
3. Enter 12
4. $\boxed{\div}$
5. Enter 3
6. $\boxed{=}$

b. $3 \times 4 + 7 = 19$
1. Enter 3
2. $\boxed{\times}$
3. Enter 4
4. $\boxed{+}$
5. Enter 7
6. $\boxed{=}$

c. $17 - 3 \times 5 = 2$
1. Enter 17
2. $\boxed{-}$
3. Enter 3
4. $\boxed{\times}$
5. Enter 5
6. $\boxed{=}$

10. *Calculator Exercise* A calculator has been used to compute the following quotients. Convert the decimal part of each quotient into a remainder. Check your answer by multiplying the divisor by the whole number part of the quotient and adding the remainder.

★   a. $47208 \div 674 = 70.041543$ b. $107253 \div 86 = 1247.1279$

      c. $13738 \div 24 = 572.41667$

11. *Approximation and Mental Calculation:* Compute each product in your head. Explain your steps.

    a. Use the fact that each number can be divided by 2 or 3 to compute the quotient.

        $84 \div 14$ $90 \div 18$ $180 \div 36$ $144 \div 16$

    b. Round off one or both numbers to approximate the quotient.

        $203 \div 50$ $82 \div 19$ $241 \div 31$ $123 \div 29$

12. During the Buddhistic period (sixth century to first century B.C.) the Hindus were particularly interested in large numbers. B. L. Van Der Waerden describes this scene from the book *Lalilavistara.** Prince Guatama (Buddha) asks the prince Dandapani for the hand of his daughter Gopa. He is required to compete with five other suitors in writing, wrestling, archery, running, swimming, and arithmetic. The great mathematician Arjuna questions him.

> Oh, young man, do you know how the numbers beyond the koti continue by hundreds?
>   I know it.
> How then do the numbers beyond the koti continue by hundreds?
>   One hundred kotis are called *ayuta,* one hundred ayutas *niyuta,* one hundred niyutas *kanikara,* one hundred kanikaras *vivara* . . .

★   a. A *koti* is $10^2 \times 10^5$ (one hundred times one hundred thousand). Compute this product and use an exponent to write your answer as a power of 10.

★   b. Prince Guatama continues through 24 stages, where koti is the first stage, ayuta is the second, niyuta is the third, etc. Write the number for the twenty-fourth stage as a power of 10.

    c. Beyond this there are eight other series. If each series increases the numbers by a multiple of 10^{48}, the total increase for all eight series will be $10^{48} \times 10^{48} \times 10^{48} \times 10^{48} \times 10^{48} \times 10^{48} \times 10^{48} \times 10^{48}$. Compute this product and write your answer as a power of ten.

*B. L. Van Der Waerden, *Science Awakening* (Groningen, Holland: P. Noordhoff, 1954), pp. 51–52.

Exercise Set 3.4: Problem Solving

1. *Repeating Pattern:* Here is a repeating pattern of dollar signs, asterisks, and number signs. If this pattern continues, which of these symbols will be in the 538th square?

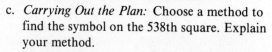

a. *Understanding the Problem:* To become more familiar with this problem, extend the pattern a few more squares. What symbol will be in the nineteenth square?

★ b. *Devising a Plan:* Because the pattern repeats itself after six squares it is suggestive of a clock with six symbols. What symbol occurs in squares 6, 12, 18, etc.? Or, you could think of the pattern as pieces of tile 6 squares long. To make the length 32 squares long, how many tiles and squares will it take?

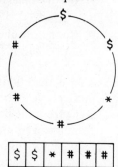

c. *Carrying Out the Plan:* Choose a method to find the symbol on the 538th square. Explain your method.

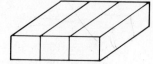

★ d. *Looking Back:* The lengths of repeating patterns and their symbols will vary. What is the 345th digit in this number if the pattern continues: 142857142857 . . . ?

★ 2. *Cake Cutting Puzzle:* After a cake was cut into three equal pieces, as shown here, it was discovered that four people each wanted an equal share of the cake. How can one more straight cut be made so that each of the four people get the same amount of cake?

★ 3. *All Four Basic Operations:* Using exactly 4 fours, and only addition, subtraction, multiplication, and division, try to write an expression that equals each of the numbers from 1 to 10. For example, $2 = (4 \div 4) + (4 \div 4)$. You do not have to use all operations, and expressions such as 44 are allowable.* (*Hint:* Guess and check.)

4. *Krypto:* Krypto is a commercially produced card game with numbers from 1 through 25. The object is to combine the numbers on 5 cards that are dealt to obtain the number on a sixth card, the target number. Any of the four basic operations may

---

*Similar equations exist for 5 fives, 6 sixes, etc. See R. Crouse and J. Shuttleworth, "Playing with Numerals," *The Arithmetic Teacher,* **21** No. 5 (May 1974), 417–19.

be used, but each of the 5 cards must be used once and only once. Each of the following sets of cards can be combined by using all four operations to obtain the target number. (*Hint:* Guess and check. Try working backward.)

Dealt Cards

| 13 | 7 | 12 | 2 | 4 |

Target Number → 1

$$13 - 7 = 6$$
$$12 \div 6 = 2$$
$$2 \times 2 = 4$$
$$4 \div 4 = 1$$

★ a. | 22 | | 19 | | 2 | | 14 | | 10 | → 7

b. | 21 | | 2 | | 3 | | 12 | | 7 | → 20

★ 5. *Faded Document:* Supply the missing digits.

```
              9 □□□
3□ | □□4□9□
       □□1
          □□□
          □1□
          2□□
          2□□
```

Exercise Set 3.4: Computers

1. The following table contains the units digits of the first, second, third, and fourth powers of the digits 1 through 9. Use Program 3.4A to compute the nth powers of a digit (page 159) to complete the table for the fifth and sixth powers.

	Digits								
	1	2	3	4	5	6	7	8	9
First Power	1	2	3	4	5	6	7	8	9
Second Power	1	4	9	6	5	6	9	4	1
Third Power	1	8	7	4	5	6	3	2	9
Fourth Power	1	6	1	6	5	6	1	6	1
Fifth Power									
Sixth Power									
Seventh Power									
Eighth Power									

a. Look for a pattern and predict the units digits for the seventh and eighth powers. Check your conjecture by computing the powers.

      b. What are the units digits for the ninth powers of the digits 1 through 9?

★   c. Which rows are symmetric about the middle number?

2. Here are the first eight powers of 3.

$$3 \quad 9 \quad 27 \quad 81 \quad 243 \quad 729 \quad 2187 \quad 6561$$

      a. Use the pattern of units digits to determine the units digit in 3^{55}.

★   b. Write a computer program to print the first eight powers of any whole number.

      c. Use the program in part **b** to obtain the first few powers of 7. Find a pattern and use it to determine the units digit in 7^{55}.

3. Use Program 3.4B for computing the cubes of numbers (page 159) to find the following numbers.

      a. There is exactly 1 four-digit perfect cube for which the sum of the digits is a perfect square.

★   b. There is exactly 1 five-digit perfect cube for which the sum of the digits is a perfect square.

      c. Are there similar numbers for two-digit and three-digit perfect cubes?

4. Write a program to compute the squares of whole numbers from 1 to N.

      a. Find a four-digit perfect square such that the sum of the digits is a perfect cube. How many such numbers are there?

★   b. Find a five-digit perfect square for which the sum of its digits is a perfect cube. Is there more than one?

3.5
PRIME AND
COMPOSITE
NUMBERS

Do you have a favorite number? The number 3 is a popular choice and there may be historical reasons for this preference. For example, in the French "très bien," which means good, "très" is derived from the word for three. One of the oldest superstitions is that odd numbers are lucky. One exception to this is the common fear of 13, called *triskaidekaphobia.* Often, hotels will not have a floor that is numbered 13, and motels will not have a room 13. Is this true in the area you live in? Some people have a particular fear of Friday, the thirteenth. These people will be unhappy to know that the thirteenth day of the month falls more frequently on a Friday than on any other day of the week.*

NUMBER MYSTICISM

The Pythagoreans (ca. 500 B.C.), a brotherhood of mathematicians and philosophers, believed that numbers had special meanings which could account for all aspects of life. The number 1 represented reason, 2 stood for opinion, 4 was symbolic of justice, and 5 represented marriage. Even numbers were weak and earthly, and odd numbers were strong and heavenly. The numbers 1, 2, 3, and 4 also represented fire, water, air, and earth, and the fact that $1 + 2 + 3 + 4 = 10$ had many meanings. When only nine heavenly bodies could be found, including the earth, sun, moon, and the sphere of stars, they imagined a tenth to "balance the earth."†

Through the ages, numbers have played an important role in astrology and in the casting of horoscopes. Since many numeration systems used letters, it was natural to substitute the number value for the letters in a name. This practice, which is called *gematria,* was popular with the ancient Hebrews and Greeks. The following table shows the letter symbols the Greeks used for numbers.

*J. O. Irwin, "Friday 13th," *Mathematical Gazette,* **55** (1971), 412–15.
†M. Kline, *Mathematics in Western Culture* (New York: Oxford University Press, 1953), p. 77.

α (1)	β 2	γ 3	δ 4	ε 5	ς 6	ζ 7	η (8)	θ 9
ι 10	κ 20	λ 30	μ (40)	ν (50)	ξ 60	ο 70	π 80	α 90
ρ 100	σ 200	τ 300	υ 400	φ 500	χ 600	ψ 700	ω 800	ⅸ 900

The Greek word for amen is αμην, and from the table this word corresponds to 1 + 40 + 8 + 50, or 99. In some old editions of the Bible the number 99 appears for the word amen. During the Middle Ages gematria was revived and used in the art of "beasting." This consisted of showing that the number of the beast, 666, could be associated with certain people by assigning numbers to the letters of their name. In one case, the Catholic theologian Peter Bungus showed that a form of the name Martin Luther was numerically equivalent to 666.

MODELS FOR FACTORS

One of the earliest number distinctions involved primes. The number 7, for example, was associated with purity and the maiden goddess Athena, because it cannot be broken down into a product of smaller factors. If one whole number divides a second whole number evenly, that is, leaving only a zero remainder, then the first number is a *factor* (or *divisor*) of the second number. Any whole number greater than 1 which has only itself and 1 as factors is called a _prime_. All other whole numbers greater than 1 are called _composite_.

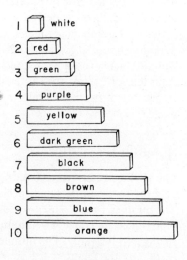

Colored rods are useful for illustrating the concepts of factor and prime and composite numbers. The rods pictured here are Cuisenaire rods. While these 10 rods can be assigned different values, they are often used to represent the whole numbers from 1 to 10, as indicated.

Numbers greater than 10 are represented by placing rods end to end. Two or more rods placed end to end are called a *train*. Here are two examples.

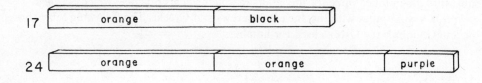

[handwritten marginal notes: "prime # test", "only have to test to sq. root of #"]

A train that is represented by only one type of rod is called a *one-color train*. The following one-color train of red rods shows that 2 is a factor of 26. Furthermore, the fact that there are 13 red rods in this train tells us that 13 is also a factor of 26. The number 3, on the other hand, is not a factor of 26 because a train of only green rods will not equal the length of the top train.

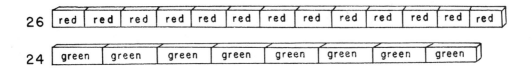

Every whole number greater than 1 can be represented by a one-color train of white rods. When a number can be represented only by a one-color train of white rods, it is a prime number. The first of the following trains represents the prime number 23. The all red, all green, and all yellow trains are not equal to the length of the top train and so 2, 3, and 5 are not factors of 23. These trains show that 22, 24, and 25 have factors other than 1, and therefore these numbers are composite.

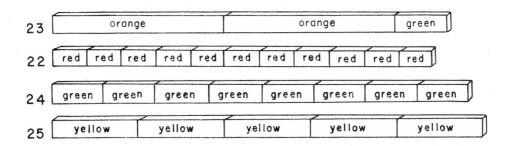

CLASSIFYING NUMBERS

Look at the following list of whole numbers and their factors. Some numbers have exactly two factors. These are *prime numbers*. The numbers with more than two factors are *composite numbers*. One is the only number with fewer than two factors and it is neither prime nor composite.

Number	Factors
1	1
2	1, 2
3	1, 3
4	1, 2, 4

Number	Factors
5	1, 5
6	1, 2, 3, 6
7	1, 7
8	1, 2, 4, 8
9	1, 3, 9
10	1, 2, 5, 10
11	1, 11
12	1, 2, 3, 4, 6, 12

There are 25 primes less than 100. Here are the first few.

$$2 \quad 3 \quad 5 \quad 7 \quad 11 \quad 13 \quad 17 \quad 19 \quad 23 \quad 29$$

There is no largest prime because there are an infinite number of prime numbers. Some very large primes have been discovered. From 1876 to 1951, this 39-digit number was the largest known prime:

$$170{,}141{,}183{,}460{,}469{,}231{,}731{,}687{,}303{,}715{,}884{,}105{,}727$$

Now with computers, someone finds a larger prime every few years. In 1983, for example, at the Cray Research Labs in Mendota Heights, Minnesota, David Slowinski found the largest known prime at that time, a number with 39,751 digits.

Prime numbers are difficult to locate because they do not occur in predictable patterns. In fact, there are arbitrarily large stretches of consecutive whole numbers with no primes! For example, between the two numbers 396,733 and 396,833 there are 99 composite numbers. It is possible to find a string of one million, one billion, etc., consecutive numbers that contain no prime numbers (see page 180, Exercise 3). No wonder mathematicians have been unable to find a formula that will give all primes less than any given number.

✳ PRIME NUMBER TEST

How can we determine whether 421 is a prime number? A natural but time-consuming approach is to try dividing by smaller whole numbers, 2, 3, 4, 5, 6, 7, 8, etc. This method can be improved by noticing that it is unnecessary to divide by the composite numbers 4, 6, 8, 9, etc. For example, if 2 is not a factor of a number, then 4, 6, and 8 will not be factors. In other words, to determine whether a number is prime or composite, it is necessary only to try dividing by smaller numbers that are prime.

A second improvement for determining whether or not a number is prime comes from the observation that composite numbers have at least one prime factor less than or equal to the positive square root of the number. Compare the prime factors of the numbers in the first column of this table to the numbers in the third column of the table. In some cases, there

n	Prime Factorization	$\sqrt{n}$
51	3×17	between 7 and 8
70	$2 \times 5 \times 7$	between 8 and 9
121	11×11	11
115	5×23	between 10 and 11
195	$3 \times 5 \times 13$	between 13 and 14
357	$3 \times 7 \times 17$	between 18 and 19

is a prime factor greater than the square root of the number, but there is always at least one prime factor that is less than this square root. Thus, to determine whether a number is prime, it is necessary only to try dividing by the primes that are less than or equal to the square root of the number.

Let's return to the question of whether or not 421 is a prime. The positive square root of 421 is between 20 and 21. Since 2, 3, 5, 7, 11, 13, 17, and 19 are not factors of 421, this number is a prime. We do not have to try dividing by primes greater than 20 because there are no prime factors of 421 which are less than 20, and the product of two primes that are both greater than 20 would be greater than 421.

SIEVE OF ERATOSTHENES

One way of finding all the primes less than a given number is to eliminate those numbers which are not prime. This method was first used by the Greek mathematician Eratosthenes (ca. 230 B.C.) and is called the *Sieve of Eratosthenes*. It is illustrated in the following array of numbers to find the primes that are less than 120.

The process begins by circling the smallest prime, 2, and crossing out all the remaining multiples of 2 (4, 6, 8, 10, . . .). Then 3 is circled and all of the remaining multiples of 3 are crossed out (6, 9, 12, 15, . . .). Continue this process by circling 5 and 7 and crossing out their multiples. With the exception of 1, the numbers that are not crossed out will be prime. Since every composite number less than 120 must have at least one prime factor less than 11 ($11 \times 11 = 121$), it is unnecessary to cross out the multiples of primes greater than 7.

FUNDAMENTAL THEOREM OF ARITHMETIC

Composite numbers can always be written as a product of primes. For this reason, prime numbers are often referred to as the building blocks of whole numbers.

One method of finding all the prime factors of a composite number is through the use of a *factor tree.* This is a series of steps in which a number is broken down into smaller and smaller factors until all the final factors are prime numbers.

EXAMPLE 1

$$36 = 2 \times 2 \times 3 \times 3$$

EXAMPLE 2

$$30 = 5 \times 2 \times 3$$

EXAMPLE 3

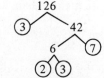

$$126 = 3 \times 2 \times 3 \times 7$$

A factor tree can be started with any two factors of a number. If the number is even, we can always begin by dividing by 2 to get a second factor. In one of the factor trees for 84 the first two factors are 2 and 42. In another factor tree for 84 the first two factors are 7 and 12.

$$84 = 7 \times 3 \times 2 \times 2 \qquad 84 = 2 \times 2 \times 3 \times 7$$

The same number may have several different factor trees, but the prime factors in each will be the same except for the order in which they appear. This fact is stated in the following important theorem.

Fundamental Theorem of Arithmetic: Every whole number greater than 1 is either a prime or a product of primes, and the product is unique, except for the order in which the factors occur.

COMPUTER APPLICATIONS

After using 440 hours of computer time at California State University at Hayward, two teenagers, Laura Nickel and Curt Noll, found a prime with 6533 digits. Looking for primes among such large numbers requires special programming techniques. However, simple programs can be written to test much smaller numbers. We will need a new command called the integer function for these programs.

2 DIVIDED BY CHALLENGE = SOLUTION

HAYWARD (UPI)—Laura Nickel had to argue with her parents and Curt Noll had to slip secretly into a closed library section to photocopy some needed research data.

But the two 18-year-olds accomplished their goal and it has been confirmed on the best authority: They have discovered the highest known prime number the world of pure mathematics has ever known.

The young mathematicians at California State University at Hayward—Noll still is technically a high school student—announced their find Tuesday and said they're "very, very excited."

Using a computer, the students found the prime number—meaning a whole number that can be evenly divided only by itself such as 1, 3, 5 or 7—to be 2 to the 21,701st power minus 1.

Written out, the number would contain 6,533 digits.

"It took them three years of diligent study and work," said the school's public affairs officer, Jane Hines.

INTEGER Function The integer function, INT(X), produces the largest whole number which is not greater than X. As examples, for any number from 5 up to but not including 6, the integer function produces 5.

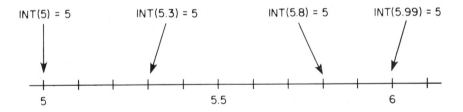

INT(5) = 5   INT(5.3) = 5   INT(5.8) = 5   INT(5.99) = 5

If the number in parentheses is a whole number M, then INT(M) = M. Otherwise, the integer function *rounds down* each number to the next lowest whole number. The next three computer programs use this function to test for divisibility.

Program 3.5A tests for divisibility by 3. In line 20 the operator types a whole number for K. If K is divisible by 3 then K/3 is an integer and it is equal to INT(K/3). If K is not divisible by 3 then K/3 is not an integer and it is not equal to INT(K/3). Follow through the steps of this program for K = 10 and K = 12.

Program 3.5A

```
10  PRINT "THIS PROGRAM DETERMINES IF A NUMBER IS DIVISIBLE
       BY 3. TYPE A NUMBER."
20  INPUT K
30  IF K / 3 =  INT (K / 3) THEN  GOTO 60
40  PRINT K" IS NOT DIVISIBLE BY 3."
50  GOTO 70
60  PRINT K" IS DIVISIBLE BY 3."
70  END
```

To determine if a number is a prime it is only necessary to try dividing by primes which are less than or equal to the square root of the number. For example, $\sqrt{100} = 10$ and therefore any number greater than 7 and less than 100 is a prime if it is not divisible by 2, 3, 5, or 7. Program 3.5B tests numbers which are greater than 7 and less than 100 to determine if they are prime. Lines 30, 40, 50, and 60 test for divisibility by 2, 3, 5, and 7 respectively.

Program 3.5B

```
10  PRINT "TYPE A NUMBER WHICH IS GREATER THAN 7 AND LESS THAN
       100."
20  INPUT K
30  IF K / 2 =  INT (K / 2) THEN  GOTO 90
40  IF K / 3 =  INT (K / 3) THEN  GOTO 90
50  IF K / 5 =  INT (K / 5) THEN  GOTO 90
60  IF K / 7 =  INT (K / 7) THEN  GOTO 90
70  PRINT K" IS A PRIME."
80  GOTO 100
90  PRINT K" IS NOT A PRIME."
100 END
```

The preceding program requires a separate line for each divisibility test. In Program 3.5C the divisibility tests are done in one line. Lines 30 through 50 form a loop in which X takes on the values from 2 through 10. Line 40 tests for divisibility by these values of X. If $K/X = INT(K/X)$ for any value of X, then the number K is not a prime and the computer is sent to line 80.

Program 3.5C

```
10  PRINT "TYPE A NUMBER WHICH IS GREATER THAN 7 AND LESS THAN
       100."
20  INPUT K
30  FOR X = 2 TO 10
40  IF K / X =  INT (K / X) THEN  GOTO 80
50  NEXT X
60  PRINT K" IS A PRIME."
70  GOTO 90
80  PRINT K" IS NOT A PRIME."
90  END
```

Activity Set 3.5 *Patterns on Grids* (Sieves for locating primes, factors, and multiples)
Just for Fun: Spirolaterals

"I tend to agree with you—especially since $6 \cdot 10^{-9} \sqrt{t_c}$ is my lucky number."

Exercise Set 3.5: Applications and Skills

1. Numerology is the study of the occult significance of numbers. It is taken lightly by most
people, who consider it to be either a recreational diversion or a fraud. Numerology buffs
sometimes choose names depending on the
number of their name, with $a = 1$, $b = 2$, etc.,
according to this chart. Use the name by which
you are best known and add the numbers corresponding to the letters.

1	2	3	4	5	6	7	8	9
A	B	C	D	E	F	G	H	I
J	K	L	M	N	O	P	Q	R
S	T	U	V	W	X	Y	Z	

a. Is your number prime or composite? Compare your number to that of a friend.
Is the sum of these numbers a prime?

b. Add the digits of your number until there is only one digit left. This final digit is called the *digital root* of the number. The digital root of 84 is 3, because $8 + 4 = 12$ and $1 + 2 = 3$. The following table lists the numbers you are most compatible with.

Compatibility with other numbers

Your number	1	2	3	4	5	6	7	8	9
Best chance of happiness	2,6	4,6	1,8	7,9	5,3	6,2	9,4	6,2	4,7

★ 2. The following sequence of numbers increases by 2, then by 4, then 6, etc. Continue this sequence until you reach the first number that is not a prime.

$$17 \quad 19 \quad 23 \quad 29 \quad 37$$

3. The numbers 2, 3, 5, 7, 11, and 13 are not factors of 173. Explain why it is possible to conclude that 173 is prime without checking for more factors.

4. Each rectangular array of squares gives information about the number of factors of a number. Two rectangles can be formed for the number 6, which shows that 6 has factors 2, 3, 1, and 6. Sketch as many different rectangular or square arrays as possible for each of the following numbers.

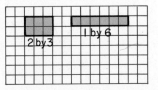

15   16   30   25   11

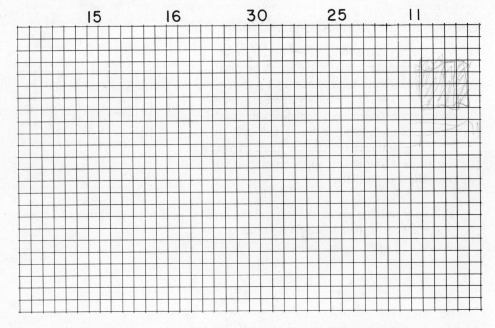

★ a. What kinds of numbers will have only one rectangular array?

b. Which three of the given numbers have an even number of factors?

★ c. Two of the given numbers have square arrays. Make a conjecture about the number of factors for square numbers.

d. *Challenge:* Find a number with nine factors. Explain how numbers with 11, 13, 15, or any other odd number of factors can be found.

5. List the factors for the numbers in this table. Include 1 and the number itself as factors.

★ a. Find the numbers having only two factors. What kind of numbers are these?

b. Which numbers have an odd number of factors? What kind of numbers are these?

Number	Factors	Number of factors
1	1	1
2	1,2	2
3	1,3	2
4	1,2,4	3
5	1,5	2
6	1,2,3,6	4
7		
8		
9	1,3,7	3
10		
11		
12		
13		
14		
15		
16		
17		
18		
19		
20		

★ 6. The first 10 prime numbers are 2, 3, 5, 7, 11, 13, 17, 19, 23, and 29. Which of these primes would you have to consider as possible factors of 367 to determine whether 367 is prime or composite?

★ 7. Which of the following numbers are prime?

a. 231   b. 227   c. 187   d. 431

8. Sketch a factor tree that contains the prime factors for each of these numbers. Express each number as a product of primes.

★ a. 924   b. 364   c. 864

★ 9. It has been estimated that life began on earth 1,000,000,000 (one billion) years ago. How can the Fundamental Theorem of Arithmetic be used to show that 7 is not a factor of this number?

10. The top train is equal in length to the one-color train of green rods.

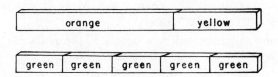

★   a. Name two other one-color trains that equal the length of this train.

     b. Name two factors of a number that is represented by seven blue rods (9 units).

★   c. If a number can be represented by an all-red train (2 units), an all-green train (3 units), and an all-black train (7 units), it has at least eight factors. Name these factors.

11. It is likely that no one will ever find a formula that will give all the primes less than an arbitrary number. The following formulas produce primes for awhile, but eventually composite numbers are obtained.

★   a. For which of the whole numbers $n = 2$ to 7 is $2^n - 1$ not a prime?

     b. The formula $n^2 - n + 41$ will give a prime for n equal to the whole numbers 1, 2, 3, . . . up to 40 but not for 41. Which of the primes less than 100 are given by this formula?

12. *Conjectures:* There are many conjectures involving primes.

★   a. Goldbach conjectured that every odd number greater than 5 is the sum of three primes. Verify this conjecture for the following numbers.
     21 =   27 =   31 =

     b. Arthur Hamann, a seventh-grade student, conjectured that every even number is the difference between two primes.* Express the following numbers as the difference of two primes.
     10 =   12 =   14 =

★   c. In 1845 the French mathematician Bertrand conjectured that between any whole number greater than 1 and its double there exists at least one prime. After 50 years this conjecture was proven true by the Russian mathematician Tchebyshev. For the numbers greater than 5 and less than 50, is it true or false that there are at least two primes between every number and its double?

---

*See M. R. Frame, "Hamann's conjecture," *The Arithmetic Teacher,* **23** No. 1 (January 1976), 34–35. A more general conjecture by A. de Polignac states that every even number is the difference of two consecutive primes in an infinite number of ways.

Exercise Set 3.5: Problem Solving

1. *Locker Room Puzzle:* In a new school built for 1000 students there were 1000 lockers that were all left closed. As the students entered the school, they decided on the following plan. The first student who entered the building opened all of the 1000 lockers. The second student closed all lockers with even numbers. The third student "changed" all lockers that were numbered with multiples of 3 ("changed" means opening those that were closed or closing those that were open). The fourth student changed all lockers that were numbered with multiples of 4, the fifth changed multiples of 5, etc. After 1000 students had entered the building and "changed" the lockers according to this pattern, which lockers were left open?

   a. *Understanding the Problem:* To better understand this problem, think about what will happen for the first few students. For example, after the first three students, will locker #6 be open or closed?

   ★ b. *Devising a Plan:* Simplifying the problem will sometimes help you to strike upon an idea for the solution. Suppose there are only 10 lockers and 10 students. We could number 10 markers and turn them upside down for open and rightside up for closed.

1	2	3	4	5	6	7	8	9	10

   Or, we could form a table showing the state of each locker as each student passes through.

Lockers

	1	2	3	4	5	6	7	8	9	10
1	O	O	O	O	O	O	O	O	O	O
2		C		C		C		C		C
3										

Student

   Which of the 10 lockers will be left open?

   c. *Carrying Out the Plan:* Solving the problem for small numbers of students and lockers will help you to see a relationship between the number of each locker and whether it is left open or closed. What types of numbers will be on the lockers that are left open?

   ★ d. *Looking Back:* How many times will a locker be opened or closed if it has a prime number? Half of the students will open or close only one locker. Which students?

2. *Grandfather's Age:* Joan's age was a factor of her grandfather's age for six consecutive years. What were her grandfather's ages during this time? (*Hint:* Guess and check.)

★ 3. *How Many Zeros?* How many zeros are there at the end of the number which is the product of whole numbers from 1 to 100?

$$1 \times 2 \times 3 \times 4 \times 5 \times 6 \times 7 \times \ldots \times 98 \times 99 \times 100$$

4. *Consecutive Composite Numbers:* There are arbitrarily long sequences of consecutive whole numbers with no primes. For example, the following five numbers are not prime. Explain why without computing the products.

$$2 \times 3 \times 4 \times 5 \times 6 + 2 \qquad 2 \times 3 \times 4 \times 5 \times 6 + 3$$
$$2 \times 3 \times 4 \times 5 \times 6 + 4 \qquad 2 \times 3 \times 4 \times 5 \times 6 + 5$$
$$2 \times 3 \times 4 \times 5 \times 6 + 6$$

   a. Construct a sequence of 10 consecutive whole numbers with no primes.

★ b. Explain how to construct a sequence of 1 million consecutive whole numbers with no primes.

Exercise Set 3.5: Computers

1. Write a computer program to determine if an arbitrary number K is divisible by 7. (*Hint:* See Program 3.5A on page 174.)

★ 2. Revise Program 3.5C (page 174), which determines if a number is prime, to test numbers greater than 100 and less than 10,000. Which of the following numbers are prime?

   a. 2477   b. 5663   c. 7189   d. 8317

3. The Italian mathematician Tartaglia (ca. 1499–1557) claimed that numbers of the form $2^N - 1$ are alternately prime and composite for whole numbers $N = 3, 4, 5, \ldots$.

   a. Write a program to print out the values of $2^N - 1$ for $N = 3$ to 12.

★ b. Use the program in part **a** and the program in Exercise 2 to determine if Tartaglia's conjecture holds for these numbers.

4. The prime which was found by David Slowinski in 1983 is equal to $2^{132,049} - 1$.

   a. Revise Program 3.5C, which determines if a number is prime, to test numbers between 1000 and 1,000,000.

★ b. For which values of $N = 13$ to 19 is $2^N - 1$ a prime number?

Census Clock, United States Department of Commerce Building, Washington, D.C.

3.6 FACTORS AND MULTIPLES

✱

factor ÷ into # — finite
multiple ÷ by # — infinite

The census clock pictured above is located in the lobby of the U.S. Department of Commerce Building and is regulated by the Bureau of Census. This clock shows the estimated population of the United States at any given moment. Four illustrated clock faces display the components of population change: births, deaths, immigration, and emigration. Plus and minus signs above the clock faces light up to show the inputs of these components. In 1981 there was a birth every 10 seconds, a death every 16 seconds, the arrival of an immigrant every 81 seconds, and the departure of an emigrant every 15 minutes.

factor same as divisor
& no greatest common multiple

The times when these lights flash together can be determined by using multiples of the different time periods of each clock. The plus sign above the birth clock lights at 10-second intervals, and the minus sign for the death clock lights every 16 seconds. If these two indicators both flashed at the same time, then they would flash together again after 80 seconds.

Birth clock intervals 0 10 20 30 40 50 60 70 80

Death clock intervals 0 16 32 48 64 80

MODELS FOR FACTORS AND MULTIPLES

Greatest Common Factor   For any two numbers there is always a number that is a factor of both. The numbers 24 and 36 both have 6 as a factor. When a number is a factor of two numbers, it is called a *common factor*. There are several common factors for both 24 and 36: 1, 2, 3, 4, 6, and 12. Among the common factors of two numbers there will always be a largest number, which is called the *greatest common factor*. The greatest common factor of 24 and 36 is 12. This is sometimes written, g.c.f. (24,36) = 12.

The concept of greatest common factor can be illustrated by forming trains of rods. The following multicolored trains, representing 15 and 24, have the same lengths as two single-colored green trains. This shows that 3 is a factor of both 15 and 24. Since the green rods are the longest rods that can be used for making one-color trains whose lengths equal the lengths of the trains for 15 and for 24, the greatest common factor of these two numbers is 3.

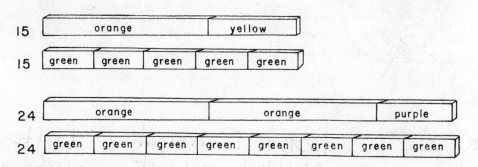

Least Common Multiple   When one number is a factor of a second, the second number is called a *multiple* of the first. Every number has an infinite number of multiples. Here are a few multiples of 5.

5   10   15   20   25   35   70   300   9000

A number is called a *common multiple* of two numbers if it is a multiple of both.

The following numbers are multiples of 7, and the ones that are circled are multiples of both 7 and 5.

7   14   21   28   (35)   42   (70)   (280)   (7000)

Every pair of numbers has an infinite number of common multiples. For instance, 35, 70, 280, and 7000 are only a few of the multiples of both 5 and 7. One of these multiples is just the product of the two numbers. The smallest multiple of both 5 and 7 is 35. This can be illustrated on the number line by intervals or jumps of 5 and of 7. Beginning at 0, jumps of 5 and jumps of 7 do not coincide at the same point again until after 35 intervals.

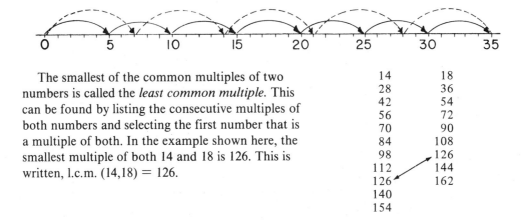

The smallest of the common multiples of two numbers is called the *least common multiple*. This can be found by listing the consecutive multiples of both numbers and selecting the first number that is a multiple of both. In the example shown here, the smallest multiple of both 14 and 18 is 126. This is written, l.c.m. (14,18) = 126.

14	18
28	36
42	54
56	72
70	90
84	108
98	126
112	144
126	162
140	
154	

The least common multiple of two numbers can be illustrated by forming two different one-color trains of equal length. The train of yellow rods (5 units) and the train of dark green rods (6 units) shown next have the same length. Therefore, the number represented by these trains is a multiple of both 5 and 6. Since there are no shorter one-color trains of yellow and dark green rods which equal the same length, 30 is the least common multiple of both 5 and 6.

30 | yellow | yellow | yellow | yellow | yellow | yellow |

30 | dark green | dark green | dark green | dark green | dark green |

COMPUTING G.C.F. AND L.C.M.

The greatest common factor of two numbers can be found from the prime factors of the two numbers. The factor trees at the right show that 2 occurs as a factor twice and 3 occurs as a factor once in *both* 60 and 72. Therefore, their greatest common factor is 2 × 2 × 3, or 12. For any two numbers the greatest common factor can be found by using the prime factors the greatest number of time they occur in both numbers.

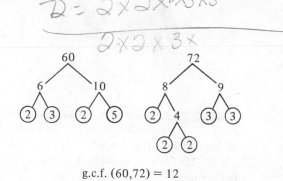

g.c.f. (60,72) = 12

Two numbers such as 24 and 35, which have no prime factors in common, are called *relatively prime.* In other words, two numbers are relatively prime if their greatest common factor is 1.

The least common multiple of two numbers can also be found from the prime factors of the numbers. For example, 2, 3, 5, and 7 are factors of 210, and so any multiple of 210 must have these prime factors occurring at least once. Similarly, any multiple of 198 must have at least one factor of 2, two factors of 3, and one factor of 11.

l.c.m. (210,198) = 2 × 3 × 3 × 5 × 7 × 11 = 6930

The smallest number satisfying these conditions for both 210 and 198 is 2 × 3 × 3 × 5 × 7 × 11 or 6930. That is, l.c.m. (210,198) = 6930. In general each prime factor of the least common multiple must occur the maximum number of times it occurs in the two given numbers.

Once you have found the prime factors of two numbers, you may find the following schemes helpful for determining the greatest common factor and least common multiple. Notice that the common factors of 198 and 210 are placed under each other for both schemes.

```
        198 = 2 × 3 × 3          × 11        198 = 2 × 3 × 3          × 11
        210 = 2 × 3     × 5 × 7               210 = 2 × 3     × 5 × 7
        g.c.f. = 2 × 3                        l.c.m. = 2 × 3 × 3 × 5 × 7 × 11
```

DIVISIBILITY TESTS

During the gasoline shortage of 1974, the state of Oregon adopted the "odd and even system." A car owner with an odd-numbered license plate could get gasoline on the odd-numbered days of the calendar, and those with even-numbered plates got gasoline on the even-numbered days. Some people whose license plate numbers ended in zero were confused as to whether their numbers were odd or even. The solution to this problem is in the way odd and even numbers are defined. If a whole number is divided by 2 and there is a remainder of 1, the number is *odd,* and if the remainder is 0, the number is *even.* Would the owner of the following license plate have purchased gasoline on the odd- or even-numbered days?

FX1080

FEB   OREGON   74

There are a few simple divisibility tests for determining whether or not a number is divisible by 2, 3, 4, 5, 6, or 9, without carrying out the division. These tests and some indications as to why they work are given in the following paragraphs.

Divisibility by 2 or 5   A number is divisible by 2 or 5 if the number represented by the units digit is divisible by 2 or 5. This means that a number is divisible by 2 if its units digit is 0, 2, 4, 6, or 8, and it is divisible by 5 if its units digit is 0 or 5.

To understand why these tests work, let's look at the expanded form for 5273.

$$5273 = \underbrace{5 \times 10^3 \ + \ 2 \times 10^2 \ + \ 7 \times 10}_{} \ + \ 3$$

Since 2 divides 10^3, 10^2, and 10, it will divide the bracketed portion of this equation. Therefore, 2 will divide the right side of the equation if and only if it divides the units digit, 3. Since 2 does not divide 3, it does not divide 5273.

Similarly, 5 divides the bracketed portion of the previous equation, but it does not divide 3. Therefore, 5 does not divide 5273.

Divisibility by 3 or 9   A number is divisible by 3 if the sum of its digits is divisible by 3. For example, 2847 is divisible by 3 since 3 divides $2 + 8 + 4 + 7$. Let's examine this test by looking at the expanded form for 2847.

$$
\begin{aligned}
2847 &= 2 \times 10^3 \; + \; 8 \times 10^2 \; + \; 4 \times 10 \; + \; 7 \\
&= 2 \times (999 + 1) \; + \; 8 \times (99 + 1) \; + \; 4 \times (9 + 1) \; + \; 7 \\
&= \underbrace{2 \times 999 \; + \; 8 \times 99 \; + \; 4 \times 9} \; + \; (2 + 8 + 4 + 7)
\end{aligned}
$$

The bracketed portion of this last equation is divisible by 3. Why? Therefore, to determine if 2847 is divisible by 3, it is necessary only to determine if the remaining portion of the equation, $2 + 8 + 4 + 7$, is divisible by 3.

The expanded form for 2847 also shows why the same test works for divisibility by 9. Since 9 divides the bracketed portion of the previous equation, it will divide 2847, if it divides $2 + 8 + 4 + 7$. In this case, 9 does not divide this sum and so it does not divide 2847.

Divisibility by 6   If a number is divisible by both 2 and 3, then it is divisible by 6, and if it is not divisible by both 2 and 3, then it is not divisible by 6. Apply this test to the following numbers. The first of these numbers is odd, so it is not divisible by 2. Therefore, it is not divisible by 6. The test shows that the second number is not divisible by 6 because it is not divisible by 3. Test the remaining two numbers for divisibility by 6.

$$4{,}328{,}561{,}785 \qquad 1{,}000{,}000{,}000 \qquad 2{,}100{,}000{,}000 \qquad 123{,}090{,}534$$

Divisibility by 4   If the number represented by the last two digits of a number is divisible by 4, the number will be divisible by 4. The number 65,932 is divisible by 4 because 4 is a factor of 32. The expanded form shows why this test works for 65,932.

$$65{,}932 = \underbrace{6 \times 10^4 \; + \; 5 \times 10^3 \; + \; 9 \times 10^2} \; + \; 3 \times 10 \; + \; 2$$

Since 4 is a factor of 10^4, 10^3, and 10^2, it will divide the bracketed portion of the expanded form. Thus, whether or not 4 divides 65,932 depends only upon whether or not it divides 32.

CASTING OUT NINES

Before the advent of calculators, it was a customary business practice to check the sums of columns of numbers by "casting out nines." This procedure was also used by schoolchildren to check computations involving addition, subtraction, and multiplication.

Casting out nines is the process of determining the remainder when a number is divided by 9. This remainder will be called the *nines excess* of the number. The nines excess can be easily found by using the test for divisibility by 9. For example, the nines excess for 47,281 is 4, because $4 + 7 + 2 + 8 + 1$ leaves a remainder of 4 when divided by 9.

The nines excess can be used to check the accuracy of any computation involving only addition, subtraction, and multiplication. For example, if addition is carried out with the nines excesses in place of the original numbers, the nines excess of this sum will equal the nines excess of the sum of the original numbers. A similar check works for subtraction and multiplication. Here are three examples.

	Excess
4728	3
3147	6
2096	8
+ 4731	+ 6
14702	23

The nines excess for both 14,702 and 23 is 5.

	Excess
2865	3
X 778	X 4
2228970	12

The nines excess for both 2,228,970 and 12 is 3.

	Excess
43891	7
− 6237	− 0
37654	7

The nines excess for both 37,654 and 7 is 7.

If there is a mistake in a computation, the chances are fairly good that it can be found by casting out nines. It is possible, however, for a mistake to occur and not be detected by this method. This will happen when the computed result differs by a multiple of 9 from the correct answer. In the example shown here, casting out nines does not indicate an error because the correct product, 41667, and the answer shown, 41217, differ by 450, which is a multiple of 9.

	Excess
731	2
X 57	X 3
41217	6

The nines excess for both 41217 and 6 is 6.

Reversing the order of two or more digits in a number is a common type of error. Suppose that a clerk receives $138.45 for a sale but rings up $183.45 on the cash register. At the end of the day, if no other mistakes are made, the cash register's total will be $45 greater than the amount of money taken in. Whenever a mistake is made by reversing two or more digits in a number, the difference between the correct number and the incorrect number is divisible by 9. Why? Because of this fact, if the amount in a cash register does not agree with the cash register's recorded intake, one approach to locating the error is to compute the difference to see if it is divisible by 9. If so, one strong possibility for the source of error is that two or more digits have been reversed.

$183.45
−138.45
$ 45.00

COMPUTER APPLICATIONS

The Pythagoreans, a society of Greek mathematicians, classified a number as *perfect* if it equaled the sum of its proper factors. The *proper factors* of a number are all its factors except the given number. For example, here are the proper factors of 28:

1   2   4   7   14

Notice that 28 is a perfect number because 1 + 2 + 4 + 7 + 14 = 28.

It is easy to write a program in which the computer finds the proper factors of a number; merely have the computer divide by all the whole numbers which are less than the number, as in Program 3.6A.

Program 3.6A

```
10   PRINT "THIS PROGRAM PRINTS THE PROPER FACTORS OF A NUMBER.
        TYPE A WHOLE NUMBER."
20   INPUT K
30   FOR X = 1 TO K - 1
40   IF K / X =  INT (K / X) THEN  GOTO 60
50   GOTO 70
60   PRINT X" ";
70   NEXT X
80   END
```

Greatest Common Factor   Program 3.6B determines the greatest common factor of two whole numbers A and B. This is done by beginning with the number $N = B$ in line 22 and successively dividing both A and B by the numbers from N down to 1 (see lines 22, 24, and 26). The first value of N which divides into both A and B is their g.c.f. and the program ends at this point. If 1 is the only whole number which divides both A and B, then A and B are *relatively prime*. Notice that the hypothesis of the conditional command in line 24 requires that both B and A be divisible by N.

Program 3.6B

```
10   PRINT "THIS PROGRAM FINDS THE G.C.F. OF TWO NUMBERS. TYPE
        TWO WHOLE NUMBERS, SEPARATED BY A COMMA."
20   INPUT A,B
22   FOR N = B TO 1 STEP  - 1
24   IF B / N =  INT (B / N) AND A / N =  INT (A / N) THEN  GOTO
        30
26   NEXT N
30   PRINT "THE G.C.F. OF "A" AND "B" IS "N"."
40   END
```

SUPPLEMENT (Activity Book)

Activity Set 3.6 Factors and Multiples with Cuisenaire Rods

Just for Fun: Star Polygons

Exercise Set 3.6: Applications and Skills

1. Some skyscrapers have double-deck elevators to minimize the number of necessary elevator shafts. People entering the buildings will use the street level if they are going to odd-numbered floors, or take a moving stairway to a mezzanine for even-numbered floors.

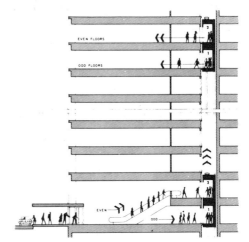

   a. Suppose you had to deliver packages to floors 11, 26, 35, and 48. How can this be done by doing the least amount of elevator riding and walking only one flight of stairs?

   ★ b. Describe an efficient scheme for delivering to any number of odd- and even-numbered floors.

2. The United States Census Clock has flashing light signs to indicate gains and losses in the population. Here are the 1984 time periods of these flashes in seconds: birth, 8; death, 16; immigrant, 81; and emigrant, 900. If these lights all started flashing at the same moment, the times when they flash together can be computed by using the concept of least common multiple.

   ★ a. If you saw the birth and immigrant signs light up at the same time, how many seconds would pass before they would both light together again? 648

   b. Suppose you saw the immigrant and emigrant signs both light at the same time. What is the shortest time before they will both be lighted together again? 800

   ★ c. If all four lights flash at the same time, what is the shortest time before they will all be lighted at the same time again? 32,400

3. The trains for both 24 and 18 are each equal to the lengths of all-red trains (a red rod represents 2).

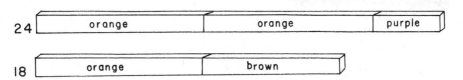

   ★ a. Name another type of rod that can be used to represent both 24 and 18 by one-color trains.

   b. What kind of rods can be used to form one-color trains that represent both 24 and 18?

★   c. What can be said about the relationship between two numbers if the only one-color trains representing these numbers contain white rods (a white rod represents 1)?

4. The following all-green train equals the length of the all-yellow train. Does this give information about common factors or common multiples?

★   a. If the purple rods (4 units) and the black rods (7 units) are used to form the shortest possible one-color trains that both have the same length, how many purple rods and how many black rods will be required?

   b. If an all-brown train (brown represents 8 units) equals the length of an all-orange train (orange represents 10 units), what can be said about the number of brown rods?

   c. If the two trains in part **b** are the shortest possible, how many brown rods will be required?

5. Circle the numbers that are divisible by 3. Can the test for divisibility by 3 be used to determine the remainder when dividing by 3?
   a. 465,076,800   b. 100,101,000   c. 907,116,341

6. Which number in Exercise 5 is divisible by 9?
★   a. If a number is divisible by 3, is it divisible by 9?
   b. If a number is divisible by 9, is it divisible by 3?

★ 7. Which of the following numbers are divisible by 4? How can the last two digits be used to determine the remainder when dividing by 4?
       47,382   512,112   14,710   4,328,104,292

8. To test for divisibility by 11, alternately add and subtract the digits from right to left beginning with the units digit, that is, units digit minus tens digit plus hundreds digit, etc. If the result is divisible by 11, then the original number will be divisible by 11. Use this test on the following numbers.
   a. 63,011,454   b. 19,321,488   c. 4,209,909
★   d. Will the test for divisibility by 11 work if the digits are alternately added and subtracted from left to right?

190   CHAPTER 3   WHOLE NUMBERS AND THEIR OPERATIONS

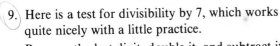 divisor = factor

9. Here is a test for divisibility by 7, which works quite nicely with a little practice.

Remove the last digit, double it, and subtract it from the remaining number. Continue this process until one or two digits remain. If the remaining number is divisible by 7, then the original number will be also.

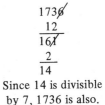

Since 14 is divisible by 7, 1736 is also.

Try this test on the following numbers. Check your answers by dividing each number by 7.

a. 3941   b. 14021   c. 17976

10. Draw factor trees for the numbers in parts **a** and **b**, and compute the greatest common factor for each pair of numbers.

★   a.   198   165   b.   280   168

g.c.f. (198,165) = _____   g.c.f. (280,168) = _____

11. Draw factor trees for these numbers and compute the least common multiple for each pair of numbers.

★   a.   30   42   b.   22   56

l.c.m. (30,42) = _____   l.c.m. (22,56) = _____

12. *Casting Out Nines:* In each of the two examples in parts **a**, **b**, and **c**, one answer is correct and one is incorrect. Use casting out nines to determine the incorrect example.

★   a.   64796   324862
             $\underline{\times\ 1560}$ $\underline{\times\ 316}$
          101181760   102656392

     b.   496742317   463798260
             341264682   281417863
         $\underline{+\ 284651906}$ $\underline{+\ 945516792}$
        1122658905   1680732915

     c.   681728906   361972482
         $\underline{-\ 419847268}$ $\underline{-\ 129865396}$
          261481638   232107086

★   d. In positional numeration, 2^{64} has 20 digits. Use casting out nines to determine the remainder when this number is divided by 9.

Exercise Set 3.6: Problem Solving

1. *Floor Tiles:* A rectangular floor has 94 tiles on one side and 118 tiles on the other side. How many tiles will be crossed by a diagonal from one corner of the floor to the opposite corner?

     a. *Understanding the Problem:* The diagonal of a square-tiled floor crosses a tile if it partitions the tile into two regions. In this 4 by 6 rectangle of tiles, the diagonal does not cross squares A and B. How many squares does it cross?

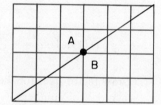

     b. *Devising a Plan:* Simplifying and drawing pictures (using graph paper, if available) is a natural way to approach this problem. By examining many special cases and looking for a relationship between the dimensions of the rectangle and the numbers of squares that are crossed, we may discover a method for solving this type of problem. How many squares are crossed in each of the following rectangles? (*Note:* The g.c.f. of the length and width of each of the rectangles 1 through 4 is 1; and the g.c.f. of each of the rectangles 5 through 8 is 2.)

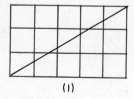
(1)

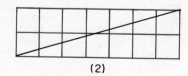
(2)

(3)

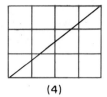
(4)

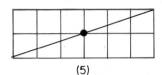
(5)

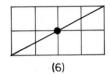
(6)

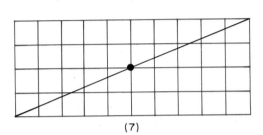
(7)

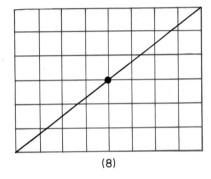
(8)

c. *Carrying Out the Plan:* Form a table like the one shown here and record the number of squares crossed for simplified versions of the problem. Circle the numbers in the table for which the g.c.f. of the length and width is 1. How is the number of squares crossed in these cases related to the length and width of the rectangle? Then look for a relationship when the g.c.f. of the length and width is 2, 3, etc. Use these observations to form a conjecture about the number of squares that will be crossed in larger rectangles. How many squares will be crossed in a 94 by 118 rectangle?

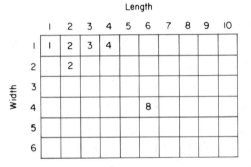

Length

	1	2	3	4	5	6	7	8	9	10
1	1	2	3	4						
2		2								
3										
4						8				
5										
6										

Width

★ d. *Looking Back:* Let's see why the g.c.f. enters into the solution of this problem. Here is an 8 by 12 rectangle. The g.c.f. (8,12) = 4. Notice how the diagonal contains several vertex points. What do the small rectangles have in common? How many tiles are crossed in each? What is the total number of squares crossed in the 8 by 12 rectangle?

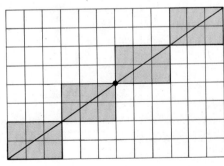

★ 2. *Card-Sorting Problem:* There are machines for sorting cards into piles. On one occasion a machine operator obtained the following curious result. When a box of cards was sorted into 7 groups there were 6 cards left over. When the box of cards was sorted into 6 groups there were 5 left over. For 5 groups, 4 groups, 3 groups, and 2 groups, there were 4, 3, 2, and 1 cards left over, respectively. What is the smallest possible number of cards in the box before the sorting took place? (*Hint:* What would have happened if the box of cards had contained one more card?)

3. *Card-Sorting Variation:* Here is a variation of the card-sorting problem. When a box of cards was sorted into 7 groups, there was one card left over. When the box of cards was sorted into 6 groups, there was 1 card left over. Similarly, for 5, 4, 3, and 2 groups, there was one card left over in each case. What is the smallest possible number of cards in the box before the sorting took place?

4. *Divisibility Problem:* Determine all the whole numbers n for which $2^n + 1$ is divisible by 3.*

5. *Hit and Run:*

HIT AND RUN

DID YOU GET HIS LICENSE NUMBER?

YES! HIS LICENSE WAS IN TWO PARTS ... A TWO DIGIT NUMBER AND A THREE DIGIT NUMBER. THE TWO DIGIT NUMBER WAS PRIME AND THE SUM OF THE TWO DIGITS WAS A TWO-DIGIT PRIME. THE TENS DIGIT WAS LARGER THAN THE UNITS DIGIT... IN THE THREE DIGIT PART, THE DIGITS WERE ALL ODD AND DIFFERENT. THE SUM OF THE THREE DIGITS WAS PALINDROMIC. THE SUM OF THE FIRST AND THIRD DIGIT WAS ONE-HALF THE SUM OF THE FIRST AND SECOND.

THAT'S ALL I REMEMBER!

*Jozsef Kurschak, *Hungarian Problem Book I* (New York, N.Y.: Random House, 1963), p. 11.

Exercise Set 3.6: Computers

☞ 1. Until 1952, there were only 12 known perfect numbers. With the aid of computers the list has grown to 23. The first two perfect numbers are 6 and 28. Use Program 3.6A for finding the proper factors of a number (page 188) to determine which of the following numbers are perfect.

    a. 254   b. 326   c. 496   d. 582

☞ ★ 2. The Pythagoreans classified a number as *deficient* if the sum of the proper factors was less than the number. A number was classified as *abundant* if the sum of the proper factors was greater than the number. Use Program 3.6A for finding the proper factors of a number (page 188) to determine which of the following numbers are deficient and which are abundant.

    a. 730   b. 920   c. 845   d. 1000

  3. Follow through Program 3.6B for determining the greatest common factor of two numbers (page 188) for $A = 12$ and $B = 30$. How many times will the computer go through lines 22, 24, and 26 before it is sent to line 30 to print the g.c.f? Use this program to find the g.c.f. of the following pairs of numbers.

    a. 180, 231   b. 227, 238   c. 209, 253

★ 4. The least common multiple of two whole numbers A and B can be found by multiplying A times B and dividing by g.c.f.(A,B). Use this fact to write a program for determining the l.c.m. of two numbers. (*Hint:* Revise the program which finds the g.c.f. of two numbers by making changes in lines 10 and 30.)

Geometric Figures

It is nature herself, and not the mathematician,
who brings mathematics into natural philosophy.

IMMANUEL KANT

Crystals of calcite

4.1
PLANE FIGURES

The assorted hexagonal prisms pictured above are crystals of calcite. The sharp edges and corners of these six-sided figures occur naturally in this mineral. The flat, thin crystals in the photo at the top of the next page are of the mineral wulfenite. This mineral usually forms these thin plates with beveled edges. The angles between the beveled edges and the flat faces of this crystal are always equal.

The study of the relationships between lines, angles, surfaces, and solids is part of geometry. This is one of the earliest branches of mathematics. More than 4000 years ago the Egyptians and Babylonians were using geometry in surveying and in architecture.

These ancient mathematicians discovered geometric facts and relationships by experimentation and inductive reasoning. Because of this approach they could never be sure of their conclusions, and in some cases their formulas for area and volume were inaccurate. The Greeks, on the other hand, viewed points, lines, and figures as abstract concepts about which they could reason deductively. They were willing to experiment in order to formulate ideas, but final acceptance of a mathematical statement depended on proof by deductive reasoning.

Crystals of wulfenite

POINTS, LINES, AND PLANES

The fundamental notion in geometry is that of a *point*. All geometric figures are sets of points. Points are abstract ideas which we illustrate by dots, corners of boxes, and tips of pointed objects. These concrete illustrations have width and thickness, but points have no dimensions. The following description of a point, from *Mr. Fortune's Maggott,* by Silvia Townsend Warner, indicates some of the pitfalls associated with this concept.*

> Calm, methodical, with a mind prepared for the onset, he guided Lueli down to the beach and with a stick prodded a small hole in it.
> "What is this?"
> "A hole."

*J. R. Newman, *The World of Mathematics,* **4** (New York: Simon and Schuster, 1956), p. 2254.

"No, Lueli, it may seem like a hole, but it is a point."

Perhaps he had prodded a little too emphatically. Lueli's mistake was quite natural. Anyhow, there were bound to be a few misunderstandings at the start.

He took out his pocket knife and whittled the end of the stick. Then he tried again.

"What is this?"

"A smaller hole."

"Point," said Mr. Fortune suggestively.

"Yes, I mean a smaller point."

"No, not quite. It is a point, but it is not smaller. Holes may be of different sizes, but no point is larger or smaller than another point."

A *line* is a very specific set of points which we sometimes intuitively describe as "straight" and by saying it extends indefinitely in both directions. The two parallel rows which Peter is drawing in the previous cartoon, the edges of boxes, and taut pieces of string or wire are all models of lines. The above line passes through points A and B and is denoted by $\overleftrightarrow{AB}$. The arrows indicate that the line continues indefinitely in both directions. If two or more points are on the same line, they are called *collinear*.

A *plane* is another basic set of points which we will not define precisely. Intuitively, we describe a plane as being "flat," like a sheet of paper and extending indefinitely. The tops of tables and the surfaces of floors and walls are common models for planes.

Points, lines, and planes are undefined terms in geometry which are used to define other terms and geometric figures. The following paragraphs contain some of the more common definitions.

LINE SEGMENTS, RAYS, AND ANGLES

Line Segments   A *line segment* consists of two points on a line and all the points between them. The line segment with *endpoints* A and B is denoted by $\overline{AB}$. To *bisect* a line segment means to divide it into two parts of equal length. The *midpoint C* bisects $\overline{AB}$.

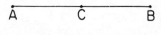

Rays   A *ray* consists of a point on a line and all the points on one side of this point. The ray which begins at D and contains point E is denoted by $\overrightarrow{DE}$. The point D is the *endpoint* of $\overrightarrow{DE}$.

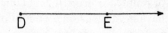

Angles An *angle* is the union of two rays having a common endpoint. This endpoint is called the *vertex,* and the rays are called the *sides of the angle.* The angle with vertex G, whose sides contain points F and H, is denoted by $\angle FGH$. Point K is in the *interior* of $\angle FGH$.

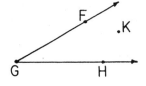

Referring to the accompanying photo, which of these geometric terms are illustrated in this art form by Kenneth Snelson?

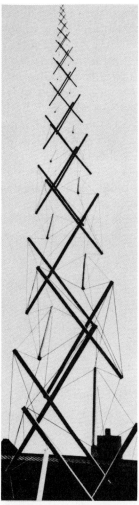

Needle Tower—Aluminum and stainless steel rods and wires, by Kenneth Snelson

Measure of Angles The ancient Babylonians devised a method for measuring angles by dividing a circle into 360 equal parts. Each angle has a measure, called *degrees,* which is some number between 0 and 360. The circular protractor in the diagram on the right shows that there are 20 degrees (20°) in ∠*KET.* To measure any angle, the vertex is placed at the center of the protractor with one side of the angle lying on the centerline. If an angle has a measure of 90°, its sides are *perpendicular* and it is called a *right angle.* If it is less than 90° it is called an *acute angle,* and if it is greater than 90° and less than 180° it is called an *obtuse angle.* Occasionally, we will need measures of angles which are greater than 180°. For example, angles in polygons are sometimes greater than 180°, as shown by ∠*ABC* in this figure. Such an angle is called a *reentrant angle.* To indicate an angle greater than 180°, a circular arc will be drawn to connect the two sides of the angle. If the sum of two angles is 90°, the angles are called *complementary;* if their sum is 180°, they are called *supplementary.*

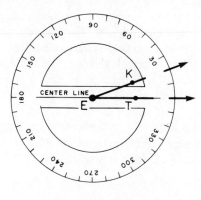

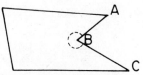

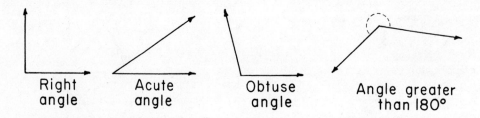

Right angle   Acute angle   Obtuse angle   Angle greater than 180°

CURVES AND CONVEX SETS

A *curve* is a set of points which can be drawn by a single continuous motion.

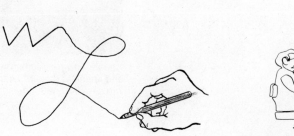

"You don't see many drawings made with one continuous line anymore."

Several types of curves are shown following this paragraph. Curve A is called a *simple curve* because it starts and stops without intersecting itself. Curve B is a *simple closed curve* because it is a simple curve which starts and stops at the same point. Curve C is a *closed curve,* but since it intersects itself it is not a simple closed curve.

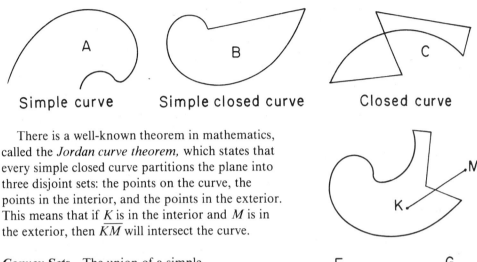

Simple curve   Simple closed curve   Closed curve

There is a well-known theorem in mathematics, called the *Jordan curve theorem,* which states that every simple closed curve partitions the plane into three disjoint sets: the points on the curve, the points in the interior, and the points in the exterior. This means that if K is in the interior and M is in the exterior, then $\overline{KM}$ will intersect the curve.

Convex Sets   The union of a simple closed curve and its interior is called a *plane region.* These regions can be classified as nonconvex and convex. You may have heard the word "concave," rather than "nonconvex." An object is concave if it is "caved in,"

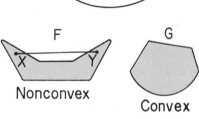

Nonconvex   Convex

such as set *F,* and convex if it is not, such as set *G.* To be more mathematically precise, we say that a set is *convex* if the line segment joining any two points of the set lies completely in the set. Set *F* is *nonconvex* because $\overline{XY}$ is not completely in the set. An intuitive way of testing for convexity of plane regions is to think about stretching a rubber band around the boundary of the set. If it touches all points on the boundary (as it will for set *G*), the set is convex, and if not, as for set *F,* the set is nonconvex.

A *circle* is a special case of a simple closed curve whose interior is a convex set. Each point on a circle is the same distance from a fixed point called the *center.* This distance is the *radius* of the circle, and twice this distance is the *diameter* of the circle. A line segment from a point on the circle to its center is also called a *radius,* and a line segment from two points on the circle which passes through the center is a *diameter.* The distance around the circle is the *circumference.* The union of a circle and its interior is called a *disc.*

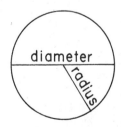

diameter

radius

POLYGONS

A *polygon* is a simple closed curve which is the union of line segments. Polygons are classified according to the number of line segments. Here are a few examples.

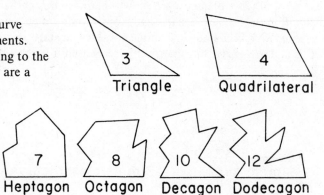

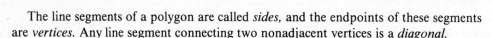

The line segments of a polygon are called *sides,* and the endpoints of these segments are *vertices.* Any line segment connecting two nonadjacent vertices is a *diagonal.*

Triangles have been the most thoroughly studied of all the polygons. One reason for this is the importance of triangular supports in buildings, bridges, and other structures. A triangle is more "rigid" than the other polygons, in that its shape is determined by its three sides. This can be illustrated by linkages, such as those shown here. The shapes of all these polygons can be changed except for that of the triangle. For example, the pentagon can be made nonconvex by pushing one of its vertex points into the interior of the polygon. Similarly, the hexagon in the illustration can be reshaped into a convex hexagon. The triangle is the only one of these linkages whose shape cannot be changed.

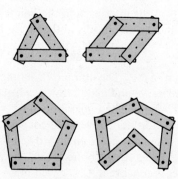

Another reason for the extensive study of triangles is that every polygon of four or more sides can be subdivided into triangles by drawing diagonals. Polygon $ABCDE$ is subdivided into three triangles by diagonals $\overline{BE}$ and $\overline{CE}$. Information about these triangles, such as angle sizes or area, can be used to obtain similar information about the polygon.

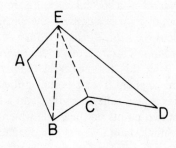

Several types of triangles and quadrilaterals occur often enough to be given special names.

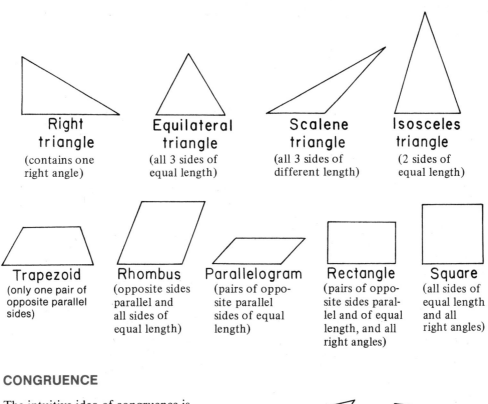

Right triangle
(contains one right angle)

Equilateral triangle
(all 3 sides of equal length)

Scalene triangle
(all 3 sides of different length)

Isosceles triangle
(2 sides of equal length)

Trapezoid
(only one pair of opposite parallel sides)

Rhombus
(opposite sides parallel and all sides of equal length)

Parallelogram
(pairs of opposite parallel sides of equal length)

Rectangle
(pairs of opposite sides parallel and of equal length, and all right angles)

Square
(all sides of equal length and all right angles)

CONGRUENCE

The intuitive idea of congruence is quite simple: Two plane figures are congruent if one can be placed on the other so that they coincide. Another way to describe congruent plane figures is to say that they have the same size and shape. (See Section 9.1 for more details on congruence.)

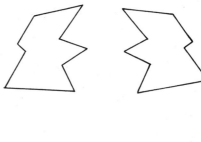

For line segments and angles we can be more precise about congruence. Two *line segments are congruent* if they have the same length, and two *angles are congruent* if they have the same number of degrees.

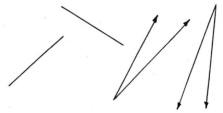

COMPUTER APPLICATIONS

Most of the geometric figures we have seen in this section can be drawn by giving the turtle commands in Logo. The command HOME is very helpful in drawing these figures because regardless of where the turtle is on the screen, this command will send the turtle back to its start position to complete a closed curve.

The following commands instruct the turtle to draw a right triangle. The two legs of the triangle are drawn first and then the hypotenuse is formed by sending the turtle home.

```
RT 90
FD 35
LT 90
FD 50
HOME
```

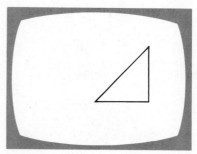

Right Triangle

Any right triangle can be drawn by using these commands and varying the length of the two legs. These lengths are replaced in the next procedure by variables. To use this procedure for drawing a triangle, type RTTRIANGLE and numbers for LEG1 and LEG2. The triangle shown below was obtained by typing RTTRIANGLE 80 50.

```
TO RTTRIANGLE :LEG1 :LEG2
  RT 90
  FD :LEG1
  LT 90
  FD :LEG2
  HOME
END
```

RTTRIANGLE 80 50

The next set of commands instructs the turtle to draw an isosceles triangle whose base angles are 38° and whose sides have length 70. Since each base angle is 38°, we begin by having the turtle make a 52° right turn (52 is the complement of 38) to draw the left side of the triangle. At the top of the triangle the turtle makes a right turn of 76° (76 is the supplement of 104). Trace out the isosceles triangle as you follow through these commands.

```
RT 52
FD 70
RT 76
FD 70
HOME
```

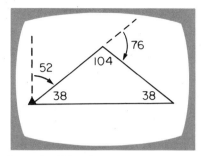

Isosceles Triangle

Now that we know how to write the commands for a specific isosceles triangle, it is easy to write a procedure for any isosceles triangle. The next procedure will draw an isosceles triangle for an arbitrary base angle and an arbitrary length for the two congruent sides. To use this procedure type ISOSTRIANGLE and numbers for ANGLE and SIDE. Why must the number for ANGLE be less than 90?

```
TO ISOSTRIANGLE :ANGLE :SIDE
 RT 90 - :ANGLE
 FD :SIDE
 RT 2 * :ANGLE
 FD :SIDE
 HOME
END
```

SUPPLEMENT (Activity Book)

Activity Set 4.1 Rectangular and Circular Geoboards
Just for Fun: Tangram Puzzles

Exercise Set 4.1: Applications and Skills

1. This picture of a cross section of natural sapphire shows angles that each have the same number of degrees.

   a. Are these angles acute or obtuse?

★ b. Approximately how many degrees are there in these angles?

Natural sapphire

2. The next picture shows three crystals of the mineral staurolite. The crystal on the far right is known as the "Fairy Stone" of the Appalachian Mountains. These stones are found in all parts of the world. They are especially common in the Shenandoah Valley. This form and the one on the left are often imitated by jewelers.

Crystals of staurolite

   a. The ridges on the top of the Fairy Stone form lines that intersect in equal angles. How many degrees are in each of these angles?

★ b. The ridges on the top of the crystal on the left form three lines that intersect in six equal angles. How many degrees are in each of these angles?

3. For the points A, B, C, and D there are six line segments which have these points as endpoints.

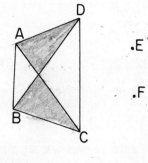

★ a. Draw line segments from E to A, B, C, and D. How many new line segments are there? What is the total number of line segments for these five points?

★ b. How many more line segments will there be if point F is used? What is the total number of line segments having six points as endpoints?

   c. Find a pattern and complete the table to show the number of line segments which have the given numbers of points as endpoints.

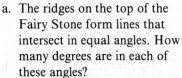

$\dfrac{n(n-1)}{2}$

n = # of points

Number of points	2	3	4	5	6	7	8	9	10	15
Number of segments	1	3	6	10	15	21				

   d. You probably recognize the numbers 1, 3, 6, etc., as being the triangular numbers (see page 6). The nth triangular number is $n(n + 1)/2$. Use this formula to find the 98th triangular number.

CHAPTER 4   GEOMETRIC FIGURES

4. The number of line segments between a given number of points has many practical applications. One of these involves the remarkable accomplishments of the Bell System. The fundamental problem was that of connecting two subscribers who wanted to talk. This was done by cords and plugs.

★   a. In 1884 Ezra T. Gilliland devised a mechanical system that would allow 15 subscribers to reach each other without the aid of an operator. How many line segments are needed to connect 15 points?

   b. In 1891 Almon B. Strowger patented a dial machine which connected up to 99 subscribers. How many different two-party calls does this permit?

5. Three lines in a plane may intersect in 0, 1, 2, or 3 points.

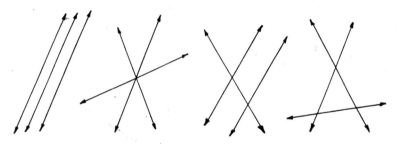

★   a. What are the possible numbers of points of intersection for four lines in a plane?

   b. What are the possible numbers of points of intersection for five lines in a plane?

6. Figure (a) has three disjoint regions which are bounded by polygons. This figure has seven line segments and five vertices. Fill in the table for Figures (b) through (e), and find a formula relating the numbers of regions, sides, and vertices. Draw a few more figures and check your conclusion.

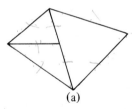

(a)

	(a)	(b)	(c)	(d)	(e)
Regions	3				
Line Segments	7				
Vertices	5				

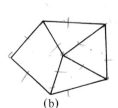

(b)

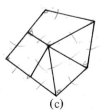

(c)

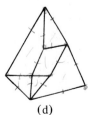

(d)

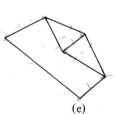

(e)

7. The white path in this ornament from the Middle Ages is a curve. Is it a simple curve? Is it closed?

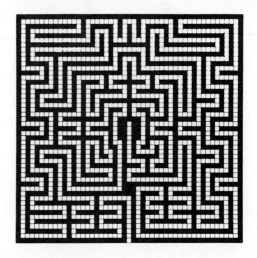

8. Use the angles in these polygons to answer the questions that follow. Which angles are:

★   a. Acute?   b.   Obtuse?

★   c. Right angles?

    d. Which angle is a reentrant angle (greater than 180°)?

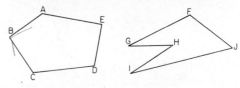

9. *Utilities Puzzle:* A, B, and C are houses and E, G, and W represent sources of electricity, gas, and water. Try connecting these houses with each utility by drawing lines or curves so that they do not cross each other. It is possible to make only 8 of the 9 connections. Draw these 8 connections.

| A | B | C |
| E | G | W |

★   a. Some of your connections will form a simple closed curve for which the remaining unconnected house and utility are on opposite sides of this curve. Find this curve and mark it with dark lines.

    b. How does the Jordan curve theorem show that all nine connections cannot be completed?

10. *Dissection Puzzle:* Trace and cut
    out the five pieces of this polygon.
    Reassemble these pieces to form a
    square.

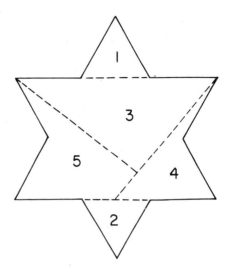

"*I am not sure it is of great value in life to know how many diagonals an n-sided
figure has. It is the method rather than the result that is valuable.*" W. W. Sawyer

Exercise Set 4.1: Problem Solving

1. *Polygons and Diagonals:* How many diagonals has a 15-sided polygon?

    a. *Understanding the Problem:* A
       diagonal is a line segment con-
       necting any two nonadjacent
       vertices of a polygon. Quadri-
       lateral $ABCD$ has diagonals $\overline{AC}$
       and $\overline{BD}$. Does a nonconvex
       quadrilateral have two diagonals?

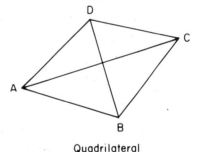

Quadrilateral

★ b. *Devising A Plan:* One approach is to simplify the problem by drawing a few
      polygons and counting the number of diagonals. By listing these in a table we
      may be able to discover a pattern. How many diagonals are there in the following
      polygons?

Pentagon   Hexagon   Heptagon

c. *Carrying Out the Plan:* Fill in the first few boxes of this table and look for a pattern. Use the pattern and inductive reasoning to complete the table. How many diagonals are there in a 15-sided polygon?

Number of sides	3	4	5	6	7	8	9	10	11	12	13	14	15
Number of diagonals	0	2	5										

d. *Looking Back:* Another approach to this problem is to use the result from page 208, Exercise 3 in Exercise Set 4.1A. How many line segments has a 15-sided polygon, including sides and diagonals? Explain how the formula for the triangular numbers can be used to obtain the number of diagonals in a 15-sided polygon.

★ 2. *Intersecting Lines:* The four lines shown here intersect in six points. What is the maximum number of points of intersection for 12 lines? (*Hint:* Draw pictures for special cases. Form a table and look for a pattern.)

$\dfrac{n(n-1)}{2}$

$\dfrac{12(12-1)}{2}$ $\dfrac{12(11)}{2}$

66

3. *Lines and Regions:* Three lines partition a plane into seven regions. What is the maximum number of regions that a plane can be partitioned into by 12 lines?

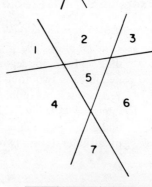

4. *Congruent Regions:* In how many ways can a rectangular region be cut into eight congruent pieces?

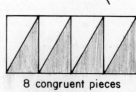

8 congruent pieces

★ 5. *Matchstick Puzzle:* Using only four more matchsticks, divide this region into four congruent regions. (*Hint:* Some of the matchsticks may be broken.)

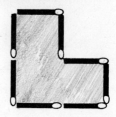

Exercise Set 4.1: Computers

1. Write the commands for drawing a scalene triangle which has an obtuse angle. Sketch the triangle.

2. Write the commands for drawing:
   a. A nonconvex hexagon
   b. A convex pentagon

★ 3. Define a procedure called PARALLELOGRAM for drawing a parallelogram with a base angle of 55 and sides of length 40 and 70. (*Hint:* Use the figure shown here.)

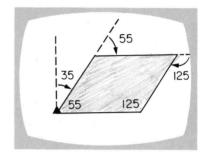

4. Define a procedure for drawing any rhombus with a variable for the base angle and a variable for the length of the sides.

TO RHOMBUS :ANGLE :SIDE

★ 5. Define a procedure called EQUITRIANGLE for drawing an equilateral triangle. Use SIDE for a variable so that different size triangles can be drawn by this procedure.

TO EQUITRIANGLE :SIDE

Cross section of cadmium sulfide crystals

"The Great Architect of the Universe now appears as a pure mathematician."

James H. Jeans

4.2
PROPERTIES OF
POLYGONS

REGULAR POLYGONS

We have become so accustomed to hearing about the regularity of patterns in Nature that it is often taken for granted. Still, it is a source of wonder to see polygons with straight edges and uniform angles such as those which occur in these photos of cadmium sulfide and tourmaline. The details and symmetry of the hexagons in the cadmium sulfide remind us of snow crystals. Equally impressive are the triangles which occur in the cross section of tourmaline. In each of these hexagons and triangles, the sides have

Cross section of the gem tourmaline

equal length and the angles have the same number of degrees. Polygons satisfying these two requirements are called *regular polygons*.

Here are the first few regular polygons.

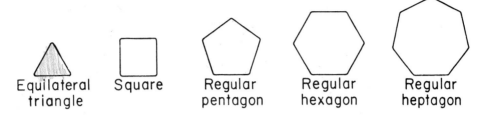

Equilateral triangle   Square   Regular pentagon   Regular hexagon   Regular heptagon

It is necessary to require both of the following conditions for regular polygons, namely: (1) sides of equal length; and (2) angles with the same number of degrees. The rhombus is not a regular polygon, because it satisfies condition 1 but not condition 2. On the other hand, although the hexagon satisfies condition 2, it is not regular because it does not satisfy condition 1.

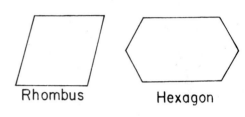

Rhombus   Hexagon

ANGLES IN POLYGONS

The angles in a polygon with four or more sides can be any size between 0° and 360°. In this hexagon, $\angle ABC$ is less than 20° and $\angle EDC$ and $\angle FAB$ are both greater than 180°. In spite of this degree of freedom, there is a relationship between the sum of the angles in a polygon and its number of sides. Let's begin by considering the triangle.

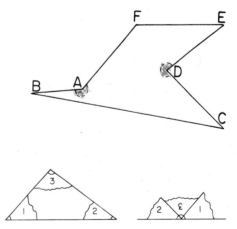

The sum of the angles in a triangle is 180°. This fact was proven by the Greek mathematicians of the fourth century B.C. One method of illustrating this theorem is to draw an arbitrary triangle and cut off its angles. When these angles are placed side by side, with their vertices at a point, they form one-half of a revolution (180°) about the point.

The sum of the angles in a polygon of four or more sides can be found by subdividing the polygon into triangles. This quadrilateral is partitioned into two triangles whose angles are numbered from 1 through 6. The sum of all six angles is 2 × 180 or 360°. The sum of the four angles of the quadrilateral is also 360° since its angles are made up of these six angles.

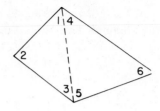

There is an infinite variety of quadrilaterals, some convex and others nonconvex. However, since each quadrilateral can be partitioned into two triangles, the sum of the angles will always be 360°. A similar approach can be used to find the sum of the angles in any polygon.

CONSTRUCTING REGULAR POLYGONS

The sum of the angles in a polygon can be used to compute the number of degrees in each angle of a regular polygon. Simply divide the sum of the angles by the number of angles. For example, the sum of the angles in a pentagon is 3 × 180, or 540, because it can be subdivided into three triangles. Therefore, each angle in a regular pentagon is 540 ÷ 5, or 108°. The regular pentagon which is

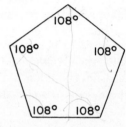

shown here was constructed by using a protractor to draw angles of 108°. Here are the first three steps in constructing a regular pentagon.

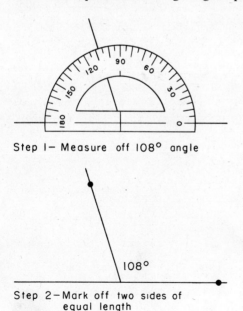

Step 1— Measure off 108° angle

Step 2—Mark off two sides of equal length

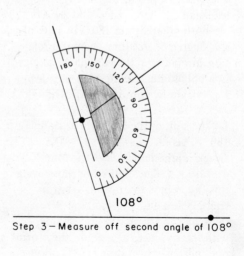

Step 3—Measure off second angle of 108°

For another approach to constructing regular polygons, we can begin with a circle and use central angles. The number of degrees in a *central angle* of a regular polygon is the number of sides of the polygon divided into 360. For a decagon, each central angle is 360 ÷ 10, or 36°.

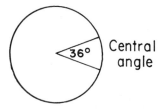

Central angle

Here is a four-step sequence for constructing a regular decagon.

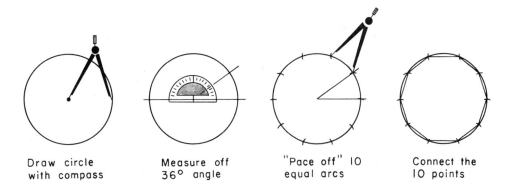

Draw circle with compass

Measure off 36° angle

"Pace off" 10 equal arcs

Connect the 10 points

TESSELLATIONS WITH POLYGONS

The hexagonal grid of the honeycomb provides another remarkable example of regular polygons in Nature. Bee eggs can be seen in the cells of the comb in the accompanying photo. These cells show that regular hexagons can be placed side by side with no uncovered gaps between them. Any arrangement of nonoverlapping figures that can be placed together to entirely cover a region is called a *tessellation*. Floors and ceilings are often *tessellated* or *tiled* with material that is square, because squares can be joined together without gaps or overlapping. Equilateral triangles are also commonly used for tessellations.

Honeycomb with bee eggs.
Courtesy of the American Museum of Natural History

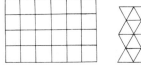 

From ancient times, tessellations have been used as patterns for rugs, fabrics, pottery, and architecture. The Moors of Spain were masters of tessellating walls and floors with colored geometric tiles. Some of their work is shown in the following picture

A room (Sala de Camas) in the Alhambra in Granada, Spain

of a room and bath in the Alhambra, a fortress palace built in the middle of the thirteenth century for Moorish kings. (There is another picture of this palace on page 254.)

The two tessellations in the center of the Alhambra photo have figures that are curved. In the following paragraphs, however, we will concern ourselves only with polygons that tessellate. The triangle is an easy case to consider first. Any triangle will tessellate by simply putting together two copies of the triangle to form a parallelogram (see shaded region). Copies of the parallelogram can then be moved horizontally and vertically. The points of tessellation at which the vertices of the triangle meet are called the *vertex points* of the

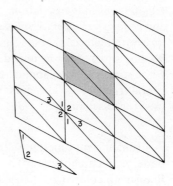

tessellation. Since the sum of the angles in a triangle is 180°, the 360° about each vertex point of the tessellation will be filled in by using each angle of the triangle twice. In the preceding tessellation of triangles, angles 1, 2, and 3 occur twice about each vertex point.

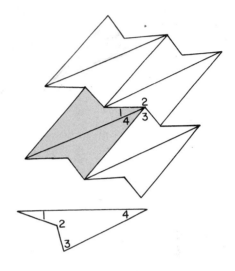

The size of the angles in a polygon and the sums of these angles will determine whether or not the polygon will tessellate. The fact that the sum of the angles in a quadrilateral is 360° suggests that quadrilaterals have the right combinations of angles to fit around each vertex point of a tessellation. In this tessellation, each angle of the quadrilateral (angles 1, 2, 3, and 4) occurs once about each vertex point of the tessellation. As in the case of triangles, two copies of the quadrilateral were matched together (see shaded region) and then moved horizontally and vertically to fill out the plane.

The quadrilateral in the preceding tessellation is nonconvex. It is quite surprising that every quadrilateral, convex or nonconvex, will tessellate. The case for polygons with more than four sides is not as strong. While there are some pentagons that will tessellate, there are also others that will not. Similarly, some hexagons will tessellate (for example, the regular hexagon), but not all of them.

If we consider only convex polygons, it can be proven that no polygon with more than six sides will tessellate. However, there are countless possibilities for tessellations of nonconvex polygons of more than six sides. The polygon in this tessellation is 12-sided.

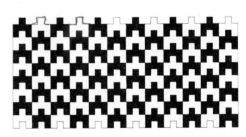

COMPUTER APPLICATIONS

Regular polygons are easy to draw using Logo commands. To draw any regular polygon the turtle will make a sequence of equal turns and equal forward moves until it has turned a total of 360°. For example, to draw the regular hexagon on page 220 the turtle makes six turns of 60° each. At the end of each forward move of 50 steps the turtle turns right 60°. Each interior angle of the polygon is 120°, the supplement of a 60° turn.

```
FD 50 RT 60 FD 50 RT 60 FD 50
RT 60 FD 50 RT 60 FD 50 RT 60
FD 50 RT 60
```

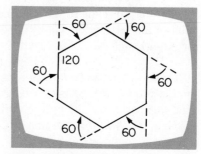

Regular Hexagon

The 12 commands for drawing this hexagon can be condensed into one command by using REPEAT. This is done in the following procedure. A variable is used for the length of the side of the hexagon. To obtain the hexagon shown above type HEXAGON 50.

```
TO HEXAGON :SIDE
  REPEAT 6 [FD :SIDE RT 60]
END
```

In general, to draw any regular polygon, the size of the turn will be 360 divided by the number of sides in the polygon. For example, a decagon will require 10 turns of $360/10 = 36$ degrees each. Here is a command for drawing a regular decagon with sides of length 15.

$$\text{REPEAT 10 [FD 15 RT 360/10]}$$

Circles and Arcs   As the number of sides in a polygon increase, the polygon becomes closer to the shape of a circle. The next procedure draws a 360-sided regular polygon which we will call CIRCLE. The "circles" shown here were obtained by using 1 and .5 for the variable SIZE.

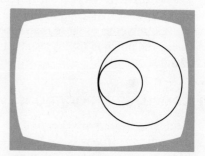

```
TO CIRCLE :SIZE
  REPEAT 360 [FD :SIZE RT 1]
END
```

CIRCLE 1 and CIRCLE .5

An arc is obtained by drawing part of a circle. The number of 1° turns the turtle makes is the number of degrees in the arc. Here is a program for producing a 90° arc with a variable size. The arc shown here was produced by letting SIZE = 1. Try this program for SIZE = .5.

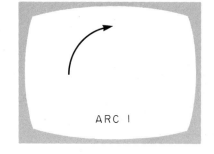

ARC 1

```
TO ARC :SIZE
  REPEAT 90 [FD :SIZE RT 1]
END
```

SUPPLEMENT (Activity Book)

Activity Set 4.2 Regular and Semiregular Tessellations
Just for Fun: Escher-type Tessellations

Exercise Set 4.2: Applications and Skills

1. This drawing of algae (sea life) is one of the unusual life forms studied by the German biologist, Ernst Haeckel (1834–1919). The polygons in the surface of this algae form the beginning of a tessellation.

★ a. There is a regular pentagon at the center. What are the polygons adjacent to this pentagon? Are they regular?

   b. There is a second ring of polygons surrounding the inner six. What kind of polygons are these?

2. This regular pentagon has been inscribed inside a circle. Angle *BAC* is a central angle of the pentagon.

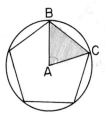

★ a. How many degrees are in each central angle of a regular pentagon?

★ b. Write the number of degrees for the central angles of the regular polygons in the following table.

	Triangle	Quadrilateral	Pentagon	Hexagon	Heptagon	Octagon	Nonagon	Decagon		
Number of sides	3	4	5	6	7	8	9	10	20	100
Central angle	120°	90°								

3. This Canadian nickel is a dodecagon (12-sided). Assume that you are to design a 12-sided one-dollar coin. Use your knowledge of central angles to inscribe a dodecagon in the given circle.

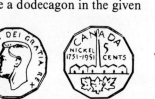

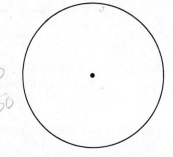

4. Use a protractor to construct a regular octagon having $\overline{AB}$ as a side.

A   B

5. This regular heptagon has been *circumscribed* about a circle.

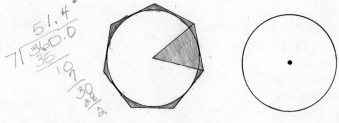

★ a. How many degrees are in the central angle of this heptagon? 51.4°

b. Construct a decagon which is circumscribed about the second circle.

6. What regular polygons will be formed by the following methods?

★ a. Tie a long rectangular strip of paper into a knot and smooth it down.

b. Cut out an equilateral triangle and fold each vertex into the center.

c. Draw a circle with a compass. With the opening of the compass equal to the radius of the circle, "pace off" this distance on the circumference and connect adjacent points.

7. Draw some figures to make conjectures about the truth or falsity of the following statements. Which three of these properties will not hold?

★ a. In any quadrilateral the sum of the opposite angles is 180°.

b. A line segment from a vertex of a triangle to the midpoint of the opposite side is called a *median*. The three medians of a triangle meet in a point.

★ c. The diagonals of a quadrilateral bisect each other.

d. If the midpoints of the sides of any quadrilateral are connected, they form a parallelogram.

★ e. A line segment from a vertex of a triangle which is perpendicular to the opposite side is called an *altitude*. The three lines containing the altitudes of a triangle meet in a point.

f. If the midpoints of the sides of any triangle are connected, they form an equilateral triangle.

8. Some regular polygons will tessellate and others will not.

a. Which of the five regular polygons in Exercise 3 on page 215 will tessellate?

b. What is very special about the sizes of the angles in regular polygons that will tessellate?

★ c. Will a regular octagon tessellate? Give a reason for your answer.

9. All triangles and quadrilaterals will tessellate, but this is not true for polygons with more than four sides. Which pentagon and which hexagon in figures **a** through **d** will tessellate? Draw a portion of the tessellation.

a. 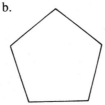   b.   c.    d.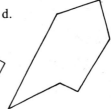

★ 10. Which of the following letters will tessellate? Draw portions of the tessellations.

a. b. c. d. e.

11. A polygon with more than six sides will not tessellate if it is convex. The following polygons have more than six sides but they are nonconvex. Sketch a portion of a tessellation for each of these polygons.

a.  b. c.

12. *A Gestalt Puzzle:* Gestalt psychology was developed in Germany in the 1930s and is concerned primarily with the laws of perception. What is represented by these polygons and their background?

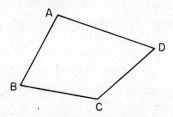

Exercise Set 4.2: Problem Solving

1. *Interior Angles:* What is the sum of the interior angles of a 50-sided polygon?

a. *Understanding the Problem:* The sum of the interior angles of any convex or nonconvex quadrilateral is 360. Explain how this can be determined.

★ b. *Devising A Plan:* Let's find the sum of the angles for a few special cases and look for a pattern. What is the sum of the angles in each of the following polygons? (*Hint:* Partition the polygon into triangles.)

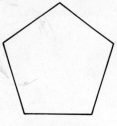
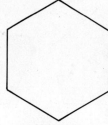
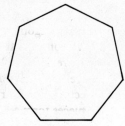

Pentagon   Hexagon   Heptagon

c. *Carrying Out The Plan:* Fill in the table for five-, six-, and seven-sided polygons. Look for a pattern and complete the table. Describe your pattern and use inductive reasoning to determine the sum of the angles in a 50-sided polygon.

Number of sides	3	4	5	6	7	8	9	10	11	12
Sum of angles	180	360	540	720						

★ d. *Looking Back:* What happens to the sum of the angles when a polygon is nonconvex? Will the results in part c still be true?

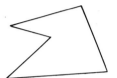

Nonconvex pentagon

2. *Regular Polygons:* There are 60° in each angle of an equilateral triangle. Find the size of each angle in regular polygons having 12 or fewer sides. What is the size of each interior angle of a regular 50-sided polygon? (*Hint:* See Exercise 1.)

★ 3. *Mirror Experiment:* This photo shows two mirrors that are hinged together with tape. The mirrors are placed in front of line *L* and a toy motorcycle to produce a regular pentagon with 5 motorcycles. Can any number of motorcycles or any regular polygon be obtained by varying the angle between the mirrors? What is the relationship between the number of motorcycles and the angle between the mirrors? (*Hint:* Trace the positions of the mirrors on the paper and measure the angles. Form a table with the measurements for each number of motorcycles.)

4. *Dissection Puzzle:* These figures show a method of dissecting a regular octagon and a regular dodecagon so that the pieces of each can be reassembled to form a square. Try it. Find a method of dissecting a regular hexagon so that the pieces form a rectangle.

a.

★ b.

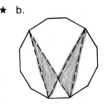

Exercise Set 4.2: Computers

1. Define a procedure for drawing these regular polygons. Use SIDE as a variable for the length of the sides.

    a. Pentagon

★   b. Octagon

    c. Dodecagon

★ 2. Define a procedure for drawing any regular *n*-sided polygon. Use NUMBER as the variable for the number of sides and SIDE as the variable for the length of the side. Here is the beginning.

    TO POLYGON :NUMBER :SIDE

3. Use CIRCLE to define a procedure called SLINKY for drawing overlapping circles. The variables NUMBER and SIZE are for the number of circles and the size of the circles.

    TO SLINKY :NUMBER :SIZE

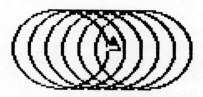

SLINKY

4. The procedure for drawing an arc (page 221) can be used twice to obtain a petal.

★   a. Define a procedure named PETAL for drawing a petal of variable size.

    b. Use PETAL to define a procedure named FLOWER to draw a flower with eight petals.

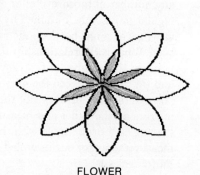

FLOWER

Cubic Space Division by M. C. Escher

"*Space is an infinite sphere whose center is everywhere and whose surface is nowhere.*" CASSIUS JACKSON KEYSER

4.3
SPACE FIGURES

In this lithograph by the Dutch artist Maurits C. Escher (1898–1972), the girders intersect at right angles to form the edges of large cubes. In this way Escher represents space as being filled with cubes of the same size. The Canadian mathematician H.S.M. Coxeter calls it the cubic honeycomb and says its perspective gives a wonderful sense of infinite space.

The notion of space in geometry is an undefined term, just as the ideas of point, line, and plane. We intuitively think of space as being 3D or three-dimensional and a plane as only 2D or two-dimensional. In his Theory of Relativity, Einstein tied together the three dimensions of space with the fourth dimension of time. He showed that space and time affect each other and give us a "four-dimensional" universe.

POLYHEDRA

The three-dimensional figures in photos (a) and (b) are crystals. Their flat sides and straight edges were not cut by people but were shaped by Nature. The crystal in (a), which is shown embedded in rock, is pyrite. It has 12 pentagonal faces, 4 of which can be seen. The crystal in (b) is zircon. Its faces are rectangles and triangles. The surfaces of space figures such as these, whose sides are polygons, are calld *polyhedra*. The polygons are called *faces* and they intersect in the *edges* and *vertices* of the polyhedron. The union of a polyhedron and its interior is called a *solid region* or *solid*.

(a) Crystal of pyrite

(b) Crystal of zircon

Regular Polyhedra The most famous of all the polyhedra are known as *regular polyhedra* or *Platonic solids*. Their faces are congruent regular polygons and there is the same arrangement of polygons at each vertex. It can be proven that there are just five such polyhedra. The *tetrahedron* has four triangular faces; the *cube* has six square faces; and the *octahedron* has eight triangular faces. These three polyhedra are found as crystals. Two of these, the cube and the octahedron, are pictured in (c). These are forms of the

(c) Crystals of pyrite

common mineral pyrite. The cube, which is embedded in rock, was found in Vermont, and the octahedron is from Peru, South America. The two other regular polyhedra are the *dodecahedron,* with 12 pentagonal faces, and the *icosahedron,* with 20 triangular faces. These cannot occur as crystals but have been found as skeletons of microscopic sea animals, called radiolarians. Posterboard models of the five regular polyhedra are shown next.

From left to right: tetrahedron, cube (hexahedron), octahedron, dodecahedron, icosahedron

Semiregular Polyhedra There is a greater variety of polyhedra once we allow two or more different types of regular polygons for faces. The faces of the boracite crystal are squares and equilateral triangles. Strange as it may seem, this crystal developed these flat, regularly shaped faces on its own, without the help of machines or people. There are only 13 polyhedra that have two or more regular polygons as faces and for which each vertex is surrounded by the same arrangement of polygons. They are called *semiregular polyhedra.* The boracite crystal is one of these. Each of its vertices is surrounded by three squares and one equilateral triangle.

Crystal of boracite

The regular and semiregular polyhedra are convex. A polyhedron is *convex,* if for any two points on its surface, the line segment connecting these points is on the surface or in the interior of the polyhedron. The polyhedra pictured next are nonconvex. They are called the Kepler-Poinsot solids. Like those of the regular and semiregular polyhedra, the faces of these polyhedra are regular polygons.

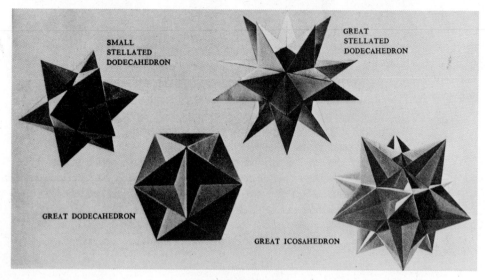

SMALL
STELLATED
DODECAHEDRON

GREAT
STELLATED
DODECAHEDRON

GREAT DODECAHEDRON

GREAT ICOSAHEDRON

Kepler-Poinsot solids

PYRAMIDS AND PRISMS

Chances are that when you hear the word "pyramid" you think, that's what the Egyptians built. Each of the Egyptian pyramids has a square base and triangular sides rising up to the vertex. This is just one type of pyramid. In general, the base of a pyramid can be any polygon but its sides are always triangular. Pyramids are named according to the shape of their base. Several pyramids with different bases are shown below. Church spires are familiar examples of pyramids. These are usually square, hexagonal, or octagonal pyramids. The spire in the photo on the right is a hexagonal pyramid.

The Bruton Steeple, Williamsburg, Virginia

Triangular pyramid
(also called tetrahedron)

Square pyramid

Pentagonal pyramid

Hexagonal pyramid

CHAPTER 4   GEOMETRIC FIGURES

Prisms are another common type of polyhedra. You probably remember from your science classes that a prism is used to produce the spectrum of colors ranging from violet to red. Due to the angle between the vertical faces of the prism, a light directed into one face will be bent when it passes out through the other face.

A *prism* has two parallel bases, an upper and a lower, which are congruent polygons. Like pyramids, prisms get their names from the shape of their bases. If the sides of a prism are perpendicular to the bases, as in the case of the triangular, quadrilateral, and hexagonal prisms, shown next, they are rectangles. Such prisms are called *right prisms,* or simply *prisms.* If some of the vertical faces are parallelograms, as in the pentagonal prism, the prism is called an *oblique prism.*

Triangular
prism

Quadrilateral
prism

Hexagonal
prism

Pentagonal
prism

The following two hexagonal prisms are crystals that grew with these flat, smooth faces and straight edges. These are oblique prisms whose faces are parallelograms.

Prisms of the crystal orthoclase feldspar

CONES AND CYLINDERS

Cones and cylinders are the circular counterparts to pyramids and prisms. Ice-cream cones, paper cups, and New Year's Eve hats are common examples of cones. A *cone* has a circular region (disc) for a base and a lateral surface that slopes to the vertex point. If the vertex lies directly above the center of the base, the cone is called a *right cone* or usually just a *cone*. Otherwise, it is an *oblique cone*.

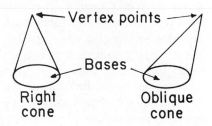

Ordinary cans are models of cylinders. A *cylinder* has two parallel circular bases (discs) of the same size and a lateral surface that rises from one circle to the other. If the center of the upper base lies directly above the center of the lower base, the cylinder is called a *right cylinder,* or simply a *cylinder,* and if it does not, it is an *oblique cylinder*. Almost without exception, the cones and cylinders we use are right cones and right cylinders.

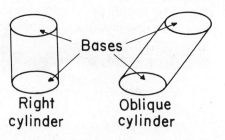

SPHERES AND MAPS

The easiest way to create a sphere is to blow a bubble. Of all the different three-dimensional figures that will hold a given amount of air, the sphere is the most economically shaped in terms of the amount of material needed.

The same laws of Nature which govern the shape of bubbles are also responsible for the spherical shape of the sun, its planets, and the moons around the planets. The accompanying photo shows a fantastic view of the earth's spherical shape. It was photographed from the *Apollo 17* spacecraft during its 1972 lunar mission. The dark areas are water. The Red Sea and the Gulf of Aden are near

Earth, as seen from *Apollo 17* during 1972 lunar mission

the top center, and the Arabian Sea and Indian Ocean are on the right. Nearly the entire coastlines of Africa and the Arabian Peninsula are visible.

Definition: A *sphere* is the set of points in space which are the same distance from a fixed point called the *center*. The union of a sphere and its interior is called a *solid sphere*.

The geometry of the sphere is especially important for navigating on the surface of the earth. The shortest distance between two points on a sphere is along an arc of a great circle. In this drawing of a sphere, G is a great circle, because its center is also the center of the sphere, and B is not a great circle. The distance between points X and Y along an arc of circle G is less than the distance between these points along an arc of circle B.

Locations on the earth's surface are often given by naming cities, streets, and buildings. A more general method of describing location is accomplished by two systems of circles. The circles that are parallel to the equator are called *parallels of latitude* [see Figure (a) shown next]. Except for the equator, these circles are not great circles. Each parallel of latitude is specified by an angle from 0° to 90°, both north and south of the equator. For example, New York City is at a northern latitude of 41°, and Sydney, Australia, is at a southern latitude of 34°. The second system of circles passes through the North and South Poles [see Figure (b) below] and are called *meridians of longitude*. These are great circles, and each is perpendicular to the equator. Since there is no natural point at which to begin numbering the meridians of longitude, the meridian that passes through Greenwich, England, is chosen as the zero meridian. Each meridian of longitude is given by an angle from 0° to 180°, both east and west of the zero meridian. The longitude of New York City is 74° West, and that for Sydney, Australia, is 151° East. These two systems of circles provide a grid or coordinate system for locating any point on earth [see Figure (c)].

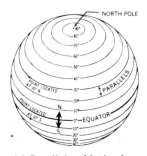

(a) Parallels of latitude

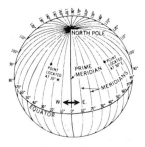

(b) Meridians of longitude

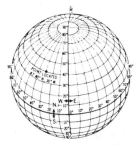

(c) Grid formed by both types of circles

The sphere cannot be placed flat on a plane without separating or overlapping some of its surface. This creates difficulties in making maps of the earth. There are three basic solutions to this problem: copying the earth's surface onto a cylinder, a cone, or a plane. These methods of copying are called *map projections*. In each case, some distortions of shapes and distances occur.

For a *cylindrical projection,* a cylinder is placed around a sphere (as shown in the next photo) and the surface of the sphere is copied onto the cylinder. Regions close to the equator are reproduced most accurately. The closer to the poles, the greater the map is distorted. The cylinder is then cut, and we have a flat map.

With a *conical projection,* a portion of the surface of a sphere is copied onto the cone. The cone can then be cut and laid flat. This type of map construction is commonly used for countries and other local regions of the earth's surface. The maps of the United States which are issued by the American Automobile Association are constructed by a conical projection.

Plane projections are made by placing a plane next to any point on a sphere and projecting the surface onto the plane. To visualize this, think of a light at the center of the sphere and the boundary of a country as being pierced with small holes. The light shining through these holes (see dotted lines) forms an image of the country on the plane. Roughly, half of the sphere's surface can be copied onto a plane by a plane projection, with the greatest distortion taking place at the outer edges of the plane.

Cylindrical projection

Conic projection

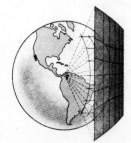

Plane projection

COMPUTER APPLICATIONS

Let's define a procedure that instructs the turtle to draw a hexagonal prism. We have a procedure called HEXAGON (page 220) which will be used to draw the upper base of the prism. This procedure uses the variable SIDE for the length of the sides of the hexagon.

Procedure 4.3A

```
TO HEXAGON :SIDE
  REPEAT 6 [FD :SIDE RT 60]
END
```

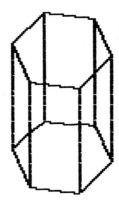

Hexagonal Prism

Next, we need some means of drawing the vertical edges of the prism. As the turtle traces each side of the lower base it must be instructed to face North at each vertex and draw an edge of the prism. To do this we need a new command that will turn the turtle in any given direction.

SETHEADING is a Logo command to set the turtle's heading in any direction. The turtle's heading can be any number of degrees from 0 to 360. For example,

SETHEADING 90

will turn the turtle toward 90°, that is, facing East. To head the turtle in the North direction for the edges of the prism, we will use SETHEADING 0.

The following procedure will draw the vertical edges of the prism. The variable HEIGHT will allow us to use different numbers for the height of the prism.

```
TO PRISMEDGE :HEIGHT
  SETHEADING 0
  FD :HEIGHT
  BK :HEIGHT
END
```

Now we are ready to define a procedure for drawing a prism. SIDE and HEIGHT are variables for the lengths of the sides of the base and the edges of the prism. There are six lines in this procedure for drawing the six sides of the lower base and the vertical edges. Notice that each right turn is 60° more than the preceding right turn. Why? The upper base is drawn by the procedure HEXAGON. Follow through the lines of this procedure and trace each corresponding part of the prism.

```
TO HEXPRISM :SIDE :HEIGHT
  HIDETURTLE
  RT 40 FD :SIDE PRISMEDGE :HEIGHT
  RT 100 FD :SIDE PRISMEDGE :HEIGHT
  RT 160 FD :SIDE PRISMEDGE :HEIGHT
```

```
RT 220 FD :SIDE PRISMEDGE :HEIGHT
RT 280 FD :SIDE PRISMEDGE :HEIGHT
RT 340 FD :SIDE PRISMEDGE :HEIGHT
FD :HEIGHT RT 40
 HEXAGON :SIDE
END
```

SUPPLEMENT (Activity Book)

Activity Set 4.3 Models for Regular and Semiregular Polyhedra
Just for Fun: Penetrated Tetrahedron

Exercise Set 4.3: Applications and Skills

1. There are six categories of illusions.* One category is called "impossible objects," because the illusions are produced by drawing three-dimensional figures on two-dimensional surfaces. Find the impossible feature in each of these pictures.

a.

Waterfall by M. C. Escher

b.

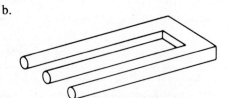

c.

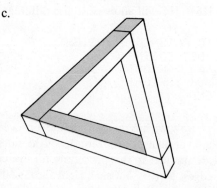

*P. A. Rainey, *Illusions* (Hamden, Conn.: The Shoe String Press, 1973), pp. 18–43.

2. A second type of illusion involves depth perception. We have accustomed our eyes to see depth in objects that are drawn on two-dimensional surfaces. Answer questions **a** and **b** by *disregarding the depth illusions.*

★ a. Is one of these cylinders larger than the others?

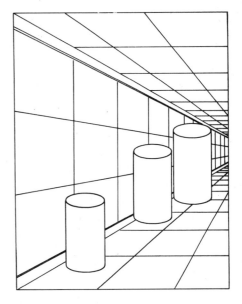

★ b. Which of the four lettered angles, *a, b, c,* or *d,* has the most degrees? Which are right angles?

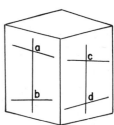

★ 3. A cube can be dissected into triangular pyramids in several ways. Pyramid *FHCA* partitions the cube into five triangular pyramids. Name the four vertices of each of the other four pyramids.

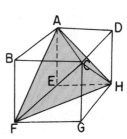

4. *F, E, G, H,* and *C* are the vertices of a square pyramid inside this cube. Name the five vertices of two more square pyramids such that the cube is partitioned into three pyramids.

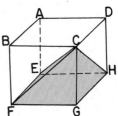

5. What are the names of each of these figures?

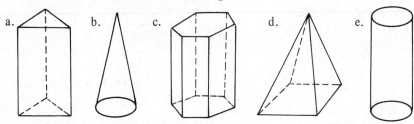

a. b. c. d. e.

6. This collection of polyhedra shows some of the limited number of forms which crystals may take. These shapes were not invented by people but were revealed to us as products of Nature. The polygons at the tops of the marked off columns are the horizontal cross sections of the polyhedra in these columns.

★   a. List the numbers of the polyhedra which are pyramids.

    b. List the numbers of the polyhedra which are prisms.

★   c. Which one of these polyhedra is most like a dodecahedron?

    d. Which one of these polyhedra is most like an octahedron?

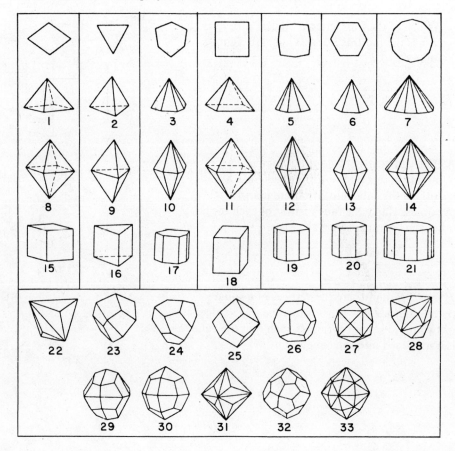

CHAPTER 4   GEOMETRIC FIGURES

7. Which of the three types of projections for making maps is most appropriate for obtaining flat maps of the following regions?
★   a. Australia.   ★   b. North, Central, and South America.
★   c. The entire equatorial region within the latitudes of 30° North and 30° South.

★ 8. Use your knowledge of the earth's coordinate system to match each of the following cities with its approximate longitude and latitude.

	Latitude	and	Longitude
Tokyo	38°N		120°W
San Francisco	56°N		4°W
Melbourne	35°N		140°E
Glasgow (Scotland)	35°S		20°E
Capetown	38°S		145°E

9. Two points on the earth's surface which are on opposite ends of a line segment through the center of the earth are called *antipodal points*. The coordinates of such points are nicely related. The latitude of one point is as far above the equator as the other is below, and the longitudes are supplementary angles (in opposite hemispheres). For example, (30°N, 40°E) is in Saudi Arabia, and its antipodal point (30°S,140°W) is in the South Pacific.

Babson College globe.
Diameter 28 feet, weight 21 tons.

★   a. The globe in the accompanying photo shows that (20°N,120°W) is a point in the Pacific Ocean just west of Mexico. Its antipodal point is just east of Madagascar. What are its coordinates?

     b. The point (30°S,80°E) is in the Indian Ocean. What are the coordinates of its antipodal point? What country is it in?

★ 10. China is bounded by latitudes of 20°N and 55°N, and by longitudes of 75°E and 135°E. It is playfully assumed that if you could dig a hole straight through the center of the earth you would come out in China. For which of the following locations is this true?

     a. Panama (9°N,80°W)   b. Buenos Aires (35°S,58°W)   c. New York (41°N,74°W)

11. Hurricane Ginger was christened on September 10, 1971, and became the longest-lived Atlantic hurricane on record. This tropical storm formed approximately 275 miles south of Bermuda and reached the U.S. mainland 20 days later.

Hurricane Ginger

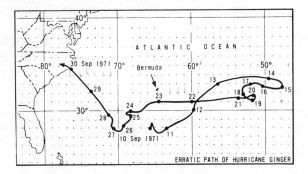

★ a. The storm's coordinates on September 10 were (28° N, 66° W). What were the coordinates for the following dates: September 15; September 23; and September 30?

b. At this latitude on the earth's surface each degree of longitude spans a distance of approximately 60 miles. About how many miles did this hurricane travel from September 10 to September 30? (*Hint:* Use a piece of string.)

Exercise Set 4.3: Problem Solving

1. *Euler's Formula:* There is a remarkable formula that relates the numbers of vertices, edges, and faces of a polyhedron. This formula was first stated by René Descartes about 1635. In 1752 it was discovered again by Leonhard Euler and is now referred to as Euler's formula. What is this formula?

a. *Understanding the Problem:* This formula holds for regular as well as nonregular polyhedra. A cube is a regular polyhedron with 6 faces and 12 edges. How many vertices does it have?

(1)

★ b. *Devising a Plan:* Count the vertices, edges, and faces of several polyhedra and form a table. Look for a relationship between the numbers for each polyhedron.

(2)

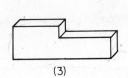

(3)

(4)

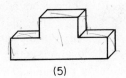

(5)

c. *Carrying Out the Plan:* Complete this table for the polyhedra in Figures 1 through 5. Find a relationship that holds for the numbers for each polyhedron. Describe this relationship by using v for the number of vertices, f for the number of faces, and e for the number of edges.

Figure	v	f	e
1			
2			
3			
4			
5			

★ d. *Looking Back:* Does Euler's formula hold for a tessellation of polygons if each polygon is counted as a face and the side of a polygon is counted as an edge?

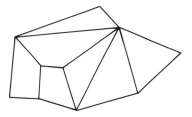

★ 2. *Regular Polyhedra:* What are the numbers of vertices, faces, and edges for each of the regular polyhedra? (*Hint:* Use Euler's formula.)

Dodecahedron   Icosahedron

	Vertices	Faces	Edges
Tetrahedron			
Cube	8	6	12
Octahedron			
Dodecahedron			
Icosahedron			

3. *Dual Polyhedra:* The centers of the faces of a cube can be connected to form a regular octahedron. Also, the centers of the faces of an octahedron can be connected to form a cube. These polyhedra are called *duals.* How is this dual relationship suggested by the table in Exercise 2? Find two other Platonic solids that are duals. Which Platonic solid is its own dual?

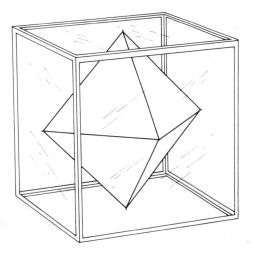

★ 4. *Patterns for Cylinders and Cones:* A rectangular sheet of paper is a pattern for a right circular cylinder without bases. Make a pattern to create each of the following figures (without bases). (*Hint:* Work backward.)

| Right circular cone | Oblique Circular cone | Oblique Circular cylinder |

★ 5. *Riddle:* Some sportspeople, having pitched camp, set out to go bear hunting. They walked 15 miles due south, then 15 miles due east, where they shot a bear. Walking 15 miles due north, they returned to their camp. What was the color of the bear? Name at least two locations on earth where the three 15-mile legs of this trip will end at the starting point. (*Hint:* Guess and check by drawing paths on a sphere.)

6. *Painted Cube:* Suppose a large cube is built from 1000 small cubes and then spray-painted on all six faces. When the large cube is disassembled, how many of the small cubes will be unpainted? Separate the remaining small cubes into groups according to the number of faces painted and compute the number in each group. Generalize this result for an *n* by *n* by *n* cube.

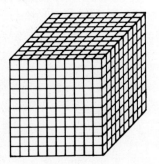

Exercise Set 4.3: Computers

1. Here are two procedures.

```
TO TRIANGLE
  RT 90 FD 80
  RT 120 FD 40
  RT 90 FD 69
END
```

```
TO TRIPRISM :HEIGHT
  HIDETURTLE
  RT 90 FD 80 PRISMEDGE :HEIGHT
  RT 210 FD 40 PRISMEDGE :HEIGHT
  RT 300 FD 69 PRISMEDGE :HEIGHT
  FD :HEIGHT TRIANGLE
END
```

a. Sketch the figure that will be drawn by TRIANGLE. What kind of a triangle is this?

b. Sketch the figure that will be drawn by TRIPRISM 80. (*Note:* PRISMEDGE is defined on page 235.)

★ 2. Use the procedure for SQUARE which is defined below, and PRISMEDGE (page 235) to define the procedure SQUARE PRISM for drawing a prism with a square base. Use HEIGHT as the variable for the height of the prism.

```
TO SQUARE
  RT 30 FD 40
  RT 90 FD 40
  RT 90 FD 40
  RT 90 FD 40
END
```

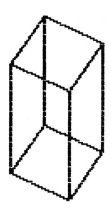

★ 3. Define a procedure (similar to the one on page 235) for drawing an octagonal prism whose base is a regular octagon. Use the variable SIDE for the length of the sides of the octagon and the variable HEIGHT for the height of the prism. (*Note:* The procedure for a regular octagon is in Exercises 1.b on page 226.)

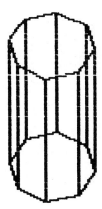

Taj Mahal, built (1630–1652) on the banks of the Jumna in Agra, India

4.4
SYMMETRIC FIGURES

The Taj Mahal is considered by many to be the most beautiful building in the world. It is made entirely of white marble and surrounded by a landscaped walled garden on the banks of the Jumna River in Agra, India. The Taj Mahal was built by the emperor Shah Jahan in memory of his wife, Mumtaz Mahal. It is an octagonal building with four of its eight faces containing massive arches rising to a height of 33 meters (108 feet). The center is an octagonal chamber containing the tombs of the emperor and his wife.

The Taj Mahal has a form and balance which can be described by saying it is symmetrical. The concept of symmetry was one of the earliest attempts of the human race to find order and harmony in a world of hostility and confusion. Perhaps the most influential factor in our desire for symmetry is the shape of the body. Even children in their earliest drawings show an awareness of body symmetry.

REFLECTIONAL SYMMETRY FOR
PLANE FIGURES

Many years before the recent trend to teach
geometric ideas in the elementary school, cutting
out symmetric figures was a routine activity. The
procedure is to fold a piece of paper and draw a
figure which encloses part or all of the crease. When
the figure is cut out and unfolded it is symmetrical.
The crease is called a *line of reflection* or *line of
symmetry,* and the figure is said to have a
reflectional symmetry.

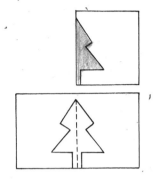

   The intuitive idea of reflectional symmetry is that
the two halves of the figure are "the same" or would
coincide if one were placed on the other. The word
"reflection" is a natural one to use because of the mirror test for symmetry. If the edge
of a mirror is placed along a line of reflection, the "half-figure" and its image will look
like the whole figure. For example, you can use this test on the front of the following
building by placing the edge of a mirror along the vertical center line of the photo. With
the mirror in this position, half of the building and its reflection will look like the whole
building. Since this is the only way the mirror can be placed so that this will happen,
the front of this building has only one line of symmetry.

Putnam Hall, University of New Hampshire

   Some figures have more than one line of reflection. To produce a figure with two such
lines, fold a sheet of paper in half and then in half again. Then cut out a figure whose

endpoints are on the creases. Where the paper is opened, the two perpendicular creases will be lines of reflection for the figure.

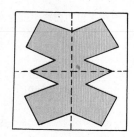

The idea of symmetry can be made more precise by adopting the terminology of "image," which is suggested by mirrors. If a line can be drawn through a figure so that each point on one side of the line has a matching point on the other side at the same perpendicular distance from the line, it is a *line of reflection.* For any two matching points, one is called the *image* of the other. A few points and their images on the house figure have been labelled. *A* corresponds to *A'*, *B* to *B'*, *C* to *C'*, and *D* to *D'*. The line segment connecting a point and its image is perpendicular to the line of reflection.

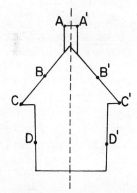

Each of the following polygons has two or more lines of reflection. The equilateral triangle has three such lines, along its medians. The square has four lines of reflection: one horizontal, one vertical, and two diagonal. The rectangle has only two lines of reflection, one horizontal and one vertical. The diagonals of the rectangle are not lines of reflection.

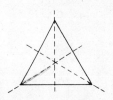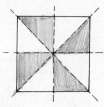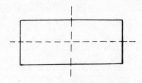

ROTATIONAL SYMMETRY FOR PLANE FIGURES

This is not a picture of a plant, as you might expect, but a type of jellyfish called Aurelia. While it seems to have the form and balance of a symmetric figure, it has no lines of reflection. It does, however, have what we call *rotational symmetry,* because it can be turned about its center so that the figure coincides with itself. For example, if it is rotated 90° clockwise, the top "arm" will move to the 3 o'clock position, the bottom "arm" will move to the 9 o'clock position, etc.

Aurelia, common coastal jellyfish

246

Let's consider another example of rotational symmetry. Trace the figure shown here and mark the center X and the arms A, B, and C. Cut it out and place it on the page so both figures coincide. By holding a pencil at point X, the top figure can be rotated clockwise so that A goes to B, B to C, and C to A. This is called a *rotational symmetry*, and X is called the *center of rotation*. Since the figure is rotated 120° (1/3 of a full turn) it has a *120° rotational symmetry*. From its original position this figure can also be made to coincide with itself after a 240° clockwise rotation, with A going to C, B to A, and C to B. This is a *240° rotational symmetry*. Since every figure can be rotated back onto itself after a 360° rotation, every figure has a *360° rotational symmetry*.

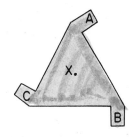

Some figures have both rotational and reflectional symmetries. The regular polygons have both types. The central angles of these polygons determine the angles for the rotational symmetries. In the regular hexagon the central angle is 360 ÷ 6, or 60°. Therefore, rotations of 60°, 120°, 180°, 240°, 300°, and 360° each move the hexagon back onto itself. Every regular hexagon has six rotational symmetries and six lines of symmetry.

Each of the hexagonal-shaped snowflakes possesses the rotational and reflectional symmetries of the hexagon. Beyond this similarity there is endless variety in their structure. These next photographs are from the more than 2200 pictures of snowflakes in the book, *Snow Crystals.**

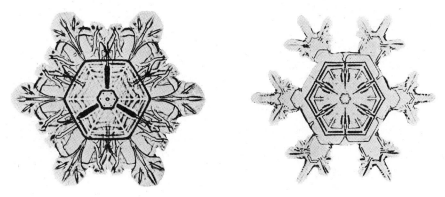

*W. A. Bentley and W. J. Humphreys, *Snow Crystals* (New York: McGraw-Hill, 1931).

REFLECTIONAL SYMMETRY FOR SPACE FIGURES

The intuitive idea of reflectional symmetry for three-dimensional objects is similar to that for plane figures. With plane figures we found lines such that one-half of the figure was the reflection of the other. With space figures there are planes of symmetry such that the two halves "look the same." Consider, for example, this antique chair. By cutting down the center of the back and across the seat to the front of the chair, it can be divided into left and right halves, both of which will be mirror images of each other. We can think of this cut as being done by a plane, called the *plane of reflection* or *plane of symmetry*. The chair is said to have *reflectional symmetry*.

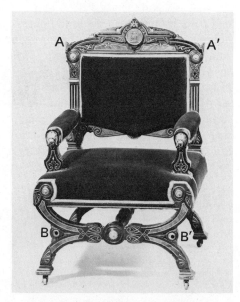

Ebonized walnut armchair, 1865–1875

This type of symmetry can be mathematically defined by requiring that for each point on the left-hand side of the chair, there is a corresponding point on the right-hand side, such that both points are the same perpendicular distance from the plane of reflection. For the antique chair, *A* corresponds to *A'* and *B* corresponds to *B'*. These points are called *images* of each other, and the segments $\overline{AA'}$ and $\overline{BB'}$ are perpendicular to the plane of reflection.

"Two-sided symmetry," such as for the antique chair, the long-horn beetle, and the zebra butterfly, is sometimes called *bilateral symmetry* or *vertical symmetry,* because

Long-horn beetle

Zebra butterfly

the plane of reflection is perpendicular to the ground. Martin Gardner offers the following explanation for the common occurrence of this type of symmetry.

"Because the earth is a sphere toward the center of which all objects are drawn by gravity, living forms have found it efficient to evolve shapes that possess strong *vertical symmetry* combined with an obvious lack of horizontal or rotational symmetry. In making objects for his use man has followed a similar pattern. Look around and you will be struck by the number of things you see that are essentially unchanged in a vertical mirror: chairs, tables, lamps, dishes, automobiles, airplanes, office buildings— the list is endless."*

Each of the objects shown above and on the preceding page has one or more planes of reflection. The chair, beetle, and butterfly each have one plane of reflection. The small square-shaped table has four planes of reflection: one from front to back; one from side to side; and one through each diagonal of the surface. The lamp has six vertical planes of reflection because its shade has six sections. The wastebasket has eight vertical planes of reflection, since its base has the shape of an octagon. If the wastebasket had a top cover to match its bottom, it would have a horizontal plane of symmetry. A sphere mounted on a pentagonal base, similar to this 1940 beechwood engraving by M. C. Escher, would have five planes of reflection because of the shape of its base.

Sphere with Fish
by M. C. Escher

*M. Gardner, The *Unexpected Hanging* (New York: Simon and Schuster, 1969), pp. 114–15.

ROTATIONAL SYMMETRY FOR SPACE FIGURES

Some three-dimensional objects, such as the table shown here, have *rotational symmetries*. By rotating this table 120° the legs will change places and the table will be moved back into the same location or position. That is, leg *A* will go to leg *B*, *B* to *C*, and *C* to *A*. In this example the table is being rotated about line *L* which passes through the center of the tabletop and its base. Line *L* is called the *axis of rotation,* and the table is said to have a *rotational symmetry*. Since the angles formed by pairs of legs of this table have 120°, the table has 120°, 240°, and 360° rotational symmetries.

   The three-legged table also has three vertical planes of reflection, one passing through each leg. It is not difficult to find objects with both planes of reflection and axes of rotation. The small table, the lamp, and the wastebasket pictured on page 249 all have both types of symmetry. Occasionally, however, you will see objects such as this paper windmill which have rotational symmetries but no plane of reflection. There are rotational symmetries of 90°, 180°, 270°, and 360° about its axis, which is the line perpendicular to this page and through the center of the windmill.

Paper windmill

COMPUTER APPLICATIONS

Symmetric figures can be obtained by interchanging all RIGHT and LEFT commands in a procedure. When the original figure is combined with the revised figure, the result will be a figure with a vertical line of symmetry. Let's see how this works.

   This procedure produces the vent-type figure shown here.

```
TO RIGHTVENT
 FD 100 RT 90 FD 80 RT 90 FD 50
 RT 90 FD 30 RT 90 FD 20 LT 90
 FD 30 LT 90 FD 70 RT 90 FD 20 RT 90
END
```

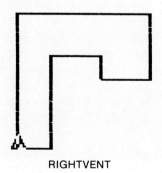

RIGHTVENT

Now, if we use the commands in RIGHTVENT, but change each "RT" to "LT" and change each "LT" to "RT", the new procedure, called LEFTVENT, will produce a vent which points to the left.

```
TO LEFTVENT
 FD 100 LT 90 FD 80 LT 90 FD 50
 LT 90 FD 30 LT 90 FD 20 RT 90
 FD 30 RT 90 FD 70 LT 90 FD 20 LT 90
END
```

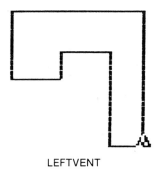

LEFTVENT

We can now instruct the turtle to draw both of these figures by typing RIGHTVENT LEFTVENT and pressing RETURN. The resulting figure has one line of symmetry, the North-South centerline of the screen.

RIGHTVENT LEFTVENT

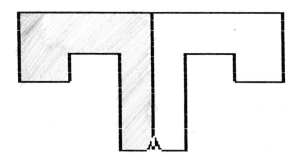

A figure with rotational symmetry can be created by rotating a given figure. The next procedure instructs the turtle to draw a flag. Then FLAG is used with the REPEAT command on the next page to create a figure with rotational symmetries.

```
TO FLAG
 FD 60
 REPEAT 4 [RT 90 FD 10]
 BK 60
END
```

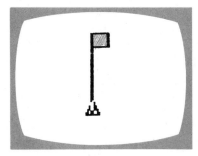

FLAG

This figure has five rotational symmetries but no lines of symmetry.

REPEAT 5 [FLAG RT 72]

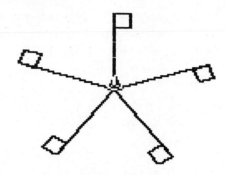

Recursion A powerful feature of computers is their ability to repeat a sequence of commands many times. This is called *recursion*. The REPEAT command is one method of obtaining recursion. Another method is to define a procedure so that one of the lines in the program calls upon the procedure itself. This is done in the next example. The procedure being defined is called STAR. After the FORWARD and RIGHT commands, the computer comes to STAR. At this point it returns to the beginning of the procedure to carry out the definition of STAR. To stop this procedure, which continues indefinitely, press CONTROL G.* This procedure draws polygons, if the number for ANGLE is chosen to be a factor of 360 (e.g., 15, 30, 60, 120, etc.). However, if ANGLE is between 120 and 180, the figure is a star. These figures have rotational symmetry and lines of symmetry.

```
TO STAR :STEP :ANGLE
  FD :STEP
  RT :ANGLE
  STAR :STEP :ANGLE
END
```

STAR 100 160   STAR 100 156   STAR 100 135

Conditional Commands One disadvantage of the recursion procedure in the preceding example is that it runs indefinitely. This problem can be avoided by using a *conditional command*. The conditional command in the following procedure stops the turtle when

*This is done by holding down the key marked CONTROL and pressing G.

its heading returns to zero.* At this point the turtle is heading North in the direction it started in and the figure is complete.

```
TO STARS :STEP :ANGLE
  FD :STEP
  RT :ANGLE
  IF HEADING = 0 STOP
  STARS :STEP :ANGLE
END
```

SUPPLEMENT (Activity Book)

Activity Set 4.4 Creating Symmetric Figures by Paper Folding

Note: In Apple Logo the action to be carried out (that is, the conclusion of the conditional command) must be enclosed in brackets. For example, IF HEADING = 0 [STOP].

The Alhambra, built in the thirteenth century for Moorish kings, Granada, Spain

Exercise Set 4.4: Applications and Skills

1. The pool, building, and fortress in this picture have a vertical plane of symmetry, about which their left sides are the reflections of their right sides.

   a. List five objects in this picture which have images about the plane of reflection.

   b. Physical objects can never be perfectly symmetric. In this scene, for example, there are several deviations from perfect symmetry. List three objects which do not have an image for the vertical plane of symmetry.

   ★ c. There are several individual items in this photo which have vertical lines of symmetry. Name an object in this picture which has a horizontal line of symmetry.

2. In 1850 gold was so plentiful that dozens of different banks and business firms minted their own coins. Some were square and others were eight-sided, such as this octagonal $50 gold piece.

★ a. How many rotational symmetries are there for a regular octagon?

b. How many degrees are there in the smallest rotational symmetry?

c. How many lines of reflection are there for a regular octagon?

Panama-Pacific octagonal $50 gold piece

3. The following organisms have lines of reflection and rotational symmetries. Determine the number of lines of reflection and rotational symmetries for each one.

★ a.   b.   ★ c.

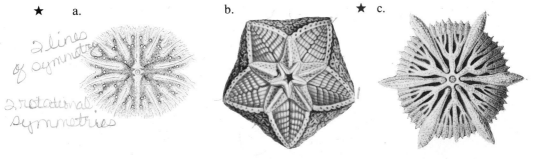

[handwritten: 2 lines of symmetry]
[handwritten: 2 rotational symmetries]

Species of algae   Extinct relative of starfish   Star coral

4. Which two of these polygons have no lines of reflection? Draw all of the lines of reflection for the remaining polygons. Find the number of rotational symmetries for each figure. The subject of beauty has been discussed for thousands of years. Aristotle felt that the main elements of beauty are order and symmetry. The American mathematician George Birkhoff (1884–1944) developed a formula for rating the beauty of objects.* Part of his formula involved counting symmetries. If only symmetry is used to rate the beauty of the polygons below, which would have the highest rating (counting all lines of reflection and rotational symmetries)? Which is the least beautiful?

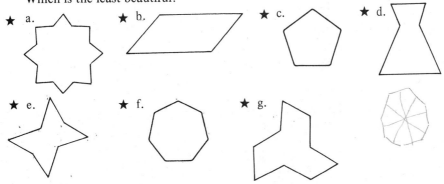

★ a.   ★ b.   ★ c.   ★ d.

★ e.   ★ f.   ★ g.

*G. D. Birkhoff, *Aesthetic Measure* (Cambridge, Mass.: Harvard University Press, 1933), pp. 33–46.

★ 5. The mirror test for symmetry is very effective when the reflecting is done with a Mira.* In addition to reflecting, the Mira allows you to see through it, so the whole figure can be seen. To find a line of symmetry, it is necessary only to move the Mira until the reflected image on the plexiglass coincides with the portion of the figure behind it. This cannot be done for several of the following figures. Which ones?

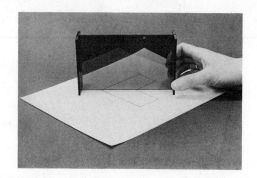

a.

b.

c.
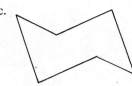

d.
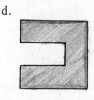

e.
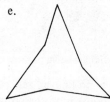

f.
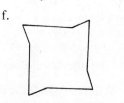

★ 6. Which letters have two lines of reflection? Which letters have two rotational symmetries but no lines of reflection?

ABCDEFGHIJKLMNOPQRSTUVWXYZ

7. If you write the letter P on a piece of paper and hold it in front of a mirror, it will look reversed.

★   a. Which capital letters will not appear reversed if you repeat this experiment? What type of symmetry do these letters have?

    b. Use the letters from part **a** to write a word such that its reflection in a mirror is also a word.

8. The following figures have been formed on a circular geoboard. Three of these figures have no lines of reflection. Sketch in all lines of reflection for the remaining

*E. Woodward, "Geometry with a Mira," *The Arithmetic Teacher,* **25** No. 2 (November 1977), 117–18.

figures. Find the number of rotational symmetries and give the number of degrees for each.

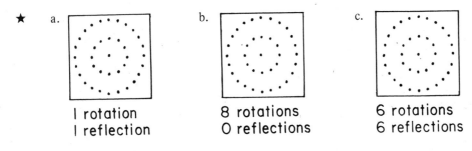

★ a.　　b.　　★c.　　d.

★ e.　　f.　　g.　　h.

9. Sketch figures on these geoboards which have the given numbers of symmetries.

★ a.

I rotation
I reflection

b.

8 rotations
O reflections

c.

6 rotations
6 reflections

10. Complete the sketches of these figures so that they are symmetric about the given line. You might want to first find the image with a mirror or Mira, as in Exercise 5.

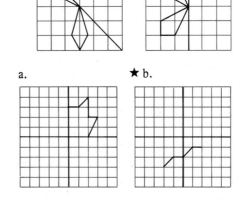

★ a.　　b.

11. Finish these figures so that they will be symmetric about the two perpendicular lines.

a.　　★ b.

12. How many planes of reflection are there for each of these objects?

13. These geometric solids are highly symmetric.

From left to right, cone, cylinder, prism, sphere, cube, pyramid (square base)

★ a. Which of these solids has a horizontal plane of reflection?

b. Does each of these solids have at least one vertical plane of reflection?

★ c. For each vertical axis of symmetry give the number of rotational symmetries for each solid.

14. These metalwork designs have many pleasing symmetries. How many rotational symmetries and lines of symmetry are there for each design?

★ a. 　　　　　　　　 b. 　　　　　　　　 c.

15. How many rotational symmetries are there for each of the Japanese crests?

★ a. b. ★c. d.

Exercise Set 4.4: Problem Solving

1. *Reflectional and Rotational Symmetries:* For every figure with reflectional symmetry there is a relationship between the number of these symmetries and the number of rotational symmetries. What is this relationship? Is it possible for a figure to have two or more reflectional symmetries and no rotational symmetry, other than a 360° rotation?

   a. *Understanding the Problem:* There are figures with both rotational and reflectional symmetries, and figures with two or more rotational symmetries but no reflectional symmetries. Name all the symmetries for the figure shown here.

   ★ b. *Devising A Plan:* Sketch figures (or form figures on geoboards) with two, three, four, and five lines of symmetry. Count the rotational symmetries, including a 360° rotation.

   c. *Carrying Out the Plan:* Record your results in this table. Is it possible to create a figure with two or more reflectional symmetries and no rotational symmetries, other than a 360° rotation? If a figure has *N* reflectional symmetries, what can be said about the number of rotational symmetries?

Number of reflectional symmetries	2	3	4	5	6	7	8	9	10
Number of rotational symmetries	4								

   d. *Looking Back:* Do the results in part **c** hold for space figures if lines of reflection are replaced by planes of reflection?

2. *Symmetries of Crystals:* Crystals are classified into different types according to their numbers of axes of rotation. This photo shows several cubes of the galena crystal.

★   a. How many axes of rotation has a cube? Describe each axis of rotation by using the figure on the left below. For example, the line which bisects segment $\overline{FG}$ and $\overline{AD}$ is an axis of rotation.

★   b. How many planes of symmetry has a cube? Describe each plane of symmetry by using the figure on the right below. For example, the plane which is shown bisects line segments $\overline{AB}$, $\overline{DC}$, $\overline{HG}$, and $\overline{EF}$.

Intersecting cubes of galena crystals

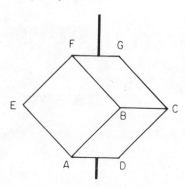

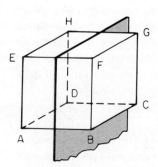

3. *Mirror Cards:** Mirror Cards were developed by Marion Walter to informally teach spatial relationships and symmetry in the early grades. The problem posed by Mirror Cards is one of using a mirror to match a pattern on one card to that on another. The mirror card at the right was used to obtain three of the four figures at the top of the next page. Draw a line on this card to indicate where the mirror should be placed to obtain these figures. Sketch four more different figures that can be obtained from this card by using a mirror.

*M. Walter, "An Example of Informal Geometry: Mirror Cards" *The Arithmetic Teacher,* **13** (October 1966), 448–52.

a.　　b.　　c.　　d.

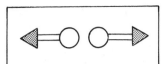

Exercise Set 4.4: Computers

1. Write a procedure called RTREE for drawing the right side of a tree. Have the turtle return to its starting position. Then define a procedure called LTREE by interchanging the "RT" and "LT" commands. Sketch the result obtained by typing HIDETURTLE RTREE LTREE.

2. Use FLAG (see page 251) to write a command for drawing a figure with the following rotational symmetry.

   ★　a. 10 rotational symmetries　　　b. 7 rotational symmetries

3. Define a procedure called HOOK for drawing the figure shown here. Return the turtle to its starting position by having it back up.

   a. Sketch the figure drawn by REPEAT 6 [HOOK RT 60].

   ★　b. Use HOOK to write a command for a figure that has 10 rotational symmetries.

4. A procedure for a drawing which has many parts, such as a face, can best be done by subprocedures. The following procedure called FACE has eight subprocedures. For each of these subprocedures, except possibly the nose (see hint below), the turtle should start from and return to its home. Design a face and define the subprocedures for drawing it. (*Hint:* Use the concept of symmetry to write subprocedures for the eyes, ears, and mouth. The turtle can be used for the nose.)

   TO FACE
   BOX　R,EYE　L,EYE　R,EAR　L,EAR　R,MOUTH　L,MOUTH　NOSE
   END

5. Some interesting figures can be obtained by increasing the variables in STAR each time the computer goes through the procedure (page 252). Experiment with the following procedures.* Which procedure produces figures with rotational symmetry?

a.
```
TO SPIRAL :STEP :ANGLE
 FD :STEP
 RT :ANGLE
 IF :STEP > 100 STOP
 SPIRAL :STEP + 10 :ANGLE
END
```

b.
```
TO INSPIRAL :STEP :ANGLE
 FD :STEP
 RT :ANGLE
 IF :ANGLE > 2000 STOP
 INSPIRAL :STEP :ANGLE + 10
END
```

*Note: To complete some inspirals a number greater than 2000 will be needed for ANGLE.

5

Measurement

Stonehenge (1900–1700 B.C.), Salisbury Plain, England

"The man who undertakes to solve a scientific question without the help of mathematics undertakes the impossible. We must measure what is measurable and make measurable what cannot be measured." GALILEO

5.1
SYSTEMS OF MEASURE

The daily rotations of the earth, the monthly changes of the moon, and our planet's yearly orbits about the sun provided some of the first units of measure. The day was divided into parts by sunrise, midday, sunset, and the tides, while the year was divided into seasons. One of the early attempts to measure the length of a year and its seasons was the prehistoric Stonehenge in southern England. This monument and some 40 or 50 others like it served

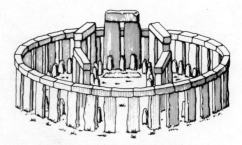

Stonehenge as it might have looked 4000 years ago

as astronomical observatories as well as temples for conducting religious and agricultural rituals. By adjusting movable stones it was possible to predict the arrival of the summer solstice, winter solstice, and occurrence of eclipses. Eventually, the sundial was invented to measure smaller periods of time, and it remained the principal method of measuring time until the nineteenth century. Today, atomic clocks measure time to within one ten-millionth of a second. (See page 400, Exercise 1.)

HISTORICAL DEVELOPMENT

It took many centuries before accurate units of measure were needed or developed. At first, rough-and-ready methods were sufficient. A distance might be described by a *day's walk* or *a number of paces.* The Chinese had an *uphill mile* and a *downhill mile,* with the uphill mile being the shorter distance. The traditional *Roman mile* was 1000 double paces of 5 feet each.

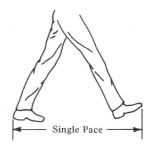

Single Pace

Many of the first units of measure were parts of the body. The early Babylonian and Egyptian records indicate that the *hand,* the *span,* the *foot,* and the *cubit* were all units of measure. The hand was used as a basic unit of measure by nearly all ancient civilizations and is still used to measure the heights of horses. The height of a horse is measured by the number of hand breadths from the ground to the horse's shoulders.

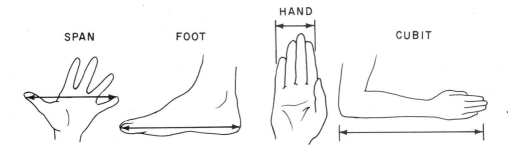

SPAN   FOOT   HAND   CUBIT

Evidence of other early units of measure still exists today. Seeds and stones were common units for measuring weight. The word "carat," which is a unit of weight for precious stones, was derived from the carob seeds of Mediterranean evergreen trees. Carat also expresses the fineness of a gold alloy. *Fourteen carat* means 14 parts of gold to 10 parts of alloy or that 14 out of 24 parts are pure gold. The *grain,* which was obtained from the average weight of grains of wheat, is another unit of weight used by jewelers. Until the past few years the *stone* was a common unit of weight in England and Canada. A newborn baby would weigh about one-half a stone.

In the English system of measure, shoe sizes are measured by an old unit called the *barleycorn.* The length of three barleycorns placed end to end is an inch, so shoe sizes vary by thirds of an inch. A size 7 shoe is 1 inch longer than a size 4 shoe. The *yard* and *fathom* are two more English system

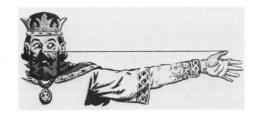

units for length which have their origins in body measurements. The yard was established by royal decree in the twelfth century by King Henry I, as the distance from his nose to his thumb. The fathom, a common unit for depths of water, is 2 yards or approximately the distance from fingertip to fingertip of a person's outstretched arms.

The *mouthful* is an ancient unit of measure for volume which was used by the Egyptians. It was also part of a unique English doubling system: 2 mouthfuls equals 1 jigger; 2 jiggers equals 1 jack; 2 jacks equals 1 jill; etc. The familiar "Jack and Jill went up the hill" nursery rhyme contains three units of volume, the *jack, jill,* and *pail,* and was a protest against the King of England for his taxation of the *jack* or

English System—Units of Volume	
2 mouthfuls = 1 jigger	
2 jiggers	= 1 jack (jackpot)
2 jacks	= 1 jill
2 jills	= 1 cup
2 cups	= 1 pint
2 pints	= 1 quart
2 quarts	= 1 pottle
2 pottles	= 1 gallon
2 gallons	= 1 pail

jackpot as it was called. The origin of the common expression, "hit the jackpot," which is used by gamblers, is due to the king's mishandling of the jackpot tax. The phrase, "broke his crown," in the nursery rhyme, refers to King Charles I and is prophetic. Not only was his crown broken but he lost his head in Britain's bourgeois revolution, not many years after the taxation of the jackpot.†

As societies evolved, weights and measures became more complex. Since most units of measure had developed independently of one another, it was difficult to change from one unit to another. As examples, in the English system there are 12 inches in a *foot;* 16 ounces in a *pound;* 2150.42 cubic inches in a *bushel;* and 5280 feet in a *mile.* By the fifteenth and sixteenth centuries the need for a simplified worldwide system of measurement had become apparent, and several proposals for a new system were made during this period.

METRIC SYSTEM

In 1790, in the midst of the French Revolution, the metric system was developed by the French Academy of Sciences. To create a system of "natural standards" they subdivided the distance from the equator to the North Pole to obtain the basic unit of length, the *meter.* The basic units for weight, the

Greek prefixes	*kilo	 1000
	hecto	 100
	deca	 10
Latin prefixes	*deci	. . . . 1/10
	*centi	. . . . 1/100
	*milli	. . . . 1/1000

gram, and for volume, the *liter,* are obtained from the meter. Smaller measures are obtained by dividing these basic units into 10, 100, and 1000 parts. Larger measures are 10, 100, and 1000 times greater than the basic units. These new measures are named by attaching prefixes to the names of the three basic units (see table above). In everyday nonscientific needs, only the above four prefixes that are marked with an asterisk are commonly used.

†A. Kline, *The World of Measurement* (New York: Simon and Schuster, 1975), pp. 32–39.

According to these prefixes, a *kilometer* is 1000 meters, a *centimeter* is one-hundredth of a meter, a *millimeter* is one-thousandth of a meter, and so on. There are more prefixes continuing in both directions for naming larger and smaller measures, with each new measure being 10 times greater or one-tenth as great as the previous one. This relationship between measures is a major advantage of the metric system. Converting from one measurement to another can be accomplished by multiplying or dividing by powers of 10, which, since we use base ten in our numeration system, is just a matter of moving decimal points.

The metric system was a radical change for the French people and met widespread resistance. Finally in 1837, the French government passed a law forbidding the use of any measures other than those of the new system. Steadily, other nations adopted the metric system. In 1866 the Congress of the United States enacted a law stating that it was lawful to employ the weights and measures of the metric system and that no contract or dealing could be found invalid because of the use of metric units. Today, less than 200 years after its creation, the metric system has been adopted by almost every country, with the last major nonmetric country, the United States, moving toward the metric system as its standard of measure. The Metric Conversion Act of 1975 set a national policy of voluntary conversion with no overall timetable.

Length   A meter is roughly the distance from the floor to the waist of an adult. If 10,000,000 meters were placed end to end, they would reach one-quarter of the distance around the world. This was the origin of the meter, but once it was established, an official meter bar of platinum was produced as the standard length. Today every country that uses the metric system has a reproduction of the French meter bar as its standard unit of length. In the United States a copy of the meter bar is maintained by the Bureau of Standards in Washington, D.C.

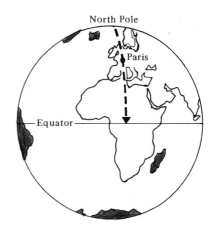

For everyday requirements, the centimeter, meter, and kilometer are sufficient for lengths and distances. The ruler shown here is marked off in *centimeters*. Our height, waist, hat size, and other such measurements are measured in centimeters.

A *meter* is 100 centimeters. The lengths of houses, racetracks, and swimming pools are measured in meters. Look at the door of the room you are in. If it is of average size, it is about 2 meters in height.

Distances between cities are measured by *kilometers*. A kilometer is shorter than a mile, approximately 3/5 as long. A person who walks 3 miles per hour will walk about 5 kilometers in an hour; and a speed limit of 30 miles per hour is equal to approximately 50 kilometers per hour.

Length	
kilometer	km
hectometer	hm
decameter	dkm
meter	m
decimeter	dm
centimeter	cm
millimeter	mm

Occasionally, you will need to measure something smaller than a centimeter. One-tenth of a centimeter is a *millimeter*. The thickness of a pencil lead is about 2 millimeters, and the width of a pencil is about 7 millimeters or 7/10 of a centimeter.

VOLUME

The most familiar unit of measure for volume or capacity is the *liter*. Compared to a quart the liter is just a "little bit" bigger. The capacities of fuel tanks, aquariums, and milk containers are measured in liters. For volumes that are less than a liter, such as some grocery store and drugstore items, the *milliliter* (1/1000 of a liter) is the common measure. Larger volumes, such as a community's reserve water supply, are measured by *kiloliters* (1000 liters).

Volume	
kiloliter	kℓ
hectoliter	hℓ
decaliter	dkℓ
liter	ℓ
deciliter	dℓ
centiliter	cℓ
milliliter	mℓ

One *milliliter* is 1/1000 of a liter and is the amount contained in a cube having a length, width, and height of 1 centimeter. A cube with these dimensions is called a *cubic centimeter*. Therefore, a liter is the amount contained in 1000 such cubes or in a container that has a length, width, and height of 10 centimeters. Converting from milliliters to liters, or from cubic centimeters to liters, is easily accomplished by dividing by 1000.

One cubic centimeter (1 cm by 1 cm by 1 cm) equals one milliliter.

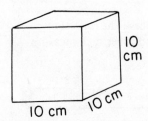

One cubic decimeter (10 cm by 10 cm by 10 cm) equals on liter.

WEIGHT

The basic unit of weight is the *gram*. This is a relatively small weight, approximately the weight of a medium-size paper clip or a dollar bill. The majority of grocery store items are measured in grams.

Heavier objects are measured in kilograms. A kilogram (about 2.2 pounds) is the common unit for people's weight. An average-size woman will weigh from 50 to 60 kilograms; and a weight-lifting record of 255 kilograms was set by Russia's Vasily Alexeyev in the 1976 Olympics.

Scientists distinguish between "weight" and "mass," with the difference in these two concepts being due to the effects of gravity. An object will weigh more at sea level than on top of a mountain, because the earth's gravity exerts a greater force on it. The same object in a spaceship would weigh practically nothing. Yet, in each of these three locations the amount of material in these objects hasn't changed! Because of this situation, it is necessary for scientific purposes to refer to the *mass* of an object as a measurement which doesn't change as it is moved further from the center of the earth. We can think of mass intuitively as the amount of matter which makes up the object. *Weight,* on the other hand, is the force that gravity exerts on the object; it varies with different locations. At sea level the mass and weight of an object are essentially equal. Since the variation in weight between sea level and our highest mountains is very small (.1% difference), the concept of weight is accurate enough for everyday measurements.

A *gram* is the weight of 1 cubic centimeter of water. (Technically, the water must be at a temperature of 4° Celsius because temperature affects volume.) This means that 1 milliliter of water weighs 1 gram. This simple relationship between weight, length, and volume gives the metric system its superiority over other systems of measure.

Weight

kilogram	kg	1000
hectogram	hg	100
decagram	dkg	10
gram	g	1
decigram	dg	1/10
centigram	cg	1/100
milligram	mg	1/1000

One cubic centimeter of water equals one milliliter of water and weighs one gram.

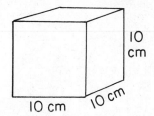

One cubic decimeter of water equals one liter of water and weighs one kilogram.

TEMPERATURE

In 1714 Gabriel Fahrenheit, a German instrument maker, invented the first mercury thermometer. The lowest temperature he was able to attain with a mixture of ice and salt he called 0°. He used the normal temperature of the human body, which he selected to be 96°, for the upper point of his scale. (We now know it is about 98.6° on the Fahrenheit scale.) On this scale of temperatures, water freezes at 32° and boils at 212°. In 1742, before the development of the metric system, the Swedish astronomer Anders Celsius devised a temperature scale by selecting 0 as the freezing point of water and 100 as the boiling point. He called his system the *centigrade* (100 grades) thermometer, but in recent years it has been changed to *Celsius* in his honor.

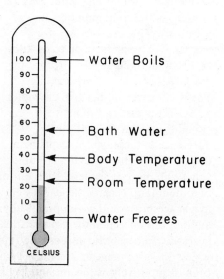

Heat is related to the motion of molecules and the faster their motion, the greater the heat. All movement stops at −273.15° Celsius. Lord Kelvin called this temperature "absolute zero" and devised the *Kelvin scale* which increase 1° for each increase of 1° Celsius. Thus, 273.15° on the Kelvin scale is 0° on the Celsius scale. Both the Celsius and Kelvin scales are part of the metric system. The Celsius scale is used for weather reports, cooking temperatures, and other day-to-day needs, and the Kelvin scale is used for particular scientific purposes.

PRECISION AND SMALL MEASUREMENTS

The objects in the picture on the next page are DNA and ribosome molecules in an active chromosome. Each DNA molecule is so small that 100,000 of them lined up side by side would fit into the thickness of this page. Measurements to this type of precision are possible with the transmission electron microscope. More recently, with the development of the field ion microscope, scientists have been able to view atoms that are ten times smaller than DNA molecules. The following scale shows eight decreasing measures from a centimeter down to an angstrom, each being one-tenth the size of the preceding one.

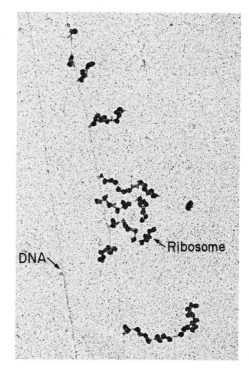

DNA and ribosome molecules

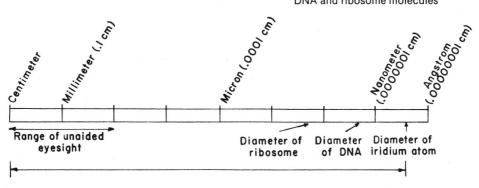

Range of microscopes

The amount of precision that is possible in taking measurements depends on the smallest unit of the measuring instrument. Using a centimeter ruler, this paper clip has a length of 3 centimeters, or we might say, just over 3 centimeters. With a ruler marked off in millimeters the length of the paper clip

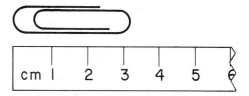

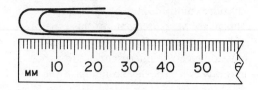

can be measured as about 32 millimeters or 3.2 centimeters. With finer
and finer measuring instruments we
might measure the length of this paper
clip to be 3.24, 3.241, or 3.2412
centimeters. It would never be possible,
however, to measure its length or the
length of any other object exactly.

If the smallest unit on the measuring instrument is a millimeter, then the
measurement can be approximated to the nearest millimeter. This means that the
measurement could be off by one-half of a millimeter, either too much or too little. In
general, the *amount of precision* is to within one-half of the smallest unit of measure
being used.

Conversely, if a measurement is given as 7.6 centimeters, we can assume that it was
measured to the nearest tenth of a centimeter and that it is closer to 7.6 centimeters
than to 7.5 centimeters or 7.7 centimeters. In other words, it is 7.6 ± 0.5 centimeters.
A measurement of 7.62 centimeters means that it has been obtained to the nearest one-
hundredth of a centimeter and may be off by as much as .005 centimeter. Sometimes
you will see a measurement such as 15.0 centimeters. This means that the measurement
is accurate to the nearest tenth of a centimeter and implies more precision than if it had
been given as 15 centimeters.

INTERNATIONAL SYSTEM
OF UNITS

The International System of Units is a modern version of the metric system which has
been established by international agreement. Officially abbreviated as SI, this system is
built on the metric units discussed previously, plus units for: time *(second);* electric
current *(ampere);* light intensity *(candela);* and the molecular weight of a substance
(mole). This system provides a logical and interconnected framework for all
measurements. To enable the type of precision needed in science today the meter is now
defined as the distance light travels in 1/299,792,458 of a second. A second of time is
defined as the deviation of 9,192,631,770 cycles of the radiation associated with the
cesium 133 atom (page 400, Exercise 1).

COMPUTER APPLICATIONS

Some of the wires in the microcircuit shown on the next page have a width of one micron,
or about 1/50th the thickness of hair. All measurements, whether large or small, are
approximate, that is, they are rounded off. Rounding off is accomplished in computer
programs by the integer function INT(X) (page 173). Here are three examples. Notice
that 4.7 and 8.5 are not rounded to the nearest whole number by the integer function.

INT(4.7) = 4   INT(3.4) = 3   INT(8.5) = 8

In order to have the integer function round off to the nearest whole number, we can add .5 to the number being rounded off. The following examples show that adding .5 ensures that a number will be rounded to the nearest integer.

$$INT(4.7 + .5) = INT(5.2) = 5$$
$$INT(3.4 + .5) = INT(3.9) = 3$$
$$INT(8.5 + .5) = INT(9) \quad = 9$$

This method of rounding off is used in Program 5.1A for converting a Fahrenheit temperature to a Celsius temperature. Follow through this program for $F = 75$.

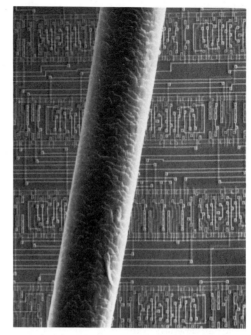

A scanning electron micrograph magnified 200 times shows a microcircuit and a human hair.

Program 5.1A

```
10   PRINT "TYPE A FAHRENHEIT TEMPERATURE. "
20   INPUT F
30   LET C = 5 * (F - 32) / 9
40   LET R =  INT (C + .5)
50   PRINT "TO THE NEAREST WHOLE NUMBER "F" DEGREES FAHRENHEIT
        EQUALS "R" DEGREES CELSIUS. "
60   END
```

Rounding Off to One Decimal Place Let's see how the integer function can be used to round off 4.72 to 4.7

$$INT(10 * 4.72)/10 = INT(47.2)/10 = 47/10 = 4.7$$

This formula, however, will not always work. For example, it will round of 4.79 to 4.7, rather than 4.8. Try it. To ensure that a number is rounded up when it should be we can use the technique of adding .5. The following equations show how 4.79 is rounded to 4.8.

$$INT(10 * 4.79 + .5)/10$$
$$= INT(47.9 + .5)/10$$
$$= INT(48.4)/10$$
$$= 48/10$$
$$= 4.8$$

Program 5.1B converts miles to kilometers and rounds each kilometer to one decimal place.

Program 5.1B

```
10  PRINT "THIS PROGRAM CONVERTS MILES TO KILOMETERS. TYPE THE
     NUMBER OF MILES."
20  INPUT M
30  LET R = 1.61 * M
40  LET K =  INT (10 * R + .5) / 10
50  PRINT M" MILES IS APPROXIMATELY EQUAL TO "K" KILOMETERS."
60  END
```

SUPPLEMENT (Activity Book)

Activity Set 5.1 Measuring with Metric Units

Just for Fun: Metric Games

Exercise Set 5.1: Applications and Skills

1. While almost all educators agree that we should not teach the metric system by converting back and forth from English units to metric units, it is sometimes helpful to make rough comparisons between the two systems.

   ★ a. Approximately what does the 55 mph speed limit equal in kilometers per hour (kph)?

   b. Jersey City to New York is 24 miles. Approximately what is this distance in kilometers?

   ★ c. It has been suggested that the speed limit be increased so that the new limit will be 100 kph. Approximately how many miles per hour is this equal to?

"All right—now convert the whole thing to metric."

2. Write the most appropriate metric units for determining the following measurements. Fill in the crossword puzzle with your answers.

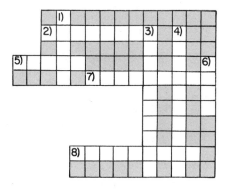

★ *Across*

2. Weight of a truck
5. Length of a building
7. Volume of a city water supply
8. Length of a river

Down

1. Volume of a gasoline tank
3. Volume of a perfume bottle
4. Width of a television screen
6. Weight of a 50-cent coin

3. Complete the statements and write your answers in the cross-number puzzle, placing one digit in each square. (Statements 4 and 9 Across and 6 and 7 Down involve volumes of water.)

Across

1. 16.5 cm = ___ mm
4. 3.15 ℓ weighs ___ g
5. .12 ℓ = ___ mℓ
8. ___ kg = 92,000 g
9. ___ mℓ weighs 7.920 kg
10. ___ m = 5.55 km

Down

2. 632,000 ℓ = ___ kℓ
3. 4.5 km = ___ m
6. ___ g is the weight of 432 mℓ
7. ___ kg is the weight of 190 ℓ
8. ___ mg = .9 g
9. .75 m = ___ cm

★ 4. Each of the following grocery store items is measured either by weight or by volume. Connect each item to the appropriate measurement.

946 mℓ   4.536 kg   567 g   384 mℓ   40 g   59 mℓ

5. A measurement to within a certain unit may be off by as much as plus or minus one-half of that unit. If a can of pineapple juice, which is labeled 1.32 ℓ, was measured to the nearest hundredth of a liter, its volume is greater than the minimum of 1.315 ℓ and less than the maximum of 1.325 ℓ. Find the minimum and maximum numbers associated with each of the following measurements.

★   a. A two-speed heavy-duty washing machine weighed 112 kg, to the nearest kg.

     b. The patient's temperature at 6:00 P.M. was 36.2°C, to the nearest tenth of a degree.

★   c. One of the speakers with a stereo 8-track player/recorder was found to have a width of 48.3 cm, to the nearest tenth of a centimeter.

     d. A three-day-old baby weighed 3.46 kg, to the nearest hundredth of a kilogram.

6. *Consumer Mathematics*

★   a. A shopper purchased the following items: tomatoes, 754 g; soup, 772 g; potatoes, 3.45 kg; sugar, 4.62 kg; raisins, 425 g; vegetable shortening, 1.361 kg; and baking powder, 218 g. What is the total weight of this purchase in kilograms?

     b. Material that is 1 m wide and 120 cm long is required to make a single window curtain. There are two curtains per window and six windows. If the curtain material comes in rolls of 1-m widths, how many meters of length will be needed to make curtains for all six windows?

★   c. The following amounts of gasoline have been charged on a credit card: 38.2 ℓ; 26.8 ℓ; 54.3 ℓ; 44.7 ℓ; and 34 ℓ. At a price of 32 cents per liter, what is the cost of this gasoline?

d. A car owner has his/her tank filled and records the odometer reading as 14368.7 (km). After a trip in the country, the tank takes 34.5 ℓ to fill and the odometer reads 14651.6. How many kilometers to the liter is this car getting?

★ e. A 24-kg bag of birdseed is priced at $16.88. If 75 g of this feed are put in a bird-feeder each day, how many days before the bag of seed will run out? Rounded off to the nearest penny, how much does it cost to feed the birds each day?

7. A fruit punch calls for these ingredients: 3.5 ℓ of unsweetened pineapple juice; 400 mℓ of orange juice; 300 mℓ of lemon juice; 4 ℓ of ginger ale; 2.5 ℓ of soda water; 500 mℓ of mashed strawberries; and a base of sugar, mint leaves and water which has a total volume of 800 mℓ.

★ a. What is the total amount of this punch in liters?

b. For a party of 30 people, how many milliliters of punch will there be per person?

c. This punch was used at a fair and each drink of 80 mℓ cost 25 cents. What was the profit on the sale of this punch if the ingredients cost $12.50?

8. Prescriptions for the antibiotic garamycin vary from 20 mg for a child to 80 mg for an adult. Garamycin is contained in a vial which has a volume of 2 cm³ (2 mℓ). The garamycin in each vial weighs 80 mg.

★ a. How many cubic centimeters of garamycin are needed for 12 injections of 24 mg each?

b. How many injections of 60 mg each can be obtained from 24 vials?

9. Roof de-icers prevent ice dams on roofs and gutter pipes. An electric heating cable is clipped to the edge of the roof in a sawtooth pattern. In the following questions, assume that this pattern is to run along the edges of a roof with a total length of 28 m.

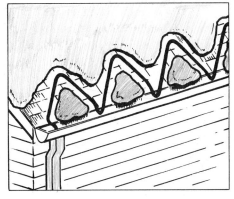

★ a. How many meters of heating cable will be needed for the edges of the roof, if each 1 m of roof-edge requires 2 m of cable?

b. There are two gutter pipes which run along the edges of the roof. Each has a length of 14 m. There are two downspouts, one from each gutter pipe to the ground. Each downspout has a length of 3.2 m. How many meters of cable will be required to go along the gutter pipes and the downspouts?

★ c. The heating cable sells for $1.20 per meter. What is the cost of the total cable?

★ 10. The National Assessment of Educational Progress (NAEP) administered the following question to 17-year-olds in both 1969 and 1973. In 1969, 36% of 17-year-olds attending high school chose the correct answer, but only 27% could do so in 1973. Select the correct answer from 1 to 5.

> *Suppose that a rubber balloon filled with air does not leak and that it is taken from earth to the moon. One can be sure that, on the moon, the balloon will have the same*

1. *size as on earth.*
2. *mass as on earth.*
3. *weight as on earth.*

4. *rate of fall as on earth.*
5. *ability to float as on earth.*
6. *I don't know.*

11. The first unit of measure that was recorded by history is the *cubit*. This is the distance from elbow to fingertips and was used more than 4000 years ago by the Egyptians and Babylonians. An ancient Egyptian cubit measuring 52.5 cm is preserved in the Louvre in Paris. Here is a picture of this cubit and its subdivisions.

Ancient Egyptian cubit

a. How does the length of the Egyptian cubit compare with your cubit?

★ b. A large dark object under the ice on Mount Ararat in Turkey may be Noah's Ark. The biblical dimensions of Noah's Ark from the sixth chapter of Genesis are contained in this table. Convert these measures to the nearest meter using the length of the Egyptian cubit.

	Cubits	Meters
Length	300	
Breadth	50	
Height	30	

12. The Celsius and Fahrenheit temperature scales are related by the following formulas:

$$C = 5(F - 32)/9 \quad \text{and} \quad F = 9C/5 + 32$$

Write in the missing two temperatures for each thermometer to the nearest tenth of a degree.

★ a. Highest recorded temperature, Libya, 1922.

b. At this body temperature, see a doctor.

★ c. At this temperature, check your car's antifreeze.

d. Lowest recorded temperature, Antartica, 1960.

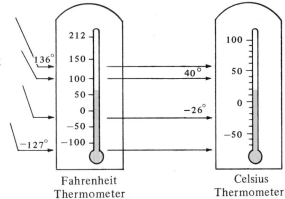

Fahrenheit
Thermometer

Celsius
Thermometer

13. On october 20, 1983, the General Conference on Weights and Measures used the speed of light to define the length of a meter (page 272).

a. Use this definition to determine the speed of light in kilometers per second.

b. In England in 1956, the speed of light was measured as 299,792.4 ± .11 kilometers. Is the speed of light that was used by the General Conference on Weights and Measures within this range?

Exercise Set 5.1: Problem Solving

1. *Bouncing Ball:* A special ball which is dropped perpendicular to the floor rebounds to half its previous height on each bounce until the height of its bounce is less than 1 centimeter. On its fifth bounce it reaches a height of 6 centimeters. What is the total distance the ball has traveled when it hits the floor after its fifth bounce?

a. *Understanding the Problem:* A diagram will help to visualize the problem. Each bounce can be represented by a vertical line which is half the height of the preceding line. Here is a diagram for the original distance the ball is dropped and the height of its first bounce. Explain why the height of each bounce must be doubled when computing the total distance the ball travels. Draw the complete diagram for this problem.

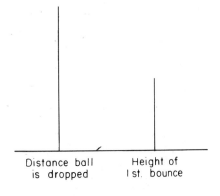

Distance ball
is dropped

Height of
1st. bounce

★ b. *Devising a Plan:* Working backward is a natural strategy for solving this problem. What is the height of the fourth bounce? What is the total distance traveled during the fourth bounce?

c. *Carrying Out the Plan:* Continue working backward to get the height of each bounce and the original distance the ball is dropped. What is the total distance the ball has traveled?

★ d. *Looking Back:* In general, as long as the height of the ball's bounce is not less than 1 centimeter, the total distance the ball will travel is the distance it is dropped plus the sum of 2 times the height of each bounce. Will this general statement continue to be true if the ball does not rebound to half its previous height on each bounce?

★ 2. *The Flea and the Trains:* Train A and train B are on the same track and headed toward each other. Both trains are traveling at 75 kilometers per hour. When the trains are 300 kilometers apart, a flea flies from the front of train A to the front of train B, then back to the front of train A, etc., returning back and forth until finally being crushed by the colliding trains. What is the length of time that the flea was in flight between the two trains?

3. *Brass Weights:* This standard set of 11 brass metric weights can be used with a balance scale (page 41) to weigh any object with a whole number weight from 1 to 1600 grams. Explain how the weight of a 917 gram object can be determined.

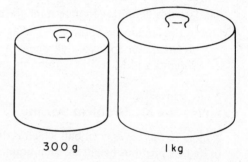

300 g   l kg

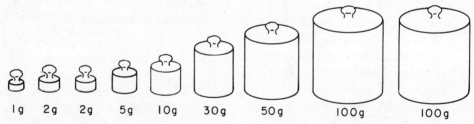

l g   2 g   2 g   5 g   10 g   30 g   50 g   100 g   100 g

★ 4. *Bookworm Puzzle:* There are four books which are numbered 1 through 4 from left to right on a shelf. Each book has 300 pages. The thickness of 300 pages is 2 centimeters and the thickness of each cover is 3 millimeters. If a worm begins at page 1 of book 1, and eats to page 300 of book 4, what is the shortest distance it could have traveled?

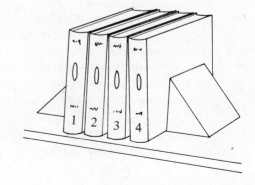

Exercise Set 5.1: Computers

1. Determine the values of the integer function.
   a. INT(14.2 + .5)   ★   b. INT(35.82 + .5)   c. INT(7.5 + .5)
   d. INT(10 * 16.37 + .5)/10   ★   e. INT(10 * 9.843 + .5)/10

2. Determine the printouts for Programs 5.1A and 5.1B on pages 273 and 274 for the following input values.
   a. F = 85 (Fahrenheit temperature of a warm day)
   ★   b. M = 412 (distance in miles from San Francisco to Los Angeles)
   ★   c. F = 98.6 (normal body temperature in Fahrenheit)
   ★   d. M = 24,874 (distance in miles around the earth)

3. Track events traditionally recorded in yards and miles are being changed to meters as shown in the table below. This change is the result of an international rule which went into effect on June 1, 1975. Using the fact that a yard is about 91.5 cm, write a computer program to convert yards to meters (rounded off to one decimal place). Use this program to determine whether each of the following metric measurements is longer or shorter than the English measurement and compute the difference in meters.

Ready or not—metric system is coming.

LOS ANGELES (AP)—The mile run and the 100-yard dash, two of track's glamor races, may soon join the horse-drawn carriage and the five-cent beer as relics of days gone by.

The United States soon will be forced to switch from measuring track meets in yards to measuring them in meters. The Amateur Athletic Union and the National Collegiate Athletic Association have long fought such a switch, but both agree it is becoming mandatory.

Under an international rule which went into effect on June 1, an athlete who runs a race in yards may not qualify for the Olympics, whether he sets a record or not.

The International Amateur Athletic Fed-

	Old Race	New Race
★ a.	100 yd	100 m
b.	220 yd	200 m
★ c.	440 yd	400 m
d.	880 yd	800 m
e.	1 mi	1500 m

Federal Reserve Bank in Minneapolis

5.2
AREA AND
PERIMETER

The design of the Federal Reserve Bank in Minneapolis is based on the mathematical curve called a *catenary*. This is the curve commonly used in the construction of bridges and is formed by suspending cables from two towers. In fact, this building can be thought of as a bridge that is 10 stories deep. There was no prototype for its structural design. Each floor has an unobstructed area of 60 by 275 feet. No occupied floors without internal columns have ever spanned such a length before.

This bank was designed to fulfill some unusual zoning restrictions. One of these was that the "coverage," or ground area occupied by the building, could be only 2.5% of the area of the city block which the bank was to be built on. To satisfy this condition, the bank is supported by two towers and its lower floor is 20 feet above the plaza. The coverage is the small amount of area which the two towers are built on. A 2.5-acre plaza runs under the building and is entirely for the public.

There are two catenaries of steel cable, one on the front side and one on the back side of the building. These cables support the weight of the bank's floors. One of these catenaries and the windows above and below can be seen in the above photograph. By counting the rectangles and parts of a rectangle which are formed by the windows, the areas above and below the catenary can be estimated (page 294).

RELATING AREA AND PERIMETER

To measure the sizes of plots of land, panes of glass, floors, walls, and other such surfaces, we need a new type of unit—one which can be used to cover a surface. Theoretically, this unit can have any shape: rectangular, triangular, etc. The number of units needed to cover a surface is called the *area*. Squares have been found to be the most convenient shape for measuring area. The area of this figure is 4 square units because it can be covered by 2 squares and 4 half-squares. Covering a region or surface with unit squares is the basic concept of area, and yet, it is often poorly understood. A Michigan State assessment found that fewer than half of the seventh graders knew the area of this figure. About 20 percent of them thought the area was 6.

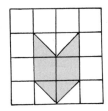

Another measure associated with a region is its *perimeter,* the length of its boundary. The perimeter of the figure on the right is 23 centimeters.

Intuitively, it seems as though the area of a region should depend on its perimeter. For example, if one person requires more fence to close in a piece of land than another person, it is tempting to assume the first person has the greater amount of land. However, this is not necessarily true. Figure (a) has an area of 4 square units and a perimeter of 8 centimeters, while Figure (b) has the same area but a perimeter of 10 centimeters.

(a)

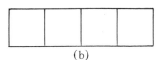

(b)

It is also possible for two figures to have the same perimeter but different areas. The perimeters of Figures (c) and (d) are each 10 centimeters, but their areas differ by 1 square.

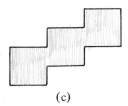

(c)

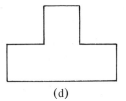

(d)

UNITS FOR AREA

The earliest units for measuring area were associated with agriculture. The amount of land that could be plowed in a day with the aid of a team of oxen was called an *acre*. In Germany, a scheffel was a volume of seed, and the amount of land that could be sown with this volume of seed became known as a *scheffel of land.* Just as for units of length, volume, and weight, more carefully defined units of area were eventually adopted.

In the metric system there is a square unit for each unit of length. The common area units are the square kilometer (km²), square meter (m²), square centimeter (cm²), and square millimeter (mm²). These units are related to one another. *The square centimeter,* for example, is equal to *100 square millimeters.*

The *are* (pronounced "air") is a metric unit for measuring the areas of house lots, gardens, and other such "medium-size" regions. The *are* is equal to the area of a square whose sides measure 10 meters each. The *hectare* is the area of a square whose sides measure 100 meters each. This is approximately the area of two football fields, including end zones.

RECTANGLES

Rectangles have right angles and pairs of opposite parallel sides, so that unit squares fit onto them quite easily. This rectangle can be covered by 24 whole squares and 6 half-squares. Its area is 27 square centimeters. This area can be obtained from the product 6 × 4.5, because there are four and one-half squares in each of six columns. In general, the *area of a rectangle* is the product of its length times its width.

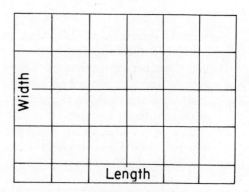

Area = length × width   ($A = lw$)

For a given perimeter, the shape of the rectangle affects its area. The same knotted piece of string in the following photos has been formed into three different rectangular shapes. By using the lengths and widths of each rectangle, you will see that their perimeters are 36 centimeters. Yet, the areas decrease from 80 square centimeters (8 × 10), to 72 square centimeters (6 × 12), to 32 square centimeters (16 × 2), as the shape of the rectangle is changed. If we continue to decrease the width of the rectangle, its area can be made as small as we please, even though the perimeter remains 36 centimeters.

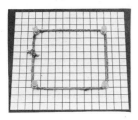

8 cm by 10 cm

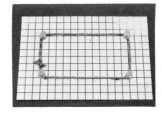

6 cm by 12 cm

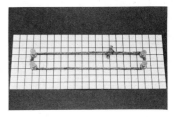

16 cm by 2 cm

PARALLELOGRAMS, TRIANGLES, AND POLYGONS

Fitting unit squares onto a figure is a good way to acquire an understanding of the concept of area. However, actually placing squares on a region is usually difficult because of the boundary, as in the case of the slanted sides of the parallelogram shown here.

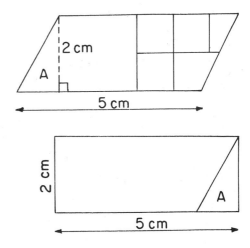

One of the basic principles in finding area is that a region can be cut into parts and reassembled without changing its area. This principle is useful in developing a formula for the area of a parallelogram. In the figure shown above, part A has been moved from the left end of the parallelogram to the right end in order to form this rectangle. The parallelogram and the rectangle both have an area of 10 square centimeters. The length (5 centimeters) and width (2 centimeters) of the rectangle are equal to the *base* and *height* (or *altitude*) of the parallelogram, respectively. This suggests the following formula for the *area of a parallelogram*. The height in this formula is the perpendicular distance between the upper and lower bases of the parallelogram.

$$\text{Area} = \text{base} \times \text{height} \quad (A = bh)$$

As in the case of a rectangle, for a given perimeter the area of a parallelogram depends on its shape. The two parallelograms in the following photo are constructed from linkages. Both have the same perimeter, but as the parallelogram is skewed more to the right, its height and area decrease. The area of the first parallelogram is approximately 66 square centimeters and the area of the second one is 40 square centimeters. The height, and consequently the area, of the parallelogram can be made arbitrarily small by further skewing the linkage, while the perimeter will stay constant.

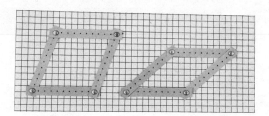

The area of the triangle in this sketch is at least 4 square centimeters. Since the unit squares do not conveniently fit inside its boundary, we will use a different approach for finding its area. By placing two copies of the triangle side by side, they form a parallelogram, as shown in the next figure.

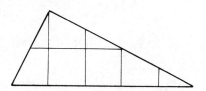

If the length of the *base* of the triangle is *b* and its *height* (or *altitude*) is *h,* the area of the parallelogram is *b* × *h*. Since the parallelogram has twice the area of the triangle, the triangle's area is one-half of the base times the height. In general, this is the formula for the area of any triangle.

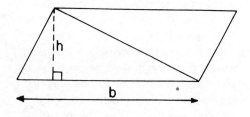

$$\text{Area} = \frac{1}{2} \times \text{base} \times \text{height} \qquad (A = \frac{1}{2} \times bh)$$

Since the preceding triangle has a base of 5 centimeters and a height of 2 centimeters, its area is 5 square centimeters. Each of the three sides of a triangle may be considered as the base, and each side has its corresponding altitude. Triangle *ABC* is drawn below in two positions showing two sets of bases and altitudes. One of these altitudes falls outside the triangle. The area of this triangle can be found by computing 1/2 × 5.8 × 2.0, or 1/2 × 4 × 2.9. In both cases, the area is the same.

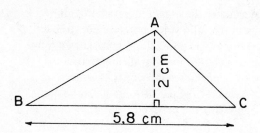

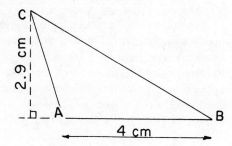

Another basic principle for finding area states that a region can be subdivided into smaller regions, and that the area of the original region is the sum of the areas of the smaller ones. Using this principle, the area of any polygon can be found by subdividing it into rectangles, triangles, or other regions whose areas can be found easily. This hexagonal region has been subdivided into one rectangle and three triangles. The area of the rectangle is 8 square centimeters, and the triangles have areas of 6, 1, and .5 square centimeters. The sum of these areas is 15.5, the approximate area of the hexagon.

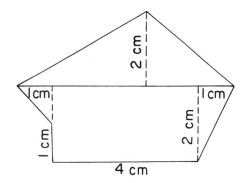

IRREGULAR SHAPES

Sometimes it is necessary to find the areas of irregular or nonpolygonal shapes. Botanists, for example, compute the areas of leaves to determine the amount of water they lose for each square centimeter of surface area. The water you see on a leaf in the early morning is often not dew condensed from the air, but water which is given off by the leaf. Some of the water which is supplied to the leaf by its system of tiny veins is lost through small openings called *stomates*. The number of stomates may range from a few thousand to over a hundred thousand for each square centimeter of surface.

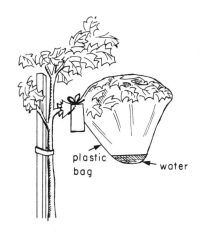

plastic bag   water

One method of finding the area of a leaf is to trace it on graph paper or place it under a transparent grid. In the following example there are 18 large squares (1 centimeter by 1 centimeter) which fall completely inside the boundary of the leaf. Thus, the leaf has a *lower bound* of at least 18 square centimeters. The total number of large squares which contain any part of the leaf is 37. Thus, the leaf has an *upper bound* of at most 37 square centimeters. For each square which intersects the boundary, the area of the square which falls inside the boundary can be estimated. These estimates can be made more accurate by subdividing the boundary squares into smaller parts. In this figure each boundary square has been subdivided into 4 smaller squares. There are 20 of these smaller squares which fall completely inside the boundary of the leaf. These 20 squares have an area of 5 square centimeters. Thus, the leaf has a new lower bound of 18 + 5, or 23 square centimeters. There are 32 of these small squares which contain some part of the boundary. These squares have an area of 8 square centimeters. Thus, the upper bound of the area of the leaf is not more than 23 + 8 or 31

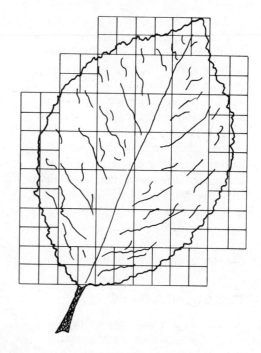

square centimeters. This process of subdividing squares into smaller squares can be continued to give lower and upper bounds which are closer to the actual area of the leaf.

We have found that the area of this leaf is between 23 and 31 square centimeters. Let's approximate its area by counting the small squares on the boundary that are partially covered by the leaf. If the leaf covers half or more than half of a small square, we will count the square. There appear to be 16 such squares. The area of these squares is 4 square centimeters. Therefore, the area of this leaf is approximately 23 + 4, or 27 square centimeters. If this leaf loses 2 milliliters of water in 24 hours, it will lose 2/27 or .074 milliliter of water per square centimeter.

Engineers and scientists sometimes use a *planimeter* for measuring area. To find the area of a figure, its boundary is traced by one arm of the planimeter while the other stays fixed. As the arm traces the boundary, a disc moves along a scale whose readings are converted into area. It is surprising that a planimeter will determine area by tracing out the perimeter of a figure,

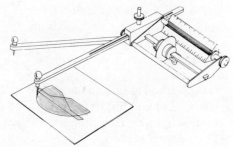

since we have seen that two figures with the same perimeter may have different areas. There is also an instrument, called a *map tracer,* for measuring the length of a curve. This device can be adjusted for different map scales.

Approximating the areas of regions is often easier than approximating the lengths of curves or the perimeters of figures. For a given region, no matter how irregular its boundary, we can always estimate its area fairly accurately by a grid. Two lines, however, can lie fairly close to each other and still differ considerably in length. For example, the following straight line is a poor approximation for the zigzag line. The line has a length of 9 centimeters. The zigzag path has a length of 12.6 centimeters, which is 40 percent longer.

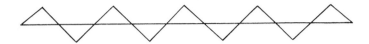

One method of approximating the length of a simple closed curve is to inscribe a polygon in its interior. This method can also be used to approximate the area of the interior by finding the area of the polygon. More accuracy can be obtained by increasing the number of sides of the polygon and drawing the lines closer to the curve.

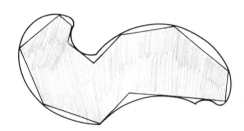

CIRCUMFERENCES OF CIRCLES

Circles are part of our natural environment. The sun, moon, flowers, whirlpools, and cross sections of trees all contain circular shapes. The nearly perfect concentric circles on this thin section of a natural pearl were formed by many layers of growth. The circular beach stones shown on page 290 were tumbled and shaped by the ocean's tides.

The circle was among the most powerful of the early symbols or signs of magic. It had neither beginning nor end and represented eternity. Making a circle with the thumb and index finger is an old good luck gesture that is still used today to mean "OK."

Concentric circles of a section of natural pearl

There is something deceptive about trying to estimate the *circumference* (perimeter) of a circle. For example, you may be surprised to know that if you place a piece of string around the edge of the pearl in the photo on page 289, the length of the string will be greater than the distance from the top line on this page to the bottom line.

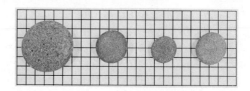

The tendency to underestimate the circumference of a circle can be dramatized by using objects such as a drinking glass, jar, or can and asking someone to compare the circumference with its height. A tennis ball can is excellent for this experiment because its height appears to be greater than its circumference. How would you answer this question?

If a piece of string just fits around a tennis ball can, as shown in this picture, how would the length of the string compare to the height of the can?

A very common but wrong answer (see photo on the right) is that the string would reach about two-thirds of the way up the can, which shows that some people tend to estimate the circumference of a circle by doubling its diameter. Actually, the distance around a circle is a little more than three times greater than the distance across. To illustrate this relationship, cut a strip of paper or a piece of string which will reach around a circular object and then fold the paper (or string) into three equal parts. The folded strip will be a little longer than the diameter of the circle. This relationship between the diameter (*d*) or radius (*r*) of a circle, and its circumference (*C*), is expressed by the following formulas:

$$C = \pi \times d \quad \text{or} \quad C = \pi \times 2r,$$

where pi (π) is approximately 3.14 or 22/7. For many estimation purposes, a value of pi equal to 3 is accurate enough.

AREAS OF CIRCLES

The area of a circle can be approximated reasonably well by the area of an inscribed polygon. As you can see from this drawing, the area of the dodecagon is just a little less than the area of the circle. The triangle formed by one side of the dodecagon and two radii of the circle has an area of $1/2 \times bh$. Since the dodecagon can be subdivided into 12 such triangles, its area is $1/2 \times 12bh$.

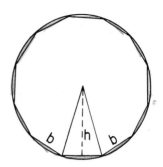

If we were to inscribe a polygon with a greater number of sides, its area would be even closer to the area of the circle. Notice that the perimeter of the dodecagon, $12b$, is very close to the circumference, C, of the circle. Also the altitude, h, of the triangle is close to the radius, r, of the circle. If $12b$ is replaced by C, and h is replaced by r, then $1/2 \times 12bh$ will be approximately equal to $1/2 \times Cr$. Since the area of the inscribed polygon can be made arbitrarily close to the area of the circle, this suggests that the area of the circle is $1/2 \times Cr$. Since the circumference is equal to $2\pi r$,

$$\frac{1}{2} \times Cr = \frac{1}{2} \times 2\pi r \times r = \pi r^2$$

In general, for any circle of radius r, its area A is π times the square of the radius.

$$\text{Area} = \pi \times (\text{radius})^2 \qquad (A = \pi r^2)$$

PRECISION

The sides of this pentagon have been measured to within .1 centimeter. This means that the side whose length is labeled 4.6 cm is accurate to within .05 centimeter. That is, its true length is between 4.55 and 4.65 centimeters. The side labeled 2.0 cm could have an actual length as small as 1.95 centimeters or as big as 2.05 centimeters. Using minimum and maximum numbers, such as these, an interval can be

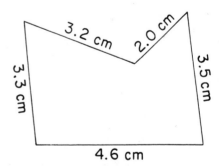

determined within which the perimeter of this pentagon is contained. The minimum and maximum numbers for the five measurements are listed on the next page. The perimeter of the pentagon is greater than the sum of the minimum numbers and less than the sum of the maximum numbers.

Minimum	Given Measurements	Maximum
3.25 cm	3.3 cm	3.35 cm
4.55 cm	4.6 cm	4.65 cm
3.45 cm	3.5 cm	3.55 cm
1.95 cm	2.0 cm	2.05 cm
+ 3.15 cm	+ 3.2 cm	+ 3.25 cm
16.35 cm	16.6 cm	16.85 cm

The sum of the given lengths of the pentagon is 16.6 centimeters. The actual length of the perimeter could be as small as 16.35 centimeters or as big as 16.85 centimeters.

The amount of precision in computing with measurements depends on the precision of the original measurements. The previous example shows that precision cannot be increased by computation. The sides of the pentagon were measured to within .1 centimeter, but the perimeter, 16.6 centimeters, is accurate to within only .25 centimeter.

A range or interval of error can also be computed for products of measurements. The length and width of this rectangle were measured to within .1 centimeter. These dimensions could be as small as 2.75 and 5.25 centimeters, respectively, or as big as 2.85 and 5.35 centimeters, respectively. The interval within which the area of this rectangle falls is determined by computing the area for these minimum numbers and then the area for the maximum numbers.

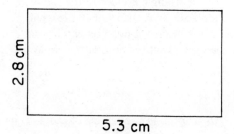

2.8 cm

5.3 cm

Minimum	Given Measurements	Maximum
2.7 5	2.8	2.8 5
× 5.2 5	× 5.3	× 5.3 5
1 3 7 5	8 4	1 4 2 5
5 5 0	1 4 0	8 5 5
1 3 7 5	1 4.8 4	1 4 2 5
1 4.4 3 7 5		1 5.2 4 7 5

The area of the rectangle, according to the given measurements (length = 5.3 centimeters and width = 2.8 centimeters), is 14.84 square centimeters. The actual area of this rectangle could be as small as 14.4375 square centimeters or as big as 15.2475 square centimeters.

Computer Applications

We have seen that computers can sometimes be programmed to aid in a systematic search for solutions to a problem. Consider this area and perimeter problem.

Problem A small rectangular pasture is to be enclosed with 96 meters of new fence. One side of the enclosure will be an existing fence. What are the dimensions of the pasture with the largest possible area?

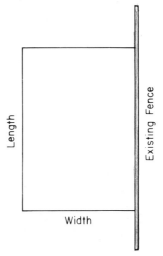

The following program computes the areas of rectangles whose length L plus two times the width W is equal to 96. The commas in lines 10 and 50 cause the printout to be spaced and lined up in three columns. The width of the rectangles varies from 1 to 47. What happens if we allow the width to be 48?

Program 5.2A

```
10   PRINT "WIDTH", "LENGTH", "AREA"
20   FOR W = 1 TO 47
30   LET L = 96 - 2 * W
40   LET A = L * W
50   PRINT W, L, A
60   NEXT W
70   END
```

Here are the first few lines of the printout for this program. Notice that as the width increases the area also increases. Will this continue to be true until W is equal to 47?

WIDTH	LENGTH	AREA
1	94	94
2	92	184
3	90	270
4	88	352
5	86	430
6	84	504
7	82	574
8	80	640
9	78	702
10	76	760

SUPPLEMENT (Activity Book)

Activity Set 5.2 Areas on Geoboards
Just for Fun: Pentominoes

Exercise Set 5.2: Applications and Skills

1. This is the skeletal structure of the Minneapolis Federal Reserve Bank. Its 10 floors and the vertical beams running down the front of the bank, partition the area above and below the catenary into rectangles. The dimensions of these rectangles are approximately 2 m by 4 m.

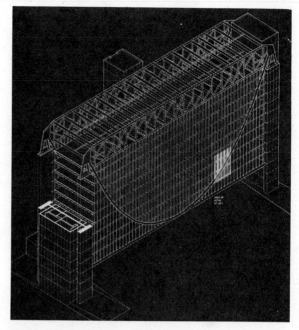

Skeletal structure of the Minneapolis Federal Reserve Bank

★ a. Use this diagram of the bank to approximate the area of the front of the bank which lies under the catenary by counting rectangles and parts of rectangles.

★ b. Use the rectangles to find the width and height of the bank's 10 stories. What is the area enclosed by these dimensions?

c. Use the areas in parts **a** and **b** to approximate the area above the catenary (not including the portion above the 10 stories).

2. The following question is taken from a mathematics test which was given to 9-year-olds by the National Assessment of Educational Progress (NAEP).

a. Which of the following figures has the same area as the 4 by 4 square?

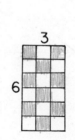
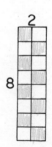

★ b. Only 44 percent chose the 8 by 2 rectangle, and almost as many selected the 3 by 5 rectangle. The selection of the 3 by 5 rectangle may indicate confusion over which two concepts of measurement? Explain why they might have made this choice.

3. *Consumer Mathematics*

★   a. A store sells two types of Christmas paper: Type A has four rolls per package and costs $2.99, and each roll is 75 cm by 150 cm; Type B has a single roll, costs $3.19, and is 88 cm by 500 cm. In which case will you get more paper for your money?

   b. An all-purpose rug costs $27.50 per square meter. What is the cost of a rug whose dimensions are 360 cm by 400 cm?

★   c. Tru-site glass for picture frames sells at the rate of $20.00 per square meter. What is the total cost for the glass in these two picture frames: Frame A, 58 cm by 30 cm; and Frame B, 40 cm by 60 cm?

   d. The length of a kitchen cupboard is 3.5 m. There are three shelves in the cupboard, each of width 30 cm. How many rolls of shelf paper will be needed to cover these shelves, if each roll is 30 cm by 3 m?

★   e. The instructions on a bag of lawn fertilizer recommend that 35 g of fertilizer be used for each square meter of lawn. How many square meters of lawn can be fertilized with a 50-kg bag?

4. Each dimension of a 30 cm by 58 cm pane of antique stained glass was measured to the nearest centimeter.

★   a. According to these measurements, what is the area in square centimeters?

   b. Assuming that these measurements could be off by as much as plus or minus .5 centimeter, what are the minimum numbers for the length and width of this glass?

★   c. What is the area of this glass for the minimum numbers in part **b**?

   d. If this glass sells for 15 cents per square centimeter, how much will it cost for the area in part **a**?

   e. If this glass has the minimum dimensions which were found in part **b**, how much money would you lose on this purchase?

5. Some humidity in homes is necessary for comfort, but too much can cause mold and peeling paint. Paint-destroying moisture can come from walls, crawl spaces, and attics.

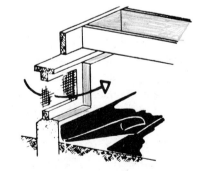

★   a. An attic should have 900 cm^2 of ventilation for each 27 m^2 of area. How many square centimeters of ventilation is needed for a 4 m by 12 m attic area?

   b. A crawl space should have 900 cm^2 of ventilation for each 27 m^2 of area, plus 1800 cm^2 for each 30 m of perimeter around the crawl space. How many square centimeters of ventilation is needed for a 10 m by 15 m crawl space?

6. Compute the areas of each of these figures in square millimeters and square centimeters.

★ a.

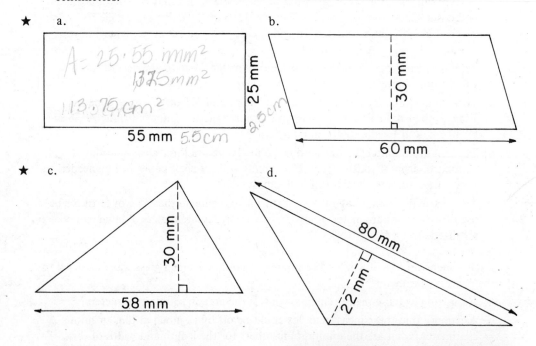

A= 25·55 mm²
      1375 mm²
113.75 cm²

25 mm

55 mm 5.5 cm

2.5 cm

b.

30 mm

60 mm

★ c.

30 mm

58 mm

d.

80 mm

22 mm

7. The 70-story cylindrical building shown here is the Peachtree Plaza Hotel in Atlanta, Georgia. It contains a 7-story central court with a half-acre lake and over 100 trees. The diameter of this building is 35.36 m (116 ft). According to its architect, John Portman, cylindrical walls were chosen rather than the more common rectangular walls because a circle encloses more area with less perimeter than any other shape.

★ a. What is the area of its cross section (that is, the area of any floor)?

b. What is the perimeter to the nearest meter of a square that encloses the same area as you found in part **a**? (*Hint:* First find the square root of the area in part **a**.)

Peachtree Plaza Hotel in Atlanta, Georgia

★   c. What is the perimeter to the nearest meter of the Peachtree Plaza Hotel?

       d. How many more meters is the perimeter of the square in part **b** than the perimeter of the hotel?

★   e. The Peachtree Plaza Hotel is 230 m (754 ft) tall. This number multiplied by your answer in part **d** will give the additional wall area that would be needed to enclose the same space if the hotel had a square base rather than a circular one. What is this area?

8. This wall has a length of 540 cm and a height of 240 cm. A few dimensions are also given around the window and fireplace.

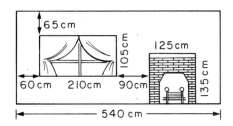

★   a. How many square centimeters of wallpaper will this wall require?

       b. A standard roll of wallpaper is 12.8 m by 53 cm, and partial rolls are not sold. How many rolls must be purchased to cover this wall?

9. The common starfish has five arms. Most species grow as large as 20 to 30 cm in diameter (greatest distance between tips of arms), but some species reach only 1 cm. The body surface contains hairlike projections capable of producing a waving motion, which carries food to the mouth and removes unwanted particles from the body. This starfish was photographed on a centimeter grid and has a diameter of approximately 9.5 cm.

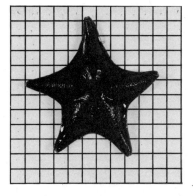

★   a. Compute an upper bound for the area of the underside of this starfish by counting every square that is covered or partially covered. (*Hint:* Count the squares which are not covered or not partially covered.)

       b. Compute a lower bound for the area of this starfish by counting only those squares which are completely covered. (*Hint:* Use the region outside the boundary of the starfish.)

c. Approximate the area of this starfish by counting only those squares which are at least half covered. (*Note:* Your answer should be between the two areas from part **a** and **b**.)

10. There are many factors that enter into the assessment of a house for determining property taxes. Once the proper category is determined, the assessment rate is per square foot or square meter. Compute the value of the following one-floor dwellings for an assessment rate of $189 per square meter.

★   a. Ranch Style: 8 m by 13.75 m

     b. L-shaped: 8.7 m by 11 m plus 8 m by 9.5 m

★   c. How much in taxes must be paid on these houses if the tax rate if $52 on every $1000 worth of assessment?

11. The sides of the square on this circle have a length equal to the radius of the circle. The area of the circle is obviously less than four of these squares. Continue placing the pie-shaped sectors of the circle on the squares (as shown) to see what part of the four squares they will cover. Estimate your answer to the nearest one-tenth of a square. How would your answer be affected if the circle were divided into smaller sectors? How should your answer be related to π?

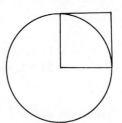

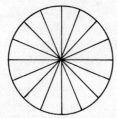 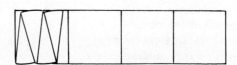

12. Physicists study cosmic radiation to learn about properties of our galaxy and levels of sun activity. The proton histogram on page 299 contains information on the intensity level of cosmic rays. Region A (shaded) under the histogram, which is called the *background area,* is compared with the total area of regions A and B. (Use a millimeter ruler for parts **a** and **b**.)

                    CHAPTER 5 MEASUREMENT

★ a. What is the approximate area
     of A in square millimeters?
   b. What is the approximate area
     of B?
★ c. Region A is what percent of the
     total area of A and B?

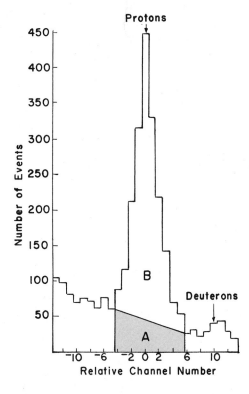

13. *Demonstration:* The vertical height of a tennis
    ball can is approximately equal to the circum-
    ference of a tennis ball. This can be illustrated
    by rolling a tennis ball along the edge of a can.
    The ball will make one complete revolution.
    Since a can holds three tennis balls, what does
    this demonstration imply about the diameter
    of a ball as compared to its circumference?

Exercise Set 5.2: Problem Solving

1. *Square-Circle-Square Problem:* Draw the largest circle possible on a square piece of
   paper. Cut the circle out, discarding the trimmings. Inside the circle, draw the largest
   square possible. Cut the square out, discarding the trimmings. What fraction of the
   original square piece of paper has been cut off and thrown away?

a. *Understanding the Problem:* The shaded portion of this diagram shows the trimmings that will be thrown away in the first step of the paper-cutting process. Inscribe a square in this circle and shade the trimmings to be thrown away in the second step.

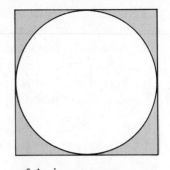

★ b. *Devising a Plan:* One approach to this problem is to compute the total area of the trimmings from the two steps. Describe the shape of the total region to be thrown away. Another approach is to derive the answer from the area of the inner square.

c. *Carrying Out the Plan:* Choose a plan from part **b** or one of your own and determine what fraction of the original square piece of paper is cut off.

d. *Looking Back:* If you answered this problem by taking measurements, then you used inductive reasoning. Explain how the diagram at the right and deductive reasoning can be used to answer the original question. (*Hint:* The vertices of the inner square are the midpoints of the sides of the outer square.)

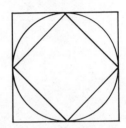

★ e. *Looking Back Again:* Suppose we begin with a circle; inscribe a square; and then inscribe a smaller circle in the square. How does the area of the small circle compare with the area of the large circle? Will the answer be the same as for the "Square-Circle-Square Problem"?

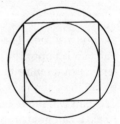

★ 2. *Cable Splicing:* Suppose a cable fits tightly around the equator of the earth. If an additional piece is to be spliced in so that the cable can be raised 2 meters above the earth (at all points), how much additional cable will be needed?

3. *Trundle Wheel:* A meter trundle wheel is a convenient device for measuring distances along the ground. Every time the wheel makes one complete revolution it has moved forward one meter. If you were to cut this wheel from a square piece of plywood, what would be the dimensions of the smallest square you could use?

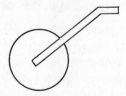

4. *Egyptian Formulas:* Many of the formulas found on ancient Egyptian scrolls (1850–1650 B.C.) are for computing land areas for purposes of taxation. Do the

following formulas produce the correct areas for circles and quadrilaterals? If not, are the results too large or too small? (*Hint:* Try these formulas on some figures. Use pi = 3.14.)

   a. The area of a circle is equal to the area of a square whose width is 8/9 the diameter of the circle.

★ b. The area of a quadrilateral with successive sides of lengths *a, b, c,* and *d* is $(a + c) \times (b + d)/4$.

5. *Wheel Contraption:* The wheel has been called the most important invention of all time. Consider the diameter of each wheel in this cartoon to be 75 cm and the distance between opposite pairs of wheels to be 300 cm. How many revolutions of each wheel will it take to turn this contraption in one complete circle?

"Just because you invented the *wheel*, it doesn't follow logically that you can go on to the *wagon.*"

Exercise Set 5.2: Computers

1. Use the computer program on page 293 to answer the following questions.

★ a. What is the printout for W = 20 to 30?

   b. Plot the widths and areas from part **a**. Use this graph and inductive reasoning to predict the dimensions of the rectangle with the largest area.

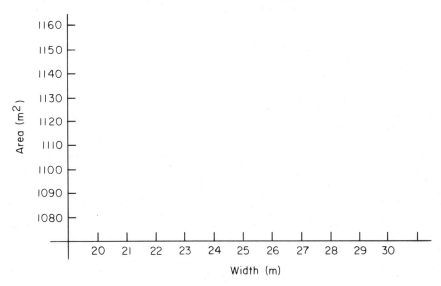

2. Of all simple closed curves of equal length, the circle encloses the largest area. Here are two figures which both have a perimeter of approximately 120 millimeters. How much greater (to the nearest square millimeter) is the area of the circle?

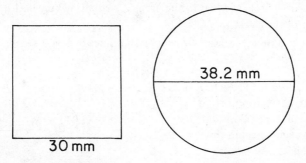

★ a. Write a program which computes the area of a square and a circle, each having a perimeter P. The printout should be:

FOR A PERIMETER OF P THE AREA OF THE SQUARE IS S
AND THE AREA OF THE CIRCLE IS C.

b. Use your program to find the areas of the square and circle which have perimeters of 120 mm.

3. This program computes the areas of a square and several rectangles which have the same perimeter. Lines 40 through 90 form a loop. The first time through the loop, K is equal to zero and L (length) and W (width) both equal P/4. Each successive time through the loop, L is increased by one and W is decreased by one so line 70 prints the area of a rectangle. What is the complete printout for P = 16?

```
10   PRINT "TYPE A NUMBER FOR THE PERIMETER."
20   INPUT P
30   LET K = 0
40   LET L = P / 4 + K
50   LET W = P / 4 - K
60   IF W < 2 THEN  GOTO 100
70   PRINT "FOR L = "L" AND W = "W" THE AREA IS "L * W"."
80   LET K = K + 1
90   GOTO 40
100  PRINT "THE WIDTH IS < 2. THIS PROGRAM IS FINISHED COMPU
     TING AREAS."
110  END
```

4. If a measurement is given to the nearest whole number then it could be off by as much as plus or minus .5. Similarly, areas which are computed by using such measurements are between minimum and maximum limits. Write a program which computes the maximum and minimum areas of rectangles when the measurements for the lengths and widths are whole numbers. Here is the first line of the program.

```
10   PRINT "TYPE TWO WHOLE NUMBERS, SEPARATED BY A COMMA, FOR
     THE LENGTH AND WIDTH OF A RECTANGLE."
```

Liquefied natural gas tanker, *Aquarius*

5.3
VOLUME AND
SURFACE AREA

This is the *Aquarius,* one of twelve liquefied natural gas tankers that was built by the Quincy Shipbuilding Division of General Dynamics. The *Aquarius* is longer than three football fields (285 meters) and carries five spherical aluminum tanks, each having a diameter of 36.58 meters. The space inside these tanks is measured by the number of unit cubes that are required to fill it. The number of such cubes is the measure which we call *volume.* The appropriate unit cube for these tanks is cubic meters (m^3), the volume of a cube whose edges have a length of 1 m. Each of these spheres holds 25,000 cubic meters of liquefied gas at ⁻165° Celsius (⁻265° Fahrenheit).

RELATING VOLUME AND SURFACE AREA

The cubic centimeter is the common unit for measuring small volumes. The first box shown here has 12 cubes on its base and will hold 12 more above these. Its volume is 24 cubic centimeters (24 cm^3).

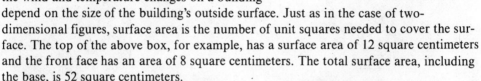

Another important measure for space objects is their amount of surface. For example, the effects of the wind and temperature changes on a building depend on the size of the building's outside surface. Just as in the case of two-dimensional figures, surface area is the number of unit squares needed to cover the surface. The top of the above box, for example, has a surface area of 12 square centimeters and the front face has an area of 8 square centimeters. The total surface area, including the base, is 52 square centimeters.

The size of the surface area of an object cannot be predicted by knowing its volume, any more than the perimeter of a figure is determined by its area. These two boxes on the right, for example, both have a volume of 18 cubic centimeters, but there is a good deal of difference in the areas of their surfaces. The top box has a surface area of 42 square centimeters, and the lower box has a surface area of 54 square centimeters. If you were going to build a box that had a volume of 18

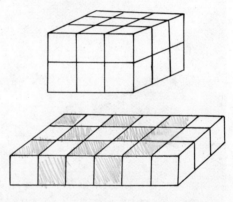

cubic centimeters, building the lower box would require about 30 percent more material than the top box. In general, for a given volume, the box that is closer to the shape of a cube will have the smaller surface area.

PRISMS

In this drawing the length (20) times the width (10) gives the number of cubes (200) on the floor of the box (or base of the rectangular prism). Since the box can be filled with six such floors of cubes, it will hold 1200 cubes. This volume of 1200 cubic centimeters can be obtained by multiplying the

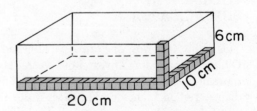

6 cm
10 cm
20 cm

three dimensions of the box. In general, a rectangular prism with length *l*, width *w*, and height *h* has the following volume:

$$\text{Volume} = \text{length} \times \text{width} \times \text{height} \quad (V = lwh)$$

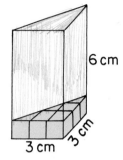

For any prism the volume can be found in a similar way. The base of this prism is a right triangle which is covered by four and one-half cubes. Since six different floors or levels of four and one-half cubes each fill the prism, its volume is 6×4.5, or 27 cubic centimeters.

The number of cubes which cover the base of this prism is the same as the area of the base. Therefore, the volume of the prism can be computed by multiplying the area of the base by the height (or altitude) of the prism. In general, for any right prism having a base of area *B* and a height of *h*, its volume can be computed by the following formula:

$$\text{Volume} = \text{area of base} \times \text{height} \quad (V = B \times h)$$

A formula for the volume of an oblique prism can be suggested by beginning with a stack of cards, and then pushing them sideways to form an oblique prism. If each file card is 12.5 centimeters by 7.5 centimeters and the stack is 5 centimeters high, its volume will be $12.5 \times 7.5 \times 5$, or 468.75 cubic centimeters. The base of the oblique prism is also 12.5 centimeters by 7.5 centimeters, and its *height* (or *altitude*), the perpendicular distance between its upper and lower bases, is 5 centimeters. Since both stacks contain the same number of cards, their volumes are both 468.75 cubic centimeters. This means that the volume of the oblique prism can be computed by multiplying the area of its base by its height. In general, the volume of any prism, right or oblique, is the area of its base times its height.

Sometimes a prism can have more than one base, as shown by the following sketches. The first prism is a right prism with a base that is a parallelogram. By turning the prism as shown in the two remaining positions, it can be classified as oblique with a rectangular base.

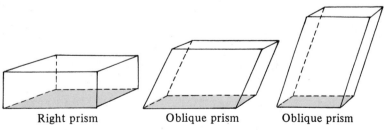

Right prism   Oblique prism   Oblique prism

The surface area of a prism is the sum of the areas of its bases and faces. For right prisms these faces are rectangles, and for oblique prisms the faces are rectangles and parallelograms.

CYLINDERS

These cylindrical buildings are part of the Renaissance Center in Detroit. Just as with the conventional rectangular-shaped buildings, architects need to know the volumes and surface areas of these glass-walled cylinders.

To compute the volumes of cylinders, we continue to use unit cubes even though they do not conveniently fit into a cylinder. The cubes in the cylinder pictured below show that more than 33 cubes are needed to cover its base. Furthermore, since the cylinder has a height of 12 centimeters, it will take at least 12 × 33, or 396, cubes to fill the cylinder.

If we were to use smaller cubes they could be packed closer to the boundary of the base, and a better approximation could be obtained to the volume of the cylinder. This suggests that the formula for the volume of a cylinder is the same as that for the volume of a prism. For a cylinder with a base area of B and a height of h, its volume is computed by this formula:

Renaissance Center in Detroit

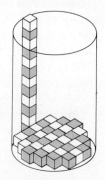

Volume = area of base × height   ($V = B \times h$)

A cylinder without bases can be formed by joining the opposite edges of a rectangular sheet of paper. The circumference of the base of the cylinder is the length of the rectangle, and the height of the cylinder is the height of the rectangle. Therefore, the surface area of the sides of a cylinder is the circumference of the base of the cylinder times its height.

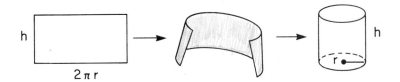

For any cylinder whose base has a radius r and whose height is h, the base has a circumference of $2\pi r$, and the surface area of the sides of the cylinder is $2\pi r h$. Adding the area of each base, πr^2, the total surface area is $2\pi r h + 2\pi r^2$.

PYRAMIDS

The Egyptian Great Pyramid, or Pyramid of Cheops, as it is called, has a height of 148 meters and a square base with a perimeter of 930 meters. The Transamerica Pyramid in San Francisco has a height of 260 meters and a square base with a perimeter of 140 meters. In spite of the height differences in these giant pyramids, the Egyptian pyramid has several times more volume.

Transamerica Pyramid in San Francisco

The shaded pyramid inscribed in the cube on the right has a square base, *EFGH,* and a height *GC.* This pyramid, together with the two pyramids having bases *EABF* and *EADH,* and vertices *C,* partition the cube into three congruent pyramids. Therefore, the volume of the shaded pyramid is one-third of the volume of the cube. That is, the volume of the pyramid is one-third of the area of the base of the cube times the height of the cube.

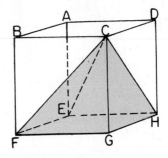

In general, if *B* is the area of the base of a pyramid and *h* is the perpendicular distance from the upper vertex to the base, the volume of the pyramid is computed by this formula:

$$\text{Volume} = \frac{1}{3} \times \text{area of base} \times \text{height} \qquad (V = \frac{1}{3}Bh)$$

Let's apply this formula to find the volume of the Great Pyramid. Its base has an area of 54056 square meters, and its height is 148 meters. Therefore, its volume is $1/3 \times 54056 \times 148$, or 2,666,763 cubic meters.

CONES

This pile of crude salt has the shape of a cone with a circular base, and it has been formed in the same way as you might build a sand castle by letting sand run through your hands. The salt has been solar-evaporated from ocean water and is being stored to await further purification. Cone-shaped piles of sand, gravel, and stone are common sights at construction companies.

The volume of a cone can be approximated by the volume of a pyramid which is inscribed in the cone. The hexagonal pyramid in this figure has a volume of 430 cubic centimeters, and the volume of the cone is approximately 523 cubic centimeters. As the number of sides of the base of the pyramid is increased, its volume becomes closer to the volume of the cone. Since the volume of the pyramid is one-third the area of its base times its height, this

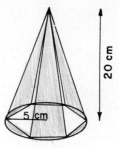

suggests that the same formula will give the volume of a cone. In general, the volume of any cone whose base has area B and whose height is h, is given by the following formula:

$$\text{Volume} = \frac{1}{3} \times \text{area of base} \times \text{height} \qquad (V = \frac{1}{3}Bh)$$

Let's apply the formula for the volume of a cone to compute the amount of salt in the photo on the previous page. The height of this cone is 12 meters, and the diameter of its base is 32 meters. Therefore, its volume equals $1/3 \times \pi \times 16^2 \times 12$, or approximately 3215 cubic meters. This will fill about 40 railroad flat cars having a capacity of 80 cubic meters each.

SPHERES

This is a view of the earth as seen from the *Apollo 10* spacecraft as it passed over the moon's surface. The earth, the planets, and their moons are all spheres, spinning and orbiting about a spherical-shaped sun. There is considerable variation in the volumes of these spheres. The earth has about 18 times more volume than the smallest planet, Mercury. The largest planet, Jupiter, is 10,900 times bigger than the earth. The sun is the largest body in our solar system with a volume which is more than a million times greater than the earth's.

Apollo 10 view of earth over moon's surface, 1969

Formulas for the volume and surface area of a sphere were known by Archimedes (287–212 B.C.) and described in his work *On the Sphere and Cylinder*. Archimedes discovered the following remarkable relationships between a sphere and the smallest cylinder containing it: The volume of the sphere is two-thirds the volume of the cylinder; and the surface area of the sphere is two-thirds the surface area of the cylinder. Archimedes rated these discoveries among his greatest accomplishments and requested that they be engraved on his tombstone.

For a sphere of radius r, the smallest cylinder to contain it has a height of $2r$. The volume of this cylinder is $\pi r^2 \times 2r$, or $2\pi r^3$. Two-thirds of this, $2/3 \times 2\pi r^3$, is $4/3 \times \pi r^3$, the formula for the volume of a sphere. If you think of π as approximately 3, the volume of the sphere is about $4r^3$, that is, four times the volume of the cube whose sides have the length of the radius of the sphere.

The surface area of this cylinder is $6\pi r^2$. Two-thirds of this, $2/3 \times 6\pi r^2$, is $4\pi r^2$, the surface area of the sphere. This shows that the surface area of a sphere is exactly four times greater than the area of a disc whose center is the center of the sphere (see shaded disc).

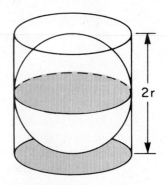

IRREGULAR SHAPES

The volume of an irregular-shaped figure can be determined quite easily by submerging it in water and measuring the volume of the water which is displaced. To illustrate this method we will find the volume in cubic centimeters of this miniature statue. Before submerging the statue, the cylinder has been filled to a height of 700 milliliters. When the statue is placed in the cylinder, the water level rises to the 800-milliliter level. This means that the volume of the statue is equal to the volume of 100 milliliters of water. Since each milliliter of water has a volume of 1 cubic centimeter, the volume of the statue is 100 cubic centimeters.

CREATING SURFACE AREA

A potato can be cooked in a shorter time if it is cut into pieces. Ice will melt faster if it is crushed, and coffee beans will provide a better coffee if they are ground before they are boiled. The purpose of crushing, grinding, cutting, or, in general, subdividing, is to increase the surface area of a substance. While you may be aware of this, it is the rate and amount of increase of surface produced that is surprising.

To illustrate how rapidly surface area can be created, consider the cube that is 2 centimeters on each edge. Its volume is 8 cubic centimeters, and its surface area is 24 square centimeters. If this cube is cut into 8 smaller cubes, the total volume is still 8 cubic centimeters, but the surface area is doubled to 48 square centimeters. This can be easily seen by looking at the small cube in the upper right-hand corner. Faces *a, b,* and *c* contributed 3 square centimeters to the area of the original cube, and after the cut, its remaining three faces were exposed, contributing 3 more square centimeters of area. Since this is true for each of the 8 smaller cubes, the total increase in surface area is 8×3 square centimeters, or 24 square centimeters.

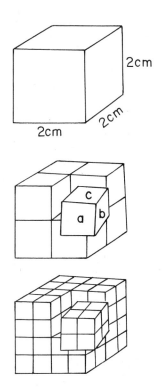

If we continue this process, by cutting each of the centimeter cubes into 8 smaller cubes, there will be 64 cubes whose edges have lengths of 1/2 centimeter. The total volume of these cubes is still 8 cubic centimeters, but the surface area is doubled to 96 square centimeters.

For the third such cut the surface area becomes $2^3 \times 24$, or 192 square centimeters, and after the twelfth such cut it has increased to $2^{12} \times 24$, or 98,304 square centimeters! During this splitting process the volume has remained the same, 8 cubic centimeters.

This process of subdividing can also be used to double the surface area of a sphere. If for example, a sphere of radius 2 centimeters is formed into 8 smaller spheres, each with a radius of 1 centimeter, the total volume will remain the same but the surface area will be doubled. As in the case of the cubes, if this subdividing is carried out far enough, a surface area can be increased many times. Consider this effect when water is sprayed into the air in a fine mist, as happens with snow-making machines. The surface area of each drop of water is increased hundreds of times, and the minute particles of water freeze into snowflakes.

Nature also divides and splits substances into smaller parts to increase surface area. Our lungs contain about 300 million tiny cavities to provide the necessary amount of surface area for oxygen. Their area is about 120 square meters, which is more than the area of the walls of a large size room! If the internal surfaces of lungs were smooth and flat, their area would be only about 1.5 square meters.

COMPUTER APPLICATIONS

We know that two figures with the same perimeter may have different areas. Similarly, two figures with the same surface area may have different volumes. Here is a problem involving volumes and surface areas of boxes. You may be surprised at what happens to the volume as the surface area decreases.

Problem An open-top box is to be made from a sheet of material which is 16 units by 16 units by cutting out the corners and folding up the edges. What are the dimensions of the box with the greatest volume?

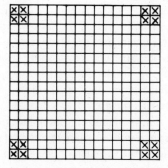

Here are three boxes with heights of 1, 2, and 3 units. The box with a height of 2 units was made by cutting 2 by 2 squares from the corners of the sheet.

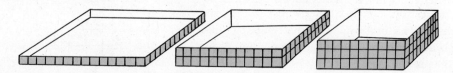

Program 5.3A determines the volumes of the boxes that are obtained by cutting out the corners of a square sheet of material of length 16. The heights of the boxes (values of H) are consecutive whole numbers beginning with 1. For each increasing height, the width of the box gets smaller and the surface area of the box decreases because larger squares are being removed from the corners of the sheet. When the height of the box is 7, the width of the box will be 2 and the last volume is computed. The commas in lines 10 and 50 cause the printout to be spaced in three columns.

Program 5.3A

```
10   PRINT "HEIGHT", "BASE", "VOLUME"
20   FOR H = 1 TO 7
30   LET W = 16 - 2 * H
40   LET V = H * W ^ 2
50   PRINT H, W" X "W, V
60   NEXT H
70   END
```

Here are the first three lines of the printout. Will the volumes of the boxes continue to increase as the surface areas decrease?

HEIGHT	BASE	VOLUME
1	14 BY 14	196
2	12 BY 12	288
3	10 BY 10	300

Activity Set 5.3 Models for Volume and Surface Area (Prisms, Pyramids, Cones, Cylinders, and Spheres)

Just for Fun: Soma Cubes

"Galileo showed men of science that weighing and measuring are worthwhile. Newton convinced a large proportion of them that weighing and measuring are the only investigations that are worthwhile."   CHARLES SINGER

Exercise Set 5.3: Applications and Skills*

1. Each of the mathematical objects listed below can be seen in the accompanying picture. Try to find the objects.

   a. Cone   ★   b. Pyramid

   c. Cylinder   d. Sphere

   e. Circle   ★   f. 30° angle

   g. Rectangle   h. Semicircle

   i. Square   ★   j. 45° angle

2. The liter is the metric unit which will replace the quart. A 1-quart milk carton has a square base of 7 cm by 7 cm and a height of 19.3 cm.

   ★   a. What is its volume?

   b. Which is greater, a quart or a liter?

Thompson Hall, University of New Hampshire

*Use $\pi = 3.14$ in this Exercise Set.

3. Thor has invented a "new wheel" but B.C. doesn't seem to be overly impressed.

★ a. What is the volume of this wheel to the nearest cubic centimeter, if it has a length and height of 1 m, a thickness of 20 cm, and an inner diameter of 46 cm?

b. If the stone this wheel is made of weighs 7 g per cubic centimeter, what is the weight of the wheel in kilograms?

4. *Consumer Mathematics*

★ a. A house with ceilings that are 2.4 m high has five rooms with the following dimensions: 4 m by 5 m; 4 m by 4 m; 6 m by 4 m; 6 m by 6 m; and 6 m by 5.5 m. Which of the following air conditioners will be adequate to cool this house: an 18,000 Btu unit for 280 m³; a 21,000 Btu unit for 340 m³; or a 24,000 Btu unit for 400 m³?

b. A woodshed has a length, width, and height of 3 m by 2 m by 2 m. If each 1.5 m³ of firewood sells for $25, how much will it cost to completely fill the shed with wood?

★ c. A catalog describes two types of upright freezers: Type A has a 60 cm by 60 cm by 150 cm storage capacity and costs $339; Type B has a 55 cm by 72 cm by 160 cm storage capacity and costs $379. Which freezer has the most cubic centimeters for each dollar?

d. A drugstore sells the same brand of talcum powder in two types of cylindrical cans: Can A has a diameter of 5.4 cm, a height of 9 cm, and sells for $1.59; Can B has a diameter of 6.2 cm, a height of 12.4 cm, and sells for $2.99. Which can is the better buy?

5. The number of fish that can be put in an aquarium depends on the amount of water, the size of the fish, and the capacity of the pump and filter system.

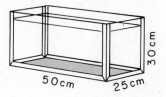

★ a. Faucet water contains chlorine and should sit for 3 days before being used for fish. How many liters of water will this tank hold?

b. The recommended number of tropical fish for this tank is 30. How many cubic centimeters does this allow for each fish?

c. Goldfish need more space and oxygen than tropical fish. Those about 5 cm in length require 3000 cm^3 of water. How many of these fish can adequately live in this tank?

6. In an NAEP (National Assessment of Educational Progress) question on volume, students were shown a sketch of a box divided into cubes, similar to this one, and asked how many cubes the box contained. Only 6% of the 9-year-olds, 21% of the 13-year-olds, and 43% of the 17-year-olds correctly answered the question.

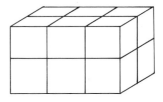

★ a. The most common wrong answer was 16. Explain how the students might have obtained this answer.

b. What two concepts of measure might they have confused?

7. Swimming pools require daily testing to determine the pH factor and the chlorine content. Pumps and filters are also necessary, and some pools have heating systems.

★ a. What is the depth of a 6 m by 12 m pool that contains 193 kℓ of water?

b. If this pool requires 112 g of chlorine every 2 days, how many kilograms of chlorine should be purchased for a 3-month period?

★ c. The Alcoa Solar Heating System for pools has 32 square panels which are 120 cm by 120 cm. Will this heating system fit onto a 5 m by 8 m roof?

d. Each panel for this system holds 5.68ℓ of water. What is the total weight of water for 32 panels?

8. This art form is Alex Lieberman's *Argo* which is at the Walker Art Center in Minneapolis. The entire display weighs about 4535 kg.

a. The cylinder shown in front is approximately 2 m tall and 1 m in diameter. What is its volume?

★ b. What is the weight of a volume of water which is equal to the volume of the front cylinder? (One cubic centimeter of water weighs approximately 1 g.)

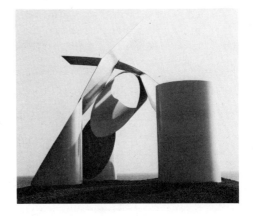

9. The concrete foundation for the office building on the corner of Congress and State streets in Boston required 496 truck loads of concrete and was formed in one continuous pouring over a 30-hour period.

   a. The concrete was poured to a depth of 1.8 m and covered an area of 2420 m². How many cubic meters of concrete did this require?   4356 m³

   b. If each truck load was the same size, what was the volume of a load of concrete?

10. Egypt's Great Pyramid, or Pyramid of Cheops, was built about 2600 B.C. and is one of the seven wonders of the ancient world. It is estimated that 100,000 men required 20 years to complete it.

   ★ a. How many times greater is the volume of the Great Pyramid than the volume of the Transamerica Pyramid in San Francisco?

   b. The heights (altitudes) of the triangular faces of the Great Pyramid and Transamerica Pyramid are 188 m and 261 m, respectively; and the bases of these triangles are 232 m and 35 m, respectively. About how many times greater is the surface area of the Egyptian pyramid (not counting the areas of their bases)?

11. Assume that a drop of unvaporized gasoline is a sphere with a diameter of 4 mm.

   ★ a. If this drop is divided into 8 smaller drops, each with a diameter of 2 mm, how many times greater is the total surface area of these 8 drops, compared to the surface area of the original drop?

   b. If each drop with a diameter of 2 mm is divided into 8 smaller drops, each with a diameter of 1 mm, how many times greater is the total surface area of the 64 drops, as compared to the surface area of the original drop?

   ★ c. If the vaporizing mechanism in a car's engine carries out this process of splitting 20 times, how many times is the surface area of the original drop increased?

12. Compute the volumes of the following figures in cubic centimeters. Figures **b** and **c** have a square base and figures **a, f,** and **g** have rectangular bases. (*Hint for Figure* **a:** Think of the front face as the base of a prism. This face can be subdivided into a 5 cm by 2 cm rectangle and two congruent triangles.)

★ a.

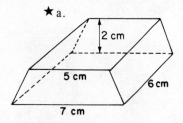

b.

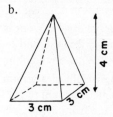

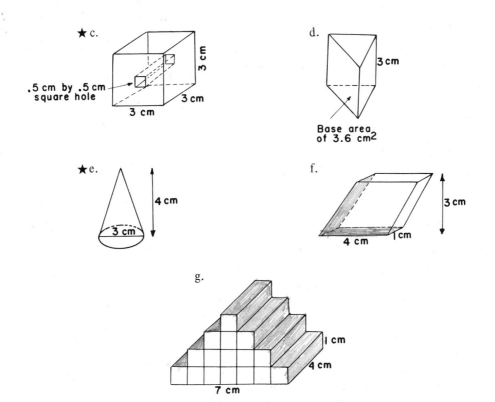

★ c.

.5 cm by .5 cm
square hole

3 cm

3 cm

3 cm

d.

3 cm

Base area
of 3.6 cm²

★ e.

4 cm

3 cm

f.

3 cm

4 cm

1 cm

g.

1 cm

4 cm

7 cm

Exercise Set 5.3: Problem Solving

1. *Rectangular Boxes:* An open-top box is to be made from a 50 cm by 30 cm
   rectangular sheet of material by cutting out squares from the corners of the sheet. It
   is required that the height of the box be a whole number of centimeters. What size
   squares should be cut out to obtain a box with maximum volume?

★   a. *Understanding the Problem:* This diagram shows how the box is to be formed.
   If 6 cm by 6 cm squares are cut from the corners, the height of the box will be
   6 cm. In this case, what is the width and length of the box?

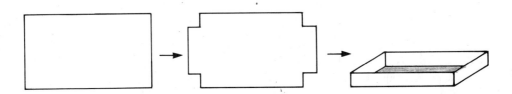

b. *Devising a Plan:* One plan for solving this problem is to systematically consider corner squares of increasing size. For squares whose dimensions are whole numbers, what is the largest possible size squares that can be cut from the corners?

c. *Carrying Out the Plan:* Complete the following table and use inductive reasoning to predict the size of the squares for obtaining the box of maximum volume.

Size of squares (cm)	2 by 2	4 by 4	6 by 6	8 by 8	10 by 10	12 by 12	14 by 14
Volume of boxes (cm^3)							

★   d. *Looking Back:* The table shows that as the size of the squares (at the corners) increases the volume of the box increases for awhile and then decreases. Try a few more sizes for the squares, using whole numbers for dimensions, to see if you can obtain a greater volume for the box.

★ 2. *Cylinders:* A cylinder without bases is to be formed from a rectangular sheet of paper by taping together two opposite edges of the paper. Assuming that the length of the rectangular sheet is greater than the width, which will have the greater volume: a cylinder whose height is the width of the sheet; or a cylinder whose height is the length of the sheet?

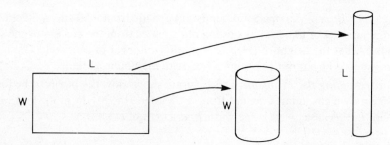

3. *Cubical Monument Legend:* There is a story about an ancient cubical monument that sits on a square plaza. Both the cube and the square plaza were constructed from the same number of smaller cubes. The plaza is twice as wide as the cube. How many cubes were needed to build the monument and the plaza?

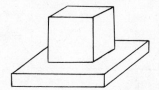

★ 4. *Footballs:* A regulation football has a length of approximately 27 centimeters and a diameter of approximately 16 centimeters. Describe two different methods of approximating its volume in cubic centimeters.

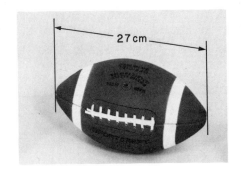

5. *Volume Puzzle:* A cylindrical can such as the one shown here is full of water. If you were to pour the water from the can, how would you know when half the water was gone if you had no measuring device?

Exercise Set 5.3: Computers

1. Use Program 5.3A on page 312 to answer these questions.

   a. What is the complete printout for this program?

★ b. What is the greatest volume of the box according to this printout?

   c. Revise the program by replacing line 20 by the following line to allow half units for the heights of the boxes. What will be the greatest volume in the printout if this line is used in the program?

   20 FOR H = 1 TO 7 STEP .5

2. Each aluminum sphere for General Dynamics' liquefied natural gas tankers has a diameter of approximately 36.6 meters and a weight of 725,750 kg (800 tons). These spheres were constructed in Charleston, South Carolina, and then towed by tug to Quincy, Massachusetts. Write a computer program to print out the volume and surface area of a sphere with an arbitrary radius. Use pi = 3.14 and round off the volume and surface area to the nearest whole number.

   a. What is the volume of the sphere in this photo?

Spheres for liquefied natural gas tankers

★　　b. A heavy external coating of insulation on the surface of the sphere enables the sphere to maintain the liquefied natural gas at ⁻165° Celsius. How many square meters of insulation are needed for one of these spheres?

3. One of the silos pictured here holds corn and the other holds hay. Chopped corn and hay are blown into the top of the silos through the pipes which can be seen. Write a program to print out the volume in cubic meters for a silo with a radius R and a height H. Also, program the computer to print out the time in hours that is required to fill a silo, if a blower can feed in one cubic meter of hay each 3 minutes. Use pi = 3.14 and round off the volume and time to the nearest whole number.

   a. The silos in this photo have a radius of 3 meters and a height of 18 meters. What is the volume to the nearest cubic meter of one of these silos?

★　　b. How many hours are needed to fill one of these silos?

Fractions and Integers

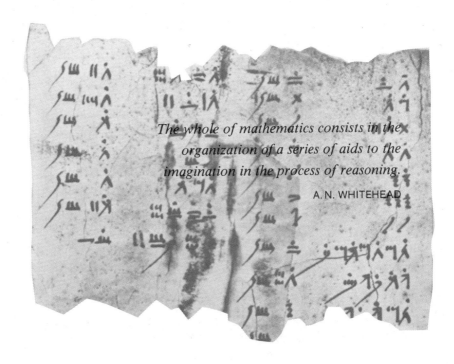

The whole of mathematics consists in the organization of a series of aids to the imagination in the process of reasoning.

A. N. WHITEHEAD

New York Stock Exchange trading floor

ratio not same as fraction

6.1
EQUALITY AND
INEQUALITY OF
FRACTIONS

The sale of stocks and bonds on the New York Stock Exchange is carried out on this three-story trading floor. Some 2000 private telephone lines connect brokers on the floor with their offices, which in turn are linked to hundreds of branch offices throughout the United States. The annunciator board in the upper left of this picture is a calling system of flashing numbers to brokers on the floor.

The prices of stocks are stated in dollars, as well as halves, fourths, eighths, and sixteenths of a dollar. Historically, when a smaller measure or unit was needed, half of it was used. For a still smaller amount, a half of a half produced a fourth. Similarly, eighths, sixteenths, and thirty-seconds resulted from repeatedly halving the original unit. The inch with its halves, fourths, eighths, etc., is a familiar example of this process. Another example is the speed of shutter openings of a camera, which is calibrated in fractions of a second: 1/2, 1/4, 1/8, 1/15, 1/30, 1/60, etc. (Exercise 1, page 335). On the other hand, when larger measures were needed, the common units were doubled (see page 266). The custom of doubling and halving seems to be a natural tendency. Karl Menninger, in *Number Words and Number Symbols,* calls them "primitive operations."*

HISTORICAL DEVELOPMENT

The use of fractions is ancient. The Egyptians were using them before 2500 B.C. With the exception of 2/3, all Egyptian fractions were *unit fractions,* that is, fractions with a numerator of 1 (1/3, 1/4, etc.).

These fractions were represented by placing the symbol ⌒, which meant "part," above their hieroglyphic numerals for whole numbers.

$$\overset{\bigcirc}{|\,|\,|} = \frac{1}{3} \qquad \overset{\bigcirc}{|\,|\,|\,|\,|} = \frac{1}{5} \qquad \overset{\bigcirc}{\cap\,|\,|} = \frac{1}{12}$$

When they used their hieratic (sacred) numerals, as shown in this scroll, a dot was placed above the numeral to represent a unit fraction. For example, ∧ represents the number 30, and ∧̇ represents the fraction 1/30. It is interesting to note that as late as the eighteenth century, over 3000 years later, the symbols ·/2 and ·/4 are found in English books for the fractions 1/2 and 1/4.

Egyptian leather scroll containing simple relations between fractions (ca. 1700 B.C.)

While the Egyptians used fractions with fixed numerators, the Babylonians (ca. 2000 B.C.) used only fractions with denominators of 60 and 60^2. The fraction 1/60 was referred to as the "first little part" and $1/60^2$ as the "second little part." These fixed

*K. Menninger, *Number Words and Number Symbols* (Cambridge, Massachusetts: MIT Press, 1969), pp. 359–60.

denominators were extensions of the Babylonian base 60 and were used in their study of astronomy. Our use of minutes, 1/60 of an hour, and seconds, $1/60^2$ of an hour, has been handed down from the Babylonians.

Gradually, fractions in which the numerator and denominator could be any number gained acceptance. These more general fractions were being used by the Hindus in the seventh century. There have been many notations for writing fractions. Our present form of writing one number above the other was used by the Hindus. The Arabs introduced

the bar between the two vertically placed numbers. The resulting notation, $\dfrac{a}{b}$, was in general use by the sixteenth century.

FRACTION TERMINOLOGY

"Fraction" is from the French "frangere," which means *to break*. The words "broken number" and "fragment" have been frequently used in the past for "fraction." These numbers were first needed for measurements that were less than a whole unit. One of the problems which occurs in the Egyptian writings from 1650 B.C.

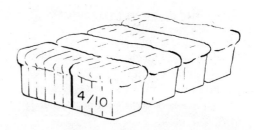

requires that 4 loaves of bread be divided equally among 10 men. If a loaf of bread is divided into 10 equal parts, the share for each man is 1/10 of a loaf. Since there are 4 loaves of bread, each man will receive 1/10 of each loaf or 4/10 of a loaf. This "broken number" tells *how much* each man is to receive.

The word "fraction" is used to refer both to a number written in the form *a/b* and to the numeral, "*a/b.*" The *a* and *b* can be any numbers, as long as *b* is not equal to zero. (See Sections 7.3 and 7.4 for fractions involving decimals and irrational numbers.) When we speak of adding two fractions, we are talking about adding numbers, not numerals (symbols). On the other hand, when we say the denominator of the fraction 5/12 has a factor of 3, we are referring to the bottom half of the numeral "5/12." You will not have to be concerned about whether or not the word "fraction" is being used as a number or as a numeral; it will be clear from its use in any particular instance.

FRACTIONS AND DIVISION

One of the major influences in the early development of fractions was the need to solve problems involving division of whole numbers. The Egyptians used fractions in cases where the quotient was not a whole number. In the example cited previously, the Egyptians needed a fraction in order to compute 4 divided by 10 (4 ÷ 10 = 4/10).

Charlie Brown's little sister Sally has a similar problem. She is trying to divide 25 by 50. Charlie Brown's comment shows that he thinks of division only in terms of *measurement,* that is, "how many 50s in 25?" However, there is another aproach to division. Remember from page 150 that division has two meanings, measurement and partitive. With the *partitive* concept, dividing by 50 means there will be 50 parts. If we picture dividing 25 objects of equal size, such as sticks of gum, into 50 equal parts, each part will be one-half of a stick. Stated in equation form, $25 \div 50 = 1/2$.

This close relationship between fractions and division is often used to define a fraction as a quotient of two numbers.

Definition: For any numbers a and b, with $b \neq 0$,

$$\frac{a}{b} = a \div b$$

MODELS FOR FRACTIONS

The most common applications of fractions involve the *part-to-whole concept,* that is, the use of a fraction to denote part of a whole. In the fraction a/b the bottom number b indicates the number of equal parts in a whole and the top number a indicates the number of parts being considered. As examples, the surface of the earth can be divided into three parts of equal area such that two of these parts are water; and an iceberg can be divided into nine parts of equal volume such that eight of these parts are under water.

2/3 of the earth's surface area is
covered by water.

8/9 of the volume of an iceberg
is underwater.

Regions   Here are four examples of regions. The fraction under each figure indicates the part of the total region which is marked by shading or letters. We can also refer to the unmarked part of these regions. For example, 5/8 of the pie model is not shaded.

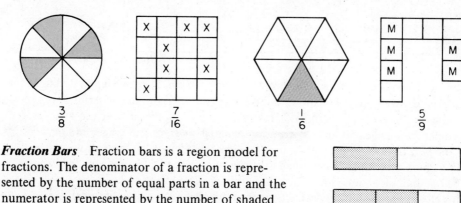

$$\frac{3}{8} \qquad \frac{7}{16} \qquad \frac{1}{6} \qquad \frac{5}{9}$$

Fraction Bars   Fraction bars is a region model for fractions. The denominator of a fraction is represented by the number of equal parts in a bar and the numerator is represented by the number of shaded parts. This model will be used in the following pages to illustrate equality and inequality of fractions and the four basic operations with fractions.

$$\frac{1}{2}$$
$$\frac{2}{3}$$
$$\frac{3}{4}$$

Discrete Objects   The part-to-whole concept of a fraction also occurs in describing part of a set of individual objects. For example, 2/8 of the objects shown here are circles and 5/8 of these objects are squares.

Another common application of fractions involves the *ratio concept* (page 424). We use this concept when comparing two different measurements. For example, the San Andreas Fault is 3/4 of the length of California's coastal region. The ratios of lengths of objects are illustrated by the next model.

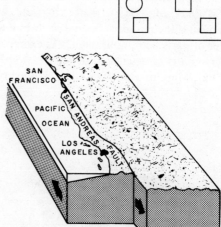

The San Andreas Fault runs 3/4 of the length of California's coastal region.

Cuisenaire Rods   Cuisenaire rods are used as models for fractions by comparing the lengths of 2 rods. It takes 3 red rods to equal the length of 1 dark green rod, so the length of the red rod is 1/3 the length of the dark green rod. If the length of the dark green rod is chosen as the unit, then the red rod represents 1/3. If a different unit is selected, the red rod will represent a different fraction. For example, if the yellow rod is the unit length, then the red rod represents 2/5, because the length of the red rod is 2/5 the length of the yellow rod. Depending on the choice of unit there are several different rods which can represent the same fraction. One-third was represented above by a red rod, but if the unit is the length of a blue rod then the green rod represents 1/3.

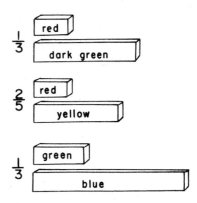

Number Line Fractions are represented on the number line by selecting a unit length and dividing the interval from 0 to 1 into parts of equal length. To locate the fraction *r/s,* subdivide the interval into *s* equal parts, and, beginning at 0, count off *r* of these parts. Here are some examples showing the sixths and tenths between 0 and 1.

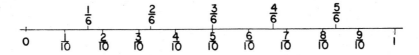

EQUALITY

Equality of fractions can be illustrated by comparing the shaded amounts of regions. This chart shows that 1/3, 2/6, and 4/12 are equal fractions. Another combination of regions shows that

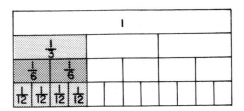

$$\frac{2}{3} = \frac{4}{6} = \frac{8}{12}$$

Different equalities can be illustrated by dividing the unit regions into different numbers of parts. In this chart we see that

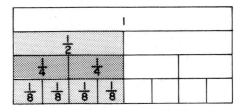

$$\frac{1}{2} = \frac{2}{4} = \frac{4}{8}$$

What other equalities can you see?

Equality of fractions can also be illustrated by sets of objects. Three out of 12 or 3/12 of the points shown here are circled. Viewed in another way, 1/4 of the points are circled because there are four rows containing the same number of points each, and one row is circled. So, 3/12 and 1/4 are equivalent fractions representing the same amount.

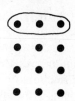

For every fraction there is an infinite number of other fractions which represent the same number. The bars on the right show how to obtain two fractions equal to 3/4. Each part of the first bar has been split into two equal parts to show that 3/4 = 6/8. We see that doubling the number of parts in a bar also doubles the number of shaded parts. This is equivalent to multiplying the numerator and denominator of 3/4 by 2. Similarly, splitting each part of a 3/4 bar into three equal parts triples the number of parts in the bar and triples the number of shaded parts. This has the effect of multiplying the numerator and denominator of 3/4 by 3 and shows that 3/4 is equal to 9/12.

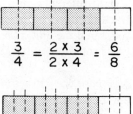

$$\frac{3}{4} = \frac{2 \times 3}{2 \times 4} = \frac{6}{8}$$

$$\frac{3}{4} = \frac{3 \times 3}{3 \times 4} = \frac{9}{12}$$

In general, two fractions are equal if one can be obtained from the other by multiplying its numerator and denominator by a whole number greater than 0.

Definition: For any fraction $\frac{a}{b}$ and whole number $k > 0$,

$$\frac{a}{b} = \frac{ka}{kb}$$

Lowest Terms When the numerator and denominator of a fraction have a common factor greater than 1, the fraction can be written in simplified form. This is done by dividing the numerator and denominator by their greatest common factor. For example, both 6 and 9 in the fraction 6/9 can be divided by 3 to get the fraction 2/3. When the only common factor of the numerator and denominator is 1, the fraction is said to be in *lowest terms*.

$$\frac{6}{9} = \frac{6 \div 3}{9 \div 3} = \frac{2}{3}$$

$$\frac{2}{3} \quad \frac{7}{12} \quad \frac{5}{8} \quad \frac{4}{9}$$

Fractions in lowest terms

COMMON DENOMINATORS

One of the more important skills in the use of fractions is that of replacing two fractions with different denominators by two fractions having equal denominators. The fractions 1/6 and 1/4 have different denominators, and the bars representing these fractions have different numbers of parts (below). If each part of the 1/6 bar is split into 2 equal parts and each part of the 1/4 bar is split into 3 equal parts, both bars will have 12 equal parts. The fractions for these new bars, 2/12 and 3/12, have a common denominator of 12.

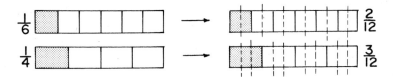

Obtaining the same number of equal parts for two bars is a visual way of illustrating common denominators. Another method for finding common denominators of two fractions is to list the multiples of their denominators. The arrows shown here point to the common multiples of 6 and 4. The least common multiple (l.c.m.) of 6 and 4 is 12. This is also the smallest common denominator of 1/6 and 1/4. In general, the *smallest common denominator* of two fractions is the least common multiple of their denominators.

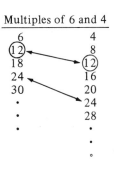

Multiples of 6 and 4

Once a common denominator is found, two fractions can be replaced by fractions having the same denominator.

$$\frac{1}{6} = \frac{2 \times 1}{2 \times 6} = \frac{2}{12} \qquad \frac{1}{4} = \frac{3 \times 1}{3 \times 4} = \frac{3}{12}$$

Sometimes only one fraction has to be replaced in order to obtain two fractions with the same denominator. For the fractions 2/3 and 11/18, 18 is the least common multiple of the two denominators. Therefore, it is necessary only to replace 2/3 by 12/18 to obtain two fractions with equal denominators.

INEQUALITY

Charts, such as the one pictured here, show many different inequalities between fractions. Place the edge of a piece of paper on this chart to compare the regions for 4/5 and 7/9. You will see that 7/9 is just a little less than 4/5.

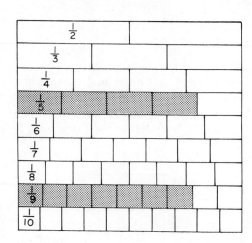

We can also see the pattern of decreasing inequalities,

$$\frac{1}{2} > \frac{1}{3} > \frac{1}{4} > \frac{1}{5} > \frac{1}{6} > \frac{1}{7} > \frac{1}{8} > \frac{1}{9} > \frac{1}{10}$$

and that of the increasing inequalities,

$$\frac{1}{2} < \frac{2}{3} < \frac{3}{4} < \frac{4}{5} < \frac{5}{6} < \frac{6}{7} < \frac{7}{8} < \frac{8}{9} < \frac{9}{10}$$

One of the reasons for finding a common denominator for two fractions is to be able to determine the greater fraction. For the fractions 5/8 and 3/5, it is difficult to know which is greater without first replacing them by fractions having a common denominator. The least common multiple of 8 and 5 is 40. Replacing both 5/8 and 3/5 by fractions having a denominator of 40, we see that 5/8 is the greater fraction, but that they differ by only 1/40.

$$\frac{5}{8} = \frac{5 \times 5}{5 \times 8} = \frac{25}{40} \qquad \frac{3}{5} = \frac{8 \times 3}{8 \times 5} = \frac{24}{40}$$

Any two fractions, a/b and c/d, can be compared by replacing them with fractions having a common denominator, ad/bd and bc/bd. An equality or inequality for these fractions can then be determined by comparing the numerators, ad and bc, of these fractions.

Definition: For any fractions a/b and c/d,
(1) $a/b < c/d$ if and only if $ad < bc$ and
(2) $a/b = c/d$ if and only if $ad = bc$.

DENSITY OF FRACTIONS

The whole numbers are evenly spaced on the number line, and for any whole number there is "the next" whole number. However, for fractions this is not possible. There is no fraction which is the next one greater than 1/2. We

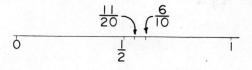

know that 6/10 is greater than 1/2 and differs from it by 1/10. But 11/20 is also greater than 1/2 and differs from it by only 1/20. We could continue with the fractions 16/30, 21/40, 26/50, etc., all greater than 1/2 and each one closer to 1/2.

Similarly, there is no fraction next to 0. Stated in another way, there is no smallest fraction greater than 0. The following sequence of fractions gets closer and closer to 0, but no matter how far we go in this sequence, these fractions will always be greater than 0.

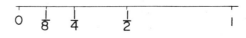

$$\frac{1}{2} \quad \frac{1}{4} \quad \frac{1}{8} \quad \frac{1}{16} \quad \frac{1}{32} \quad \frac{1}{64} \quad \frac{1}{128} \cdots$$

These examples are special cases of the more general fact that between any two fractions there is always another fraction. We refer to this by saying that the fractions are *dense*. As a result of this fact, there are an infinite number of fractions between any two fractions. Thus, it appears that the number line is just about filled up with fractions. However, in Section 7.4 you will see that there are lots of points on the number line which do not have corresponding fractions.

FRACTIONS AND CALCULATORS

One method of comparing two fractions to determine whether or not they are equal, or which is the greater, is to find their decimal representations by dividing the numerator by the denominator. For example, 7/12 and 9/16 are both very close to 1/2, but which is the greater fraction? The top calculator display was obtained by dividing 7 by 12 and shows that 7/12 is about 58 hundredths. The second display is from dividing 9 by 16 and shows that 9/16 is about 56 hundredths. Therefore, 7/12 is greater than 9/16 by about 2 hundredths.

7/12   `0.58333333`

9/16   `0.56250000·`

For those fractions which are represented by *infinite repeating decimals,* such as 7/12 = .5833333333 . . . , the number in the calculator display is an approximation because it shows only a few of the digits. For most applications this is sufficient accuracy, and we will not need to be concerned over the fact that the decimal is not exactly equal to the fraction. The decimal .5625, on the other hand, is exactly equal to 9/16 because it is a *terminating decimal* whose digits are contained on the calculator display. For more details on infinite repeating decimals and terminating decimals see page 394.

MIXED NUMBERS AND IMPROPER FRACTIONS

Historically, a fraction stood for part of a whole and represented numbers less than 1. The idea of a fraction such as 4/4 or 5/4 having a numerator greater than or equal to the denominator was uncommon even as late as the sixteenth century. Such fractions are sometimes called *improper fractions*.

When these fractions are written as a combination of whole numbers and fractions, they are called *mixed numbers*. The numbers 1 1/5, 2 3/4, and 4 2/3 are examples of mixed numbers. Placing a whole number and a fraction side by side, as in mixed numbers, indicates the sum of the two numbers. For example, 1 1/5 means 1 + 1/5. The following number lines are labeled with fractions, mixed numbers, and whole numbers.

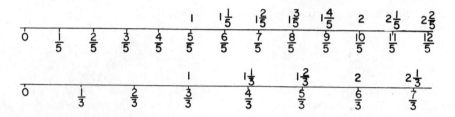

APPROXIMATION AND MENTAL CALCULATION

The smaller of two unit fractions can be determined by inspection; the fraction with the larger denominator is the smaller fraction. This can be illustrated with the fraction bar model by noting that the more parts in a bar, the smaller the parts. More generally, for any set of unit fractions, the smallest fraction is the one with the greatest denominator and the largest fraction is the one with the smallest denominator. For example, 1/20 is the smallest of the following fractions and 1/5 is the largest.

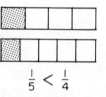

$$\frac{1}{12} \quad \frac{1}{5} \quad \frac{1}{9} \quad \frac{1}{20} \quad \frac{1}{11}$$

It is also easy to mentally classify a fraction as either less than 1/2, equal to 1/2, or greater than 1/2. If the numerator is less than half the denominator, the bar for the fraction is less than half-shaded and the fraction is less than 1/2. If the numerator is equal to half the denominator, the bar for the fraction is half-shaded and the fraction is equal to 1/2. If the numerator is greater than the denominator, then more than half of the bar for the fraction is shaded and the fraction is greater than 1/2.

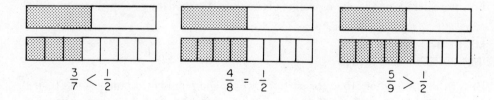

An inequality for two fractions can sometimes be determined quickly by comparing each fraction to 1/2. For example, since 4/11 is less than 1/2 and 5/9 is greater than 1/2, we know that 4/11 < 5/9. By similar reasoning, 5/7 > 3/8.

$$\frac{4}{11} < \frac{5}{9}$$

$$\frac{5}{7} > \frac{3}{8}$$

Rounding Off Rounding off fractions and mixed numbers to the nearest whole number involves comparing fractions to 1/2. If a fraction is less than 1/2, it is rounded off to 0, and if it is greater than or equal to 1/2, it is rounded off to 1. Similarly, a mixed number is rounded down or up depending on whether the fraction rounds off to 0 or 1. Here are some examples.

$$\frac{2}{3} \xrightarrow{\text{rounds off to}} 1 \qquad 2\frac{5}{9} \xrightarrow{\text{rounds off to}} 3$$

$$\frac{1}{5} \xrightarrow{\text{rounds off to}} 0 \qquad 4\frac{3}{7} \xrightarrow{\text{rounds off to}} 4$$

$$\frac{1}{2} \xrightarrow{\text{rounds off to}} 1 \qquad 7\frac{3}{6} \xrightarrow{\text{rounds off to}} 8$$

COMPUTER APPLICATIONS

Here are all the different fractions from 0 to 1 with denominators of 5 or less which are in lowest terms.

$$\frac{0}{1} \quad \frac{1}{1} \quad \frac{1}{2} \quad \frac{1}{3} \quad \frac{2}{3} \quad \frac{1}{4} \quad \frac{3}{4} \quad \frac{1}{5} \quad \frac{2}{5} \quad \frac{3}{5} \quad \frac{4}{5}$$

We can see that there are two fractions with a denominator of 1; three fractions with a denominator of 2 or less; five fractions with a denominator of 3 or less; seven fractions with a denominator of 4 or less; and eleven fractions with a denominator of 5 or less. These numbers of fractions are the first four prime numbers. Will the number of fractions in this sequence be a prime for any denominator?

Let's look at the program for printing all the fractions with a denominator of 6 or less. This program has two loops: one for the denominators and one for the numerators. First, in line 10, B takes on a value of 1 for the denominator, and in lines 20, 30, and 40, A takes on the values of 0 and 1 for the numerator. Then B takes on a value of 2 from line 10 for the denominator and this time A is assigned the values of 0, 1, and 2, etc. Notice that the division symbol in line 30 is in quotes. This causes the slash to be printed for the fractions. Without the quotation marks the computer would divide A by B and print out a decimal.

Program 6.1A

```
10  FOR B = 1 TO 6
20  FOR A = 0 TO B
30  PRINT A"/"B" ";
40  NEXT A
50  NEXT B
60  END
```

RUN

```
0/1 1/1 0/2 1/2 2/2 0/3 1/3 2/3 3/3 0/4 1/4 2/4 3/4 4/4
0/5 1/5 2/5 3/5 4/5 5/5 0/6 1/6 2/6 3/6 4/6 5/6 6/6
```

This printout contains all the fractions with a denominator of 6 or less but some of the fractions are not in lowest terms. This problem is corrected by the next program. Lines 22, 24, and 26 (Similar lines were also used on page 188 for finding the greatest common factor of two numbers.) form a loop in which the numerator and denominator of the fraction are tested to see if they have a common factor greater than 1. If they do, line 24 sends the computer to line 40 and the fraction in not printed.

Program 6.1B

```
10  FOR B = 1 TO 6
20  FOR A = 0 TO B
22  FOR N = 2 TO B
24  IF B / N =  INT (B / N) AND A / N =  INT (A / N) THEN
      GOTO 40
26  NEXT N
30  PRINT A"/"B" ";
40  NEXT A
50  NEXT B
60  END
```

RUN

```
0/1 1/1 1/2 1/3 2/3 1/4 3/4 1/5 2/5 3/5 4/5 1/6 5/6
```

Notice that this printout has 13 fractions and so the pattern of primes continues to hold.

SUPPLEMENT (Activity Book)

Activity Set 6.1 Models for Equality and Inequality (Fraction Bars and Cuisenaire Rods)
Just for Fun: Fraction Games

Exercise Set 6.1: Applications and Skills

1. This is a shutter speed knob on top of a 35-mm camera. The numerals on this knob determine the amount of time the camera's shutter stays open. The settings of "4" and "2," following "B," open the shutter for 4 and 2 seconds, respectively. The remaining numerals 1, 2, 4, 8, etc., represent 1, 1/2, 1/4, 1/8, etc., of a second. The fastest opening on this camera is 1/1000 of a second.

35-mm camera shutter speed knob

★   a. The less light that is available, the longer the shutter must stay open for the film to be properly exposed. Will a shutter setting of 15 allow more or less light than a setting of 60?

     b. If a shutter setting of 250 doesn't allow quite enough light, what number should the dial be set on?

2. There are only a few fractions with different denominators which occur frequently in newspapers, magazines, sales ads, etc.

     a. Name the different denominators for the fractions in the following collage.

     b. There are many mixed numbers in the collage. Name ten of them.

     c. Write your mixed numbers in part **b** as improper fractions.

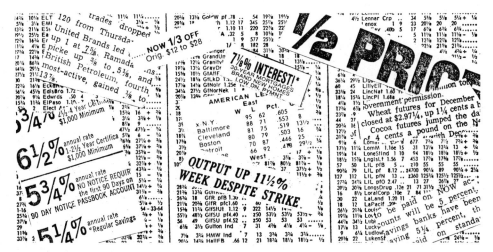

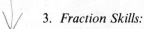

3. *Fraction Skills:*

   a. Write each of these fractions as mixed numbers or as whole numbers.

   $$\frac{5}{3} \qquad \frac{8}{8} \qquad \bigstar \; \frac{25}{6} \qquad \frac{21}{7} \qquad \frac{17}{5}$$

   b. Write each of these mixed numbers as fractions.

   $$1\frac{3}{4} \qquad \bigstar \; 2\frac{1}{5} \qquad 4\frac{2}{3} \qquad \bigstar \; 2\frac{5}{6} \qquad 1\frac{3}{7}$$

   c. Write the missing numbers for these fractions.

   $$\frac{7}{8} = \frac{28}{32} \qquad \bigstar \; \frac{360}{3} = \frac{40}{24} \qquad \frac{2}{3} = \frac{12}{72} \qquad \bigstar \; \frac{5}{} = \frac{20}{24}$$

   d. Write each fraction as a fraction with a denominator of 9.

   $$\frac{4}{18} \qquad \frac{12}{27} \qquad \bigstar \; \frac{4}{12} \qquad \frac{16}{24}$$

4. Complete the equations so that each pair of fractions has the smallest common denominator. → needed for adding & subtracting

   a. $\frac{2}{3} =$ b. $\frac{1}{6} =$ ★ c. $\frac{3}{15} =$ d. $\frac{5}{8} =$

   $\frac{4}{5} =$ $\frac{7}{12} =$ $\frac{5}{6} =$ $\frac{3}{10} =$

5. Split the parts of these fraction bars to illustrate the given equalities.

   ★ a. 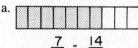   b. 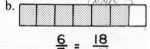   c.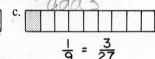

   $$\frac{7}{10} = \frac{14}{20} \qquad\qquad \frac{6}{7} = \frac{18}{21} \qquad\qquad \frac{1}{9} = \frac{3}{27}$$

6. Cuisenaire rods represent various fractions, depending on the choice of the unit rod. If the unit rod is brown, then the purple rod represents 1/2.

    ★ a. What is the unit rod if the purple rod represents 2/3? dark green

   b. If the unit rod is the orange rod, what fraction is represented by the black rod?

    ★ c. If the dark green rod represents 3/4, what is the unit rod? brown

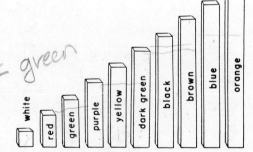

60 < 147

336

7. This news clipping shows the five
most active stocks and the twenty
top stocks in dollars and fractions
of a dollar for a given day. The first
column of numbers contains the
daily closing prices. The second
column shows the fraction of a
dollar which the stock was up (u),
or down (d), or unchanged (unc.)
from the previous day's price.

★ a. Find the cost in dollars and
cents for these stocks.

    Dow Ch   Gen Elec   Gulf Oil

★ b. Find the daily price change in
cents for these stocks.

    RCA   Xerox   Tyco Lb

5 MOST ACTIVE			
Occid. Pt	17 3/8	d	1/8
Dow Ch	44 3/8	d	1/4
Gn Mot	67 1/4	u	5/8
Aetna Lf	31 1/2	unc.	
Brist My	74	u	3/8

20 TOP STOCKS			
Am T&T	59	u	1/4
RCA	27 1/8	u	1/8
Data Genl	47 5/8	u	3/8
Nat Gyp	14 3/8	unc.	
P Sv Eg	20 3/4	u	1/8
Con Foods	25 1/4	d	1/8
Reyn Ind	60	u	1/4
Pub S.N.H.	20 1/2	unc.	
Xerox	63 3/8	u	3/16
U.S. Steel	47 3/4	d	1/2
Exxon	51 7/8	u	1/2
Whel Fry	22 1/8	d	1/4
Gen Elec	52 3/4	u	1/4
Gulf Oil	25 7/8	d	1/4
Polaroid	37 7/8	u	1/4
Unit Tech	33 1/4	u	1/16
Con Edis	18 7/8	d	1/8
Tyco Lb	12 5/8	d	1/4
Ca Pw Pf	27 7/8	unc.	
IBM	271 3/4	u	1/2

8. Here are five daily reports of stock
prices with the lowest and highest
costs per share for the day.

★ a. Which of these five stocks had
the lowest cost for the day?

b. Which stock had the highest
cost for the day?

★ c. Alaska Airlines finished the day
at 8 and 3/16 dollars. Is this
greater than or less than the
day's high price for Canadian
Homestead?

Company	Low Price	High Price
Alaska Airlines	$7\frac{5}{8}$	$8\frac{3}{4}$
Canadian Homestead	$7\frac{5}{16}$	$8\frac{1}{4}$
Mobile Home Indiana	$23\frac{5}{8}$	$24\frac{1}{2}$
Ranger O Can	$19\frac{1}{2}$	$20\frac{3}{8}$
Vintage Enterprise	$28\frac{3}{8}$	$29\frac{3}{4}$

9. *Calculator Exercise:* Use a calculator to rewrite these fractions in increasing order.

$$\frac{16}{20} \qquad \frac{19}{34} \qquad \frac{38}{52} \qquad \frac{21}{25} \qquad \frac{11}{17}$$

10. *Mental Calculations:* An inequality for each of the following pairs of fractions can be found mentally without getting a common denominator. Determine each inequality and explain the reason for your answer.

   a. $\dfrac{1}{9}$ $\dfrac{1}{12}$ ★ b. $\dfrac{2}{5}$ $\dfrac{4}{6}$ c. $\dfrac{5}{9}$ $\dfrac{5}{12}$ d. $\dfrac{1}{5}$ $\dfrac{1}{3}$

11. *Rounding Off:* Round off each fraction or mixed number to the nearest whole number.

   a. $\dfrac{4}{10}$ b. $1\dfrac{1}{3}$ ★ c. $3\dfrac{1}{2}$ d. $\dfrac{4}{7}$ e. $2\dfrac{3}{4}$ ★ f. $\dfrac{2}{5}$

12. *Word Problems:*

   a. Magnesium is lighter than iron but stronger. One-fiftieth of the earth's crust is magnesium, and 1/20 of the earth's crust is iron. Is there more iron or more magnesium?

★ b. Pure gold is quite soft. To make it more useful for such items as rings and jewelry it is mixed with other metals, such as copper and zinc. Pure gold is marked 24k (24 karat). What fraction of a ring is pure gold if it is marked 14k? Write your answer in lowest terms.

   c. Some health authorities say that we should have 1 gram of protein a day for each kilogram of our weight. There are about 40 grams of protein in a liter of milk. A liter of fat-free milk weighs about 1040 grams. What fraction of milk's weight is protein? Write your answer in lowest terms.

Exercise Set 6.1: Problem Solving

1. *Poetry and Precision:* In Lord Tennyson's poem, "The Vision of Sin," there is a verse which reads,

> Every minute dies a man,
> Every minute one is born.

In response to these lines the English engineer Charles Babbage wrote a letter to Tennyson in which he noted that if this were true, the population of the world would be at a standstill. He suggested that the next edition of the poem should read:

> Every moment dies a man,
> Every moment 1 1/16 is born.

The world population birth and death rates in 1983 were 43 live births every 10 seconds and 16 deaths every 10 seconds.* What mixed number should be used in this poem for these rates in place of 1 1/16?

*Mary Kent, *1983 World Population Data Sheet* (Washington, D.C.: Population Reference Bureau, 1983).

a. *Understanding the Problem:* Let's determine the birth and death rates at the time of Babbage's calculation. If 1 1/16 people are born each moment then in 16 moments 17 people will be born. That is, a sum with 1 1/16 recurring 16 times equals 17.

$$1\frac{1}{16} + 1\frac{1}{16} + 1\frac{1}{16} + 1\frac{1}{16} + 1\frac{1}{16} + 1\frac{1}{16} + 1\frac{1}{16} + \cdots + 1\frac{1}{16} = 17$$

For every 17 that are born, how many will die?

★ b. *Devising a Plan:* Simplifying a problem by using more convenient numbers will often help to illustrate the concepts involved. For example, what number should be used in place of 1 1/16 if: 2 people were born for every person that died?; 3 people were born for every 2 people that died?

c. *Carrying Out the Plan:* Use your method of getting the correct answer in part **b** on the 17 births and 16 deaths in part **a**. Do you get 1 1/16? What number should be used in the poem for the world's 1983 birth and death rates?

d. *Looking Back:* What can be said about a number to replace 1 1/16 if the world's death rate should become greater than the birth rate?

★ 2. *Damaged Plates:* Mr. Hash bought some plates at a yard sale. After arriving home he found that 2/3 of the plates were chipped, 1/2 were cracked, and 1/4 were chipped and cracked. Only two plates were without chips or cracks. How many plates did he buy in all? (*Hint:* Use Venn diagrams and try guessing and checking.)

3. *Fibonacci Numbers:* The Fibonacci numbers, 1, 1, 2, 3, 5, 8, 13. . . . , occur as the numerators and denominators of fractions that are associated with patterns of leaf arrangements on trees and plants.

a. Begin with a leaf from a pear tree and count the number of leaves until you reach a leaf just above the first. This number of leaves, not counting the first (zero leaf), is the Fibonacci number 8. Furthermore, in passing around the branch from a leaf to the one directly above, you will make three complete turns. This means that each leaf is 3/8 of a turn from its adjacent leaves. This fraction, which is called *phyllotaxis* (or *leaf divergence*), can be found for trees by counting the leaves (or branches) and the turns around the stem. Fill in the missing numbers for phyllotaxis in the following table. (*Hint:* Look for a relationship between the numerator and denominator in each of the first three fractions.)

Pear Tree Stem

Tree	Apple	Pear	Beech	Almond	Elm	Willow
Phyllotaxis	$\dfrac{2}{5}$	$\dfrac{3}{8}$	$\dfrac{1}{3}$	$\dfrac{5}{}$	$\dfrac{}{2}$	$\dfrac{3}{}$

★ b. The seeds of sunflowers and daisies, the scales of cones and pineapples, and the leaves on certain vegetables are arranged in two spirals. In these cases the numerator and denominator of the fraction for phyllotaxis gives the number of clockwise and counterclockwise spirals. Look for a relationship between the numerator and denominator of each fraction and fill in the missing numbers for phyllotaxis in this table.

Pine cone

Plant	White Pine Cone	Pineapple	Daisy	Cauliflower	Celery	Medium Sunflower	Large Sunflower
Phyllotaxis	$\dfrac{5}{8}$	$\dfrac{8}{13}$	$\dfrac{21}{34}$	$\dfrac{}{3}$	$\dfrac{1}{}$	$\dfrac{55}{}$	$\dfrac{}{144}$

4. *Tower of Bars:* This tower of bars represents fractions with denominators 2 through 12. Each geometric pattern corresponds to a pattern of fractions. Explain how the tower illustrates the following facts. Complete each pattern of fractions for the tower of bars.

a. $\dfrac{1}{2} = \dfrac{2}{4} = \dfrac{3}{6} = \cdots$

b. $\dfrac{1}{3} = \dfrac{2}{6} = \cdots$

★ c. $\dfrac{1}{3} < \dfrac{2}{4}, \dfrac{2}{4} < \dfrac{3}{5}, \dfrac{3}{5} < \dfrac{4}{6}, \cdots$

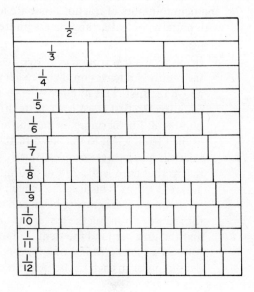

d. $\dfrac{1}{2} = \dfrac{1}{3} +$ half of $\dfrac{1}{3}$

$\dfrac{2}{4} = \dfrac{2}{5} +$ half of $\dfrac{1}{5}$

$\dfrac{3}{6} = \dfrac{3}{7} +$ half of $\dfrac{1}{7}$

$\vdots$

5. *Three Hungry Men:** Three tired and hungry men had a bag of apples. When they were asleep one of them awoke, ate 1/3 of the apples and went back to sleep. Later a second man awoke, ate 1/3 of the remaining apples, and went back to sleep. Finally, the third man awoke and ate 1/3 of the remaining apples, leaving 8 apples in the bag. How many apples were in the bag originally?

Exercise Set 6.1: Computers

1. Revise Program 6.1B (page 334), which prints fractions in lowest terms, so that it will print out all the fractions with denominators less than or equal to K. This can be done by inserting a new line 5 and changing line 10.

   5 INPUT K
   10 FOR B = 1 TO K

   a. Use the revised program to complete the following table.
   b. How far do the primes continue?

Denominators less than or equal to:	1	2	3	4	5	6	7	8	9	10	11	12
Number of fractions	2	3	5	7	11	13						

2. A sequence of increasing fractions in lowest terms from 0 to 1 whose denominators are less than or equal to n is called the *n-th Farey sequence.* Here is the 4-th Farey sequence.

$$\dfrac{0}{1} \qquad \dfrac{1}{4} \qquad \dfrac{1}{3} \qquad \dfrac{1}{2} \qquad \dfrac{2}{3} \qquad \dfrac{3}{4} \qquad \dfrac{1}{1}$$

*"October Calendar," *The Mathematics Teacher,* 76 No. 6 (October 1983), pp. 502–3.

These fractions have the following *Farey property:* if we take any three in a row, say 1/4, 1/3, and 1/2, the middle fraction can be obtained by adding the numerators and denominators of the other two fractions.

$$\frac{1+1}{4+2} = \frac{2}{6} = \frac{1}{3}$$

a. List the 11 fractions in the 5-th Farey sequence. Does the Farey property hold for any three fractions in a row?

★ b. List the 13 fractions in the 6-th Farey sequence. Does the Farey property hold for any three fractions in a row?

c. Does the Farey property hold if we use the fractions in the order they are printed by the program in Exercise 1?

3. Select two unequal fractions. Form a new fraction by adding the numerators to get a new numerator and adding the denominators to get a new denominator. In the example shown here, the new fraction is 4/7 and it is between 1/2 and 3/5. If this method is used on any two nonequal fractions, will the new fraction be between the two fractions? Write a computer program to help you investigate this question. Here are the first few lines. $\frac{1}{2} < \frac{1+3}{2+5} < \frac{3}{5}$

```
10   PRINT "TYPE THE NUMERATOR AND DENOMINATOR OF THE SMALLER
         FRACTION. SEPARATE THEM BY A COMMA."
20   INPUT A, B
30   PRINT "TYPE THE NUMERATOR AND DENOMINATOR OF THE GREATER
         FRACTION. SEPARATE THEM BY A COMMA."
40   INPUT C, D
```

★ 4. There are occasions when it is convenient to have a computer program which checks a fraction to see if it is in lowest terms, and if not, to write an equivalent fraction in lowest terms. Write a program to do this. (*Hint:* Use lines 22, 24, and 26 from Program 3.6B on page 188. Let N vary from B to 2.)

United States space shuttle, *Enterprise*

6.2
OPERATIONS WITH FRACTIONS

Fractions are often used to indicate how much smaller a diagram or picture is than the original object. A number which describes the relative size from an object to its representation is called a *scale factor*. Scale factors may be greater than 1 (see Section 9.2) or less than 1. If a model or drawing is bigger than the actual object, the scale factor is greater than 1. Models of buildings, planes, and ships have scale factors that are less than 1. The scale factor from the space shuttle orbiter *Enterprise* to the model of the space shuttle which is shown here is approximately 1/50. This model is in a wind tunnel which simu-

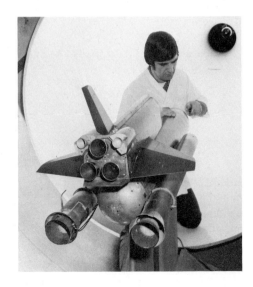

lates space conditions at 4 1/2 times the speed of sound and up to 32 miles above the earth. The rocket engines in this model actually fire during tests.

In each of the following examples the scale factor from the object to its picture is a fraction less than 1. The picture of the foxhound is 32 times smaller than the actual dog, and so the scale factor from the dog to its picture is 1/32. The reductions in the size of the trumpeter and turkey buzzard are also indicated by fractions.

Foxhound

Trumpeter

Turkey buzzard

ADDITION

The concept of addition of fractions is the same as that for whole numbers. The addition of whole numbers is illustrated by "putting together" or "combining" two sets of objects. Similarly, the addition of fractions can be illustrated by combining two regions. By placing two bars end to end, the total shaded amount represents the sum of the two fractions.

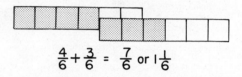

$$\frac{4}{6} + \frac{3}{6} = \frac{7}{6} \text{ or } 1\frac{1}{6}$$

Addition of fractions can also be illustrated on the number line by placing arrows for the fractions end to end.

$$\frac{5}{8} \quad + \quad \frac{6}{8} \quad = \quad \frac{11}{8} \text{ or } 1\frac{3}{8}$$

0   1 $1\frac{3}{8}$ 2

Unlike Denominators The difficulty in adding fractions occurs when the denominators are unequal. The 2/5 bar and the 1/3 bar show that 2/5 + 1/3 is greater than 3/5 but less than 4/5. To

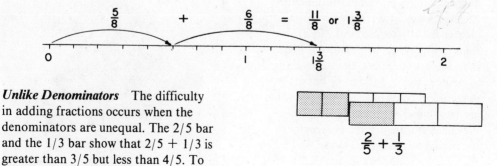

$$\frac{2}{5} + \frac{1}{3}$$

344

determine this sum exactly, the two fractions must be replaced by fractions having the same denominator.

Since the smallest common denominator of 2/5 and 1/3 is 15, these two fractions can be replaced by 6/15 and 5/15, respectively. The sum of these two fractions is 11/15, so 2/5 + 1/3 = 11/15.

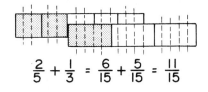

$$\frac{2}{5} + \frac{1}{3} = \frac{6}{15} + \frac{5}{15} = \frac{11}{15}$$

Multiplying the denominators of two fractions by each other will always yield a common denominator, but not necessarily the smallest one. Once two fractions have a common denominator, their sum is computed by adding the numerators and retaining the denominator.

Definition: For any fractions $\frac{a}{b}$ and $\frac{c}{d}$,

$$\frac{a}{b} + \frac{c}{d} = \frac{ad}{bd} + \frac{bc}{bd} = \frac{ad + bc}{bd}$$

Mixed Numbers Mixed numbers are combinations of whole numbers and fractions. The sum of two such numbers can be found by adding the whole numbers and the fractions separately. If the denominators of the fractions are unequal, as in this example, a common denominator must be found before adding.

$$2\frac{1}{4} = \quad 2\frac{3}{12}$$
$$+1\frac{2}{3} = +1\frac{8}{12}$$
$$\overline{\qquad\qquad 3\frac{11}{12}}$$

SUBTRACTION

The concept of subtraction for fractions is the same as for subtraction of whole numbers. That is, we can think of subtraction as "take away" or in terms of "missing addends." These two bars, for example, show that 1/2 take away 1/6 is 2/6, and that 2/6 must be added to 1/6 to get 1/2.

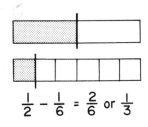

$$\frac{1}{2} - \frac{1}{6} = \frac{2}{6} \text{ or } \frac{1}{3}$$

To illustrate subtraction on the number line, the fraction to be subtracted is represented by an arrow from right to left, just as was done when subtracting whole numbers on the number line. In this example, 11/12 − 7/12 = 4/12, or 1/3.

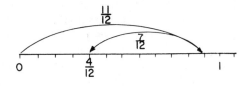

Unlike Denominators   These bars show that the difference between 5/6 and 1/4 is greater than 3/6 and less than 4/6. To compute this difference we can replace these fractions by fractions having a common denominator.

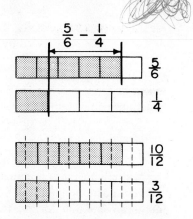

The smallest common denominator of 5/6 and 1/4 is 12, so these fractions can be replaced by 10/12 and 3/12, respectively. The difference between these two fractions is 7/12, so $5/6 - 1/4 = 7/12$.

$$\frac{5}{6} - \frac{1}{4} = \frac{10}{12} - \frac{3}{12} = \frac{7}{12}$$

The general rule for subtracting fractions is always stated by using a common denominator that is the product of the two denominators, even though this may not be the smallest common denominator. Once two fractions have a common denominator, their difference is computed by subtracting the numerators and retaining the denominators.

Definition: For any fractions $\frac{a}{b}$ and $\frac{c}{d}$,

$$\frac{a}{b} - \frac{c}{d} = \frac{ad}{bd} - \frac{bc}{bd} = \frac{ad - bc}{bd}$$

Mixed Numbers   The difference between two mixed numbers can be found by subtracting the whole number parts and the fractions separately. Sometimes before this can be done, borrowing or regrouping is necessary. In the accompanying example, the 4 1/5 is replaced by 3 6/5 before subtracting.

$$
\begin{array}{rcl}
4\frac{1}{5} & = & 3\frac{6}{5} \\[2mm]
- 1\frac{2}{5} & = & -1\frac{2}{5} \\[2mm]
\hline
& & 2\frac{4}{5}
\end{array}
$$

$$4\frac{1}{5} = 4 + \frac{1}{5} = (3+1) + \frac{1}{5} = 3 + \left(1 + \frac{1}{5}\right) = 3 + \left(\frac{5}{5} + \frac{1}{5}\right) = 3 + \frac{6}{5} = 3\frac{6}{5}$$

If the denominators of the fractions in the mixed numbers are unequal, the fractions must be replaced by fractions having a common denominator. In this example, 3/4 and 1/3 were replaced by 9/12 and 4/12, respectively, before the numbers were subtracted. In some cases, both borrowing and changing denominators will be necessary before subtracting mixed numbers.

$$
\begin{array}{rcl}
3\frac{3}{4} & = & 3\frac{9}{12} \\[2mm]
- 1\frac{1}{3} & = & -1\frac{4}{12} \\[2mm]
\hline
& & 2\frac{5}{12}
\end{array}
$$

346

handwritten at top: of means times

$$\frac{1}{2} \div 3 = \frac{1}{2} \times \frac{1}{3}$$
reciprocal

MULTIPLICATION

In the product of two whole numbers, $m \times n$, we think of the first number m as indicating "how many" of the second number. For example, 2×3 means 2 "of the" 3s. Multiplication of fractions may be viewed in a similar manner. In the product $1/3 \times 6$, the first number tells us "how much" of the second, that is, $1/3 \times 6$ means $1/3$ "of" 6.

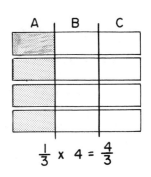

$\frac{1}{3} \times 6 = 2$

There is one major difference between multiplying by a whole number and multiplying by a fraction. When multiplying by a whole number greater than 1, the product is greater than the second number being multiplied. However, when multiplying by a fraction less than 1, the product is less than the second number being multiplied. This has always been a problem in explaining multiplication by fractions.*

Whole Number Times Fraction In this case we can interpret multiplication to mean "repeated addition," just as we did for multiplication of whole numbers in Chapter 3. The whole number indicates the number of times the fraction is to be added to itself. For any whole number, k, $k \times a/b$ is equal to a sum with a/b occuring k times. In this example, $k = 3$ and $a/b = 2/5$.

$$3 \times \frac{2}{5} = \frac{2}{5} + \frac{2}{5} + \frac{2}{5} = \frac{6}{5} \quad \text{or} \quad 1\frac{1}{5}$$

Fraction Times Whole Number The product $1/3 \times 4$ means $1/3$ "of" 4. To illustrate this we will use 4 bars and divide them into 3 equal parts. The vertical lines partition these bars into equal parts A, B, and C. Part A, which is one-third of the 4 bars, consists of 4 one-thirds, or 4 thirds. That is,

$$\frac{1}{3} \times 4 = \frac{1}{3} + \frac{1}{3} + \frac{1}{3} + \frac{1}{3}$$

$$= \frac{4}{3}$$

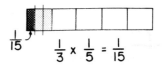

$\frac{1}{3} \times 4 = \frac{4}{3}$

Fraction Times Fraction The product $1/3 \times 1/5$ means $1/3$ of $1/5$. This can be illustrated by using a $1/5$ bar and taking $1/3$ of its shaded amount. In order to do this, the shaded part of the bar has been split into three equal parts. The double-shaded part

$\frac{1}{15}$ $\frac{1}{3} \times \frac{1}{5} = \frac{1}{15}$

*According to D. E. Smith, *History of Mathematics*, **2** (Lexington, Mass: Ginn, 1925), p. 225, even early writers (fifteenth and sixteenth centuries) on the subject of fractions expressed concern over this problem.

of the bar is 1/3 of 1/5. Each of these new parts is 1/15 of a whole bar, so $1/3 \times 1/5 = 1/15$.

To illustrate $1/3 \times 4/5$, we will use a 4/5 bar and take 1/3 of each of the shaded regions. To do this, each shaded part of the 4/5 bar is split into three equal parts. Each of these new parts is 1/15 of a whole bar. The four double-shaded parts of the bar represent 4/15, so, $1/3 \times 4/5 = 4/15$.*

The rule for computing the product of two fractions is the easiest of the algorithms for the four basic operations. This rule is suggested in the preceding illustrations of a fraction times a fraction. In each of these examples the product can be found by multiplying numerator by numerator and denominator by denominator.

Definition: For any fractions $\dfrac{a}{b}$ and $\dfrac{c}{d}$,

$$\frac{a}{b} \times \frac{c}{d} = \frac{ac}{bd}$$

DIVISION

Division of fractions can be viewed in much the same way as division of whole numbers. One of the meanings of division of whole numbers is the "repeated subtraction" or "measurement" concept (see page 150). For example, to explain $15 \div 3$, we often say, "How many times can we subtract 3 from 15?" Similarly, for $3/5 \div 1/10$ we can ask, "How many

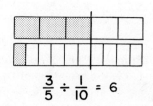

times can we subtract 1/10 from 3/5?" The bars show that the shaded amount of a 1/10 bar can be subtracted from the shaded amount of a 3/5 bar six times. Or, viewed in terms of multiplication, the shaded amount of the 3/5 bar is six times greater than the shaded amount of the 1/10 bar.

This interpretation for division of fractions continues to hold even when the quotient is not a whole number. The bars in this sketch show that the shaded amount of the 1/3 bar can be subtracted from the shaded amount of the 5/6 bar two times, and there is a remainder.

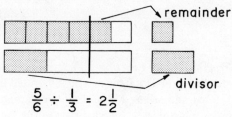

---

*Because of the distributive property, taking 1/3 of 4/5 of a bar is the same as taking 1/3 of each 1/5 of a bar: $\dfrac{1}{3} \times \dfrac{4}{5} = \dfrac{1}{3} \times \left(\dfrac{1}{5} + \dfrac{1}{5} + \dfrac{1}{5} + \dfrac{1}{5}\right) = \left(\dfrac{1}{3} \times \dfrac{1}{5}\right) + \left(\dfrac{1}{3} \times \dfrac{1}{5}\right) + \left(\dfrac{1}{3} \times \dfrac{1}{5}\right) + \left(\dfrac{1}{3} \times \dfrac{1}{5}\right) = \dfrac{1}{15} + \dfrac{1}{15} + \dfrac{1}{15} + \dfrac{1}{15} = \dfrac{4}{15}.$

Just as in whole number division, the remainder is then compared to the divisor by a fraction. In the preceding example, the remainder is 1/2 as big as the divisor, so the quotient is 2 1/2.

One of the early methods of dividing one fraction by another was to replace both fractions by fractions having a common denominator. When this is done the quotient can be obtained by disregarding the denominators and dividing the two numerators. For example,

$$\frac{3}{4} \div \frac{2}{3} = \frac{9}{12} \div \frac{8}{12} = 9 \div 8 = 1\frac{1}{8}$$

It may have been this approach, of getting a common denominator for dividing fractions by fractions, that eventually led to the present "invert and multiply" method. In this method of division, which came into general use in the seventeenth century, we invert the divisor and then multiply the two fractions. The Hindus and Arabs divided this way as early as the eleventh century, but it seems not to have been used for the next 400 years.*

Definition: For any fractions $\frac{a}{b}$ and $\frac{c}{d}$, with $\frac{c}{d} \neq 0$,

$$\frac{a}{b} \div \frac{c}{d} = \frac{a}{b} \times \frac{d}{c} = \frac{ad}{bc}$$

NUMBER PROPERTIES

The commutative, associative, and distributive properties which hold for the operations of addition and multiplication with whole numbers also hold for these operations with fractions. The rules for adding and multiplying fractions will be used to illustrate each of these properties in the following examples.

Addition Is Commutative   Two fractions that are being added can be interchanged (commuted) without changing the sum: $7/8 + 3/5 = 3/5 + 7/8$.

$$\frac{7}{8} + \frac{3}{5} = \frac{35}{40} + \frac{24}{40} = \frac{59}{40} = 1\frac{19}{40} \qquad \frac{3}{5} + \frac{7}{8} = \frac{24}{40} + \frac{35}{40} = \frac{59}{40} = 1\frac{19}{40}$$

This property is illustrated next on the number lines by placing the lengths for the fractions end to end.

*D. E. Smith, *History of Mathematics,* **2** (Lexington, Mass.: Ginn, 1925), pp. 226–28.

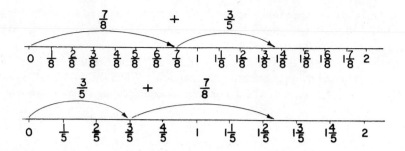

Addition Is Associative   In a sum of three fractions, the middle number may be grouped (associated) with either of the other two numbers. (The sums inside the parentheses are computed first.)

$$\left(\frac{1}{3} + \frac{1}{4}\right) + \frac{1}{6} = \frac{7}{12} + \frac{1}{6} = \frac{9}{12} \qquad \frac{1}{3} + \left(\frac{1}{4} + \frac{1}{6}\right) = \frac{1}{3} + \frac{5}{12} = \frac{9}{12}$$

Multiplication Is Commutative   Two fractions that are being multiplied can be interchanged (commuted) without changing the product: $1/2 \times 1/3 = 1/6$, and $1/3 \times 1/2 = 1/6$. A physical illustration of this property is interesting because the processes of taking $1/2$ of something and taking $1/3$ of something are quite different. To take $1/2$ of $1/3$ we begin with a $1/3$ bar, and to take $1/3$ of $1/2$ we use a $1/2$ bar.

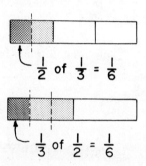

Multiplication Is Associative   In a product of three fractions the middle number may be grouped with either of the other two numbers. (The products inside the parentheses are computed first.)

$$\left(\frac{1}{2} \times \frac{3}{4}\right) \times \frac{1}{5} = \frac{3}{8} \times \frac{1}{5} = \frac{3}{40} \qquad \frac{1}{2} \times \left(\frac{3}{4} \times \frac{1}{5}\right) = \frac{1}{2} \times \frac{3}{20} = \frac{3}{40}$$

Multiplication Is Distributive over Addition   When a sum of two fractions is multiplied by a third number we can add the two fractions and then multiply, as in the first of the following equations, or we can multiply both fractions by the third number and then add, as in the second row of equations.

$$\frac{1}{2} \times \left(\frac{3}{4} + \frac{7}{10}\right) = \frac{1}{2} \times \left(\frac{15}{20} + \frac{14}{20}\right) = \frac{1}{2} \times \frac{29}{20} = \frac{29}{40}$$

$$\frac{1}{2} \times \left(\frac{3}{4} + \frac{7}{10}\right) = \left(\frac{1}{2} \times \frac{3}{4}\right) + \left(\frac{1}{2} \times \frac{7}{10}\right) = \frac{3}{8} + \frac{7}{20} = \frac{29}{40}$$

Inverses for Addition For every fraction there is another fraction, called its *negative* or *inverse for addition,* such that the sum of the two fractions is 0. The fractions $3/4$ and $^-3/4$ are inverses of each other for addition: $3/4 + {}^-3/4 = 0$.

$$\frac{3}{4} + \frac{^-3}{4} = \frac{3 + {}^-3}{4} = \frac{0}{4} = 0$$

Inverses for Multiplication For every fraction not equal to 0 there is a nonzero fraction, called its *reciprocal* or *inverse for multiplication,* such that the product of the two numbers is 1. For the fraction $3/8$, its reciprocal is $8/3$, and $3/8 \times 8/3 = 1$.

$$\frac{3}{8} \times \frac{8}{3} = \frac{3 \times 8}{8 \times 3} = \frac{24}{24} = 1$$

OPERATIONS ON CALCULATORS

The four basic operations with fractions can be carried out by first changing the fractions to decimals and then performing the given operations. Sometimes a computation with fractions is only approximately equal to one with decimals. This will occur whenever the decimal representations for the fractions or the computation exceeds the capacity of the calculator. (The symbol $\approx$ means "approximately equal to")

$$\frac{4}{5} + \frac{1}{3} + \frac{5}{8} \approx .8 + .3333333 + .625 = 1.7583333$$

On some calculators sums or differences can be computed by merely entering the fractions (as quotients of whole numbers) and the addition or subtraction operations as they occur from left to right. To compute the previous sum, if $4 \boxed{\div} 5 \boxed{+} 1$ is entered onto this type of calculator, 1 is not added to the quotient, $4 \div 5$, but it is divided by 3 when $\boxed{\div} 3$ is entered. Here are the steps for computing the sum on this type of calculator.

$$4 \boxed{\div} 5 \boxed{+} 1 \boxed{\div} 3 \boxed{+} 5 \boxed{\div} 8 \boxed{=}$$

Multiplication of fractions can be carried out on a calculator by simply entering the fractions (as quotients of numbers) and the multiplication operation in the order in which they occur. For example, here are the steps for computing $5/7 \times 2/9$.

$$5 \boxed{\div} 7 \boxed{\times} 2 \boxed{\div} 9 \boxed{=}$$

This sequence of steps produces the correct answer because

$$[(5 \div 7) \times 2] \div 9 = (5 \div 7) \times (2 \div 9)$$

Division cannot be carried out as conveniently on the calculator as can multiplication. To compute $5/7 \div 2/9$, for example, we cannot enter the fractions and the division operation in the same order as we did for multiplication, because

$$[(5 \div 7) \div 2] \div 9 \neq (5 \div 7) \div (2 \div 9)$$

One method of computing $5/7 \div 2/9$ on a calculator is to replace $2/9$ by $9/2$ and multiply.

$$\frac{5}{7} \div \frac{2}{9} = \frac{5}{7} \times \frac{9}{2} = (5 \div 7) \times (9 \div 2)$$

Another method of computing $5/7 \div 2/9$ is to replace each fraction by a decimal and then divide one decimal by the other. This can be done by first computing $2 \div 9$ and placing the decimal for this quotient in memory storage. Then compute $5 \div 7$ and divide it by the number in storage. Here are the steps and corresponding displays for this computation.

Steps	Displays
1. 2 $\div$ 9 $=$	0.22222222
2. M+	0.22222222
3. 5 $\div$ 7 $=$	0.71428571
4. $\div$	0.71428571
5. MR	0.22222222
6. $=$	3.2142857

Sometimes you will want the sum or difference of two fractions to be written as another fraction. Let's look at an example for addition. First, replace each fraction by a decimal and add the decimals.

$$\frac{3}{8} + \frac{1}{3} \approx .375 + .3333333 = .7083333$$

Next, the smallest common denominator of $3/8$ and $1/3$ is 24, so we need to solve the following equation for N.

$$.7083333 = \frac{N}{24}$$

To find N multiply $.7083333 \times 24$, and since N is the numerator of a fraction, round off your answer to the nearest whole number. In this example N is equal to 17.

$$.7083333 \times 24 = 16.999999$$

APPROXIMATION AND MENTAL CALCULATION

The sum or difference of mixed numbers can be approximated by rounding off each mixed number to the nearest whole number.

$$6\frac{1}{3} + 2\frac{3}{4} + 1\frac{1}{5} \approx 6 + 3 + 1 = 10$$

$$8\frac{1}{2} - 2\frac{3}{5} \approx 9 - 3 = 6$$

Computing a product by rounding off two mixed numbers may produce a good approximation.

$$6\frac{3}{4} \times 8\frac{1}{3} \approx 7 \times 8 = 56 \quad \left(\text{the actual product is } 56\frac{1}{4}\right)$$

or it may give a rough approximation,

$$6\frac{3}{4} \times 8\frac{1}{2} \approx 7 \times 9 = 63 \quad \left(\text{the actual product is } 57\frac{3}{8}\right)$$

A more reliable approximation can be obtained for this example by multiplying the two whole numbers, 6×8, and then adding the products of each fraction times the opposite whole number: $3/4 \times 8$ and $6 \times 1/2$.

$$6\frac{3}{4} \times 8\frac{1}{2} \approx (6 \times 8) + \left(\frac{3}{4} \times 8\right) + \left(6 \times \frac{1}{2}\right)$$

$$= 48 + 6 + 3$$

$$= 57$$

The distributive property shows why this method produces a good approximation.

$$6\frac{3}{4} \times 8\frac{1}{2} = \left(6 + \frac{3}{4}\right) \times \left(8 + \frac{1}{2}\right)$$

$$= \left(6 + \frac{3}{4}\right)8 + \left(6 + \frac{3}{4}\right)\frac{1}{2}$$

$$= (6 \times 8) + \left(\frac{3}{4} \times 8\right) + \left(6 \times \frac{1}{2}\right) + \left(\frac{3}{4} \times \frac{1}{2}\right)$$

$$= 48 + 6 + 3 + \frac{3}{8}$$

$$= 57\frac{3}{8}$$

A good approximation can still be obtained if we round off the products of the fractions and whole numbers.

$$5\frac{1}{3} \times 7\frac{3}{4} \approx (5 \times 7) + \left(\frac{1}{3} \times 7\right) + \left(5 \times \frac{3}{4}\right)$$

$$\approx 35 + 2 + 4$$

$$= 41 \quad \left(\text{the actual product is } 41\frac{1}{3}\right)$$

COMPUTER APPLICATIONS

This photo shows several monitoring screens on the American Stock Exchange trading floor with the latest computerized stock prices in fractions, mixed numbers, and whole numbers. The fractions on these screens are halves, fourths, eighths, sixteenths, and thiry-seconds.

Monitoring screens on American Stock Exchange trading floor

Since the computer normally prints whole numbers and decimals, you might wonder how it can be programmed to perform operations with fractions and print the results as fractions. The following program shows how the sum of two fractions can be computed and printed in lowest terms. Lines 22, 24, and 26 are the same as those used on page 188 for determining the greatest common factor of two numbers. Follow through the steps of this program for $1/2 + 3/5$.

Program 6.2A

```
10  PRINT "TYPE THE NUMERATOR AND DENOMINATOR OF A FRACTION
    AND SEPARATE THEM BY A COMMA."
12  INPUT C,D
14  PRINT "TYPE THE NUMERATOR AND DENOMINATOR OF THE NEXT
    FRACTION AND SEPARATE THEM BY A COMMA."
16  INPUT E,F
18  LET A = (D * E) + (C * F)
20  LET B = D * F
22  FOR N = B TO 1 STEP  - 1
24  IF B / N =  INT (B / N) AND A / N =  INT (A / N) THEN
    GOTO 30
26  NEXT N
30  PRINT C"/"D" + "E"/"F" = "(A / N)"/"(B / N)
40  END
```

Here's a problem that involves computing the sum of many fractions.

Rabbit in the Box Problem Suppose a rabbit looks out one end of a box and then 1/2 of a second later it looks out the other end; 1/4 of a second later it looks out the first end; 1/8 of a second later it looks out the other end; etc. If the rabbit continues to cut the time in half, how long before it will have looked out the box 12 times?

To find the total amount of time, we must compute the sum of these 11 fractions.

$$\frac{1}{2} + \frac{1}{4} + \frac{1}{8} + \frac{1}{16} + \frac{1}{32} + \frac{1}{64} + \frac{1}{128} + \frac{1}{256} + \frac{1}{512} + \frac{1}{1024} + \frac{1}{2048}$$

We could use the preceding program to compute this sum by adding two fractions at a time. However, the following program is more convenient. It begins by asking for the number of fractions to be added. Lines 30, 40, and 50 form a loop. The first time through the loop, S is equal to $0 + 1/2 = 1/2$. Notice that the S on the right side of the equal sign in line 40 is equal to zero the first time throught the loop. Then each succeeding time through the loop the next fraction is added to S. Will line 60 print the sum as a decimal or a fraction?

Program 6.2B

```
10   PRINT "HOW MANY FRACTIONS DO YOU WISH TO ADD? TYPE A WHOLE
       NUMBER."
20   INPUT N
30   FOR X = 1 TO N
40   LET S = S + 1 / (2 ^ X)
50   NEXT X
60   PRINT "THE SUM OF THE FIRST "N" FRACTIONS IS "S
70   END
```

SUPPLEMENT (Activity Book)

Activity Set 6.2 Computing with Fraction Bars

Just for Fun: Fraction Games for Operations

Exercise Set 6.2: Applications and Skills

1. The fraction bar for $1/5$ can be used to show that $1/2 \times 1/5 = 1/10$. Draw sketches of bars to visually illustrate each of the following computations.

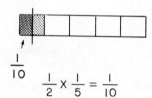

$$\frac{1}{2} \times \frac{1}{5} = \frac{1}{10}$$

   a. $\dfrac{3}{10} + \dfrac{2}{5} = \dfrac{7}{10}$ ★ b. $\dfrac{5}{6} - \dfrac{1}{3} = \dfrac{3}{6}$ ★ c. $\dfrac{1}{3} \times \dfrac{1}{4} = \dfrac{1}{12}$

   d. $\dfrac{2}{3} \div \dfrac{1}{6} = 4$ e. $\dfrac{1}{4} \times 3 = \dfrac{3}{4}$ f. $\dfrac{1}{3} \times \dfrac{1}{6} = \dfrac{1}{18}$

2. Place the edge of a piece of paper on the eighths' line and mark off the length 1 1/8. Then place the beginning of this marked-off length at the 1 1/5 point on the fifths' line and approximate the sum 1 1/5 + 1 1/8. What number on the fifths' line is this sum closest to?

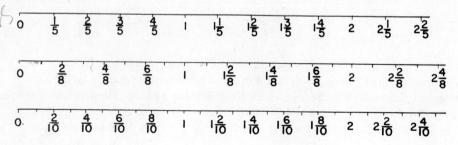

a. Use the preceding number lines to approximate the following sums. Write the number from the given line that the sum is closest to.

$$\frac{4}{5} + 1\frac{3}{8} \text{ (fifths' line)} \qquad \frac{5}{8} + \frac{7}{10} \text{ (eighths' line)} \qquad \frac{3}{10} + 1\frac{4}{5} \text{ (tenths' line)}$$

★ b. Compute these sums and compare them to your approximations.

$$\begin{array}{c} \frac{4}{5} \\ +1\frac{3}{8} \\ \hline \end{array} \qquad\qquad \begin{array}{c} \frac{5}{8} \\ +\frac{7}{10} \\ \hline \end{array} \qquad\qquad \begin{array}{c} \frac{3}{10} \\ +1\frac{4}{5} \\ \hline \end{array}$$

3. Place the edge of a piece of paper of the fifths' line in Exercise 2 and mark off the length 3/5. Then place this marked-off length at the point 1 7/8 on the eighths' line and approximate the difference, 1 7/8 − 3/5. What number on the eighths' line is this difference closest to?

a. Use the number lines of Exercise 2 to approximate the following differences. Write the number from the given line that the difference is closest to.

$$1\frac{4}{5} - \frac{5}{8} \text{ (fifths' line)} \qquad 1\frac{1}{8} - \frac{7}{10} \text{ (eighths' line)} \qquad \frac{9}{10} - \frac{3}{5} \text{ (tenths' line)}$$

★ b. Compute these differences and compare them with your approximations.

$$\begin{array}{c} 1\frac{4}{5} \\ -\frac{5}{8} \\ \hline \end{array} \qquad\qquad \begin{array}{c} 1\frac{1}{8} \\ -\frac{7}{10} \\ \hline \end{array} \qquad\qquad \begin{array}{c} \frac{9}{10} \\ -\frac{3}{5} \\ \hline \end{array}$$

4. *Fraction Skills:* Perform the operations. Replace all improper fractions by whole numbers or mixed numbers and write all fractions in lowest terms.

a. $\frac{2}{3} + \frac{3}{4}$ ★ b. $\frac{1}{6} + \frac{3}{8}$ c. $\frac{7}{8} - \frac{1}{3}$ d. $\frac{3}{4} - \frac{2}{5}$

e. $\frac{2}{3} \times 6$ f. $\frac{3}{4} \times \frac{2}{5}$ g. $\frac{3}{4} \div \frac{1}{10}$ h. $\frac{2}{3} \div \frac{1}{5}$

i. $2\frac{1}{4} + 1\frac{1}{3}$ j. $1\frac{5}{6} + 3\frac{1}{2}$ ★ k. $3\frac{1}{4} - 1\frac{1}{8}$ l. $5\frac{1}{3} - 2\frac{1}{2}$

m. $2\frac{1}{4} \times 3\frac{1}{2}$ ★ n. $14\frac{1}{2} \div 2\frac{1}{4}$ o. $3 \div \frac{1}{5}$ ★ p. $4 \times 5\frac{1}{8}$

5. *Error Analysis:* In computing with fractions there are several types of errors which frequently occur. In this example, 3 2/5 should have been replaced by 2 7/5. Instead, the 1 which was borrowed from the 3 was placed in the numerator to form 2 12/5. Find plausible reasons for the errors in the following examples.

$$3\frac{2}{5} = 2\frac{12}{5}$$
$$-\frac{4}{5} = -\frac{4}{5}$$
$$\overline{\qquad\qquad}$$
$$2\frac{8}{5} = 3\frac{3}{5}$$

★ a. $\frac{1}{4} + \frac{5}{6} = \frac{6}{10}$ b. $\frac{7}{8} - \frac{2}{3} = \frac{5}{5}$ c. $\frac{1}{2} \times \frac{3}{8} = 68$ d. $\frac{3}{4} \div \frac{1}{5} = \frac{4}{15}$

★ e. $\qquad 3\frac{1}{4}$ f. $\qquad 5\frac{3}{8}$ ★ g. $\frac{1}{3} + \frac{2}{5} = \frac{6}{8} + \frac{5}{8} = \frac{11}{8}$

$$\qquad -2\frac{3}{4} \qquad\qquad\qquad -2\frac{2}{3}$$
$$\qquad \overline{\;\;1\frac{2}{4}\;\;} \qquad\qquad\qquad \overline{\;\;3\frac{1}{5}\;\;}$$

6. This table contains high and low prices for six companies during one day.

★ a. General Plywood's low price was 1/8 of a dollar less than its high price for the day. What was its low price?

b. The day's high price for Sea Container was 3/16 of a dollar above its low price. What was the high price?

★ c. What is the difference between the high price and low price for each of these stocks?

Drew National   MEM Company   Old Town

d. How much money would a person have saved if he or she had purchased 1000 shares of MEM Company stock at the low price rather than the high price for the day?

	High Price	Low Price
Canadian Homestead	$8\frac{1}{4}$	$7\frac{5}{16}$
Drew National	$11\frac{1}{8}$	$9\frac{7}{8}$
General Plywood	$3\frac{3}{4}$	
MEM Company	$26\frac{1}{4}$	$25\frac{5}{8}$
Old Town	$7\frac{3}{4}$	$6\frac{3}{8}$
Sea Container		$16\frac{1}{2}$

7. This picture of a blue shark has a length of 45 millimeters. The scale factor from an average size blue shark to this picture is 1/120.

★ a. How long is an average blue shark?

b. A right whale has a length of approximately 12 meters. If the scale factor from this whale to a picture of the whale is 1/300, how long is the picture?

8. State the number property that is being used in each of these equalities.

★ a. $\frac{3}{7} + \left(\frac{2}{9} + \frac{1}{3}\right) = \frac{3}{7} + \left(\frac{1}{3} + \frac{2}{9}\right)$

b. $\frac{3}{7} + \left(\frac{2}{9} \times \frac{9}{2}\right) = \frac{3}{7} + 1$

★ c. $\frac{2}{9} + \left(\frac{3}{7} + \frac{1}{3}\right) = \left(\frac{2}{9} + \frac{3}{7}\right) + \frac{1}{3}$

d. $\frac{5}{6} \times \left(\frac{3}{4} + \frac{1}{2}\right) = \left(\frac{3}{4} + \frac{1}{2}\right) \times \frac{5}{6}$

★ e. $\frac{3}{4} \times \frac{5}{6} + \frac{1}{2} \times \frac{5}{6} = \left(\frac{3}{4} + \frac{1}{2}\right) \times \frac{5}{6}$

9. *Calculator Exercise:* Here are sequences of calculator steps for computing a product, quotient, and sum of fractions. Which of these sequences will not produce the correct answer? Explain why.

a. 80 × 3/5
   1. Enter 80
   2. ☒
   3. 3 ÷ 5
   4. =

b. 80 ÷ 3/5
   1. Enter 80
   2. ÷
   3. 3 ÷ 5
   4. =

c. 80 + 3/5
   1. Enter 80
   2. +
   3. 3 ÷ 5
   4. =

10. *Approximation and Mental Calculation:* Round off each mixed number to the nearest whole number and compute the sum.

Think				Approximate Sum
⑧	③	④	⑥	
$7\frac{2}{3}$	$2\frac{3}{4}$	$4\frac{1}{8}$	$5\frac{9}{10}$	21
a. $1\frac{2}{3}$	$3\frac{1}{4}$	$1\frac{1}{2}$	$2\frac{1}{10}$	
★ b. $3\frac{4}{5}$	$2\frac{1}{6}$	$3\frac{2}{5}$	$4\frac{1}{2}$	
c. $10\frac{1}{3}$	$5\frac{1}{2}$	$2\frac{7}{8}$	$5\frac{1}{6}$	
★ d. $6\frac{1}{2}$	$2\frac{3}{4}$	$1\frac{1}{4}$	$2\frac{5}{6}$	

11. *Approximation and Mental Calculation:* Approximate the product of each pair of mixed numbers by multiplying the whole numbers and adding the products of the fractions times the opposite whole numbers (page 353). Show each step for obtaining the answer.

   a. $4\frac{1}{3} \times 6\frac{1}{2}$ ★   b. $5\frac{1}{4} \times 8\frac{2}{5}$

   c. $3\frac{1}{4} \times 4\frac{2}{3}$ ★   d. $10\frac{1}{3} \times 6\frac{1}{4}$

12. A taxpayer is going to file his federal income tax report by using the income averaging method. The eight steps of this method and part of the computation are shown below. Compute the missing amounts.

"You'll be happy to know that nobody in the government is out to get you, nobody's reported you for the finder's fee, nor have we received any anonymous tips. You're here only because we think you've been cheating on your return."

1. 1/3 of base income of $22,000 . . . . . . _____
2. 1/5 of averageable income of $12,000. . _____
3. Line 1 plus line 2 . . . . _____
4. Tax on line 3 . . . . $2,318
5. Tax on line 1 . . . . $1,630
6. Line 4 minus line 5 . . . . _____
7. Line 6 multiplied by 4 . . . . _____
8. Line 4 plus line 7 (Total tax) . . . . . . _____

13. Musical notes which are produced from two strings of equal diameter and tension will vary according to the lengths of the strings. Different fractions of the length of the unit string (see lower string) can be used to produce the notes C, D, E, F, G, A, B, and c (do, re, mi, fa, so, la, ti, do). In particular, if one string is half as long as another, its tone or note will be an octave higher than the longer string.*

*See C. F. Linn, *The Golden Mean* (New York: Doubleday, 1974), pp. 9–13, for an elementary explanation of the origin of these fractions.

1/2		c (one octave higher than C)
128/243		B
16/27		A
2/3		G
3/4		F
64/81		E
8/9		D
(unit string)	1	C

★ a. Proceeding from the unit string to the top string, all of the strings except two are 8/9 of the length of the previous string. For example, the G string is 8/9 the length of the F string since $8/9 \times 3/4 = 2/3$. Which strings are 8/9 of the length of the preceding strings?

b. The white piano keys pictured here are the notes of the scale from C to c. All but two of these keys are separated by black keys. How is this observation related to the answer for part **a**?

Exercise Set 6.2: Problem Solving

1. *Candy Jar Problem:* Five jars which are numbered 1 through 5 contain a total of 92 candy bars. If each jar contains two more candy bars than the previous jar, how much candy is in each jar?

a. *Understanding the Problem:* Sometimes a problem such as this can be solved by guessing and then adjusting the next guess. Even if no solution is found, you may obtain a better understanding of the problem. Try a few numbers and describe what you learn by guessing.

★   b. *Devising a Plan:* The guessing in part **a** suggests that the answers are not whole numbers. Let's simplify the problem by changing the number of candy bars. How many bars will there be in each jar if there is a total of 100 bars?

    c. *Carrying Out the Plan:* Continue solving the problem for 100 candy bars for each of the following cases.

Each jar has 3 more than the previous jar   _____   _____   _____   _____   _____

Each jar has 4 more than the previous jar   _____   _____   _____   _____   _____

Each jar has 5 more than the previous jar   _____   _____   _____   _____   _____

Look for a pattern in your answers. What general approach to solving this type of problem is suggested? What is the solution to the original problem with 92 candy bars?

★   d. *Looking Back:* Solve this problem for 7 jars containing 92 candy bars, if each jar has 2 more bars than the previous jar.

2. *Fraction Patterns:* Look carefully at the fractions shown here. In each case we begin with a unit fraction whose denominator is an odd number. Each unit fraction is replaced by a new fraction which in lowest terms is nicely related to the original fraction.

    a. What is this relationship?

    b. Continue the method suggested in this table to find the new fractions for 1/9, 1/5, and 1/11. Does the relationship still hold?

★   c. Will the same relationship hold for unit fractions whose denominators are even numbers?

Unit Fraction	New Fraction
$\dfrac{1}{3}$	$\dfrac{1+3}{3+3} = \dfrac{4}{6} = \dfrac{2}{3}$
$\dfrac{1}{7}$	$\dfrac{1+7}{7+7} = \dfrac{8}{14} = \dfrac{4}{7}$
$\dfrac{1}{15}$	$\dfrac{1+15}{15+15} = \dfrac{16}{30} = \dfrac{8}{15}$
$\dfrac{1}{9}$	
$\dfrac{1}{5}$	
$\dfrac{1}{11}$	

3. *Rhind Papyrus:* The Rhind Papyrus is an ancient Egyptian mathematics text for solving practical problems. These problems are preceded by a table for fractions with a numerator of 2 and denominators which are odd numbers from 5 to 101. Each of these fractions is written as a sum of unit fractions with different denominators. Use the method suggested in the following equations to write 2/7, 2/15, and 2/35 as a sum of unit fractions. Check your results.

$$\frac{2}{5} = \frac{2}{(1)(5)} = \frac{2}{6}\left(\frac{1}{1} + \frac{1}{5}\right) = \frac{1}{3} + \frac{1}{15}$$

$$\frac{2}{21} = \frac{2}{(3)(7)} = \frac{2}{10}\left(\frac{1}{3} + \frac{1}{7}\right) = \frac{1}{15} + \frac{1}{35}$$

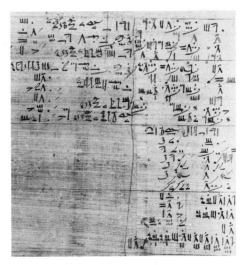

Rhind Papyrus—mathematical text
(ca. 1650 B.C.)

4. *Tower of Bars:* Explain how this tower of bars illustrates the following facts. Complete each pattern of fractions for the tower of bars and check the equations.

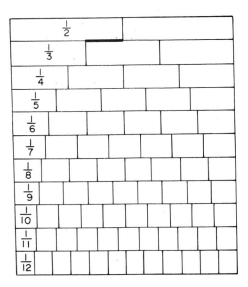

a. $\dfrac{1}{2} - \dfrac{1}{3} = \dfrac{2}{3} - \dfrac{1}{2}$

$\dfrac{1}{3} - \dfrac{1}{4} = \dfrac{3}{4} - \dfrac{2}{3}$

$\vdots$

b. $\dfrac{2}{4} = \dfrac{2}{5} + \dfrac{1}{2}\left(\dfrac{1}{5}\right)$

$\dfrac{3}{6} = \dfrac{3}{7} + \dfrac{1}{2}\left(\dfrac{1}{7}\right)$

$\vdots$

★ c. $\dfrac{1}{2} - \dfrac{1}{3} = \dfrac{1}{2} \times \dfrac{1}{3}$ (*Hint:* See the dark line between the 1/2 bar and the 1/3 bar.)

$\dfrac{1}{3} - \dfrac{1}{4} = \dfrac{1}{3} \times \dfrac{1}{4}$

$\vdots$

5. *The Missing Fraction Puzzle:* In Farmer Brown's will, he bequeathed his 17 horses to his 3 sons in the following manner: 1/2 of his horses to the oldest son, Al; 1/3 of his horses to the middle son, Garry; and 1/9 of his horses to the youngest son, Greg. Being fairly capable with fractions, the boys computed their shares to be 8 1/2 horses, 5 2/3 horses, and 1 8/9 horses. However, each boy was disappointed at the prospect of getting parts of a horse and they fell to quarrelling about their predicament. At that point Farmer Smith rode up and, after being informed of their dilemma, proposed the following solution. First he donated his horse to make a total of 18 horses. Then he gave 1/2 of the horses to Al, 1/3 of the 18 to Garry, and 1/9 of the 18 to Greg. How many horses did each boy receive? What was the total number of these horses? Seeing their satisfaction with his solution, Farmer Smith jumped on his horse and rode away. Explain why his solution is possible.

Exercise Set 6.2: Computers

1. These questions pertain to Program 6.2B on page 356.
   a. What is the printout for N = 4?
   b. What is the sum of the first 11 fractions?
   c. How long before the rabbit will have looked out of the box 20 times?
   d. No matter how many fractions are used in this sum, the sum never exceeds some number. What is this number?

2. Suppose we slow down the speed of the rabbit in the "Rabbit in the Box Problem" by replacing the given fractions by the following fractions.

$$\frac{1}{2} + \frac{1}{3} + \frac{1}{4} + \frac{1}{5} + \frac{1}{6} + \frac{1}{7} + \frac{1}{8} + \frac{1}{9} + \frac{1}{10} + \frac{1}{11} + \cdots$$

   a. Revise Program 6.2B (page 356) to compute this sum.
   b. What is the sum of these fractions for N = 10?
   c. How long before the rabbit will have looked out of the box 20 times?
   d. By adding more and more fractions in this sum, the sum can be made as large as we please. What is the sum of the first 100 fractions? 1000 fractions?

3. Most of the problems in the Egyptian Leather Scroll and the Rhind Papyrus involve sums of unit fractions. Use Program 6.2A (page 355) to answer the following questions.
   a. Here are the three sums from the Egyptian Leather Scroll. One was computed incorrectly by the Egyptian scribes. Which one?*

$$\frac{1}{28} + \frac{1}{49} + \frac{1}{196} = \frac{1}{13} \qquad \frac{1}{25} + \frac{1}{50} + \frac{1}{150} = \frac{1}{15} \qquad \frac{1}{96} + \frac{1}{192} = \frac{1}{64}$$

*Note: The computer may take several minutes to compute the sum of two fractions.

↪ ★   b. Here are three sums from the Rhind Papyrus. Write each sum as a fraction in lowest terms.

$$\frac{1}{12} + \frac{1}{51} + \frac{1}{68} \qquad \frac{1}{20} + \frac{1}{124} + \frac{1}{155} \qquad \frac{1}{4} + \frac{1}{8} + \frac{1}{9} + \frac{1}{10} + \frac{1}{30} + \frac{1}{40} + \frac{1}{45}$$

4. Between any two fractions there is another fraction. One method of finding this fraction is to add the two fractions and divide by 2. This produces a fraction which is halfway between the given fractions. Write a program to find the fraction in lowest terms that is half way between any two given fractions.

6.3
POSITIVE AND
NEGATIVE NUMBERS

HISTORICAL DEVELOPMENT

Whole numbers and fractions were used hundreds of years before negative numbers were developed. As trading became more common, two distinctly different uses of whole numbers were needed—one to indicate credits or gains and one for debits or losses. Conventions were developed to permit the use of whole numbers in both cases. Around 200 B.C. the Chinese were computing credits with red rods and debits with black rods. Similarly, in their writing they used red numerals and black numerals.*

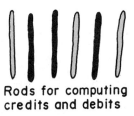

Rods for computing credits and debits

*D. E. Smith, *History of Mathematics,* **2** (Lexington, Mass: Ginn, 1925), pp. 257–58.

The custom today is to reverse the color scheme used by the Chinese. Banks often use red numerals to indicate overdrawn accounts, and we hear the phrase "operating in the red," which refers to business losses. This scoreboard for the 1976 Masters Tournament used red numerals for golf scores below par and green numerals for scores above par.

LEADERS

	HOLE	1	2	3	4	5	6	7	8	9	10	11	12	13	14	15	16	17	18
	PAR	4	5	4	3	4	3	4	5	4	4	4	3	5	4	5	3	4	4
15	FLOYD	15	15	15	14	15	15	15	15	15	15	15	16	16	16	17	17	17	
6	ZIEGLER	6	6	6	6	7	7	7	7	7	7	6	6	7	7	7	7	6	
7	NICKLAUS	7	7	6	6	6	6	6	7	6	6	6	6	7	7	7	6	6	6
5	COODY	5	6	6	6	5	5	5	6	5	5	5	5	5	5	5	4	3	
4	KITE	4	5	5	4	4	4	5	5	5	4	4	4	4	4	4	3	3	
3	GRAHAM L.	3	3	2	2	2	2	2	3	3	3	3	2	2	1	1	0	1	
4	CRENSHAW	3	3	4	4	4	4	5	6	6	6	6	6	8	9	9	9	9	
2	WEISKOPF	1	2	1	1	1	1	1	1	0	0	0	0	0	0	1	0	0	
2	CASPER	1	0	0	0	0	1	1	2	2	1	1	2	1	1	1	1	1	
1	IRWIN	1	0	0	1	0	1	1	0	0	0	0	0	1	2	2	2	3	3

It is only after hundreds of years of proven necessity that a new type of number can earn its place beside the commonly accepted older numbers. This is especially true of negative numbers. By the seventh century, Hindu mathematicians were using these numbers on a limited basis. They had symbols for negative numbers, such as ⑤ and 5̊ for ⁻5, and rules for computing with them. However, it was another thousand years before the Italian mathematician Jerome Cardan (1501–1576) gave the first significant treatment of negative numbers. Cardan called these new numbers "false" and represented each number by writing "m:" in front of the numeral. For example, he wrote "m:3" for negative three. Other writers of this period called negative numbers "absurd numbers." In 1759, the English mathematician Baron Francis Masères published *Dissertation on the Use of the Negative Sign in Algebra,* in which he expressed the following opinion about negative numbers as solutions to equations:

> ". . . they serve only, as far as I am able to judge, to puzzle the whole doctrine of equations, and to render obscure and mysterious things that are in their own nature exceeding plain and simple . . ."

The resistance to negative numbers can be seen as late as 1796 when William Frend, in his text *Principles of Algebra,* argued against their use.

APPLICATIONS

The concept of positive and negative numbers, also referred to as *signed numbers,* is useful whenever we wish to measure on both sides of a fixed point of reference. The positive numbers indicate one direction, and the negative numbers indicate the opposite direction.

Credits and Debits   One common example of "opposites" is credits, which are represented by positive numbers, and debits, which are represented by negative numbers. The graph on page 367 shows America's trade balance, which is the difference between exports and imports. It was positive from 1960 to 1970 and negative from 1976 to 1981. Its lowest balance was less than ⁻33 billion in 1978.*

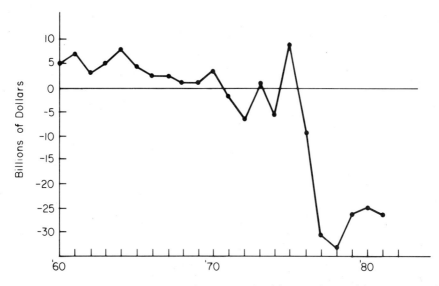

Temperature Measuring temperature is another familiar use for positive and negative numbers. The fixed reference point on the Celsius thermometer is 0, the temperature at which water freezes. On the Fahrenheit thermometer water freezes at 32°. On both scales temperatures above 0 are positive and those below 0 are negative.

Time Scientists often find it convenient to designate a given time as "zero time" and then refer to the time before and after as being negative and positive, respectively. This practice is followed in the launching of rockets. If the time with respect to blast-off is ⁻15 minutes, then it is 15 minutes before the launch. For Apollo 11, the first mission to land men on the moon, countdown began many hours before lift-off, which occurred on July 16, 1969, at 9:32 A.M. Eastern Daylight time. Before this time, minutes and hours were labeled by negative numbers: 9:32 A.M. on July 16 was time 0; and the time following 9:32 A.M. was positive.

Apollo 11 moon launch, July 16, 1969

*Source: U.S. Bureau of Economic Analysis.

Altitude Sea level is the common reference point for measuring altitudes. Charts and maps which label altitudes below and above sea level use negative and positive numbers. The following chart shows the altitudes in terms of negative numbers for the "floor" of the Atlantic Ocean between South America and Africa.

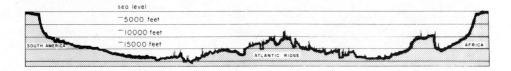

MODELS FOR POSITIVE
AND NEGATIVE NUMBERS

Number Line Model The number line with a fixed reference point labeled 0 is the common model for positive and negative numbers. For each positive number to the right of 0, there is a negative number to the left, symmetrically opposite. These pairs of numbers 2 and ⁻2, 3/4 and ⁻3/4, etc., are called *opposites*. More precisely, two numbers are called *opposites* (or *inverses for addition*) if their sum is 0. Sometimes numbers greater than 0, such as 4, 7/8, 9, etc., are labeled as ⁺4, ⁺7/8, and ⁺9, in order to emphasize that they are positive numbers. Notice the raised plus and minus signs to denote positive and negative numbers. Raising these signs helps to distinguish between the concepts of positive and negative numbers and the operations of addition and subtraction.

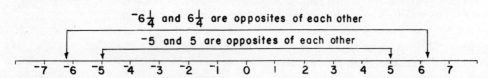

Inequality For any two numbers (positive or negative) on the number line, the number on the left is *less than* the number on the right. As examples: ⁻8 < ⁻3; ⁻5 1/4 < ⁻3 1/2; and ⁻5 < 0. This is stated more precisely in the following definition for inequality of positive and negative numbers which was stated on page 110.

> **Definition:** For any two numbers *m* and *n* (positive or negative), *m* is *less than n* (*m* < *n*) if there is a positive number *k*, such that *m* + *k* = *n*.

You may find it helpful, in thinking about inequalities of negative numbers, to recall the applications on the previous pages. Temperatures of ⁻15°C and ⁻6°C are both cold, but ⁻15°C is colder than ⁻6°C: ⁻15 < ⁻6 because ⁻15 + 9 = ⁻6. Similarly, an altitude of ⁻8000 feet is further below sea level than an altitude of ⁻5000 feet: ⁻8000 < ⁻5000 because ⁻8000 + 3000 = ⁻5000.

Integers   The whole numbers together with their opposites are called *integers*.

$$\ldots\ ^-6,\ ^-5,\ ^-4,\ ^-3,\ ^-2,\ ^-1, 0, 1, 2, 3, 4, 5, 6, \ldots$$

These numbers are also referred to as the positive integers, the negative integers, and 0. The integers are assigned points on the number line by marking off unit lengths to the right and left of 0. This model shows the integers increasing in order from left to right ($^-4 < ^-3$, $^-3 < ^-2$, $^-2 < ^-1$, $^-1 < 0$, etc.).

Black and Red Chips Model   The red and black rods which the Chinese used for positive and negative numbers form a concrete model for the integers. In place of rods we will use chips, and the color scheme will be reversed; that is, black chips will represent positive integers and red chips negative integers. By agreeing that each black chip cancels a red chip, every integer can be represented in an infinite number of ways. Two different sets for 3 are shown above.

On the following pages the elementary definitions and theorems for addition, subtraction, multiplication, and division of integers are illustrated by this model. These same definitions and theorems hold for the basic operations on all positive and negative numbers, including fractions and decimals.

ADDITION

Addition of whole numbers is illustrated by "putting together" (taking the union of) sets of objects. This same model can also be used to illustrate the addition of integers. To find the sum of two integers we take the union of sets of chips which represent these integers. In the union of the sets for $^-5$ and 2, the 2 black chips cancel 2 of the red chips, leaving 3 red chips. This shows that $^-5 + 2 = ^-3$.

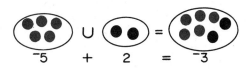

By using black and red chips for computing sums of positive and negative integers, the usual rules for addition can be discovered. For example, to add a positive and a negative integer, the sum will be positive or negative depending on whether there are more black chips or red chips. The three general cases for addition which involve negative numbers are on the following page. In each of these theorems *n* and *s* are positive numbers.

"Negative plus negative equals negative" $$^-n + {}^-s = {}^-(n + s)$$ $$^-3 + {}^-7 = {}^-(3 + 7) = {}^-10$$	"Positive plus negative equals positive, if $n > s$" $$n + {}^-s = n - s$$ $$13 + {}^-5 = 13 - 5 = 8$$	"Positive plus negative equals negative, if $n < s$" $$n + {}^-s = {}^-(s - n)$$ $$6 + {}^-11 = {}^-(11 - 6) = {}^-5$$

The number line is a more abstract model for illustrating addition of positive and negative numbers. To add two numbers we begin by drawing an arrow from 0 to the point which corresponds to the first number in the summand. Then, if the second number is positive, we move to the right on the number line, and if it is negative, we move to the left. Here are two examples.

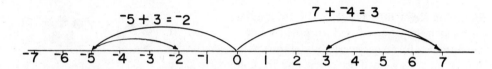

SUBTRACTION

The "take-away" model for subtraction of whole numbers also accommodates the subtraction of integers. The example shown here illustrates $^-6$ take away $^-2$. We begin by representing $^-6$ by 6 red chips and then take away 2 red chips.

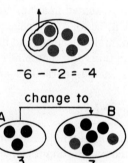

$$^-6 - {}^-2 = {}^-4$$

change to

Traditionally, the take-away model for subtraction is used only when one whole number is subtracted from a larger one. However, it is still possible to use this model in cases such as $3 - 5$, where the number being subtracted is the larger one. This can be accomplished by using a suitable representation for 3. For example, instead of representing 3 by black chips as in set A, 3 can be represented by 5 black chips and 2 red chips as in set B. Then, 5 black chips can be taken away, leaving 2 red chips: $3 - 5 = {}^-2$.

There are two common definitions of subtraction, *adding opposites* and *missing addends,* and both are stated in terms of addition. These definitions can be introduced through the chips model.

Adding Opposites If instead of removing 5 black chips from set B in the preceding example we put in 5 red chips, the final set will still represent $^-2$. In other words, putting in 5 red chips has the same effect as taking away 5 black chips. This suggests that subtracting 5 is the same as adding its opposite, $^-5$.

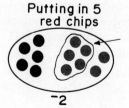

Putting in 5 red chips

$$3 - 5 = 3 + {}^-5 = {}^-2$$

This approach to subtraction is called *adding opposites*. Using this method we can compute differences by replacing them with sums. In general, the difference between any two numbers can be defined in terms of adding opposites.

Definition: For any two numbers, *a* and *b,*

$$a - b = a + \bar{}b$$

Missing Addends   The *missing addends* definition which was used to define subtraction of whole numbers can also be used to define subtraction of negative numbers. To use this approach for computing $3 - 5$, we must find the number that can be added to 5 in order to get 3. That is, we must find the missing addend for $\square$ such that $3 = 5 + \square$. On the preceding page we computed $3 - 5$ by replacing Set A by Set B so that 5 black chips could be taken away. However, before taking away the 5 black chips, set *B* shows that $3 = 5 + \bar{}2$, and therefore, the missing addend is $\bar{}2$. In general, the difference between two numbers can be defined in terms of missing addends.

Definition: For any two numbers, *a* and *b,*

$$a - b = \square \qquad \text{if and only if} \qquad a = b + \square$$

MULTIPLICATION

The familiar rules for multiplying with negative numbers, such as, "a negative times a negative is a positive number," are easy enough to remember but difficult to illustrate. There are many different approaches to this topic which attempt to justify the rules for multiplying with negative numbers in an intuitive manner. Three of the more common methods will be explained in the following paragraphs.

Extension of Patterns   This approach to multiplying with negative numbers is based on the assumption that the patterns we observe in the first few equations on the right will continue to hold. In this column of equations we begin with the products of positive integers. For each move down the column, the products on the right side of the equations decrease by 3, until we arrive at $0 \times 3 = 0$. If this pattern is continued into the negative numbers, it suggests that "a negative times a positive is a negative number."

$$4 \times 3 = 12$$
$$3 \times 3 = 9$$
$$2 \times 3 = 6$$
$$1 \times 3 = 3$$
$$0 \times 3 = 0$$
$$\bar{}1 \times 3 = \bar{}3$$
$$\bar{}2 \times 3 = \bar{}6$$
$$\bar{}3 \times 3 = \bar{}9$$

In the next column of equations we begin with the fact that "a negative times a positive is a negative number," and extend the pattern to where we are multiplying a negative number times a negative number. This time the products on the right side of the equations increase by 4. A continuation of this pattern suggests that "a negative times a negative equals a positive number."

$$
\begin{aligned}
^-4 \times\ \ 3 &= \ ^-12 \\
^-4 \times\ \ 2 &= \ \ ^-8 \\
^-4 \times\ \ 1 &= \ \ ^-4 \\
^-4 \times\ \ 0 &= \ \ \ \ 0 \\
^-4 \times\ ^-1 &= \ \ \ \ 4 \\
^-4 \times\ ^-2 &= \ \ \ \ 8 \\
^-4 \times\ ^-3 &= \ \ 12
\end{aligned}
$$

Physical Representations   The "mail carrier model" is perhaps the best known of the physical representations. This model can be illustrated with the black and red chips by thinking of each black chip as a credit of 1 dollar and each red chip as a bill for 1 dollar. Anybody whose financial situation is represented by the chips shown here would have a total worth of 4 dollars.

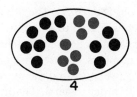

4

To create a model for multiplication let us agree that in the product $n \times s$, n tells us the number of times we *put in* or *take out* s chips. For example, if n is positive, such as in $3 \times\ ^-6$, then we will put in 6 red chips 3 times. If n is negative, such as in $^-4 \times\ ^-6$, then we will take out 6 red chips 4 times.

We are now ready to use the mail carrier model. Suppose that you have 2 bills, each for 3 dollars. These are represented by the 2 groups of 3 red chips (see figure). If the mail carrier takes away both of these bills, it represents the product $^-2 \times\ ^-3$. In this case the available funds will increase from 4 dollars to 10 dollars. That is, taking out 6 red chips is equivalent to putting in 6 black chips. This suggests that $^-2 \times\ ^-3 = 6$.

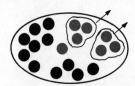

The number represented by this set increases from 4 to 10 by removing 6 red chips.

Deductive Approach   The patterns approach and the mail carrier model illustrate the reasonableness of the rules for multiplying with negative numbers. These illustrations, however, do not prove that these rules will always hold. To establish these facts we must use deductive reasoning. The following proof that $^-2 \times 3 = \ ^-6$ is essentially the same as the general proof that a negative number times a positive number is a negative number. One way of proving that $^-2 \times 3 = \ ^-6$ is to first show that $^-2 \times 3 + 6 = 0$.

$$
\begin{aligned}
\textit{Proof} \quad\quad ^-2 \times 3\ +\ 6 &= \ ^-2 \times 3\ +\ 2 \times 3 \\
&= (^-2 + 2) \times 3 \quad\quad \text{(distributive property)} \\
&= 0 \times 3 \quad\quad \text{(a number plus its opposite equals 0)} \\
&= 0 \quad\quad \text{(0 times any number equals 0)}
\end{aligned}
$$

This sequence of equations shows that $^-2 \times 3 + 6 = 0$. Since $^-6$ is the only number that can be added to 6 to give 0, $^-2 \times 3$ must equal $^-6$.

Similar but more general proofs will establish the following theorems for multiplication. In each case n and s are positive numbers.

"Positive times negative equals negative"	"Negative times positive equals negative"	"Negative times negative equals positive"
$n \times {}^-s = {}^-(n \times s)$	${}^-n \times s = {}^-(n \times s)$	${}^-n \times {}^-s = n \times s$
$5 \times {}^-2 = {}^-(5 \times 2) = {}^-10$	${}^-7 \times 3 = {}^-(7 \times 3) = {}^-21$	${}^-4 \times {}^-5 = 4 \times 5 = 20$

DIVISION

Both the partitive and measurement concepts of division will be needed in the following illustrations of division with negative integers.

To illustrate ${}^-8 \div {}^-2$, we begin with 8 red chips and then measure off or subtract as many groups of 2 red chips as possible. Since there are 4 such groups, ${}^-8 \div {}^-2 = 4$. This is the *measurement* use of division.

$$^-8 \div {}^-2 = 4$$

To illustrate ${}^-6 \div 3$, 6 red chips can be divided into 3 equal groups. Since there are 2 red chips in each group, ${}^-6 \div 3 = {}^-2$. In this illustration the divisor, 3, indicates the number of equal parts into which the set is divided. This is the *partitive* use of division.

$$^-6 \div 3 = {}^-2$$

Examples such as these can be helpful in understanding why "a negative divided by a negative equals a positive number" and "a negative divided by a positive equals a negative number." (*Note:* The black and red chips model is not convenient for illustrating a positive number divided by a negative number.)

The three theorems for division involving negative numbers are stated here. In each case n and s are positive numbers.

"Positive divided by negative equals negative"	"Negative divided by positive equals negative"	"Negative divided by negative equals positive"
$n \div {}^-s = {}^-(n \div s)$	${}^-n \div s = {}^-(n \div s)$	${}^-n \div {}^-s = n \div s$
$24 \div {}^-6 = {}^-(24 \div 6) = {}^-4$	${}^-14 \div 2 = {}^-(14 \div 2) = {}^-7$	${}^-30 \div {}^-6 = 30 \div 6 = 5$

Division with negative numbers can be defined in terms of multiplication, just as it was for whole numbers.

Definition: For any two numbers n and s, with $s \neq 0$,

$$n \div s = \square \qquad \text{if and only if} \qquad n = s \times \square$$

For example,

$$^-12 \div {^-4} = \boxed{3} \qquad \text{because} \qquad ^-12 = {^-4} \times \boxed{3}$$

It is this inverse relationship between division and multiplication which accounts for the similarity between the rules of signs for division and multiplication with negative numbers.

PROPERTIES OF INTEGERS

Inverses for Addition   Addition of integers has one property which addition does not have for whole numbers. Every integer has an inverse for addition. That is, for any integer k, there is a unique integer ^-k, such that $k + {^-k} = 0$. The integers 8 and $^-8$ are inverses of each other for addition:

$$8 + {^-8} = 0$$

Commutative Properties   The operations of addition and multiplication are commutative. In particular, these properties hold for negative integers. Here are two examples.

$$^-3 + {^-5} = {^-5} + {^-3} \qquad \text{and} \qquad ^-3 \times {^-5} = {^-5} \times {^-3}$$

Associative Properties   The operations of addition and multiplication are associative. These properties hold for any combination of three integers. For the integers, $^-7$, $^-2$, and $^-6$,

$$(^-7 + {^-2}) + {^-6} = {^-7} + (^-2 + {^-6}) \qquad \text{and} \qquad (^-7 \times {^-2}) \times {^-6} = {^-7} \times (^-2 \times {^-6})$$

Distributive Property   The distributive property of multiplication over addition holds in the set of integers. Show that this is true in the following example by computing both sides of the equation.

$$^-2 \times (3 + {^-7}) = {^-2} \times 3 + {^-2} \times {^-7}$$

OPERATIONS ON CALCULATORS

Most calculators are designed to compute with negative as well as positive numbers. Here are four steps for multiplying $^-44$ times $^-16$. These steps may not work on your calculator.

Steps	Displays
1. $\boxed{-}$ 44	$^{-}44.$
2. $\boxed{\times}$	$^{-}44.$
3. $\boxed{-}$ 16	$^{-}16.$
4. $\boxed{=}$	$704.$

On some calculators the minus sign which is used in Step 3 of the previous sequence will cancel (replace the $\boxed{\times}$ in Step 2. As a result, the calculator will evaluate $^{-}44 - 16$ and give an answer of $^{-}60$. There are several ways around this problem. One is to use the fact that a "negative times a negative is positive." That is, we can compute $^{-}44 \times {}^{-}16$ by computing 44×16.

Some calculators have a button which will *negate* (multiply by negative 1) the number in the display. The $\boxed{\text{CS}}$ and $\boxed{+/-}$ are two such buttons. Here are the steps for computing $^{-}44 \times {}^{-}16$ using the $\boxed{+/-}$ button.

Steps	Displays
1. 44 $\boxed{+/-}$	$^{-}44.$
2. $\boxed{\times}$	$^{-}44.$
3. 16 $\boxed{+/-}$	$^{-}16.$
4. $\boxed{=}$	$704.$

Negative numbers can be added by using the following relationship between addition and subtraction: $a + {}^{-}b = a - b$. that is, sums of negative numbers can be computed by subtracting positive numbers. The sum $^{-}118 + {}^{-}249 + {}^{-}403$ can be replaced by $^{-}118 - 249 - 403$. The following steps will work on most calculators.

Steps	Displays
1. Enter $\boxed{-}$ 118	$118.$
2. $\boxed{-}$	$^{-}118.$
3. Enter 249	$249.$
4. $\boxed{-}$	$^{-}367.$
5. Enter 403	$403.$
6. $\boxed{=}$	$^{-}770.$

The $\boxed{+/-}$ button is handy if you wish to subtract negative numbers. It negates the display without interfering with the minus operation, as seen in the following steps for computing $^-15 - \,^-8$.

Steps	Displays
1. 15 $\boxed{+/-}$	$^-15.$
2. $\boxed{-}$	$^-15.$
3. 8 $\boxed{+/-}$	$^-8.$
4. $\boxed{=}$	$^-7.$

APPROXIMATION AND MENTAL CALCULATION

The commutative and associative properties for addition allow us to rearrange the numbers in any sum. The following sum can be more easily computed mentally by combining the negative numbers and positive numbers separately.

$$^-30 + 40 + \,^-62 + 10 = \,^-92 + 50 = \,^-42$$

Sometimes there will be convenient combinations of positive and negative numbers which will cancel each other. In the next example, $^-80 + \,^-40$ almost cancels the 125, leaving $60 + 5$.

$$^-80 + 60 + \,^-40 + 125 = 60 + 5 = 65$$

Rounding Off   We can combine the technique of rearranging the order of the numbers with rounding off, to approximate a sum.

$$81 + \,^-32 + 21 + \,^-47 \approx (80 + 20) + (^-30 + \,^-50)$$
$$= 100 + \,^-80$$
$$= 20$$

Can you see another easy way to approximate this sum?

COMPUTER APPLICATIONS

Select any four nonnegative integers and place them at the corners of a square. Then between each pair of numbers write the difference of the larger minus the smaller and form an inner square whose corners have these differences. Continue this process of taking differences and forming inner squares. Eventually the differences will all be equal. This observation was attributed to the Italian mathematician, E. Ducci, in the 1930s.* Four inner squares (or four steps) are needed before all the differences are equal in the example shown here. Try it.

It is surprising how few steps are needed before all the differences are equal, even for large numbers.

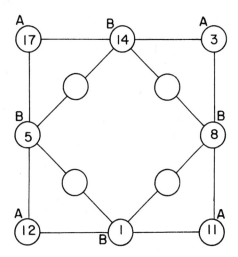

PROBLEM Find four nonnegative integers for which seven steps are needed before the differences are all equal.

To solve this problem using a computer it will be convenient to have a new function.

Absolute Value Function The absolute value function, ABS(X), computes the positive numerical value of the number in the parentheses. Here are some examples.

$$\text{ABS}(^-12) = 12 \qquad \text{ABS}(^-7.8) = 7.8 \qquad \text{ABS}(14) = 14$$

This function enables the computer to subtract consecutive numbers on the square in the preceding example and get differences which are positive (or zero) without necessarily subtracting the smaller number from the larger. For example,

$$3 - 17 = {}^-14 \qquad \text{and} \qquad \text{ABS}(3 - 17) = \text{ABS}(^-14) = 14$$

Beginning with any four nonnegative integers, the following program computes the differences until they are all equal. Lines 30, 32, 34, and 36 compute the differences and lines 70, 72, 74, and 76 set A, B, C, and D equal to each new set of differences. The printout below the program is for the square above.

Program 6.3A

```
10  PRINT "TYPE FOUR INTEGERS, SEPARATED BY COMMAS."
20  INPUT A,B,C,D
30  LET E =  ABS (A - B)
32  LET F =  ABS (B - C)
```

*Ross Honsberger, *Ingenuity in Mathematics* (Washington, D.C.: The Mathematical Association of America, 1970, 2nd Ed.) pp. 73, 80–83.

```
34    LET G =  ABS (C - D)
36    LET H =  ABS (D - A)
40    LET N = N + 1
50    PRINT "THE DIFFERENCES FOR STEP "N" ARE "E", "F", "G", "H
60    IF E = F AND F = G AND G = H THEN  GOTO 90
70    LET A = E
72    LET B = F
74    LET C = G
76    LET D = H
80    GOTO 30
90    END

RUN

TYPE FOUR INTEGERS, SEPARATED BY COMMAS.
?17, 3, 11, 12
THE DIFFERENCES FOR STEP 1 ARE 14, 8, 1, 5
THE DIFFERENCES FOR STEP 2 ARE 6, 7, 4, 9
THE DIFFERENCES FOR STEP 3 ARE 1, 3, 5, 3
THE DIFFERENCES FOR STEP 4 ARE 2, 2, 2, 2
```

Integer Function   The integer function, INT(X), which up to this point has been used only for nonnegative numbers (Sections 3.5, 3.6, and Chapter 5), is also used for rounding down negative numbers. As before, INT(X) equals the greatest integer which is not greater than X. For example, INT($^{-}$5.2) = $^{-}$6.

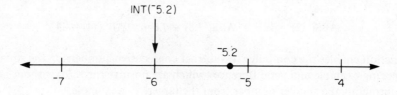

Here are some more examples.

$$INT(^{-}3.1) = ^{-}4 \qquad INT(^{-}213.2) = ^{-}214 \qquad INT(^{-}12) = ^{-}12$$

The integer function is used in the same way for rounding off negative numbers as it was for rounding off positive numbers. Here is the formula which was used in Chapter 5 for rounding off a number to its nearest tenth.

$$INT(10 * X + .5)/10$$

The following equations show step by step how this formula rounds off $^{-}$21.37 to $^{-}$21.4.

$$\text{INT}(10 * {}^-21.37 + .5)/10 = \text{INT}({}^-213.7 + .5)/10$$
$$= \text{INT}({}^-213.2)/10$$
$$= {}^-214/10$$
$$= {}^-21.4$$

SUPPLEMENT (Activity Book)

Activity Set 6.3 Models for Operations with Integers (Illustrations of the four basic operations, with black and red chips)

Just for Fun: Games for Negative Numbers

Exercise Set 6.3: Applications and Skills

1. Antarctica, the only polar continent, is centered near the South Pole and is covered by a huge ice dome reaching a height of almost 4 kilometers (about 13,000 ft). One of the hazards of South Polar exploration is the hidden crevasses in the ice. This picture shows a crevasse detector operating in the Antarctic.

Ice crevasse detector, Antarctic

★ a. Here are five daytime Celsius temperatures from an Antarctic summer: ${}^-32°$, ${}^-27°$, ${}^-24°$, ${}^-34°$, and ${}^-28°$. What is their average?

b. Winter temperatures are usually below ${}^-75°$C. What is the average of these winter temperatures: ${}^-84°$C, ${}^-72°$C, ${}^-79°$C, ${}^-81°$C, and ${}^-78°$C?

2. a. *Addition:* Sketch black and red chips to illustrate this sum. Complete the equation.

$$3 \quad + \quad {}^-7 \quad = $$

b. *Subtraction:* Remove chips from this diagram to compute the difference and then complete the equation.

$${}^-8 - {}^-3 = $$

c. *Multiplication:* Remove pairs of black chips to compute this product and then complete the equation.

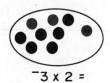

$^{-}3 \times 2 =$

d. *Division:* Divide these chips into five equivalent sets to compute this quotient. Complete the equation.

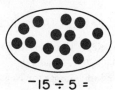

$^{-}15 \div 5 =$

3. Illustrate each sum or difference of integers by drawing arrows on the corresponding number line.

★ a. $6 + {}^{-}5 =$
   $$^{-}8 \quad {}^{-}7 \quad {}^{-}6 \quad {}^{-}5 \quad {}^{-}4 \quad {}^{-}3 \quad {}^{-}2 \quad {}^{-}1 \quad 0 \quad 1 \quad 2 \quad 3 \quad 4 \quad 5 \quad 6 \quad 7 \quad 8$$

   b. $^{-}4 + 9 =$
   $$^{-}8 \quad {}^{-}7 \quad {}^{-}6 \quad {}^{-}5 \quad {}^{-}4 \quad {}^{-}3 \quad {}^{-}2 \quad {}^{-}1 \quad 0 \quad 1 \quad 2 \quad 3 \quad 4 \quad 5 \quad 6 \quad 7 \quad 8$$

   c. $^{-}3 + {}^{-}4 =$
   $$^{-}8 \quad {}^{-}7 \quad {}^{-}6 \quad {}^{-}5 \quad {}^{-}4 \quad {}^{-}3 \quad {}^{-}2 \quad {}^{-}1 \quad 0 \quad 1 \quad 2 \quad 3 \quad 4 \quad 5 \quad 6 \quad 7 \quad 8$$

   d. $7 - 5 =$
   $$^{-}8 \quad {}^{-}7 \quad {}^{-}6 \quad {}^{-}5 \quad {}^{-}4 \quad {}^{-}3 \quad {}^{-}2 \quad {}^{-}1 \quad 0 \quad 1 \quad 2 \quad 3 \quad 4 \quad 5 \quad 6 \quad 7 \quad 8$$

★ e. $4 - 7 =$
   $$^{-}8 \quad {}^{-}7 \quad {}^{-}6 \quad {}^{-}5 \quad {}^{-}4 \quad {}^{-}3 \quad {}^{-}2 \quad {}^{-}1 \quad 0 \quad 1 \quad 2 \quad 3 \quad 4 \quad 5 \quad 6 \quad 7 \quad 8$$

   f. $^{-}5 - 2 =$
   $$^{-}8 \quad {}^{-}7 \quad {}^{-}6 \quad {}^{-}5 \quad {}^{-}4 \quad {}^{-}3 \quad {}^{-}2 \quad {}^{-}1 \quad 0 \quad 1 \quad 2 \quad 3 \quad 4 \quad 5 \quad 6 \quad 7 \quad 8$$

4. *Negative Number Skills:*

   a. Compute the following products and quotients.

   ★ $^{-}6 \times 2 =$ $^{-}8 \times {}^{-}3 =$ ★ $1/3 \times {}^{-}12 =$
      $24 \div {}^{-}6 =$ ★ $^{-}20 \div 4 =$ $^{-}6 \div 1/3 =$

   b. Find the missing number for each equation.

   ★ $4 + \square = {}^{-}10$ $^{-}3 + \square = {}^{-}11$
   ★ $6 - \square = 10$ $^{-}4 - \square = 7$
   ★ $^{-}6 \times \square = {}^{-}12$ $^{-}1/5 \times \square = 1$
   ★ $^{-}15 \div \square = {}^{-}3$ $24 \div \square = {}^{-}8$

   c. A number k is defined to be less than m if there is a positive number $\square$ such that $k + \square = m$. Find the replacement for $\square$ to show that each of the following inequalities is true.

   ★ $^{-}3 < {}^{-}2$ $^{-}14 < 3$ $^{-}7 < 1$
     $^{-}3 + \square = {}^{-}2$ $^{-}14 + \square = 3$ $^{-}7 + \square = 1$

CHAPTER 6   FRACTIONS AND INTEGERS

★ d. For each of the numbers in the table on the right, write its negative (inverse for addition) and its reciprocal (inverse for multiplication).

Number	$\frac{7}{8}$	‾4	$\frac{-1}{2}$	10
Negative				
Reciprocal				

5. *Approximation and Mental Calculation:* Round off each number to its leading digit and mentally approximate the sum of the numbers in each row.

Think
100   ‾200   ‾200   500

				Approximate Sum
118	‾235	‾190	485	200
a. ‾123	207	‾315	186	
b. ‾238	175	‾103	‾214	
c. 78	‾41	19	‾38	
d. ‾23	51	‾48	‾82	

★ b., ★ d.

6. This table contains the percentage increases (positive numbers) and decreases (negative numbers) from one year to the next in energy consumption for petroleum products, natural gas, and coal from 1970 to 1981.*

a. Find three consecutive years in which the consumption of petroleum products decreased. What was happening to the consumption of coal during these years?

★ b. Over the 12-year period covered by this table, there was one year in which the use of petroleum decreased and the use of natural gas increased. What year was this?

YEAR	Energy consumption			
	Total [2]	Refined petroleum products	Natural gas	Coal
1970	4.8	4.9	6.7	1.3
1971	2.2	3.5	3.1	−5.1
1972	4.9	7.8	1.0	3.7
1973	4.2	5.7	−.8	6.8
1974	−2.5	−4.0	−3.5	−3.2
1975	−2.8	−2.2	−8.2	−.5
1976	5.4	7.5	2.0	7.1
1977	2.4	5.5	−2.1	1.7
1978	2.4	2.3	.4	−.8
1979	.9	−2.2	3.4	9.1
1980	−3.9	−7.9	−1.4	2.3
1981	−2.8	−6.4	−3.7	3.7

*Source: U.S. Energy Information Administration.

7. Which number property of the integers is being used in each of the following equalities?

★ a. $^-4 \times (16 + {^-9})/({^-6} + 13) = {^-4} \times (16 + {^-9})/(13 + {^-6})$

b. $^-4 \times ({^-3} + 3) + {^-17} = ({^-4} \times {^-3}) + ({^-4} \times 3) + {^-17}$

★ c. $({^-8} + 7) + 2 \times ({^-6} \times {^-5}) = ({^-8} + 7) + (2 \times {^-6}) \times {^-5}$

d. $^-3 \times (16 \times {^-5})/({^-14} + 2) = {^-3} \times ({^-5} \times 16)/({^-14} + 2)$

8. Extend the patterns in each of these columns of equations by writing the next three equations. What multiplication rule for negative numbers is suggested by the last few equations that you have written in each column?

a. $5 \times 3 = 15$
   $5 \times 2 = 10$
   $5 \times 1 = 5$
   $5 \times 0 = 0$

b. $3 \times 6 = 18$
   $2 \times 6 = 12$
   $1 \times 6 = 6$
   $0 \times 6 = 0$

c. $^-3 \times 3 = {^-9}$
   $^-3 \times 2 = {^-6}$
   $^-3 \times 1 = {^-3}$
   $^-3 \times 0 = 0$

9. For each number written above the number line, locate its reciprocal (inverse for multiplication).

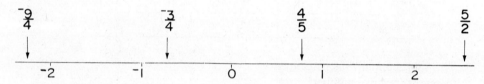

★ a. What general statement can be made about the reciprocals of nonzero numbers in the interval from $^-1$ to 1?

b. What general statement can be made about the reciprocals of numbers outside of the interval from $^-1$ to 1?

c. Does 0 have a reciprocal?

10. *Calculator Exercise:* These two sequences of steps use memory storage, [M+], and memory recal, [MR], for subtracting negative numbers. Which one of these sequences computes $^-15 - {^-6} = {^-9}$? What number will show in the other display?

a.	Steps	Displays
1.	$-$ 15 $=$	$^-15.$
2.	M+	$^-15.$
3.	C	0.
4.	$-$ 6 $=$	$^-6.$
5.	$-$	$^-6.$
6.	MR	$^-15.$
7.	$=$	?

b.	Steps	Displays
1.	$-$ 6 $=$	$^-6.$
2.	M+	$^-6.$
3.	C	0.
4.	$-$ 15 $=$	$^-15.$
5.	$-$	$^-15.$
6.	MR	$^-6.$
7.	$=$	?

c. Write the steps and the corresponding displays for computing 15 × ⁻6 by using the ⊞ button.

11. NASA's *Voyager 1* achieved its closest approach to Jupiter (about 280,000 km) on March 5, 1979. In December, 1978, ⁻80 days before its closest approach, the spacecraft swivelled its narrow-angle television camera to begin its "observatory phase." The inner squares shown next are the camera's fields of view at four different approach times.

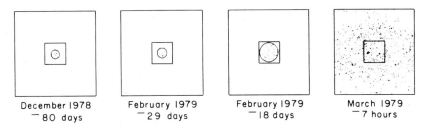

| December 1978 | February 1979 | February 1979 | March 1979 |
| ⁻80 days | ⁻29 days | ⁻18 days | ⁻7 hours |

★   a. How much time elapsed between the ⁻18-day view and the ⁻7-hour view?

b. As *Voyager 1* moved away from Jupiter, it examined its moons: Io at ⁺3 hours, Europa at ⁺5 hours, and Ganymede at ⁺14 hours. Thirty-six hours after the ⁻7-hour view it examined Callisto. How many hours was this after its closest approach to Jupiter?

12. *Magic Squares:* A square array of numbers in which the sum of numbers in any horizontal row, vertical column, or diagonal is always the same is called a *magic square.* Write the numbers below in the 3 by 3 array to produce a magic square.

⁻10, ⁻8, ⁻6, ⁻4, 0, 2, 4, 6

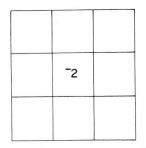

Exercise Set 6.3: Problem Solving

1. *Neighbor Integers:* The integers from ⁻4 to 4 can be placed around a circle so that each integer from ⁻10 to 10 can be obtained by adding two or more neighbor numbers (numbers that are next to each other) from around the circle. How can this be done?

a. *Understanding the Problem:* Here is a simplified version of the problem for the integers from ‾3 to 3. The neighbor numbers 1, 0, ‾3, and ‾2 have a sum of ‾4. How can 4 be obtained as the sum of two or more neighbor numbers?

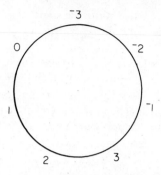

★ b. *Devising a Plan:* This type of problem can be solved by guessing and checking the results. However, the requirement that we be able to get sums from ‾10 and 10 is a strong condition on the placement of the integers. Explain why. For trying different placements of numbers you may find it helpful to move numbered markers around the circle.

c. *Carrying Out the Plan:* Find the placement for the integers from ‾4 to 4 and show how each number from ‾10 to 10 can be obtained as a sum of neighbor numbers.

d. *Looking Back:* Each of the integers from ‾4 to 4 can be routinely obtained by adding all the numbers on the circle except the negative of the given number. For example, to get ‾4 add all the numbers except 4. Explain why this works.

★ 2. *Neighbor Integers (Extension):* The integers from ‾5 to 5 can be placed around a circle so that each integer from ‾15 to 15 can be obtained by adding two or more neighbor numbers. Show how this can be done. The integers from ‾5 to 5 can be routinely obtained by adding all the numbers except the negative of the given number. Show how each of the remaining numbers from ‾15 to 15 can be obtained as the sum of two or more neighbor numbers.

3. *Nine-Integer Problem:* Here are the positive and negative values of the integers from 1 to 9.

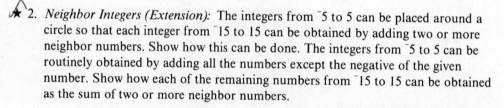

The sum of the nine positive integers is 45 and the sum of the nine negative integers is ‾45. Which of the remaining integers from ‾45 to 45 can be obtained by adding either the positive or negative value of each of these nine integers exactly once. For example, here are nine integers whose sum is 17.

$$1 + ‾2 + ‾3 + ‾4 + ‾5 + 6 + 7 + 8 + 9 = 17$$

★ 4. *Eight-Integer Problem:* Here are the positive and negative values of the integers from 1 to 8.

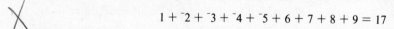

Which of the integers from ⁻36 to 36 can be obtained by adding either the positive or negative value of each of these eight integers exactly once? What general conclusion can be drawn regarding this type of problem?

5. *Plus and Minus Signs:* Keeping the single-digit numbers from 1 to 9 in order,

$$1 \quad 2 \quad 3 \quad 4 \quad 5 \quad 6 \quad 7 \quad 8 \quad 9$$

it is possible to insert seven plus or minus signs to obtain a sum of 100.*

$$1 + 2 + 3 - 4 + 5 + 6 + 78 + 9 = 100$$

   a. Show how three plus or minus signs can be inserted to obtain a sum of 100.
★ b. Suggest an extension to this problem.

Exercise Set 6.3: Computers

1. *Professor Ducci's Observation:*
   a. Determine the printout for Program 6.3A on page 377 if the first four numbers are A = 4, B = 7, C = 2, D = 1.
   b. Find four nonnegative integers for the corners of a square for which seven steps are needed before the differences are all equal.
☛★ c. Answer part **b** for four nonnegative integers that require eight steps. (*Hint:* This can be done for four numbers which are each less than or equal to 12.)

2. The formula INT(10 * C + .5)/10 rounds off a number to its nearest tenth.
   a. Show step by step how this formula rounds off ⁻18.362.
★ b. Write a computer program to convert a Fahrenheit temperature to a Celsius temperature which is rounded off to the nearest tenth of a degree: C = 5(F − 32)/9.

☛ c. The coldest place on earth is the Pole of Cold in Antarctica. Its average annual temperature is ⁻72° Fahrenheit. Revise the program in part **b** to print out all the Fahrenheit temperatures from ⁻60 to ⁻80 and their equivalent Celsius temperatures to the nearest tenth.

3. In the "Rabbit in the Box Problem" (page 364) we saw that the sum of the reciprocals of the whole numbers can be made as large as we please by using enough fractions. Let's investigate what happens when we negate every other fraction in that sum.

*B. A. Kordemsky, *The Moscow Puzzles,* (New York: Charles Scribner's Sons, 1972) p. 23.

$$\frac{1}{2} + \frac{^-1}{3} + \frac{1}{4} + \frac{^-1}{5} + \frac{1}{6} + \frac{^-1}{7} + \frac{1}{8} + \frac{^-1}{9} + \cdots$$

a, Revise Program 6.2B on page 356 to compute this sum (*Hint:* Let $S = S + (1/x)(^-1)^x$.)

b, What is the sum of the first: 10 fractions? 20 fractions? 100 fractions?

c, No matter how many fractions are used in this sum, the sum is always less than some number. What is this number to the nearest tenth?

Decimals: Rational and Irrational Numbers

He is unworthy of the name of man who is ignorant of the fact that the diagonal of a square is incommensurable with its side.

PLATO

Circular patterns of atoms in an iridium crystal,
magnified more than a million times by a field ion microscope

7.1
DECIMALS AND
RATIONAL NUMBERS

Each dot in this remarkable picture is an atom in an iridium crystal. These circular patterns show the order and symmetry governing atomic structures. The diameters of atoms, and even the diameters of electrons contained in atoms, can be measured by decimals. Each atom in this picture has a diameter of .000000027 centimeter, and the diameter of an electron is .00000000000056354 centimeter.

The use of decimals is not restricted to that of describing small objects. The gross national product (GNP) and the national income (NI) for four 5-year periods are expressed to an accuracy of tenths of a billion dollars in the following table.*

	1965	1970	1975	1980
GNP (billions)	$691.1	$992.7	$1,549.2	$2,631.7
NI (billions)	$572.4	$810.7	$1,239.4	$2.116.6

HISTORICAL DEVELOPMENT

The person most responsible for our use of decimals is Simon Stevin, a Dutchman. In 1585 Stevin wrote *La Disme,* the first book on the use of decimals. He not only stated the rules for computing with decimals but also pointed out their practical applications. Stevin showed that business calculations can be performed as easily as if they involve only whole numbers. He recommended that the government adopt the decimal system and enforce its use.

As decimals gained acceptance in the sixteenth and seventeenth centuries, there were a variety of notations. Many writers used a vertical bar in place of a decimal point. Here are some examples of how 27.847 was written during this period.

27 | 847   27(847)   27 |847   27 847

27847 . . . ③   27,8^{i}4^{ii}7^{iii} 27847   27⓪8①4②7③

Today, there are still variations in the use of decimal notation. The English place their decimal point higher above the line than it is written in the United States. In other European countries a comma is used in place of a decimal point. A comma and raised numeral denote a decimal in Scandinavian countries.

United States	England	Europe	Scandinavian countries
82.17	82·17	82,17	82,^{17}

MODELS FOR DECIMALS

Decimal squares, the abacus, and number lines provide three models for decimals with increasing levels of abstraction.

Decimal Squares   This model illustrates the part-to-whole concept of decimals. Squares are divided into 10, 100, and 1000 equal parts and the decimal for a square tells what part of the square is shaded. As examples, 3 shaded parts out of 10 represents the decimal .3, 35 shaded parts out of 100 represents .35, and 375 shaded parts out of a thousand represents .375.

*Bureau of Economic Statistics, *Handbook of Basic Economic Statistics* (Washington, D.C.: Economics Statistics Bureau, 1983), pp. 224–25, 230–31.

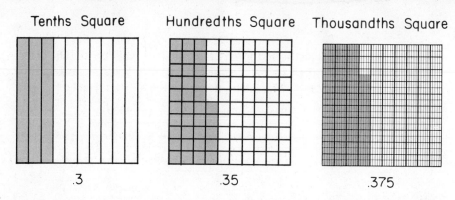

Tenths Square	Hundredths Square	Thousandths Square
.3	.35	.375

Equalities of decimals are illustrated by equal shaded amounts. By comparing the squares above you can see that

$$.3 = .30 = .300$$

because a tenths square with 3 shaded parts, a hundredths square with 30 shaded parts, and a thousandths square with 300 shaded parts all have the same amount of shading. The fact that 10 parts out of 100 equals 1 part out of 10, and 10 parts out of 1000 equals 1 part out of 100 is used in regrouping, as you will see in the following pages.

$$.10 = .1 \text{ and } .010 = .01$$

Abacus   Decimals can be illustrated on an abacus by using columns for powers of 10 and for reciprocals of powers of 10. From left to right the columns shown on this abacus represent thousands, hundreds, tens, units, tenths, hundredths, thousandths, and ten-thousandths. The 2 markers on the 1/10 column represent 2/10 or .2, the 8 markers on the $1/10^2$ column represent 8/100 or .08, and the 6 markers on the $1/10^3$ column represent 6/1000 or .006.

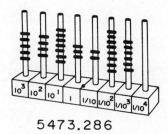

5473.286

A marker from any column can be replaced by putting 10 markers on the column to the right. Conversely, 10 markers from any column can be replaced by 1 marker on the column to the left. For example, 10 markers on the $1/10^3$ column can be regrouped or replaced by putting 1 marker on the $1/10^2$ column. This is visually illustrated by the decimal squares at the top of this page and is shown in the following equations by using fractions.

$$10 \times \frac{1}{10^3} = \frac{10}{10^3} = \frac{1}{10^2}$$

Number Line   The number line is a common model for illustrating decimals. We can locate .372 by subdividing the unit interval into tenths, hundredths, and thousandths. First, the interval from 0 to 1 is divided into 10 equal parts. These are tenths of a unit, and .372 is between .3 and .4. Second, the interval from .3 to .4 is divided into 10 equal parts. These are hundredths of a unit, and .372 is between .37 and .38. Finally, if the interval from .37 to .38 is divided into 10 equal parts, each will be one-thousandth of a unit (see enlarged segment below number line). The second one-thousandth mark in this interval corresponds to .372.

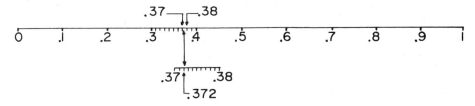

This process of repeatedly subdividing intervals into 10 parts can be used to locate any terminating decimal in a finite number of steps. Infinite repeating decimals cannot be located in this manner. For example, subdividing intervals into 10 parts to locate .333 . . . could go on indefinitely without producing the point on the number line which corresponds to this decimal. There is, however, another way of locating infinite repeating decimals on the number line. Every infinite repeating decimal can be written as a rational number of the form r/s (see Examples 1 and 2 on page 396). The point for r/s can be located by dividing the unit interval into s equal parts and counting off r of these parts. The decimal .3333 . . . is equal to 1/3, and this can be located by dividing the unit interval into three equal parts.

Decimals are used for negative as well as positive numbers. This graph contains increasing and decreasing changes from month to month in the Producer Price Index in tenths of a percent for 1983.* The change was ⁻1.1 in January, ⁻.3 in March, ⁻.2 in November, and there was no change in April.

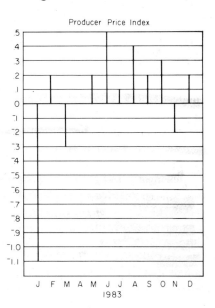

For each positive decimal, whether terminating or infinite repeating, there is a corresponding negative decimal, such that the sum of the two decimals is zero. When two numbers have a sum of zero they are called *opposites* (or *inverses for addition*) of each other.

*Source: U.S. Department of Labor, Producer Price Index office.

The November and December price changes in the graph on page 391 are opposites. Several decimals and their opposites are shown on the following number line.

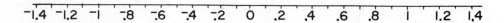

INEQUALITY OF DECIMALS

The definition for inequalities of decimals is the same as the definition that was stated for the inequality of positive and negative numbers on page 368. For any two decimals if there is a positive number which can be added to the first decimal to get the second, then the first decimal is less than the second. Consider .372 and .6. Since .372 + .228 = .6, .372 is less than .6. This definition, however, is not a practical test for determining inequality. It is much easier to see that .372 is less than .6 by comparing the tenths digits in these decimals: .3 is less than .6 (3/10 < 6/10). The "72" in ".372" represents 72/1000, which is less than .1 and not great enough to affect the inequality. Remember how we located .372 on the number line on the previous page, by subdividing smaller and smaller intervals? The "72" in ".372" shows us that .372 lies between .3 and .4 on the number line, and therefore .372 lies to the left of .6.

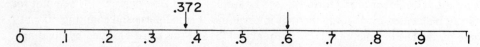

In general, for positive decimals that are both less than 1, the decimal with the greater digit in the tenths place will be the greater number. If these digits are equal, this test is applied to the hundredths digits, etc.

DECIMAL TERMINOLOGY
AND NOTATION

The word "decimal" comes from the Latin "decem," meaning ten. Technically, any number written in base 10 positional numeration is a decimal. However, "decimal" is more often used to refer only to numbers such as 17.38 or 104.5, which are expressed with decimal points. The number of digits to the right of the decimal point is called the *number of decimal places*. There are two decimal places in 17.08 and one decimal place in 104.5.

The positions of the digits to the *left* of the decimal point represent increasing powers of 10 (1, 10, 10^2, 10^3, . . .). The positions to the *right* of the decimal point represent decreasing powers of 10 (10^{-1}, 10^{-2}, 10^{-3}, . . .) or reciprocals of powers of 10 (1/10, $1/10^2$, $1/10^3$, . . .). In the decimal 5473.286 the "2" represents 2/10, the "8" denotes 8/100, and the "6" stands for 6/1000.

Just as in the case of whole numbers, decimals can be written in expanded form in several ways.

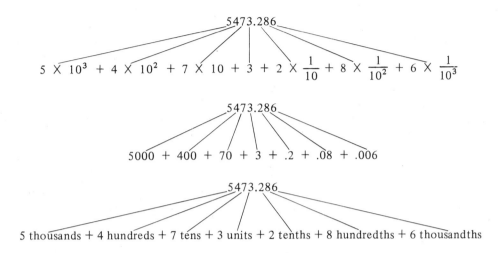

5473.286

$$5 \times 10^3 + 4 \times 10^2 + 7 \times 10 + 3 + 2 \times \frac{1}{10} + 8 \times \frac{1}{10^2} + 6 \times \frac{1}{10^3}$$

5473.286

$$5000 + 400 + 70 + 3 + .2 + .08 + .006$$

5473.286

5 thousands + 4 hundreds + 7 tens + 3 units + 2 tenths + 8 hundredths + 6 thousandths

Reading and Writing Decimals The digits to the left of the decimal point are read as a whole number (see page 71) and the decimal point is read as "and." The digits to the right of the point are also read as a whole number, and then followed by the name of the place value of the last digit. For example, 1208.0925 is read, "one thousand, two hundred eight and nine hundred twenty-five ten-thousandths."

1 2 0 8 . 0 9 2 5

One thousand, two hundred eight

and

nine hundred twenty-five ten-thousandths

When writing an amount of money, the decimal part of a dollar is in hundredths. Notice that on this bank check it is unnecessary to write "dollars" or "cents." The amount is in terms of dollars, and this unit is printed at the end of the line upon which the amount of money is written. Some

people write the decimal part of a dollar as a fraction. For example, the amount on this check might have been written as "one hundred seventy-seven and 24/100."

ROUNDING OFF

To round off a decimal at a given place value or to a given number of decimal places, locate the given place value and check the digit to its right. If the digit to the right is 5 or greater, then all digits to the right are dropped off and the place value is increased by

1. Otherwise, all digits to the right of the given place value are merely dropped off. Here are some examples.

1. Rounding to two decimal places (rounding to hundredths)

Hundredths
$1.6825 \longrightarrow 1.68$

2. Rounding to one decimal place (rounding to tenths)

Tenths
$1.6825 \longrightarrow 1.7$

3. Rounding to three decimal places (rounding to thousandths)

Thousandths
$1.6825 \longrightarrow 1.683$

RATIONAL NUMBERS

Decimals, percents, fractions, and ratios are different types of notation for the same numbers. The following four numerals represent the same rational number.

.25	25%	1/4	1:4
Decimal	Percent	Fraction	Ratio

These examples show that a rational number can be represented by several different types of numerals. Regardless of the notation, if a number can be written as the quotient of two integers, it is called a rational number.

Definition: A rational number is a number that can be written in the form a/b, where a and b are integers and $b \neq 0$.

Whenever the denominator of a/b is 1, the rational number equals an integer: $0/1 = 0$, $6/1 = 6$, $^-4/1 = ^-4$, etc. Therefore, the rational numbers contain the integers. Here are some more examples of rational numbers. Each is equal to a quotient of integers, a/b.

$$.45 = \frac{9}{20} \qquad ^-9 = \frac{^-9}{1} \qquad 3 = \frac{3}{1} \qquad 62\% = \frac{62}{100} \qquad \frac{1.6}{5} = \frac{8}{25}$$

A decimal with a finite number of digits is called a *terminating decimal*. Here are some rational numbers and their terminating decimals.

$$\frac{3}{4} = .75 \qquad \frac{1}{8} = .125 \qquad \frac{17}{80} = .2125 \qquad \frac{46}{25} = 1.84$$

When a decimal has an infinite number of digits which repeat a pattern, it is called an *infinite repeating decimal*. The pattern of digits which is repeated is called the *period of the decimal*. This repeating pattern is indicated by three dots or by placing a bar over the numerals that repeat. As examples, 1/12 has a period of length 1; 1/11 has a period of length 2; and 3781/9999 has a period of length 4.

$$\frac{1}{12} = .08333 \ldots \qquad \frac{1}{11} = .090909 \ldots \qquad \frac{3781}{9999} = .37813781 \ldots$$

$$= .08\overline{3} \qquad\qquad\qquad = .\overline{09} \qquad\qquad\qquad = .\overline{3781}$$

CONVERTING BETWEEN FRACTIONS AND DECIMALS

Every number of the form a/b can be written as either a terminating or an infinite repeating decimal, and conversely, every terminating or infinite repeating decimal can be written in the form a/b, where a and b are integers and $b \neq 0$.

From Fractions to Decimals Every rational number of the form a/b can be changed to a decimal by dividing the numerator by the denominator. For example, when the numerator of 3/8 is divided by its denominator, the division algorithm shows that the decimal terminates after three digits.

$$
\begin{array}{r}
.375 \\
8\,\overline{)\,3.000} \\
\underline{2\ 4}\ \ \\
60 \\
\underline{56} \\
40 \\
\underline{40}
\end{array}
$$

On the other hand, when the numerator of 4/7 is divided by the denominator of 4/7, the quotient does not terminate but repeats the same arrangement of six digits (571428) over and over. In this case, the decimal is infinite repeating. This can be seen by the two circled 40s, which show that after six steps in the division process, all of the numbers will be repeated.

$$
\begin{array}{r}
.5714285 \\
7\,\overline{)\,4.0000000} \\
\underline{3\ 5} \\
50 \\
\underline{49} \\
10 \\
\underline{7} \\
30 \\
\underline{28} \\
20 \\
\underline{14} \\
60 \\
\underline{56} \\
\textcircled{40} \\
\underline{35}
\end{array}
$$

It is fairly easy to see why every rational number r/s can be represented either by an infinite repeating decimal or by a terminating decimal. When r is divided by s, the remainders are always less than s. If a remainder of zero occurs in the division process, as it does when dividing 3 by 8, then the decimal terminates. If there is no zero remainder then eventually a remainder will be repeated, in which case the digits in the quotient will also start repeating.

From Decimals to Fractions Terminating decimals can be written as fractions whose denominators are powers of 10 ($.92 = 92/10^2$ and $.4718 = 4718/10^4$). The power of 10 will be the number of decimal places in the decimal. The expanded form of .378 shows why this decimal equals 378/1000. The common denominator of the fractions in the top equation is 1000, and their sum is 378/1000.

$$.378 = \frac{3}{10} + \frac{7}{100} + \frac{8}{1000}$$

$$= \frac{300}{1000} + \frac{70}{1000} + \frac{8}{1000}$$

$$= \frac{378}{1000}$$

Infinite repeating decimals can be converted to fractions by following a sequence of steps similar to those in the following examples.

EXAMPLE 1

1. Represent the decimal by x.

$$x = .656565 \cdots$$

2. Multiply both sides of the equation by 10^2. (It is necessary to move the decimal point past the first repeating pattern.)

$$100x = 65.6565 \cdots$$

3. Subtract Equation 1 from Equation 2. (By doing this the infinite repeating part of the decimal is subtracted off.)

$$99x = 65$$

4. Solve for x.

$$x = \frac{65}{99}$$

EXAMPLE 2

1. Represent the decimal by x.

$$x = .2222 \cdots$$

2. Multiply both sides of the equation by 10.

$$10x = 2.222 \cdots$$

3. Subtract Equation 1 from Equation 2.

$$9x = 2$$

4. Solve for x.

$$x = \frac{2}{9}$$

Both of these examples can be checked by dividing the numerators by their denominators. In each case you will obtain the original decimal.

Since a rational number can always be represented as the quotient of two integers, a/b, or as a decimal, we have two different types of numerals that represent the same numbers. In the following paragraphs we will refer to both types of numerals, fractions and decimals, and you may think of them as interchangeable.

DENSITY OF RATIONAL NUMBERS

Between any two terminating or infinite repeating decimals on the number line, there is always another such decimal. It follows from this, that between any two decimals there are an infinite number of decimals. This crowding together of decimals is referred to by saying that the rational numbers are *dense*.

One method of locating a decimal that is between two given decimals is to add the two decimals and divide by 2. If we begin with 1.4 and 1.82, their sum divided by 2 is 1.61. Next, 1.4 + 1.61, divided by 2, is a decimal between 1.4 and 1.61. If this process is continued, the resulting sequence of decimals will get closer and closer (but never equal) to 1.4.

In spite of the abundance of rational numbers, there are still an infinite number of points left on the number line which do not correspond to these numbers (see Section 7.4, Irrational and Real Numbers).

DECIMALS AND CALCULATORS

To enter a decimal onto a calculator the keys are pressed for the digits from the highest to lowest place values with the decimal point being entered between the units and tenths digits. The digits usually appear in the right side of the display and are pushed to the left as each new digit is entered.

Some calculators automatically round off decimals which exceed the full display of the calculator. On these calculators if 2 is divided by 3, the decimal 0.66 · · · 667 will show in the display. Almost all calculators which round off at a digit before 5 will increase this digit, as described previously. For example, 55/99 is equal to the infinite repeating decimal .5555 . . . . If 55 is divided by 99 on a calculator which rounds off, 0.55 · · · 556 will show in the display.

There are calculators that use two or three more digits than shown in their display. If you compute 1 ÷ 17 on a calculator with 8 places for digits and the calculator does not round off, you will get 0.0588235. The next three digits, 2, 9, and 4, are kept in the calculator for computing but are not displayed. To find the first "hidden digit," multiply by 10. The display will then show 0.5882352. To find the next hidden digit, multiply by 10 and subtract 5. The display will then show 0.8823529. To find the last hidden digit multiply by 10 and subtract 8. The display will show 0.8235294.

Sometimes a decimal may appear to terminate or repeat at a certain point when actually it doesn't. If you compute 1/17 on a calculator, the first seven decimal places are .0588235 but this last "5" is not the beginning of a repeating pattern.

Let's see how a calculator that displays eight digits can be used to find more digits in the decimal representation of 1/17. The long-division algorithm shows the remainder is 9 after obtaining six decimal places in the quotient. This remainder can be found by multiplying .058823 by 17 and subtracting the result from 1.

$$.058823 \times 17 = 0.999991$$

```
         .058823
17 ) 1.000000
       85
       ──
       150
       136
       ───
        140
        136
        ───
         40
         34
         ──
          60
          51
          ──
Remainder  9
```

This product, 0.999991, differs from 1 by .000009, which shows that the next step in the division algorithm is to divide 9 by 17. (*Note:* To avoid problems due to hidden digits or rounding off, 17 was multiplied by .058823 rather than by .0588235. Similarly, the last digit in each calculator display will not be used in the following steps.)

Dividing 9 by 17 on the calculator gives 0.5294117, the next seven digits in the decimal representation for 1/17. Since there is no repeating pattern of decimals, we will multiply .529411 by 17 to get the next remainder.

$$17 \times .529411 = 8.999987$$

This product differs from 9 by .000013, so the next remainder is 13. Dividing once more by 17, $13 \div 17$ will show where the pattern begins to repeat.

$$13 \div 17 = 0.7647058$$

Thus, the decimal for 1/17 is an infinite repeating decimal with a period of 16.

$$\frac{1}{17} = .\overline{0588235294117647}$$

COMPUTER APPLICATIONS

The study of periodic decimals is a "natural" for student curiosity and discovery. Look at the first few decimals for fractions with denominators of 9 and 11. Will these patterns continue to hold?

$$\frac{1}{9} = .111111 \ldots \qquad \frac{1}{11} = .090909 \ldots$$

$$\frac{2}{9} = .222222 \ldots \qquad \frac{2}{11} = .181818 \ldots$$

$$\frac{3}{9} = .333333 \ldots \qquad \frac{3}{11} = .272727 \ldots$$

The decimals for these fractions are infinite repeating. What other types of fractions will have infinite repeating decimals? The following computer program will be useful for investigating these questions. It prints out the proper fractions for a given denominator and their decimals. All but one of the proper fractions with a denominator of 6 have infinite repeating decimals.

Program 7.1A

```
10   PRINT "TYPE A WHOLE NUMBER FOR THE DENOMINATOR OF A
     FRACTION."
20   INPUT D
30   FOR N = 1 TO D - 1
40   PRINT N"/"D" = "N / D"."
50   NEXT N
60   END

RUN

TYPE A WHOLE NUMBER FOR THE DENOMINATOR OF A FRACTION.
?6
1/6 = .166666667.
2/6 = .333333333.
3/6 = .5.
4/6 = .666666667.
5/6 = .833333333.
```

Here are two fractions whose decimals are infinite repeating with periods of maximum length. That is, the length of the period for each decimal is one less than the denominator of the fraction.

$$\frac{1}{7} = .\overline{142857}$$

$$\frac{1}{17} = .\overline{0588235294117647}$$

Since the denominators of these fractions are primes, it is natural to wonder if there are other primes whose reciprocals have decimals with periods of maximum length. To investigate this question, it will be convenient to have a program that prints out any number of digits in a decimal. The next program does this. Line 30 asks you how many digits you would like in the decimal representation of a fraction. The loop in lines 50, 60, 70, and 80 performs K steps in the long-division algorithm to obtain the K digits in the decimal. The printout shows the first 50 digits in the decimal for 1/23. What is the period of this decimal?

Program 7.1B

```
10   PRINT "TYPE THE NUMERATOR AND DENOMINATOR OF A FRACTION.
        SEPARATE THESE NUMBERS BY A COMMA. "
20   INPUT N,D
30   PRINT "HOW MANY DECIMAL PLACES DO YOU WANT? TYPE A WHOLE
        NUMBER. "
40   INPUT K
50   FOR X = 1 TO K
60   LET N = 10 * (N -  INT (N / D) * D)
70   PRINT  INT (N / D)" ";
80   NEXT X
90   END

RUN

TYPE THE NUMERATOR AND DENOMINATOR OF A FRACTION. SEPARATE
THESE NUMBERS BY A COMMA.
? 1, 23
HOW MANY DECIMAL PLACES DO YOU WANT? TYPE A WHOLE NUMBER.
? 50
0 4 3 4 7 8 2 6 0 8 6 9 5 6 5 2 1 7 3 9 1 3 0 4 3 4 7 8 2 6 0
8 6 9 5 6 5 2 1 7 3 9 1 3 0 4 3 4 7 8
```

SUPPLEMENT (Activity Book)

Activity Set 7.1 Decimal Squares

Exercise Set 7.1: Applications and Skills

1. Intervals of time can be measured by this cesium-beam atomic clock to an accuracy of .0000000000001 of a second. (This is equivalent to an accuracy of within 1 second every 300,000 years.) This strange clock, which is 6 m long, is operated at Boulder, Colorado, by the National Bureau of Standards. The frequency of cesium waves is 9,192,631,770 cycles per second. If the clock is adjusted to emit a pulse every 9,192,632 cycles (9,192,631,770 rounded off to the nearest thousand and divided by one thousand) it provides time intervals of .001 (one-thousandth) of a second.

Cesium-beam atomic clock

★ a. What is 9,192,631,770 rounded off to the nearest million? The answer divided by one million is the number of cycles the clock is adjusted to in order to provide intervals of .000001 (one-millionth) of a second.

   b. How many cycles should the clock be adjusted to in order to provide intervals of .000000001 (one-billionth) of a second?

2. The first book written about decimals was by Simon Stevin in 1585. He used small circled numerals between the digits. Among his examples he defined 27 + 847/1000 to be 27⓪8①4②7③. Explain how the decimal point could have evolved from this notation.

   a. In England this decimal is written as 27 · 847 and in the United States as 27.847. Which of these notations is closer to that used by Simon Stevin?

   b. What advantage is there in the English location of the decimal point as compared to the decimal point location which is used in the United States?

3. Use decimal squares to explain why:

   a. .8 = .80

★ b. .25 = .250

   c. .45 < .6

★ d. .3 > .125

4. Regroup the markers on this abacus so that there are fewer than ten on each column. Then write the number that is represented.

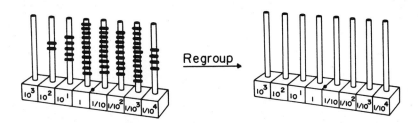

5. The sketch that follows is from a study showing the influence of surface temperature on air currents.

★ a. What is the highest surface temperature on this graph?

   b. What is the difference between the highest and lowest surface temperatures?

★ c. What is the average surface temperature (sum of the six temperatures divided by 6)?

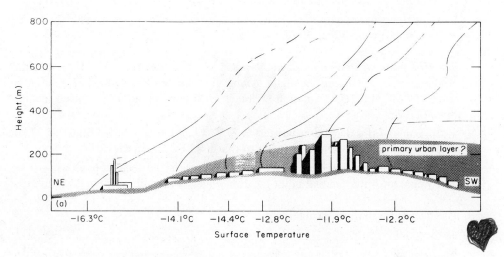

6. Draw an arrow from the numbers to their corresponding points on the number line. For each point that is labeled with a letter, write the number which corresponds to the point.

.07   .72   1.40   1.68

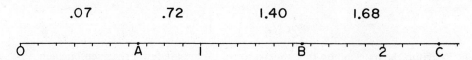

7. Write the names of these numbers.
★ a. 360.2866   b. .99999   c. 7.060   d. 49.00158
★ e. $347.96 _____ Dollars
   f. $23.50 _____ Dollars

8. Express the following decimals as fractions.
★ a. .837   b. .5̄ repeating   ★ c. .6̄4̄   d. 4.1̄
              so 5/9

9. Round off these decimals to the given number of decimal places.
★ a. 2 places   b. 3 places   ★ c. 2 places   d. 4 places
   .4372   .7272···   3.715   .06999

10. *Calculator Exercise:* The fraction 1/19 is represented by an infinite repeating decimal that has 18 digits before the pattern repeats. Use a calculator and the method described in the text to find this repeating pattern of digits.

decimals are fractions

11. *Calculator Exercise:* The rational numbers are dense, which means that between any two rational numbers there is always another one.

★   a. Use a calculator to find the number that is halfway between .4172 and .436 by adding these decimals and dividing by 2.

     b. Add your answer in part **a** to .436 and divide by 2.

★   c. Add your answer in part **b** to .436 and divide by 2.

     d. Add your answer in part **c** to .436 and divide by 2.

★   e. If you continue obtaining numbers in this manner, they will get closer and closer to what number?

12. *Calculator Exercise:* Every decimal except 0 has a *reciprocal*. The product of the decimal times its reciprocal is 1. The reciprocal of 2.318 is 1/2.318, which in decimal form to 7 places is .4314064. Use a calculator to compute the reciprocals of the following decimals. Multiply each number by its reciprocal.

★   a. 2.4   b. .48   c. .0086

★ 13. The 1980 edition of the *Guinness Book of World Records* contains an evolution of sports records for the twentieth century. Two categories of competition with updated records are shown below in the table. For each distance, replace the feet and inches by feet to two decimal places and the minutes and seconds by minutes to two decimal places. (*Hint:* Divide the inches by 12 and the seconds by 60.)

	Start of the Century	Middle of the Century	Recent Records
High Jump	M. Sweeney 6′ 5 5/8″ U.S. 1895	Lester Steers 6′ 11″ U.S. 1941	Zhu Jianhua 7′ 9 1/4″ China 1983
Fastest Mile	W. G. George 4m 12.8s U.K. 1886	Gunder Hägg 4m 1.3s Sweden 1945	Sebastian Coe 3m 47.3s G.B. 1981

Exercise Set 7.1: Problem Solving

1. *Infinite Repeating Decimals:* What types of fractions cannot be written as terminating decimals.

   a. *Understanding the Problem:* Here are two fractions with terminating decimals and two with infinite repeating decimals.

$$\frac{3}{8} = .375 \quad \frac{7}{20} = .35 \quad \frac{5}{6} = .8\overline{3} \quad \frac{7}{12} = .41\overline{6}$$

Which of the following fractions have infinite repeating decimals?

$$\frac{4}{11} \quad \frac{2}{15} \quad \frac{5}{16} \quad \frac{1}{18}$$

This is ridiculous!

★   b. *Devising a Plan:* One approach is to systematically list fractions with denominators of 2, 3, 4, 5, etc. and note which fractions have infinite repeating decimals. Why is it only necessary to consider fractions in lowest terms?

       c. *Carrying Out the Plan:* Choose an approach for solving this problem. How can we predict which fractions will have terminating decimals and which will have infinite repeating decimals?

★   d. *Looking Back:* If you formed your conclusion in part **c** by looking at examples then you used inductive reasoning. Let's use deductive reasoning to solve this problem. Suppose a fraction a/b is in lowest terms and has a terminating decimal with n digits.

$$\frac{a}{b} = .k_1 k_2 k_3 \cdots k_n$$

Multiplying both sides of this equation by 10^n we obtain

$$\frac{a}{b} \times 10^n = k_1 k_2 k_3 \cdots k_n$$

Since the right side of this equation is a whole number, the left side must also be a whole number and so b must divide into 10^n. Why? What does this tell us about the prime factors of b?

2. *Purely Periodic:* Some infinite repeating decimals have a nonrepeating part. That is, digits that occur before the repeating pattern of digits. As examples, the decimal for $1/22$ has a nonrepeating part with one digit and the decimal for $5/12$ has a nonrepeating part with two digits.

$$\frac{1}{22} = .0\overline{45} \qquad \frac{5}{12} = .41\overline{6}$$

       a. What types of fractions will have infinite repeating decimals with a nonrepeating part?

★   b. If a decimal does not have a nonrepeating part, it is called *purely periodic.* What types of fractions will have infinite repeating decimals that are purely periodic?

3. *Odometer Readings:* A truck driver noticed that the mileage on the odometer was 72927, a palindromic number. Four hours later the driver was surprised to find the odometer had another palindromic number. What was the truck's average speed during this four-hour period?

Exercise Set 7.1: Computers

↻ 1. Write a computer program to print out the first few decimals for fractions of the form $1/N$, with $N = 2$ to 30. Use this program to determine the fractions with terminating decimals. What do these fractions have in common?

★ 2. Use Program 7.1A (page 399) to investigate the decimals for fractions of the form N/9, M/99, K/999, and T/9999. What can be said about the periods of the decimals for each of these types of fractions?

3. Use Program 7.1B (page 400) to obtain the decimals for fractions of the form $1/p$, where p is a prime greater than 5 and less than 30.

★ a. Compute the sum of the digits in the period of each of these decimals. What do these sums have in common?

b. The digits in the period of the decimal for 1/7 are 142857. If the first half of these digits is placed under the second half, as shown here, their sum contains all 9's.

$$
\begin{array}{r}
142 \\
+\,857 \\
\hline
999
\end{array}
$$

Does this hold for all the decimals for fractions of the form $1/p$, where p is a prime greater than 5 and less than 30?

A phototimer finish of a one-mile race between Marty Liquori and Jim Ryun

7.2
OPERATIONS WITH
DECIMALS

Decimals to hundredths and thousandths of a second are used in electronic timers for athletic competition. The above picture shows a phototimer finish of a one-mile race in which Marty Liquori of Villanova beat Jim Ryun of the Oregon Track Club at the 1971 International Freedom Games at Philadelphia's Franklin Field. Both runners were officially clocked in 3:54.6 seconds by officials using hand-operated stopwatches. However, the phototimer clocked Liquori in 3:54.54 seconds and Ryun in 3:54.75 seconds, a difference of .21 second. What you see is not a simultaneous photograph of the two runners but a continuous photograph of the finish line showing each runner as he crossed it.

ADDITION

The concept of addition of decimals is the same as that for whole numbers or fractions; it involves "putting together" or "combining" two amounts. The total shaded amount of the two decimal squares shown here is 75 parts. Since each of these parts is one-hundredth of a whole square, the sum of .45 and .30 is .75.

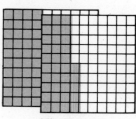

.40 + .35 = .75

Abacus Addition can be illustrated on the abacus by representing the numbers to be added on the columns. In this example, 430.353 is represented on the upper portion of the columns and 145.472 beneath. The sum is computed by counting the total number of markers in each column. When there are 10 or more markers in a column, regrouping is necessary. The total number of markers in the hundredths column in this example is 12. We know (from the decimal squares model) that 10 hundredths is equal to 1 tenth. So, 10 markers on the hundredths column can be replaced by 1 marker on the tenths column. This leaves 2 markers on the hundredths column. Continue computing the sum by using this abacus.

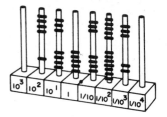

$$430.353 + 145.472$$

Pencil-and-Paper Algorithm In this method for addition of decimals, the digits are aligned, tenths under tenths, hundredths under hundredths, etc., similar to the vertical alignment of the markers on the abacus. When the sum of the digits in any column is 10 or greater, regrouping or "carrying" is necessary. In this example the sum of the digits in the hundredths column is 12, so a 1 is regrouped and added to $3 + 4$ in the tenths column. On the abacus this regrouping is accomplished by removing 10 markers from the $1/10^2$ column and carrying 1 marker to the $1/10$ column. This "1" which is regrouped represents $10/100$, or $1/10$. The following equations show the sum in the hundredths column of this example and the regrouping.

$$
\begin{array}{r}
1 \\
430.353 \\
+\ 145.472 \\
\hline
575.825
\end{array}
$$

$$\frac{5}{100} + \frac{7}{100} = \frac{12}{100} = \frac{10}{100} + \frac{2}{100} = \left(\frac{1}{10}\right) + \left(\frac{2}{100}\right)$$

"1" carried to tenths column   "2" recorded in hundredths column

SUBTRACTION

The concept of subtraction of decimals, as with whole numbers or fractions, can be viewed as "take away" or in terms of "missing addends." This decimal square, for example, shows that $.75 - .35 = .40$ or that .35 must be added to .40 to get .75.

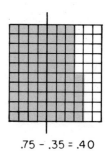

$.75 - .35 = .40$

Abacus To illustrate subtraction on the abacus, only one number is represented on the columns. Then we take away the appropriate number of markers from each column to compute the difference. For example, to compute 77.6528 − 65.9386, we begin by removing 6 markers from the $1/10^4$ column. Then 8 markers should be removed from the thousandths column, but it contains only 2 markers. We know (from the Decimal Squares model) that each hundredth is equal to 10 thousandths. So, we can replace one of the markers on the hundredths column by 10 markers on the thousandths column. Continue computing the difference by using this abacus.

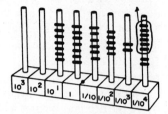

77.6528 − 65.9386

Pencil-and-Paper Algorithm In this method for subtraction of decimals, the digits are aligned as they are for addition of decimals. Subtraction then takes place from right to left, with thousandths subtracted from thousandths, hundredths from hundredths, etc. When regrouping or "borrowing" is necessary, it is done just as it is for subtracting whole numbers. In this example, the "2" in the thousandths column is changed to a "12" by borrowing 1 from 5. On the abacus this regrouping is done by replacing 1 marker

$$
\begin{array}{r}
4 \\
77.6\cancel{5}28 \\
-\ 65.9386 \\
\hline
42
\end{array}
$$

from the $1/10^2$ column by 10 markers on the $1/10^3$ column. This can be done because $1/100 = 10/1000$. The following equations show the change in the hundredths column and the regrouping.

$$
\frac{5}{100} = \frac{4}{100} + \frac{1}{100} = \left(\frac{4}{100}\right) + \left(\frac{10}{1000}\right)
$$

"4" is left in the hundredths column   "10" is regrouped to the thousandths column

To continue this example, regrouping is also needed to subtract 9 from 6 in the tenths column. In this case, 1 is regrouped from the 7 in the units column and used as 10 tenths in the tenths column. These equations show the change in the units column and the regrouping.

$$
\begin{array}{r}
6\ \ \ 4 \\
7\cancel{7}.6\cancel{5}28 \\
-\ 65.9386 \\
\hline
11.7142
\end{array}
$$

$$
7 = 6 + 1 = \left(6\right) + \left(\frac{10}{10}\right)
$$

"6" is left in the units column   "10" is regrouped to the tenths column

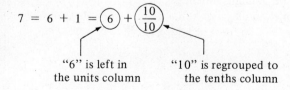

✳MULTIPLICATION

The product of a whole number times a decimal can be illustrated by "repeated addition." Each of these decimal squares, for example, has 7 shaded parts. In all there is a total of $2 \times 7 = 14$ shaded parts. This is 4 more parts than a whole square, so $2 \times .7 = 1.4$.

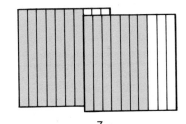

Know blocks

$$\begin{array}{r} .7 \\ \times\ 2 \\ \hline 1.4 \end{array}$$

✳   The product of a decimal times a decimal, such as $.2 \times .3$, means .2 "of" .3. This can be illustrated by using a decimal square for .3 and taking .2 of its shaded part. To do this the shaded part of the decimal square for .3 has been split into 10 equal parts. The 6 double-shaded parts of the square represent .2 of .3. Each of these new parts is one-hundredth of a whole square, so $.2 \times .3 = .06$

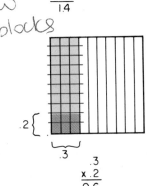

$$\begin{array}{r} .3 \\ \times\ .2 \\ \hline 0.6 \end{array}$$

Abacus   The product of a whole number times a decimal can be illustrated on the abacus by multiplying the numbers of markers on each column separately. To multiply by 2 the number of markers on each column is doubled. The regrouping can be done as you multiply or after the multiplication has been carried out. In this example, the markers for 352.46 have been doubled. Which columns require regrouping? Use this abacus to finish computing the product.

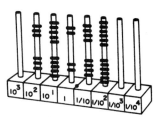

$$\begin{array}{r} 3\,5\,2.4\,6 \\ \times\ 2 \\ \hline 7\,0\,4.9\,2 \end{array}$$

Pencil-and-Paper Algorithms   One of the notations for locating decimal points in the sixteenth century was that of writing circled indices to the right of the numerals. For example, 27.487 was represented as 27487 . . . ③, and 9.21 was 921 . . . ②. This notation is especially convenient when computing the product of two decimals. The whole numbers are multiplied, and then the numbers in circles are

$$\begin{array}{r} 27487 \ldots ③ \\ \times\ 921 \ldots ② \\ \hline 27487 \\ 54974\ \ \\ 247383\ \ \ \ \\ \hline 25315527 \ldots ⑤ \end{array}$$

added to determine the location of the decimal point. This particular notation points out the close relationship between computing products of whole numbers and

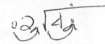

computing products of decimals. Using our present notation the decimal point should be placed between the "3" and "1" in this product.

The current pencil-and-paper algorithm for multiplying decimals is essentially the same as the sixteenth-century example. Two numbers are multiplied as though they were whole numbers and then the decimal point is located in the product. The digits *do not* have to be positioned so that units are above units, tenths above tenths, etc., as they are for addition and subtraction of decimals. The number of decimal places in the answer is the total number of decimal places in the original two numbers.

$$
\begin{array}{r}
2\,7.4\,8\,7 \\
\times\ \ 9.2\,1 \\
\hline
2\,7\,4\,8\,7 \\
5\,4\,9\,7\,4\ \ \\
2\,4\,7\,3\,8\,3\ \ \ \ \\
\hline
2\,5\,3.1\,5\,5\,2\,7
\end{array}
$$

The grid (or graph paper) model which was used for multiplying one-digit and two-digit whole numbers (pages 134 and 135) can be used to illustrate the products of mixed decimals. The product 2.3 × 1.7 is represented by the total area of the rectangle shown below. The four regions of this rectangle correspond to the four partial products of 2.3 × 1.7.

$$
\begin{array}{r}
1.7 \\
\times 2.3 \\
\hline
.21 \\
.3 \\
1.4 \\
2 \\
\hline
3.91
\end{array}
\qquad
\begin{array}{l}
(.3 \times .7) \\
(.3 \times 1) \\
(2 \times .7) \\
(2 \times 1)
\end{array}
$$

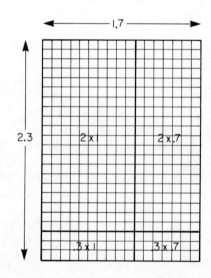

DIVISION

There are two concepts of division, the measurement and the partitive (page 150), and both are useful for illustrating division with decimals. The measurement concept involves repeatedly measuring off or subtracting one amount from another. For example, to compute .90 ÷ .15 we must determine how many times .15 can be subtracted from .90, or how many times .15 "goes into" .90. The decimal square for .90 has been marked off to show that the quotient is 6.

division is repeated subtraction

 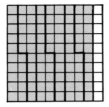 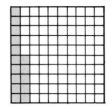

.90 ÷ .15 = 6

To illustrate the division of a decimal by a whole number, we can use the partitive concept. In this case the divisor is the number of equal parts into which a set or region is divided. The shaded part of the decimal square for .80 has been divided into four equal parts to illustrate .80 ÷ 4. Since each part has 20 hundredths, the quotient is .20.

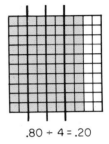

.80 ÷ 4 = .20

Abacus   When a decimal which is represented on the abacus is divided by a whole number, the division can be performed separately on the numbers of markers in each column. In this example, 97.56 is divided by 3, beginning with the 9 markers on the tens column.

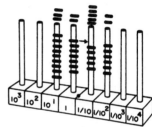

97.56 ÷ 3 = 32.52

1. 9 tens ÷ 3 = 3 tens, so 3 markers are placed above the tens column to represent 30.

2. 7 ÷ 3 = 2, with 1 remaining, so 2 markers are placed above the units column. The remaining marker is regrouped as 10 markers on the tenths column.

3. There are now 15 tenths: 15 tenths ÷ 3 = 5 tenths, so 5 markers are placed above the tenths column.

4. 6 hundredths ÷ 3 = 2 hundredths, so 2 markers are placed above the hundredths column.

Notice that in this example on the abacus we did not make any special use of the location of the decimal point. That is, the division process and regrouping are the same as though we were dividing a whole number by 3.

Pencil-and-Paper Algorithm   Four steps, similar to those above, are carried out when computing $97.56 \div 3$ by the division algorithm. Division is performed without regard to the decimal point, just as though we were computing $9756 \div 3$. Then the decimal point is placed in the quotient directly above the decimal point in 97.56. The following equations show why this quotient can be computed by dividing a whole number by a whole number. In the first equation the decimal point is removed; in the second equation, 9756 is divided by 3; and in the last equation the decimal point is replaced.

```
      3 2.5 2
   3/9 7.5 6
     9
     ─
     7
     6
     ─
     1 5
     1 5
     ──
       6
       6
       ─
```

$$\frac{97.56}{3} = \frac{9756}{3} \times \frac{1}{10^2}$$

$$= 3252 \times \frac{1}{10^2}$$

$$= 32.52$$

In the long-division algorithm for dividing with decimals we actually never do divide by a decimal. Before dividing, a slight adjustment is made so that the divisor is always a whole number. For example, the algorithm shown here indicates that we are to divide 1.504 by .32. Before dividing, however, the decimal points in .32 and 1.504 are moved two places to the right. This has the effect of changing the divisor and the dividend so that we are dividing 150.4 by 32, as shown in the lower algorithm.

$$.32 \,\overline{\big)\, 1.504}$$

changed to

```
          4.7
   32. /150.4
        128
        ───
        224
        224
        ───
```

The rule for dividing by a decimal is to count the number of decimal places in the divisor and then move the decimal points in the divisor and the dividend this many places to the right. In the previous example the decimal points in .32 and 1.504 were moved two places to the right because .32 has two decimal places. The justification for this process of shifting decimal points is illustrated in the following equations.

$$\frac{1.504}{.32} = \frac{1.504 \times 10^2}{.32 \times 10^2} = \frac{150.4}{32.}$$

These equations show that the answer to $1.504 \div .32$ is the same as that for $150.4 \div 32$. No further adjustment is needed as long as we shift the decimal points in both the divisor and the dividend by the same amount. In this manner, dividing a decimal by a decimal can always be carried out by dividing a decimal by a whole number.

APPROXIMATION AND MENTAL CALCULATION

There are times when approximate computations are as useful as exact computations. To make a decision regarding a purchase, for example, all we may need is a "rough idea" of the cost. Suppose you are interested in the total cost of a stereo system with these components and prices: tape deck, $219.50; turntable with cartridge, $179; two speakers, $284; and receiver $335.89. An approximate computation will save time and can be done in your head. Here is the exact sum and the approximate sum for numbers rounded off to the hundreds place.

Exact Sum	Approximate Sum
$ 219.50	$ 200
$ 179.00	$ 200
$ 284.00	$ 300
$ 335.89	$ 300
$1018.39	$1000

Whenever exact answers are required, approximate computations can serve as a guide for detecting large errors. One common source of error is that of misplacing a decimal point in the answer. This can usually be discovered by rounding off each number to its highest nonzero place value. For example,

$$342.8 \times .046 = 15.7688$$

can be checked by rounding off 342.8 to 300 and .046 to .05. The product of 300 and .05 is 15, which is a good approximation to the original product of 15.7688. If the decimal point in 15.7688 had been misplaced, such as in 1.57688 or in 157.688, the relatively large difference between these numbers and 15 would indicate an error.

Another source of error is that of misplacing the decimal point when a number is entered into a calculator. Suppose the previous product is computed on a calculator and "342.8" is mistakenly entered as "34.28." In this case, the product

$$34.28 \times .046 = 1.57688$$

is sufficiently different from the approximate answer of 15 to indicate an error.

Rounding off will help to detect large errors but not small ones. Here are three more examples of approximate computations from rounding off numbers to their highest place value.

Exact	Approximation		Exact	Approximation
430.48	400		194.15	200
92.81	→ 90		− 76.62	→ − 80
+ 160.36	+ 200		117.53	120
683.65	690			

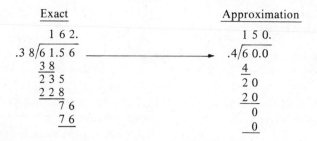

	Exact		Approximation

```
        Exact                              Approximation
          1 6 2.                               1 5 0.
     .3 8/6 1.5 6        ──────────▶        .4/6 0.0
          3 8                                    4
          ───                                   ──
          2 3 5                                  2 0
          2 2 8                                  2 0
          ───                                   ──
            7 6                                    0
            7 6                                    0
            ───                                   ──
```

OPERATIONS ON CALCULATORS

The steps for carrying out the four basic operations with decimals on a calculator are the same as for whole numbers. A pair of numbers and their operations are entered in the order in which they appear (from left to right) when written horizontally. However, when addition and subtraction are combined with multiplication and division, care must be taken regarding the order of the operations. As in the case of whole numbers, multiplication and division are performed before addition and subtraction. Consider the following example of computing income tax. According to Schedule Y in the 1983 Internal Revenue Service forms, the tax on earnings greater than $11,900 and less than $16,000 is $1,149 plus 17% of the amount over $11,900. Therefore, the tax on $13,600 is

$$\$1149 + .17 \times \$1700$$

On some calculators this cannot be computed by entering the numbers and operations into the calculator as they appear from left to right, because the .17 would be added to the 1149. One method of determining this tax is to compute $.17 \times 1700$ and then add 1149. Another method is to place 1149 in storage, compute $.17 \times 1700$, and then perform the addition. Here are the steps for finding the answer by this method.

Steps	Displays
1. Enter 1149	1149.
2. $\boxed{M+}$	1149.
3. $.17 \boxed{\times} 1700 \boxed{=}$	289.
4. $\boxed{+} \boxed{MR}$	1149.
5. $\boxed{=}$	1438.

COMPUTER APPLICATIONS

Personal loans for cars, boats, home improvements, etc. are called *consumer* or *install-ment* loans. To *finance* a loan over 12 months means that the loan plus the finance charge will be paid off in installments during the 12 months. If the consumer wishes to pay off the loan early, the *Rule of 78ths* is one method of determining what part of the finance charge should be refunded. The sum of whole numbers from 12 to 1 is used to determine the refund.

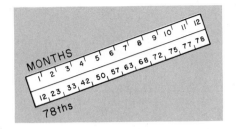

The "Rule of 78 ths"

$$12 + 11 + 10 + 9 + 8 + 7 + 6 + 5 + 4 + 3 + 2 + 1 = 78$$

If the consumer wishes to pay off the loan in the first month, 12/78 of the total finance charge must be paid. If the loan is paid off in the second month, 23/78 (12 + 11 = 23) of the finance charge must be paid, etc.

Here is a computer program which determines the amount of finance charge to be paid based on the Rule of 78ths. Lines 50, 60, and 70 form a loop which computes the appropriate sum of numbers, beginning with 12, for the numerator S in line 80. Line 90 rounds off the finance charge to hundredths of a dollar. Follow through lines 50, 60 and 70 for K = 1, 2 and 3.

Program 7.2A

```
10    PRINT "TYPE THE AMOUNT OF THE FINANCE CHARGE."
20    INPUT X
30    PRINT "THIS PROGRAM COMPUTES THAT PART OF $"X" WHICH MUST
        BE PAID AFTER THE FIRST K MONTHS. TYPE A NUMBER FOR K."
40    INPUT K
50    FOR N = 1 TO K
60    LET S = S + (13 - N)
70    NEXT N
80    LET Z = S / 78 * X
90    LET R =  INT (Z * 100 + .5) / 100
100   PRINT "THE AMOUNT OF FINANCE CHARGE TO BE PAID AFTER "K"
        MONTHS IS $"R"."
110   END
```

SUPPLEMENT (Activity Book)

Activity Set 7.2 Decimal Operations on the Abacus
Just for Fun: Decimal Tricks with Calculators

Exercise Set 7.2: Applications and Skills

1. Misplacing a decimal point can be a costly mistake.

   a. This article says that a .05-cent difference was used rather than a half-cent difference. What is the decimal for one-half of a cent?

   ★ b. If a bid for 650,000 cartons is .05 of a cent per carton higher than another bid, how many dollars greater is the bid?

   c. If a bid for 650,000 cartons of milk is .5 of a cent per carton higher than another bid, how many dollars greater is the bid?

   ★ d. How much money did the school lose by misplacing the decimal point? (*Hint:* Use the answers from parts **b** and **c**.)

Subtraction error to cost Rochester schools $3,000

By MARK C. BUDRIS
Rochester Bureau Chief

ROCHESTER—An arithmetic error may end up costing the School Department almost $3,000 next year.

School Board Chairman Roland Roberge said Thursday night the subtraction error during a comparison of milk bids led the board to accept a bid it thought was only $298 higher than a second. It was actually $2,986 higher.

"Well, the decimal point was put in the wrong place," Roberge told the board.

He said the error was in turning a half-cent difference in milk prices into a .05-cent difference, which was then multiplied out over the more than 650,000 cartons of milk used in a year by the School Department.

2. Explain how decimal squares can be used to illustrate each of the following equalities. *use grids*

   a. $3 \times .625 = 1.875$
   b. $10 \times .37 = 3.7$
   ★ c. $.1 \times .1 = .01$
   d. $.3 \times .4 = .12$
   ★ e. $.75 \div .05 = 15$
   f. $.60 \div 10 = .06$

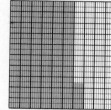

625

★ 3. a. The markers on the top half and bottom half of this abacus represent 35.654 and 125.347 respectively. The sum of these numbers can be obtained by regrouping the markers so that there are fewer than ten on each column. Show this regrouping on the other abacus. What is this sum?

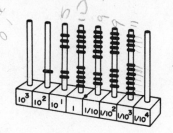

Regrouping
for addition →

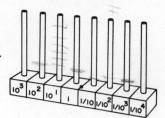

b. The abacus on the left represents 84.0273. In order to compute
84.0273 − 6.5409, regrouping is necessary several times. Show this regrouping
by drawing markers on the abacus on the right. Then compute 84.0273 − 6.5409
by circling the markers to be removed from this abacus. What is this difference?

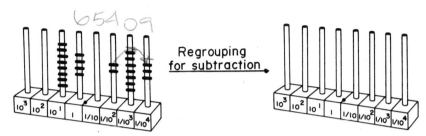

Regrouping
for subtraction

4. In the example shown here, a "1" is regrouped from the hundredths column to the
tenths column. This can be explained by the equations

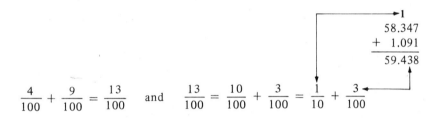

$$\frac{4}{100} + \frac{9}{100} = \frac{13}{100} \quad \text{and} \quad \frac{13}{100} = \frac{10}{100} + \frac{3}{100} = \frac{1}{10} + \frac{3}{100}$$

$$\begin{array}{r} 1 \\ 58.347 \\ + \ \ 1.091 \\ \hline 59.438 \end{array}$$

In each of the following exercises, there is one column for which regrouping is
needed. Mark this column and use equations to explain how the regrouping takes
place.

★ a. $\begin{array}{r} 4.821 \\ + 61.73 \\ \hline \end{array}$ b. $\begin{array}{r} .367 \\ .015 \\ + .509 \\ \hline \end{array}$ c. $\begin{array}{r} 7.00048 \\ + \ \ .1738 \\ \hline \end{array}$

★ d. $\begin{array}{r} 66.43 \\ - 41.72 \\ \hline \end{array}$ e. $\begin{array}{r} .046 \\ - .018 \\ \hline \end{array}$ f. $\begin{array}{r} 5.63 \\ - \ .17 \\ \hline \end{array}$

5. Use a grid (or graph paper) to illustrate these products.
   a. $2.5 \times 3.7 = 9.25$ b. $1.8 \times 4.6 = 8.28$

6. *Error Analysis:* There are many types of errors that can occur in computation, even when a student knows the basic operations with single digit numbers. Determine each type of error.

a.
```
  .4
+ .8
 .12
```

★ b.
```
  99.4
- 27.86
  71.66
```

c.
```
 21.8
×  .4
 87.2
```

★ d.
```
      9.62
  4 ) 38.6
      36
      26
      24
       2
```

7. *Approximation and Mental Calculation:* Do each computation mentally by rounding off the numbers to more convenient numbers. Show each rounded off number. Compare the approximate and actual computations.

			Approximate	Actual
Example	258.35 + 32.70		290	291.5
a.	161.92 + 47.8			
★ b.	481.3 − 168.6			
c.	614.5 × 3.82			
★ d.	4.32 ÷ .82			

(Example row shows 260 above 258.35 and 30 above 32.70)

8. *Calculator Exercise:* This is a geometric sequence.

$$2, \quad 1.7, \quad 1.445, \quad 1.22825,$$

★ a. What is the next number in this sequence?

b. How many numbers will there be in this sequence before reaching a number that is less than .1?

c. If this sequence is continued it gets closer and closer to some number. What is this number?

9. *Payroll Deductions:* A job that pays $5.50 an hour amounts to a gross income of $220 for a 40-hour week. Compute the deductions in parts **a** through **c** and enter them into the table.

★ a. The federal income tax on $220 for a single person is .118 times $220.

b. The state income tax is .029 times $220.

Total gross earnings $220.00

Federal income tax _____

State income tax _____

★   c. The FICA or social security tax   FICA tax   _____
         is .067 times $220.
                                            Net pay   _____

     d. Compute the sum of the three
        deductions in parts **a, b,** and **c.**
        How many working hours, at $5.50 per hour, are needed to pay for these
        deductions?

★   e. Subtract the sum of the deductions in parts **a, b,** and **c** from the gross earnings
        in order to find the net or "take-home" pay.

10. *Credit Cards:* Fill in the missing amounts on this credit card by completing parts **a**
     through **e.**

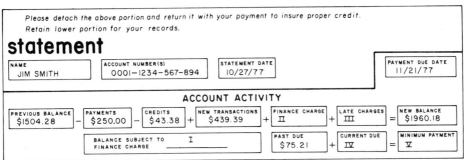

★   a. The balance subject to the finance charge (box I) is the difference between the
        previous balance and the sum of the payments and credits. Compute
        $250 + $43.38, and subtract this from the balance of $1504.28. Write this
        amount in box I.

     b. The monthly finance charge on the amount in box I is determined by the
        following rule: .0125 times the first $500; .0095 times the next $500; and .0083
        times the amount over $1000, which in this example is .0083 times $210.90.
        Compute these three products and, if necessary, round off to the nearest penny.
        Add these products and record the sum in box II.

★   c. The late charges (box III) are .05 times the amount past due, $75.21. Compute
        this product and round your answer to the nearest penny. Enter this amount in
        box III.

     d. If you add the previous balance, $1504.28; the new transactions, $439.39; the
        finance charge in box II; and the late charges in box III; and subtract the
        payments, $250.00, and the credits, $43.38, then you should get the new balance
        of $1666.80. The current amount due is .05 times $1666.80. Compute this
        product and enter it into box IV.

★   e. The minimum payment (box V), which is due within 25 days after the date of
        this statement, is the past amount due, $75.21, plus the amount currently due
        from box IV. Compute this sum and enter it into box V.

11. *Electricity Costs:* The basic unit for measuring electricity is the *kilowatt-hour* (kWh). This is the amount of electrical energy required to operate a 1000-watt appliance for 1 hour. For example, it takes 1 kilowatt-hour of electricity to light ten 100-watt bulbs for 1 hour. This table contains the average number of kilowatt-hours for operating the appliances for 1 month.

Appliance	Kilowatt-hours per month	Cost per month
Microwave oven	15.8	_____
Range with oven	97.6	_____
Refrigerator	94.7	_____
Frostless refrigerator	152.4	_____
Water heater	400.0	_____
Radio	7.5	_____
Television (black-white)	29.6	_____
Television (color)	55.0	_____
Total		Sum

★ a. What is the sum of the kilowatt-hours for operating these eight appliances for 1 month? Record this at the bottom of the table.

b. At a cost of $.04 for each kilowatt-hour, what is the monthly cost for operating each of these appliances? Round off your answers to the nearest penny and record the amounts in the table.

★ c. Compute the sum of the monthly costs in part **b** for the eight appliances.

d. Compute the difference in the monthly costs of electricity for black and white television and color television.

e. How much less does it cost to operate a refrigerator for 1 year than a frostless refrigerator?

12. The greatest record spree ever, occurred in the 1976 Olympic swimming competition when world records were set in 22 out of 26 events. In one of these events, an East German, Petra Thümer, set a world record in the 400-meter freestyle, winning in 4:09.89 (4 minutes and 9.89 seconds).

★ a. Thümer's time was 1.87 seconds faster than the old record. What was the time for the old record?

b. In the 200-meter freestyle a world record was set by another East German, Kornelia Ender, in the time of 1:59.26. If this rate of speed could be maintained, how long would it take to swim the 400-meter event? Compare this with Thümer's time for the 400-meter event.

c. In the 1964 Olympics in Tokyo, Don Schollander, an American, had a winning time of 4:12.2 in the 400-meter freestyle. How many seconds faster was Thümer's time for the 400-meter event?

Exercise Set 7.2: Problem Solving

1. *Exchange Rates:* In the summer of 1984, Holly and Kathy traveled to Canada. Holly exchanged her U.S. money for Canadian money before leaving. For each 82 cents she received $1 in Canadian money. Kathy exchanged her money in Canada. For each U.S. dollar she received $1.20 in Canadian money. Who had the better rate of exchange?

   ★ a. *Understanding the Problem:* Let's answer a few easy questions to become more familiar with the problem. If Holly received $100 in Canadian money, what did it cost her in U.S. money? If Kathy exchanged $50 in U.S. money, how much did she receive in Canadian money?

   b. *Devising a Plan:* It is tempting to conclude that Kathy had the better rate of exchange because it looks like she "gained" 20 cents while Holly only "gained" 18 cents. Let's check this reasoning by exaggerating the exchange rate. Suppose that for each 50 cents Holly received $1 in Canadian money and that for each U.S. dollar Kathy received $1.50 in Canadian money. At these rates, how much U.S. money will it cost each person to buy an item which costs $3 in Canadian money?

   c. *Carrying Out the Plan:* Using the simplified rates in part **b,** it is easy to see that the amount of Canadian money Holly received is 2 times greater than her U.S. money, and the amount of Canadian money Kathy received is 1.5 times her U.S. money.

   Holly's rate

   $$\frac{\$1}{50\cent} = 2$$

   Kathy's rate

   $$\frac{\$1.50}{\$1} = 1.5$$

   Use this approach to determine who had the better rate for the original problem.

   ★ d. *Looking Back:* We have not only solved the original problem, but we have learned something more general. Suppose one person pays $1 - t$ cents in U.S. money for each Canadian dollar, and a second person pays $1 in U.S. money for $1 + t$ cents in Canadian money. Are these rates equal?

2. *Five and Ten Store:* Ken bought some items at the Five and Ten store. All the items were the same price and he bought as many items as the number of total cents in the cost of each item. His bill was $6.25. How many items did he buy?

3. *Reduced Price:* A certain make of ballpoint pen was priced at $.50 in a store but found few buyers. When the store reduced its price, the remaining stock was sold for $31.93. What was the reduced price?*

★ 4. *Stack of Paper:* A piece of paper is cut in half and one piece is placed on top of the other. Then the two pieces are cut in half and one half is placed on top of the other,

---

*"Problems of the Month," *The Mathematics Teacher,* 76 No. 5 (May 1983), p. 310.

forming a stack with four pieces. If this process is continued 25 times and the original piece of paper is .003 of an inch thick, what is the height of the stack to the nearest tenth of a mile?

Exercise Set 7.2: Computers

1. When Program 7.2A on page 415 is run, the computer will print question marks for lines 20 and 40 to request numbers for X and K. Determine the printout for this program for the following values of X and K.
   a. X = 120 and K = 4.
   b. X = 120 and K = 6.

2. If a loan is financed for 18 months, the sum of the whole numbers from 18 to 1 is used to determine the amount of the finance charge to be refunded.

$$18 + 17 + 16 + 15 + 14 + \ldots + 3 + 2 + 1 = 171$$

   For example, if the loan is repaid in the third month, 51/171 of the finance charge must be paid.
   a. Write a computer program which determines the amount of the finance charge to be paid on an 18-month loan after K months.
★ b. What will be the printout of the program in part **a** if the finance charge is 300 and the loan is repaid after the third month?

3. A palindromic decimal is one that reads the same from right to left as from left to right. For example, 37.73 is a palindromic decimal but 5.65 and 988.9 are not. In the example shown here the process of repeatedly adding a number to its reverse is carried out three times before obtaining a palindromic number.

$$
\begin{array}{r}
9.7 \\
+\ 7.9 \\
\hline
17.6 \\
+\ 6.71 \\
\hline
24.31 \\
+\ 13.42 \\
\hline
37.73
\end{array}
$$

★ a. Program 3.1B on page 111 was used for adding a whole number to its reverse. Can this program be used for decimals? Try N = 6.7.
   b. In the examples shown here, 149 took two steps to produce a palindromic number but 1.49 only took one step. By inserting a decimal into a whole number, will the number of steps needed to produce a palindromic number always be less than or equal to the number of steps required for the whole number?

$$
\begin{array}{r}
149 \\
+\ 941 \\
\hline
1090 \\
+\ 0901 \\
\hline
1991
\end{array}
\qquad
\begin{array}{r}
1.49 \\
+\ 94.1 \\
\hline
95.59
\end{array}
$$

★ c. It takes 24 steps to produce a palindromic number if we begin with 89. How many steps will it take for 8.9?

Juxtaposition of photos of planets, each taken by a different spaceship.

7.3
RATIO, PERCENT,
AND SCIENTIFIC
NOTATION

An astronomical unit is the average distance from the earth to the sun. The distance from each planet to the sun in astronomical units is the ratio of that planet's distance to the sun divided by the earth's distance to the sun. For example, the earth's average distance from the sun is 93,003,000 miles and Jupiter's average distance from the sun is 483,881,000 miles. The ratio of 5.2 is Jupiter's distance from the sun in astronomical units.

$$\frac{483,881,000}{93,003,000} \approx 5.2$$

RATIOS AND RATES

Ratio is one of the most useful ideas in everyday mathematics. A *ratio* is a pair of positive numbers that is used to compare two sets. For example, in the United States in 1981 the ratio of the number of women arrested to the number of men arrested was 4 to 21. This tells us that for every 4 women arrested, 21 men were arrested. This ratio gives the relative sizes of these sets but not the actual number of women or men.

There are two common notations for ratios. The ratio of the number of women to the number of men in the preceding example can be written as 4:21 or 4/21 and is read as "4 to 21."

Definition: For any two positive numbers, *a* and *b,* the ratio of *a* to *b* is *a/b*. This is sometimes written as *a:b*.

Notice that *a* and *b* do not have to be whole numbers. You will see examples of ratios of decimals in the paragraphs on proportion on pages 425 and 426.

Comparing the relative size of large sets by the use of small numbers is the primary use of ratios. There are 8 teeth on the small gear shown here and 40 teeth on the large gear. This is a ratio of 1 to 5 from the number of teeth on the small gear to the number of teeth on the large gear, because 8/40 = 1/5. Every time the small gear revolves 5 times, the large gear revolves once. This ratio of turns from the small gear to the large gear is 5 to 1 or 5/1.

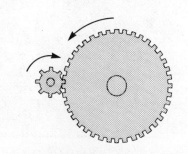

Gears such as these which step-down the number of revolutions per minute were used by farmers as they operated windmills for pumping water. Between 1880 and 1930 more than 6.5 million of these windmills were built in the United States.

Sometimes the two numbers of a ratio are divided to obtain one number. As an example, there is a federal law that the ratio of the length to the width of an official United States flag must be 1.9. This means that no matter what the size of the flag, if the length is divided by the width, the result should be 1.9.

Rates A rate is a special kind of ratio in which the two sets being compared have different units. Kilometers per hour, cost per hour, and births per day are all examples of rates. This graph shows a rate of 3 cents for each 2 kilowatt-hours. For 2 kilowatt-hours it costs 3 cents; for 4 kilowatt-hours it costs 6 cents; etc.

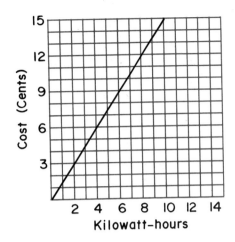

PROPORTIONS

Each ratio gives rise to infinitely many pairs of numbers all of which are *equal ratios*. For example, in 1980 the ratio of truck accidents to car accidents was 11 to 46. This means that for every 11 truck accidents there were 46 car accidents, for every 22 there were 92, etc.*

Truck Accidents	Car Accidents
11	46
22	92
33	138
44	184
.	.
.	.
.	.

An equality of two ratios is called a *proportion*. In the previous example, 11/46 and 22/92 are equal ratios, and 11/46 = 22/92 is an example of a proportion.

Definition: For any two equal ratios, a/b and c/d, $a/b = c/d$ is called a *proportion*.

We can use the test for equality of fractions to determine whether we have a proportion. That is,

$$\frac{a}{b} = \frac{c}{d} \qquad \text{if and only if} \qquad ad = bc$$

For example, 11/46 = 22/92 because $11 \times 92 = 46 \times 22$.

Proportions are useful in problem solving. Typically, three of the four numbers in a proportion are given and the fourth is to be found. For this reason the rule for solving

*Source: National Safety Council, Chicago, Illinois.

proportions has historically been called the "Rule of Three." This rule was so highly prized by merchants of the past that it was referred to as the *golden rule*. Here is a typical example. If 2.3 kilograms of flour cost $1.20, how much will you save buying 9 kilograms of flour at $4.45? Let's compute the cost of 9 kilograms of flour based on the rate of 2.3 kilograms for $1.20. Using ratios of "dollars to kilograms," and x for the unknown cost, the price for 9 kilograms of flour can be found by solving the following proportion.

$$\frac{1.20}{2.3} = \frac{x}{9}$$

This proportion can be changed to $2.3x = 10.80$ by using the preceding test for equality of ratios. This equation is satisfied by $x \approx \$4.70$. Therefore, based on the cost of 2.3 kilograms of flour, 9 kilograms will cost $4.70. The $4.45 price represents a savings of 25 cents.

A proportion may be set up in many different ways and still produce the correct answer. The previous problem may be solved by any one of the following proportions. The important thing is that we remain consistent on both sides of the equation. The same value of $x \approx \$4.70$ satisfies each of these equations.

Dollars to kilograms	Dollars to kilograms	Kilograms to dollars	Kilograms to dollars	Kilograms to kilograms	Dollars to dollars
$\dfrac{1.20}{2.3} = \dfrac{x}{9}$		$\dfrac{2.3}{1.20} = \dfrac{9}{x}$		$\dfrac{2.3}{9} = \dfrac{1.20}{x}$	

PERCENT

Percent has evolved as an outgrowth of the use of fractions with denominators of 100. The Roman method of taxation involved hundredths, or fractions which could easily be changed to hundredths. As examples, there was a sales tax of $1/100$ on merchandise and a tax of $1/25$ on slaves.

The word "percent" comes from the Latin "per centum," meaning "out of a hundred," and the symbol % has evolved from the seventeenth century use of "per ⸺." Percent was first used in the fifteenth through seventeenth centuries for computing interest, profits, and losses. Currently, it has much broader applications as illustrated by the accompanying news clippings.

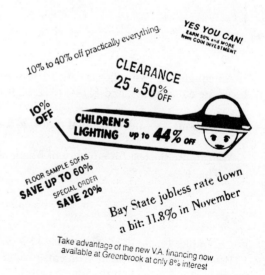

A percent can be changed to a decimal in two easy steps. First write the percent as a fraction with a denominator of 100, as shown in the following examples. (Notice the similarity between the percent symbol "%" and the numeral "100.") Second, a fraction with a denominator of 100 can be written as a decimal by moving the decimal point in the numerator 2 places to the left.

$$42\% = \frac{42}{100} = .42 \qquad 5.2\% = \frac{5.2}{100} = .052 \qquad 132\% = \frac{132}{100} = 1.32$$

Conversely, to write a decimal as a percent, first write it as a fraction with a denominator of 100 and then as a percent.

$$.647 = \frac{64.7}{100} = 64.7\% \qquad .1735 = \frac{17.35}{100} = 17.35\%$$

Most computations with percents fall into two categories. In one case we are given a percent in order to determine a part of the whole. Consider the percentages of injuries to different parts of the body which were obtained from a study of National Football League players. In games and practices, 1169 players were injured. The number of players with a particular type of injury can be found by converting the percent to a decimal and multiplying by 1169. The following computation shows the number of players with shoulder injuries.

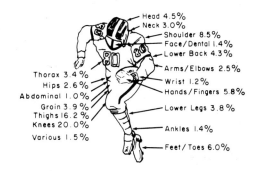

National Football League 1974 Injuries

$$8.5\% = \frac{8.5}{100} = .085 \qquad \text{and} \qquad .085 \times 1169 = 99.365$$

Rounded off to the nearest whole number, there were 99 players with shoulder injuries. In the second case we determine a percent by comparing a part to the whole. In the preceding football study, 189 players suffered injuries to the thigh. Thus, the fractional part of players with thigh injuries was 189/1169. This fraction can be converted to a decimal and then to a percent.

$$189 \div 1169 \approx .161676 \qquad \text{or approximately} \qquad 16.2\%$$

SCIENTIFIC NOTATION

It has been remarked that the human race appears to be about "half-way in size" between many of the smallest and largest things we can measure. Through microscopes we look at objects that are many times smaller than people, and through telescopes we see stars and galaxies that are many times bigger. For example, the diameter of hydrogen atoms is .0000000000106 meters, and the diameter of the largest stars is 2,770,000,000,000 meters. Such very small and very large numbers can be expressed by using powers of 10. The diameter of this star in meters can be written as

$$2.77 \times 1,000,000,000,000 = 2.77 \times 10^{12}$$

The diameter of the atom in meters can be written as

$$\frac{1.06}{100,000,000,000} = \frac{1.06}{10^{11}} = 1.06 \times 10^{-11}$$

because $10^{-11} = 1/10^{11}$. In general, for any numbers x and n, with $x \neq 0$,

$$x^{-n} = \frac{1}{x^n}$$

Any positive number can be written as the product of a number from 1 to 10 and a power of 10. This method of writing numbers is called *scientific notation* or *scientific form*. The number from 1 to 10 is called the *mantissa* and the exponent of 10 is called the *characteristic*. The product 2.77×10^{12} is a number in scientific notation. The mantissa is 2.77 and the characteristic is 12. The following table contains five examples of numbers written in positional numeration and scientific notation.

	Positional Numeration	Scientific Notation
Years since age of dinosaurs	150,000,000	1.5×10^8
Seconds of half-life of U-238	142,000,000,000,000,000	1.42×10^{17}
Wave length of gamma ray (m)	.0000000000003048	3.048×10^{-13}
Size of viruses (cm)	.000000914	9.14×10^{-7}
Orbital velocity of earth (kph)	41290	4.129×10^4

COMPUTING WITH SCIENTIFIC NOTATION

This graph shows the world's rapidly increasing population. It wasn't until 1825 that the population reached 1 billion (1×10^9), and by 1985 it was 4.7 billion (4.7×10^9). Since there are about 3.1×10^3 square meters of cultivated land (land with crops) per person, the total amount of this land in square meters is

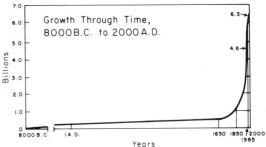

World Population Growth

Growth Through Time, 8000 B.C. to 2000 A.D.

Billions

Years

$$(4.7 \times 10^9) \times (3.1 \times 10^3)$$

Rearranging these numbers and using the rule for adding exponents, this product can be written as

$$(4.7 \times 3.1) \times 10^{12} = 14.57 \times 10^{12} = 1.457 \times 10^{13}$$

Finally, the product of the two mantissas (4.7×3.1) is computed, and the answer is written in scientific notation.

$$(4.7 \times 3.1) \times 10^{12} = 17.108 \times 10^{12} = 1.7108 \times 10^{13}$$

This example shows that the products of numbers in scientific notation can be found by multiplying numbers from 1 to 10 (the mantissas) and adding exponents of 10 (the characteristics). This method of computing products holds whether the exponents are positive or negative. Let's consider an example with a negative exponent. There is a type of bacteria, called *Escherichia coli,* which is sometimes detected in swimming pools. These bacteria multiply rapidly, and under ideal conditions one bacteria would be replaced by a population of 4.8×10^8 after 30 hours. Since each bacteria weighs 2×10^{-12} grams, the total weight of the population in grams would be

$$(4.8 \times 10^8) \times (2 \times 10^{-12}) = (4.8 \times 2) \times 10^{8 + ^-12} = 9.6 \times 10^{-4}$$

This is approximately $1/1000$ of a gram or 1 milligram.

The convenience of scientific notation for computing is due primarily to the rules for adding and subtracting exponents. These rules, which were stated for whole numbers on page 157, can be extended to fractions and decimals. For any rational number b, and any integers n and m,

$$b^n \times b^m = b^{n+m} \quad \text{and} \quad b^n \div b^m = b^{n-m}, \text{ for } b \neq 0$$

PERCENT AND SCIENTIFIC
NOTATION ON CALCULATORS

One method of determining the percentage of a given amount with a calculator, whether or not the calculator has a percent key, is to convert the percent to a decimal and multiply times the given number. To illustrate this, 6% of the 1169 National Football League injuries (see page 427) were foot injuries. Here are the steps for determining 6% of 1169.

Steps	Displays
1. Enter .06	0.06
2. $\boxed{\times}$	0.06
3. Enter 1169	1169.
4. $\boxed{=}$	70.14

The most common occurrences of percent in day-to-day affairs are in discounts and sales taxes. In the case of discounts we want to subtract a certain percent of the original cost. What is the cost, for example, of an object which is listed for $36.50 with a 15% discount? Many calculators are designed to compute this cost using the steps that follow. When the 15% is entered in Step 3, the display shows the amount that is to be subtracted from $36.50. Pressing $\boxed{=}$ in Step 4 shows the discounted price.

Steps	Displays
1. Enter 36.50	36.50
2. $\boxed{-}$	36.50
3. Enter 15 $\boxed{\%}$	5.475
4. $\boxed{=}$	31.025

The cost of an object plus the sales tax is computed in a similar way with the minus sign in Step 2 replaced by a plus sign. The cost of the previous item ($31.03) plus a 6% sales tax is found by the following steps.

Steps	Displays
1. Enter 31.03	31.03
2. $\boxed{+}$	31.03
3. Enter 6 $\boxed{\%}$	1.8618
4. $\boxed{=}$	32.8918

Calculators which add a given percentage, as in the previous example, make it easy to compute bank interests on saving accounts. Suppose you deposit $500 in a bank that pays 6% interest compounded annually. This means that at the end of each year you earn 6% on the money which was in the savings for the past year. After the first year (or beginning of the second year) you will have $500 plus 6% of $500, which according to Step 2 in the following calculator sequence is $530. Step 3 shows the amount of your savings at the beginning of the third year, and each succeeding step shows the annual amount in the savings account.

Steps					Displays	
1.	Enter 500				500.	
2.	+	6	%	=	530.	(Beginning of 2nd year)
3.	+	6	%	=	561.8	(Beginning of 3rd year)
4.	+	6	%	=	595.508	(Beginning of 4th year)

Scientific Notation   Some calculators have a special button for displaying numbers in scientific notation. These may be labeled $\boxed{EE}$ or $\boxed{EEX}$ or something similar, where E stands for exponent. If 749,300,000 is entered on such a calculator and the buttons for scientific notation and equality are pressed, the mantissa, 7.493, and the characteristic or exponent, 8, will appear in the display to denote the number for 7.493×10^8. The base 10 will not appear in the display. On some calculators the mantissa, 7.493, and the characteristic, 8, must be entered separately.

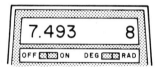

Calculators with scientific notation will automatically display numbers in this form whenever the number is too large or too small for the display. Suppose, for example, you computed 473,200 times 639,000, which is 302,374,800,000. If the calculator does not represent numbers in scientific notation, the display will be exceeded and flashing lights or some other type of signal will appear. A calculator with scientific notation will represent this product, which equals 3.023748×10^{11}, by the display shown here.

3.023748	11

Similarly, if the number is too small for the standard display, it will be represented by a mantissa and a negative power of 10. Consider the product .0004 times .000006, which is .0000000024, or 2.4×10^{-9} in scientific notation. If you compute this on a calculator whose display has only 8 places for digits and no scientific notation, it will show a product of 0. On a calculator with scientific notation, a mantissa of 2.4 and a characteristic of ¯9 will appear in the display, as shown here.

2.4	¯9

COMPUTER APPLICATIONS

This table shows the growth of $100 compounded once a year at 10% interest for 50 years. At the end of 50 years the $100 has increased to approximately $11,700. Here is a formula for determining the amount A which results from an investment of principal P over N compounding periods with interest rate R.

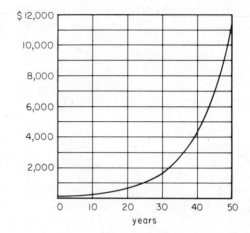

$$A = P(1 + R)^N$$

Let's use this formula to determine the exact amount of the investment in the preceding example. The interest is compounded once a year, so $N = 50$. The principal P is $100 and the interest rate R is 10%. Replacing the letters in the formula by these numbers we obtain

$$A = 100(1 + .1)^{50} = \$11,739.09$$

Program 7.3A prints the amount A at the end of each year for an investment of $100 at 10% interest compounded once each year. Line 40 rounds off this amount to the nearest penny (hundredth of a dollar). The commas in lines 10 and 50 cause the printout to be lined up in two columns. The first 10 lines of the printout, as well as the graph above, show that the money accumulates slowly in the first few years.

Program 7.3A

```
10    PRINT "COMPOUNDING PERIOD", "AMOUNT"
20    FOR N = 1 TO 50
30    LET A = 100 * (1 + .1) ^ N
40    LET D =  INT (100 * A + .5) / 100
50    PRINT N, D
60    NEXT N
70    END

RUN

COMPOUNDING PERIOD    AMOUNT
1                     110
2                     121
3                     133.1
4                     146.41
5                     161.05
6                     177.16
```

7	194.87
8	214.36
9	235.79
10	259.37

The more frequently that an investment is compounded the faster it grows. An example of this can be seen by using the following program. The number of years N is multiplied by the number of compounding periods per year, and the interest rate R is divided by the number of compounding periods per year.† Try this program for $100 at 10% interest for 50 years and compare the results from each type of compounding.

Program 7.3B

```
10   PRINT "THIS PROGRAM COMPUTES THE AMOUNT A WHICH RESULTS
        FROM AN INVESTMENT P OVER N YEARS WITH AN INTEREST RATE R,
        FOR DIFFERENT TYPES OF COMPOUNDING. TYPE NUMBERS FOR P, N,
        AND R. "
20   INPUT P,N,R
30   LET A = P * (1 + R) ^ N
40   PRINT "$"P" COMPOUNDED ANNUALLY FOR  "N" YEAR(S) BECOMES
        $"A". "
50   LET B = P * (1 + R / 2) ^ (2 * N)
60   PRINT "$"P" COMPOUNDED SEMIANNUALLY FOR "N" YEAR(S) BECOMES
        $"B". "
70   LET C = P * (1 + R / 4) ^ (4 * N)
80   PRINT "$"P" COMPOUNDED QUARTERLY FOR "N" YEAR(S) BECOMES
        $"C". "
90   LET D = P * (1 + R / 12) ^ (12 * N)
100  PRINT "$"P" COMPOUNDED MONTHLY FOR "N" YEAR(S) BECOMES
        $"D". "
110  END
```

SUPPLEMENT (Activity Book)

Activity Set 7.3 Computing with the Calculator

Just for Fun: Number Search

Exercise Set 7.3: Applications and Skills

1. In his will dated July 17, 1788, Benjamin Franklin stated that he wished "to be useful even after my death if possible," and to this end left 1000 pounds sterling (about $4570) to the inhabitants of Boston.*

   a. Franklin's will stipulated that not more than 60 pounds (about $274) was to be loaned to apprentices at a 5% annual interest rate. What is the interest on this amount for one year?

*J. Bigelow, *The Life of Benjamin Franklin, Vol. 3* (Philadelphia: J.B. Lippincott, 1893), pp. 470–89.
†Note: The interest rates which are given here and in exercises are *annual interest rates*.

b. The will also required that "each borrower shall be obliged to pay, with yearly interest, one-tenth part of the principal." Under this condition, how much of the principal plus interest would be paid back at the end of the first year on a $274 loan?

2. This public service ad in the *Eugene Register-Guard* points out the need for car pooling to reduce traffic.

a. According to this ad, what fraction of the car seats are empty during the morning's rush hour?

★ b. In a city the size of Los Angeles, there would be 9,000,000 empty seats during the rush hour. How many seats are filled?

In this morning's rush hour, empty seats outnumbered full seats 4 to 1.

In a city the size of Los Angeles, that's 9,000,000 empty seats in cars jammed up on the freeways.
Think about that while you're sitting in traffic.

Share the ride with a friend. It sure beats driving alone.

Ad Council

Presented as a public service by
Eugene Register-Guard
Daily and Sunday

★ 3. Fill in the missing numbers in this table. Use these numbers to determine each planet's average distance from the sun in astronomical units.

| | Average Distance from Sun in Miles | | |
Planet	Scientific Notation	Positional Numeration	Astronomical Units
Mercury	3.6002×10^7		
Venus		67,273,000	
Earth	9.3003×10^7		
Mars		141,709,000	
Jupiter	4.83881×10^8		
Saturn		887,151,000	
Uranus	1.784838×10^9		
Neptune		2,796,693,000	
Pluto	3.669699×10^9		

4. a. Write the following numbers in scientific notation.

   ★ Size of minute   Length of day   Number of years
     insects in centimeters in seconds   since earth's formation
       .032   86,400   3,250,000,000

   b. Write the following numbers in positional numeration.

   ★ Wavelength of   Average length of   ★ Total number of
     X rays in centimeters   solar year in seconds   possible bridge hands
       3.048×10^{-9} 3.15569×10^{7} 6.35×10^{11}

5. Write the answers to parts **a, b,** and **c** in scientific notation.

★   a. The velocity of a jet plane is 1.76×10^{3} kph, and the escape velocity of a rocket
      from earth is 22.7 times faster. Find the rocket's velocity by computing
      $1.76 \times 10^{3} \times 22.7$.

     b. The earth travels 10^{9} km each year about the sun, in approximately 9×10^{3} hr.
      Compute $10^{9} \div (9 \times 10^{3})$ to determine its speed in kilometers per hour.

★   c. A *light-year,* the distance that light travels in 1 year, is 9.4488×10^{13} km. The
      sun is 2.7×10^{4} light-years from the center of our galaxy. Find this distance in
      kilometers by computing $9.4488 \times 10^{13} \times 2.7 \times 10^{4}$.

     d. At one point in *Voyager 1*'s journey to Jupiter, its radio waves traveled
      7.44×10^{8} km to reach the earth. These waves travel at a speed of 3.1×10^{5}
      kilometers per second. Compute $7.44 \times 10^{8} \div 3.1 \times 10^{5}$ to determine the
      number of seconds for these signals to reach the earth. How many minutes is
      this?

★ 6. *Discounts:* The cost of a $9.85 item which is being discounted at 12% can be
     determined by subtracting 12% of $9.85 from $9.85. The following equations show
     that the cost of this item can also be found by taking 88% of $9.85. What number
     property is used in the first of these two equations? distributive
     property

$$9.85 - (.12 \times 9.85) = (1 - .12) \times 9.85 = .88 \times 9.85$$

     Use one of these two methods for computing the cost after discount of these items.
★   a. Portable typewriter $209.50 (15% off)
     b. Backpacker sleeping bag $153.95 (20% off)
     c. Snowshoes $86 (28% off) $61.92

7. *Sales Taxes:* The total cost of a $15.70 item plus a 6% sales tax can be determined by adding 6% of $15.70 to $15.70. The total cost can also be found by multiplying 1.06 times $15.70, as shown by these equations. What number property is used in the first of these two equations?

$$15.70 + (.06 \times 15.70) = (1 + .06) \times 15.70 = 1.06 \times 15.70$$

Use one of these two methods for computing the cost plus the sales tax of these items.

★ a. Fishing tackle outfit $48.60 (4% tax) $50.54

   b. Ten-speed bike $189 (5% tax) $198.45

   c. Cassette tape recorder $69.95 (6% tax)   74.15

8. The controversy over unit pricing has existed for many years. Without a price per unit, such as cost per gram or cost per ounce, consumers are often confused about the better buy. For example, would you pay more per ounce for the $1.68 package mix pictured here or for the $1.26 package? Since 168 divided by 48 is 3.5, each ounce of the larger package costs 3.5 cents. How much does each ounce of the smaller package cost? Which package is the better buy?

48 ounces
for 63¢

32 ounces
for 46¢

   For each of the following pairs of packages, which is the better buy?

★ a. Betty Crocker Complete Buttermilk
      Small Size   40 ounces at $2.35
      Large Size   56 ounces at $3.14

   b. Bisquick Variety Baking Mix
      Small Size   20 ounces at $1.17   5.9
      Large Size   32 ounces at $1.99   6.2

   c. If the large box of Hungry Jack has enough mix for 115 4-inch pancakes, how many 4-inch pancakes can be made from the small box?

76

9. This 100-foot-high windmill was built by NASA near Sandusky, Ohio. It is designed to generate 100 kW, enough electricity to power 25 homes.

★ a. The rotor is designed to rotate at a constant 40 revolutions per minute. It drives a generator which turns 1800 rpm. This increase in speed is accomplished by step-up gearing. Write the ratio of 40 to 1800 using the smallest positive whole numbers.

b. *Tip-speed ratio* is a comparison of the speed of the blade tips to the speed of the wind. Ratios of 5 to 1 and 8 to 1 are not uncommon for high-speed windmills. The blade-tip speed of NASA's rotor is 180 mph in a 20 mph wind. What is this tip-speed ratio?

9 : 1

Windmill near Sandusky, Ohio
for generating electricity

10. The diagram on page 438 shows each step in the nuclear fuel cycle, from mining to the reactor.

★ a. It takes 125,900 tons of uranium ore to produce 239 tons of uranium oxide (yellowcake). The weight of the yellowcake is what percent of the weight of the uranium ore?

b. It takes 300 tons of uranium hexafluoride gas to produce 42.3 tons of enriched uranium. The weight of the enriched uranium is what percent of the weight of the hexafluoride gas?

★ c. The enriched uranium supplies 28.3 tons of power-plant fuel when it is converted back to a solid. This supply loses 4% of its weight in running the nuclear reactor for 1 year. What is the weight of the uranium which is left over for reprocessing?

Mining	Conversion	Fuel assembly
It takes 25,900 tons of uranium ore to produce one year's fuel for a 1,000-megawatt reactor	The yellowcake, combined with fluorine, produces 300 tons of uranium hexafluoride gas	Converted back to a solid the enriched uranium supplies 28.3 tons of power-plant fuel

Milling	Enrichment	Reactor
The ore yields 239 tons of uranium oxide or yellowcake	The gas yields 42.3 tons of enriched uranium leaving 257.5 tons of depleted uranium	Running a reactor for a year the 28.3 tons of fuel loses 4% of its weight

11. *Calculator Exercise:* One thousand dollars is placed in a savings account which pays 6% interest compounded annually.

★   a. How much money is in the account at the end of 1 year? $10,060

     b. If the original deposit and the money earned from interest are left in the bank for 5 years, what will be the total savings? $1338.23

★   c. In how many years will the $1000 deposit be doubled? $2012.20

     d. If $500 had been placed in this savings account rather than $1000, in how many years would it be doubled?

12. *Calculator Exercise:* Costs due to an annual inflation rate of 4% can be determined by repeatedly multiplying the cost of an item by 1.04. Use this rate of inflation to answer the following questions.

     a. How much will the cost of a $545 washing machine increase after 5 years, due to inflation?

★   b. In how many years will the cost of the washing machine in part **a** be doubled?

     c. If a $16,500 annual salary is increased 4% each year to keep pace with inflation, what will this salary be after 3 years?

★ 13. *Calculator Exercise:* This table contains the numbers of public elementary school students and teachers in 1980 for several states. The student-teacher ratio for each state is the number of students divided by the number of teachers. Compute these ratios to the nearest tenth.

     a. Which state has the best student-teacher ratio?

     b. Which has the poorest?

State	Number of Teachers	Number of Students	Student-Teacher Ratio
Alabama	17,300	528,000	
Florida	38,200	1,042,000	
Hawaii	4,700	110,000	
Iowa	15,200	351,000	
Maine	6,900	153,000	
Missouri	24,700	567,000	
Oregon	14,900	319,000	
Wyoming	3,100	70,000	

14. *Calculator Exercise:* Population density is the ratio of the number of people to each unit of land area. The population density of California in 1980 was 151.4 people per square mile. This is an increase of 23.8 people per square mile from 1970 to 1980.

California Population in 1980		Square Miles in California		Population Density
23,667,902	÷	156,361	=	151.4

Compute the population densities for the 1980 populations of the states in the following table.

State	1980 Population	Land Area in Square Miles	1970 Population Density	1980 Population Density
California	23,667,902	156,361	127.6	151.4
New Jersey	7,364,823	7,521	953.0	
Texas	14,229,191	262,134	42.7	
Alaska	401,851	566,432	.5	
Montana	786,690	145,587	4.8	
Rhode Island	947,154	1,049	905.4	

★   a. Which of these six states had the greatest increase in the number of people per square mile from 1970 to 1980?

     b. In 1970 New Jersey became the most densely populated state, overtaking Rhode Island which had held the title since the first census in 1790. How many more people did New Jersey have per square mile than Rhode Island in 1980? Round your answer to the nearest whole number.

     c. The U.S. population in 1980 was 226,545,805 and the 50 states have a land area of 3,536,855 square miles. What was the U.S. population density in 1980?

Exercise Set 7.3: Problem Solving

1. *Discounts and Sales Taxes:* Suppose an item is on sale at a 20% discount but there is a 5.5% sales tax. Is the consumer better off if the discount is computed first and then the tax or if the tax is computed first and then the discount?

     a. *Understanding the Problem:* If the discount is taken first then the sales tax will be computed on an amount that is less than the original price. If the sales tax is computed first then the discount will be taken on an amount that is more than the original price. Is one method better for the consumer than the other? Make an intuitive guess.

b. *Devising a Plan:* One approach to this problem is to try a few different prices, compare the results, and use inductive reasoning. What happens when the two methods are used for an item that costs $25?

c. *Carrying out the Plan:* Use the plan suggested in part **b** or one of your own to solve this problem.

★ d. *Looking Back:* It may have occurred to you to compute the final cost of the item by taking a discount of 14.5% (20% discount minus 5.5% tax). Will this method produce the correct result?

e. *Looking Back Again:* The two methods described in the original problem result in different amounts of sales tax to be paid to the state. Which method would the owner of the business prefer: discount then tax? or tax and then discount?

★ 2. *One Hundred Percent Savings?:* In recent years people have become more aware of the need to conserve energy. If one new engine device will save 20% on fuel, a second will save 30%, and a third saves 50%, what is the total percent of fuel that can be saved if all three devices are used?

3. *Oscillating Paychecks:* For 6 years in a row the amount of Mr. Little's paycheck varied. At the end of the first year it was decreased 20%; the second year it was increased 20%; the third year it was decreased 20%; etc., alternating between 20% decreases and increases for 6 years.

a. What was the overall effect on his original salary?

b. What would have been the overall effect on his salary if the 6 years of alternating changes had begun with a 20% increase?

4. *Gear Systems:* This gear system has five gears. Gear A has 40 teeth; gear B 320; gear C 120; gear D 30; and gear E 75. Gear B and gear C are bolted together so that they both turn the same number of revolutions. If gear A is turned 8 revolutions, how many revolutions will gear E turn?

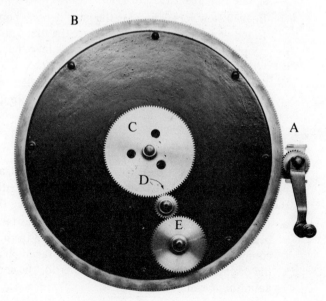

Exercise Set 7.3: Computers

1. In Benjamin Franklin's will (see page 433, Exercise 1) he predicted that the 1000 pounds he was leaving to the inhabitants of Boston would grow to 131,000 pounds in 100 years if loaned at 5% compounded annually.

   a. Write a program to determine the amount to the nearest pound that will be obtained by investing 1000 pounds for 100 years at 5% compounded annually.

   b. The will further stipulated that at the end of 100 years, ¾ of the money was to be spent on "public work" (the Franklin Institute was built) and the remaining ¼ was to be loaned for another 100 years. The ¼ balance was approximately $150,000. What is $150,000 at 5% compounded annually for 100 years?

2. The graph on page 432 shows the result of annually compounding $100 at a 10% interest rate for 50 years. Use Program 7.3B (page 433) to determine the amounts from such an investment if it is compounded:

   a. Semiannually   b. Quarterly   c. Monthly

★ 3. Most banks now compound interest daily. Daily interest which is computed by dividing the annual interest rate by 365 is called *exact interest*.* Write a computer program to print out the amount A which results from investing P dollars at R% interest compounded daily for N years. Program the computer to round off A to the nearest hundredth of a dollar. Use this program to answer Exercise 2 for daily compounding.

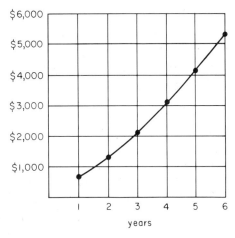

4. Few people realize the cumulative power of regular deposits and compound interest working together. This graph shows the results of depositing $50 a month for 6 years at 12% interest compounded monthly. Here's the formula for determining the amount A when P dollars is deposited monthly for N years at R% interest compounded monthly.

$$A = P \times \frac{(1 + R/12)^{12N + 1} - (1 + R/12)}{R/12}$$

   a. Write a program to determine the amount A when P dollars is deposited monthly for N years at R% interest compounded monthly. Round off A to the nearest hundredth of a dollar.

   b. Use the program in part **a** to determine the amount from investing $50 a month for 6 years at 12% interest compounded monthly.

*Note: Dividing by 360 is called *ordinary interest*.

"We have reason to believe Bingleman
is an irrational number himself."

7.4
IRRATIONAL AND
REAL NUMBERS

IRRATIONAL NUMBERS

The following number line shows the locations of a few rational numbers. Each rational number corresponds to a point on the number line, and between any two such numbers, no matter how close, there is always another rational number. Crowded together as they are, it would seem that there is no room left for any new types of numbers.

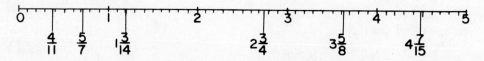

There are, however, points on the number line which correspond to numbers that are not rational. The $\sqrt{2}$ is such an example. This is the number which when multiplied by itself equals 2. The following equations show that $\sqrt{2}$ is just a little bigger than 1.4.

$$(1.4)^2 = 1.96$$
$$(1.41)^2 = 1.9881$$
$$(1.414)^2 = 1.999396$$
$$(1.4142)^2 = 1.99996164$$

The beginning of the decimal representation for $\sqrt{2}$ is 1.4142135, but no matter how many places are computed there will be no repeating pattern as in the case of rational numbers.* Such infinite nonrepeating decimal numbers are called *irrational numbers*. It is easy to exhibit examples of this type of decimal. In the following numeral, each "7" is preceded by one more zero than the previous "7": .07007000700007 . . . . While there is a pattern here, there is no repeating pattern of digits, as in the case of a rational number. Therefore this is an irrational number.

Numbers which are not rational were first recognized by the Pythagoreans of the fifth century B.C. Previously, they had thought that the length of any line segment was a rational number, and most of their proofs depended on this assumption. Of even greater importance to them was the belief that all practical and theoretical affairs could be explained by whole numbers and ratios of whole numbers. The discovery of line segments whose lengths were not expressible by rational numbers caused a logical scandal which threatened to destroy the Pythagorean philosophy. According to one legend, they attempted to keep the matter a secret by taking its discoverer, Hippacus, on a sea voyage from which he never returned.

The Greeks used the word "expressible" if a line segment had a rational length and "inexpressible" if it did not. Our words "rational" and "irrational" are derived from these words. It wasn't until the nineteenth century, more than 2000 years after the discovery of inexpressible lengths, that our number system was extended to include irrational numbers. This was accomplished by the German mathematician Richard Dedekind (1831–1916) who carefully defined irrational numbers and removed the ambiguities surrounding them.

PYTHAGOREAN THEOREM

It is assumed by some scholars that the discovery of irrational numbers arose in connection with the Pythagorean theorem. This theorem states that for any right triangle, the sum of the squares of the legs is equal to the square of the hypotenuse. The *legs* of a right triangle are the perpendicular sides, and the *hypotenuse* is the side opposite the 90-degree angle. For a

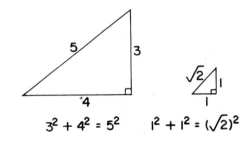

right triangle with legs of lengths 3 and 4, the hypotenuse has length 5, because $3^2 + 4^2 = 5^2$. If the legs are both 1, the hypotenuse is the irrational number $\sqrt{2}$ since $1^2 + 1^2 = (\sqrt{2})^2$.

---

*For a proof that $\sqrt{2}$ is irrational, see R. Courant and H. Robbins, *What Is Mathematics* (New York: Oxford University Press, 1941), pp. 59–60.

The first proof of the Pythagorean theorem is thought to have been given by Pythagoras (ca. 540 B.C.). According to legend, when Pythagoras discovered this theorem he was so overjoyed that he offered a sacrifice of oxen. The theorem was used, however, many years before by the Babylonians and Egyptians. It is illustrated on this 4000-year-old Babylonian tablet, which contains a square and its diagonals. The numbers on this tablet show that the Babylonians had computed $\sqrt{2}$ to be 1.414213, which is the correct value to 6 decimal places.

Babylonian stone tablet with approximation of $\sqrt{2}$

There are many proofs of the Pythagorean theorem. *The Pythagorean Proposition,* a book by E. S. Loomis, contains 370 such proofs. The proof suggested by the following squares was known by the Greeks and may have been the one given by Pythagoras. The triangles with sides a, b, and c are all right triangles. The relationship, $a^2 + b^2 = c^2$, can be proven by the following observations. Square I has an area of $a^2 + b^2 + 4T$, where T is the area of each triangle. Square II has an area of $c^2 + 4T$. Both squares have sides of length $a + b$, and therefore their areas are equal. Setting these areas equal to each other, $a^2 + b^2 + 4T = c^2 + 4T$, shows that $a^2 + b^2 = c^2$.

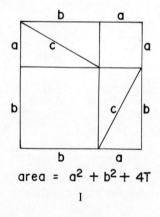

area = $a^2 + b^2 + 4T$

I

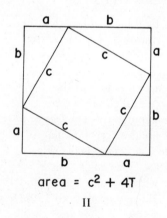

area = $c^2 + 4T$

II

The converse of the Pythagorean theorem also holds. If the sum of the squares of two sides of a triangle equals the square of the third side, then the triangle must be a right triangle. This means that if you used a rope with 30 knotted intervals of equal length and formed a triangle of sides 5, 12, and 13, it would be a right triangle. This fact was undoubtedly known by the ancient Egyptians and used by them to form right triangles. The top part of the next photo shows Egyptian surveyors, called "rope stretchers," with a piece of rope. Mark off a piece of string with 30 equal intervals and try forming a triangle whose sides have lengths of 5, 12, and 13.

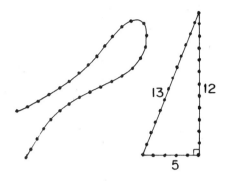

An Egyptian painting (about fifteenth century B.C.) depicting the needs of an advanced society. The upper part shows surveyors with rope.

SQUARE ROOTS AND OTHER IRRATIONAL NUMBERS

The square root of a number n is defined as the number which when multiplied by itself equals n. The square root of 5.29 is 2.3, since $2.3 \times 2.3 = 5.29$. The symbol for the square root of a number n is $\sqrt{n}$. The term "square root" comes from the word "root." It was first represented by "r" then by $\checkmark$ and finally by $\sqrt{}$.

$$5.29 \text{ cm}^2 \qquad \sqrt{5.29} \text{ cm}$$

$$\sqrt{5.29} = 2.3$$

   The square roots of square numbers 1, 4, 9, 16, 25, etc. are always integers. The square roots of all other whole numbers greater than 0 are irrational: $\sqrt{2}, \sqrt{3}, \sqrt{5}, \sqrt{6}, \sqrt{7}, \sqrt{8}, \ldots$. These numbers all have infinite nonrepeating decimals.

Even though we cannot write the complete decimal for an irrational number, these numbers should not be thought of as mysterious or illusive. They are the lengths of line segments, as illustrated by the following triangles. The legs of the first triangle each have a length of 1, and so, by the Pythagorean theorem, the hypotenuse is $\sqrt{2}$. The legs of the second triangle are 1 and $\sqrt{2}$, and the hypotenuse is $\sqrt{3}$. Next, we can use legs of length $\sqrt{2}$ and $\sqrt{3}$ to obtain a hypotenuse of $\sqrt{5}$. The lengths of these hypotenuses have been transferred to the number line below the triangles. Line segments of lengths $\sqrt{6}$, $\sqrt{7}$, $\sqrt{8}$, etc., can be constructed in a similar manner.

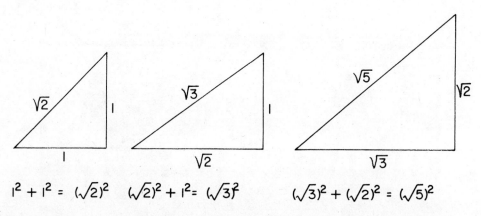

$$1^2 + 1^2 = (\sqrt{2})^2 \qquad (\sqrt{2})^2 + 1^2 = (\sqrt{3})^2 \qquad (\sqrt{3})^2 + (\sqrt{2})^2 = (\sqrt{5})^2$$

In this manner, line segments whose lengths are irrational numbers can be constructed and used to locate these numbers on the number line.

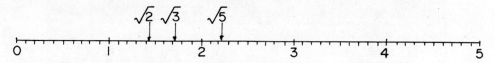

The cube root of a number n is denoted by $\sqrt[3]{n}$. This is the number s such that $s \times s \times s$ equals n. If a cube has a volume of 15 cubic centimeters then each side of the cube has length $\sqrt[3]{15}$. The cube roots of cube numbers 1, 8, 27, 64, 125, etc., are whole numbers. The cube roots of all other whole numbers greater than zero are irrational numbers. For example, the cube roots of 4, 10, and 35 are all infinite nonrepeating decimals. Try cubing the following decimals to see how close you get to 4, 10, and 35.

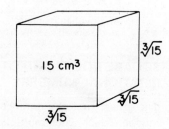

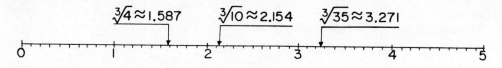

The most celebrated of all irrational numbers is pi, which is denoted by π. This number is the ratio of the circumference of a circle to its diameter. To four decimal places it is 3.1416. Pi has had a long and interesting history. In the ancient Orient, π was frequently taken to be 3, and this value also occurs in the Bible (I Kings, Chapter 7, and II Chronicles, Chapter 2). There have been many attempts to compute π. Archimedes computed π to two decimal places, and in 1841, Zacharias Dase computed π to 200 places. In 1873 William Shanks of England computed π to 707 places. In 1946 D. F. Ferguson of England discovered errors starting with the 528th place in Shank's value for π, and a year later he gave a corrected value of π to 710 places. In recent years, electronic computers have calculated π to more than one million places. Among the curiosities connected with π are the word devices for remembering the first few decimal places. In the following sentence the number of letters in each word is a digit in π.

3. 1   4 1   5   9   2   6

May I have a large container of coffee?*

REAL NUMBERS

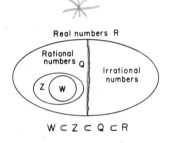

The irrational numbers together with the rational numbers form the set of real numbers. The accompanying diagram shows the relationships between the familiar sets of numbers. The set of rational numbers and the set of irrational numbers are disjoint and their union is the set of real numbers, R. The rational numbers, Q, contain the whole numbers, $W = \{0, 1, 2, 3, \ldots\}$, and the integers, $Z = \{0, \pm 1, \pm 2, \pm 3, \ldots\}$. Viewed in another way, the sets W, Z, Q, and R form an increasing sequence of subsets: The set of whole numbers is contained in the set of integers, $W \subset Z$; the set of integers is contained in the set of rational numbers, $Z \subset Q$; and the set of rational numbers is contained in the set of real numbers, $Q \subset R$. The next diagram shows another way to picture the various relationships between these types of numbers.

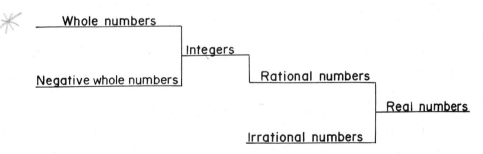

*This mnemonic and others are given by H.W. Eves, *An Introduction to the History of Mathematics,* 3rd ed. (New York: Holt, Rinehart and Winston, 1969) p. 94.

PROPERTIES OF REAL NUMBERS

The commutative, associative, and distributive
properties which were stated in Chapters 3 and 6
hold for the whole numbers, integers, rational
numbers, and real numbers. There are other
properties which do not hold for all of these number
systems. The integers have inverses for addition
(because they have negative numbers) but the whole
numbers do not. The rational numbers have
inverses for multiplication (because they have

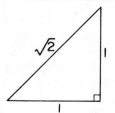

reciprocals) but the integers do not. The real numbers have the property called
completeness, but the rational numbers do not have this property. Intuitively,
completeness means that for every line segment there is a real number which equals its
length. If we limit ourselves to the rational numbers, this is not true. For example, we
have seen that there is no rational number corresponding to the length of the
hypotenuse of a right triangle whose legs have lengths of 1 unit.

Expressed in a slightly different way, completeness of the real numbers means that
there is a one-to-one correspondence between the real numbers and the points on a line.
Because of this relationship, a line called the *real number line* is used as a model for the
real numbers. Once a zero point is labeled and a unit is selected, each real number can
be assigned to a point on the line. Each positive real number is assigned a point to the
right of zero such that the real number is the distance from this point to the zero point.
The negative of this number corresponds to a point symmetrically to the left of zero.
Here are a few examples.

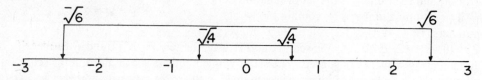

Here are 12 properties which hold for the real number system.

Closure for Addition   For any two real numbers, their sum is another real number.
This is stated by saying that *addition is closed* on the set of real numbers. Closure can
be illustrated best by a set of numbers on which addition is not closed. Consider the set
of odd numbers, {1, 3, 5, 7, . . .}. The sum of two odd numbers is not another odd
number, so addition is not closed on this set. Addition is closed on the sets of whole
numbers, integers, and rational numbers.

Closure for Multiplication   For any two real numbers, their product is another real
number. That is, *multiplication is closed* on the set of real numbers. Multiplication may
or may not be closed, depending on the given set. The product of any two negative
numbers is not another negative number, so multiplication is not closed on the set of
negative numbers. Multiplication is closed on the sets of whole numbers, integers, and
rational numbers.

Addition Is Commutative   For any real numbers r and s, $r + s = s + r$.

Multiplication Is Commutative   For any real numbers r and s, $r \times s = s \times r$.

Addition Is Associative   For any real numbers, r, s, and t, $(r + s) + t = r + (s + t)$.

Multiplication Is Associative   For any real numbers r, s, and t, $(r \times s) \times t = r \times (s \times t)$.

Identity for Addition   For any real number r, $0 + r = r$. Zero is called the *identity for addition* because when it is added to any number, there is "no change." That is, the identity of the number is left unchanged.

Identity for Multiplication   For any real number r, $1 \times r = r$. One is called the *identity for multiplication* because multiplying by 1 leaves the identity of each number unchanged.

Inverses for Addition   For any real number r, there is a unique real number ^-r, called its *negative* or *inverse for addition*, such that $r + {}^-r = 0$.

Inverses for Multiplication   For any real number r, not equal to 0, there is a unique real number $1/r$, called its *reciprocal* or *inverse for multiplication*, such that $r \times 1/r = 1$.

Multiplication Is Distributive over Addition   For any real numbers r, s, and t, $r \times (s + t) = r \times s + r \times t$.

Completeness Property   For any line segment there is a real number which equals its length.

OPERATIONS WITH IRRATIONAL NUMBERS

At first it is difficult to imagine how to perform arithmetical operations with numbers which cannot be expressed exactly in decimal notation. One solution is to replace irrational numbers by rational approximations. Thus, for example, we see both 3.14 and 3.1416 being used for π. For a circle whose diameter is 2.8 centimeters, its circumference will be approximately 8.8 centimeters. This is sufficient accuracy for most purposes.

Another solution is to write products and sums without doing the computing. For example, the sides of the square on this centimeter dot grid each have a length of $\sqrt{2}$ centimeters. The perimeter is denoted by $4\sqrt{2}$, which means 4 times $\sqrt{2}$. It is fairly easy to prove that $4\sqrt{2}$ is an irrational number. The argument is as follows. We know by closure for multiplication of real numbers that $4\sqrt{2}$ is a real number, so it is either rational or irrational. Let's suppose it is a rational number and denote it by r. That is, $r = 4\sqrt{2}$. Multiplying both sides of this equation by $1/4$, we get $\sqrt{2} = r \times 1/4$. Now by closure for multiplication of rational numbers, $r \times 1/4$ is a rational number. However, this can't be true because $\sqrt{2}$ is an irrational number. Since the assumption that $4\sqrt{2}$ is rational leads to a contradiction, $4\sqrt{2}$ must be irrational. A similar argument can be used to prove that the product of any nonzero rational number with an irrational number is an irrational number.

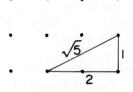

As another example, the triangle whose sides have lengths 1, 2, and $\sqrt{5}$ has $3 + \sqrt{5}$ as a perimeter. This is also an irrational number, as can be proven by an argument similar to the previous one. In general, the sum of a rational number and an irrational number is an irrational number.

The perimeter of the square, $4\sqrt{2}$, and the triangle, $3 + \sqrt{5}$, can be approximated by substituting a rational approximation for $\sqrt{2}$ and $\sqrt{5}$, respectively. There are cases, however, in which we can compute with irrational numbers without replacing them by decimal approximations. The area of the square whose sides have length $\sqrt{2}$ (see page 449) is $\sqrt{2} \times \sqrt{2}$. By the definition of square root, $\sqrt{2} \times \sqrt{2} = 2$. We can also see that the area of this square is 2, by dividing it into 4 smaller half-squares.

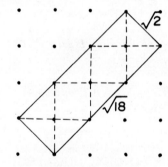

Let's consider another example of a product of two irrational numbers. The rectangle in this figure has a length of $\sqrt{18}$ and a width of $\sqrt{2}$. Its area, using the formula for the area of a rectangle, is $\sqrt{18} \times \sqrt{2}$. Using a second method, this area can be seen to be 6 square units by dividing the rectangle into 12 small half-squares. These two methods indicate that $\sqrt{18} \times \sqrt{2} = 6$. Since $6 = \sqrt{36} = \sqrt{18 \times 2}$, we see that $\sqrt{18} \times \sqrt{2} = \sqrt{18 \times 2}$. In this example the product of the square roots of two numbers is equal to the square root of the product of the two numbers. This result is stated in the following theorem.

Theorem: For any positive numbers a and b, $\sqrt{a} \times \sqrt{b} = \sqrt{a \times b}$.

Often, as in the preceding examples, $\sqrt{a}$ and $\sqrt{b}$ will be irrational numbers but their product will be a rational number. In addition to computing the products of square roots, this theorem is useful for simplifying square roots. For example, $\sqrt{18} = \sqrt{9 \times 2} = \sqrt{9} \times \sqrt{2} = 3\sqrt{2}$. A square root can always be simplified if the number under the root sign has a factor that is a square number.

IRRATIONAL NUMBERS
AND CALCULATORS

Some calculators have buttons for irrational numbers. One such button is for π. However, since the decimal representation for an irrational number has an infinite number of digits, the calculator computes with an approximate value for π. Pressing $\boxed{\pi}$ on a calculator with 10 places for digits will give the number shown in the display on this calculator.

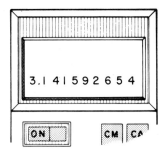

Many calculators have a square root key, $\boxed{\sqrt{x}}$. If the square root of a number is irrational, the decimal which appears in the display will be an approximation. Here are the steps for obtaining $\sqrt{2}$.

Steps	Displays
1. Enter 2	2.
2. Press $\boxed{\sqrt{x}}$	2.
3. $\boxed{=}$	1.414213562

By definition, $\sqrt{2}$ is that number which when multiplied by itself produces 2. If the decimal in the previous display is multiplied by itself, $(1.414213562)^2$, the product is a decimal with 19 digits, whose first 12 digits are 1.99999999894. On a calculator which has 10 positions for digits and which automatically rounds off, 1.999999999 will appear in the display. If, however, the calculator has 9 or fewer positions for digits and automatically rounds off, the product will be rounded off to the number 2.

Some calculators have a button for finding *any* root (cube root, fourth root, etc.) of a positive number. One common button for this is $\boxed{\sqrt[x]{y}}$. Here are the steps for finding the cube root of 12 by using this button. The number in Step 4 is only an approximation because $\sqrt[3]{12}$ is an irrational number.

Steps	Displays
1. Enter 12	12.
2. Press $\boxed{\sqrt[x]{y}}$	12.
3. Enter 3	3.
4. $\boxed{=}$	2.289428485

If this decimal is cubed, the result may or may not be equal to 12. This will depend on the number of places for digits in the display and whether or not the calculator rounds off.

COMPUTER APPLICATIONS

For over 2000 years architects and artists have been fond of using a rectangle called the *golden rectangle*. The length of a golden rectangle divided by its width is an irrational number, which when rounded to six decimal places is 1.618033. This irrational number is called the *golden ratio*. The Fibonacci numbers (page 4) are related to the golden ratio. Here are the first 10 Fibonacci numbers.

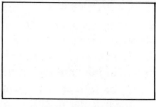

Golden Rectangle

$$1 \quad 1 \quad 2 \quad 3 \quad 5 \quad 8 \quad 13 \quad 21 \quad 34 \quad 55$$

As we use larger and larger consecutive Fibonacci numbers their ratio gets closer and closer to the golden ratio. For example,

$$13 \div 8 = 1.625 \text{ and } 55 \div 34 \approx 1.6176471$$

How far must we go in this sequence before the ratio of two consecutive Fibonacci numbers is approximately 1.618033? The following program prints numbers in the Fibonacci sequence if 1 is typed for A and for B when the program is run. Lines 40 through 90 form a loop which computes the terms of the sequence from the third term on and prints these numbers. The semicolons in lines 30 and 50 cause the printout to be printed on one line until that line is full. Line 80 uses the absolute value function (page 377) to compare the ratio of consecutive numbers from the sequence to the golden ratio. When the difference in these ratios is less than one-millionth, the computer is sent to line 100 to print the final ratio. Begin with A = 1 and B = 1 and follow through the lines of this program to see if you can obtain the first few Fibonacci numbers.

Program 7.4A

```
10   PRINT "TYPE THE FIRST TWO NUMBERS OF A SEQUENCE. SEPARATE
        THEM BY A COMMA."
20   INPUT A,B
30   PRINT A" "B" ";
40   LET F = A + B
50   PRINT F" ";
60   LET A = B
70   LET B = F
80   IF  ABS (B / A - 1.618033) < .000001 THEN  GOTO 100
90   GOTO 40
100  PRINT "THE RATIO OF THE LAST TWO NUMBERS IS "B / A"."
110  END
```

Activity Set 7.4 Irrational Numbers on the Geoboard
Just for Fun: Golden Rectangles

There once was a lady named Lou,
Who computed the square root of 2.
  When no pattern repeated
  She gave up defeated,
Two million digits is all she would do.

$\sqrt{2} = 1.41421356241933916628197598871307959868$
$34890650961931894324235266142798191004$
$5554661670432543765054609450559457028$
$25327193147647412885462978\,4\ldots$

Exercise Set 7.4: Applications and Skills

1. For each number in the left-hand column, determine what type of number it is and put checks in the appropriate columns. For example, $^-3$ is an integer, a rational number, and a real number.

	Whole numbers	Integers	Rational numbers	Real numbers
$^-3$		✓	✓	✓
★ $\frac{1}{8}$				
★ $\sqrt{3}$				
π				
14				
$\frac{16}{4}$				
$.\overline{82}$				

2. We know that $\sqrt{a} \times \sqrt{b} = \sqrt{ab}$ for all positive numbers a and b. A similar condition for the sums of square roots would be: $\sqrt{a} + \sqrt{b} = \sqrt{a+b}$. Find two numbers for a and b to show that this equation does not hold.

3. Simplify the following square roots so that the smallest possible whole number is left under the square root symbol.
   ★   a. $\sqrt{45}$ b. $\sqrt{48}$ ★   c. $\sqrt{60}$

4. Compute each of the following products. Write your answers as whole numbers.
   ★   a. $\sqrt{20} \times \sqrt{5}$ b. $\sqrt{6} \times \sqrt{24}$ ★ c. $\sqrt{7} \times \sqrt{28}$ d. $\sqrt{3} \times \sqrt{12}$

5. Use the Pythagorean theorem to find the missing length for each of the right triangles. Compute each answer to one decimal place.

★ a.

b.

★ c.

d.

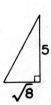

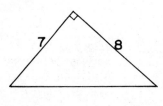

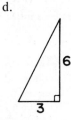

6. Any three whole numbers a, b, and c, such that $a^2 + b^2 = c^2$, are called *Pythagorean triples*. The following equations can be used to find such numbers, by using whole numbers for u and v.

$$a = 2uv, \quad b = u^2 - v^2, \quad c = u^2 + v^2$$

For example, if $u = 5$ and $v = 3$, then a, b, and c have the values shown in the table. Use the values of u and v in the table to find the remaining values for a, b, and c. Check your answers by showing that $a^2 + b^2 = c^2$.

u	v	a	b	c
2	1			
3	2			
4	3			
4	2			
5	3	30	16	34

7. *Rationalizing the Denominator:* Quotients of real numbers, such as $2 \div \sqrt{3}$, are often written as fractions, $2/\sqrt{3}$. The process of replacing an irrational denominator by a rational denominator is called *rationalizing the denominator*.

$$\frac{2}{\sqrt{3}} = \frac{2 \times \sqrt{3}}{\sqrt{3} \times \sqrt{3}} = \frac{2\sqrt{3}}{3}$$

Rationalize the following denominators of these fractions.

★ a. $\dfrac{3}{\sqrt{7}}$

b. $\dfrac{3}{2\sqrt{6}}$

★ c. $\dfrac{5}{\sqrt{5}}$

d. $\dfrac{^-1}{\sqrt{2}}$

8. *Calculator Exercise:* Use a calculator with a square root key to carry out Steps 1 through 4. Write in the numbers for the display in Step 5.

Steps	Displays
1. Enter 2	2.
2. Press $\boxed{\sqrt{x}}$	1.4142135

3. Press $\sqrt{x}$ 1.1892070

4. Press $\sqrt{x}$ 1.0905076

5. Press $\sqrt{x}$ _____

★   a. If you continue to press the square root key in this example, all the numbers in the display will eventually be equal. What is this number?

★   b. What number will eventually show in the display of a calculator if you enter a number less than 1 and repeatedly press the square root key?

9. Which of the following operations are closed for the given sets? If an operation is not closed, provide an example to show this.

★   a. Subtraction on the set of whole numbers.

     b. Division by nonzero numbers on the set of rational numbers.

     c. Multiplication on the set of irrational numbers.

★   d. Addition on the set of integers.

10. State the property of the real numbers which is being used in each equality.

★   a. $3(cb)y + 3c(2bz) = 3(cb)y + 3(c2)bz$

     b. $3c(by) + 3c(2bz) = 3c(by + 2bz)$

★   c. $4[2k + 5(k + y)] = 4[5(k + y) + 2k]$

     d. $(2k + 5) + (6a + 7)/bc = 1(2k + 5) + (6a + 7)/bc$

     e. $4 + [2k + 5(k + y)] = (4 + 2k) + 5(k + y)$

11. This spiral of right triangles, somewhat resembling a cross section of the seashell of the chambered nautilus (see page 4), shows the square roots of consecutive whole numbers. The first triangle has two legs of unit length and a hypotenuse of $\sqrt{2}$. This hypotenuse is the leg of the next triangle which has a hypotenuse of $\sqrt{3}$. Each triangle uses the hypotenuse of the preceding triangle as a leg.

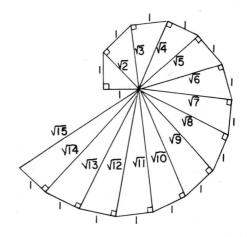

     a. For each whole number n on the horizontal axis of the grid on the following page, plot a point $\sqrt{n}$ units above the axis. The lengths of $\sqrt{n}$ can be measured from the spiral of triangles.

★ b. Will this graph continue to rise as *n* increases?

★ c. Connect the points of the graph with a curve. Use the graph to approximate $\sqrt{7.5}$. Could this graph be used to approximate the square root of any non-negative number less than 12?

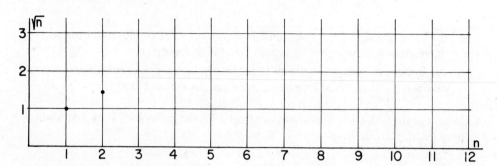

12. *Calculator Exercise:* Here is a brief chronology of some early approximations for π. Divide the numerators of these fractions by their denominators and compare the decimals with the following value of π: to 15 decimal places, $\pi = 3.141592653589793$. Which one of these fractions is closes to the value of π? Which two of these fractions are equal?

★ a. Archimedes (240 B.C.) $\dfrac{223}{71}$

b. Claudius Ptolemy (A.D. 150) $\dfrac{377}{120}$

★ c. Tsu Ch'ung-chih (A.D. 480) $\dfrac{355}{113}$

d. Aryabhata (A.D. 530) $\dfrac{62832}{20000}$

★ e. Bhāskara (A.D. 1150) $\dfrac{3927}{1250}$

Exercise Set 7.4: Problem Solving

1. *Infinity of Squares:* The inner square shown here was obtained by connecting the midpoints of the sides of the outer square. If this process of forming smaller inner squares by connecting the midpoints of the sides of the preceding square is continued, what will be the dimensions of the twentieth square?

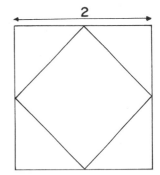

★ a. *Understanding the Problem:* The outer square is 2 by 2 and has an area of 4. The inner square is $\sqrt{2}$ by $\sqrt{2}$. Explain why. What is the area of the inner square? The third square will be formed by connecting the midpoints of the second square.

b. *Devising a Plan:* One approach is to use the Pythagorean theorem to compute the length of the side of each new square. Another approach is to use the area of each new square to find the length of the side of that square. How is the length of the side of a square related to its area?

c. *Carrying Out the Plan:* Complete the following table and look for patterns. Use these patterns to predict the length of the side of the twentieth square.

	Square 1	Square 2	Square 3	Square 4	Square 5	Square 6
Length of side	2 by 2	$\sqrt{2}$ by $\sqrt{2}$				
Area	4	2				

★ d. *Looking Back:* This problem is related to the "Square-Circle-Square" problem (page 299). In that problem a circle was inscribed in a square and then a second square was inscribed in the circle. If this process of inscribing an inner circle and an inner square is continued, what will be the area of the twentieth square?

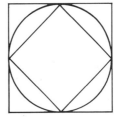

2. *Square Root Inequalities:* Is the square root of any positive number other than 1 always less than the number? If not, what are the conditions under which this statement is true?

★ 3. *Constructing Square Roots:* Describe a method of constructing line segments whose lengths are $\sqrt{2}, \sqrt{3}, \sqrt{5},$ and $\sqrt{7}$. Use the line segments to locate points for these numbers on this number line.

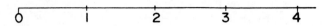

4. *Patterns for Pyramids*:* For any polygon (*ABCD*) and any point *P* in its interior, describe a procedure for constructing triangular "flaps" on the sides of the polygon so that the pattern can be folded to form a pyramid with point *P* at the foot of its altitude. (*Hint:* The outer vertices of the triangular flaps will lie on lines which are perpendicular to the sides of the polygon and pass through point *P*. The altitude, which can be any length, will be needed together with the Pythagorean Theorem to locate the vertices of the flaps.)

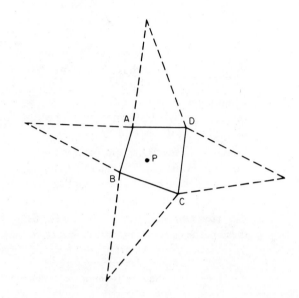

Exercise Set 7.4: Computers

1. Use Program 7.4A on page 452 to answer these questions.

   a. If 1 is typed for A and for B in line 20, what is the ratio B/A in line 80 the first time through the program?

   b. When this program ends, B/A is approximately 1.618033. What are the values of B and A when this happens?

*George Polya, "On Learning, Teaching, and Learning Teaching," *American Mathematical Monthly, 70,* June–July 1963, pp. 605–619.

2. A *Fibonacci-type sequence* is any sequence of positive numbers in which each succeeding number is obtained by adding the two preceding numbers.

   a. What are the next six terms in the Fibonacci-type sequence that begins with 1.2 and 1.5?

   b. Use the sequence from part **a** to determine two consecutive numbers whose ratio is approximately 1.618033. (*Hint:* Run Program 7.4A on page 452 and type 1.2 for A and 1.5 for B.)

3. There are many paths which may be taken in traveling from one planet to another, but the one which takes the least amount of fuel is a looping course called a *Hohmann Transfer*. Let's imagine a space trip using this course. In the following formulas D is the distance in millions of miles and T is the time in days. The distances from each planet to the sun, R_1 and R_2, are also in millions of miles.* These distances are given in the table on page 434.

$$D = \frac{\sqrt{R_1^2 + R_2^2}}{.45} \qquad T = \frac{(\sqrt{R_1 + R_2})^3}{13.8888}$$

   a. Write a program which prints out the distance D the spaceship will travel and the time T for the trip.

   ★ b. The average distance from the earth to the sun is 93.003 million miles and the average distance from Mars to the sun is 141.709 million miles. Let $R_1 = 93.003$ and $R_2 = 141.709$ to find the distance and time for a trip from Earth to Mars.

4. The time it takes a satellite to orbit the earth depends on its apogee and perigee. The *apogee* (A) of a satellite is its greatest distance from the center of the earth and the *perigee* (P) is its least distance. The formula for the time in hours (T) of one orbit is:

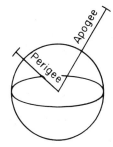

$$T = \frac{(A + P)\sqrt{A + P}}{501186}$$

   a. Write a program using the apogee A and perigee P of a satellite to compute the time in hours for one orbit. Round off the time to the nearest hour.

   ★ b. Satellites are often placed in circular orbits with apogees and perigees of approximately 26,300 miles. Why is this distance chosen?

*These formulas are based on simplifications. The actual formulas are more complicated. They do, however, give close approximations.

8

Algebra and Functions

The advancement and perfection of mathematics are intimately connected with the prosperity of the State.

NAPOLEON

"If only he could think in abstract terms."

8.1
SOLVING EQUATIONS

Algebra originated in Babylonia and Egypt more than 4000 years ago. At first there were no equations and words were used for variables rather than letters. The Egyptians used "heap" and "aha" for unknown quantities in their word problems. Here's a problem from the *Rhind Papyrus,* which was written by the Egyptian priest Ahmes about 1650 B.C.:

Heap and one-seventh of heap is 19. What is heap?

Today we would use a letter for the unknown quantity in the Egyptian problem and express the given information in an equation. Can you solve this equation?

$$x + \frac{1}{7}x = 19$$

VARIABLES

A variable is a letter or symbol that is used to denote an unknown quantity. The concept of a variable is often misunderstood. One source of confusion is that a letter may be used to represent *a number of objects* or as a label which represents *the object itself.* The following example, taken from a study of 3000 secondary school students in

England, shows that more than half of the students thought that the variables c and d represented *the objects* rather than *the numbers of objects:*[*]

> Cabbages cost 8 pence each and turnips cost 6 pence each.
> If c stands for the *number* of cabbages bought
> and t stands for the *number* of turnips bought,
> what does $8c + 6t$ stand for?
> What is the total number of vegetables bought?

Fewer than 5 percent of the students answered either question correctly; 52 percent thought that $8c + 6t$ meant "8 cabbages and 6 turnips;" and 72 percent wrote 14 for the total number of vegetables.

One method of introducing variables in the elementary school is to use geometric shapes such as $\triangle$ and $\square$. To replace a variable by a number the student can write the number in the geometric shape, like filling in a blank.

$$3\square + 8 = 14$$
$$\triangle - 7 = 3$$

Another method of providing readiness for variables is to use words such as "STEP," "WIDTH," and "NUMBER" for the unknowns, as is done in the computer language Logo (see Section 2.3).

SOLVING EQUATIONS

A balance scale is a common model for introducing equations. If each ball on the scale has the same weight, the weight on the left side of the scale *equals* (is the same as) the weight on the right side. Similarly, the sum of numbers on the left side of the equation *equals* (is the same as) the number on the right side.

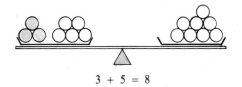

$$3 + 5 = 8$$

An *algebraic equation* is an equation with one or more variables. A number that can be used to replace a variable so that the equation becomes a true statement is called a *solution* of the equation. There is one solution for the first equation shown below, two solutions for the second equation, and infinitely many solutions for the third equation.

Equation 1 $2x + 7 = 15$ *Equation 2* $3x^2 - 5 = 7$

Solution $x = 4$ *Solutions* $x = 2$

                                                        $x = {}^-2$

---

[*]Dietmar Kuchemann, "Children's Understanding of Numerical Variables," *Mathematics in Schools,* 7 No. 4 (September 1978), 23–26.

Equation 3 $x + 3y = 24$

Solutions $x = 0,\ y = 8$

$x = 3,\ y = 7$

$x = {}^{-}3,\ y = 9$

$x = 6,\ y = 6$

$x = 24,\ y = 0$

$$\begin{matrix} \cdot & & \cdot \\ \cdot & & \cdot \\ \cdot & & \cdot \end{matrix}$$

The series of steps for finding the solution(s) to an equation is called *solving the equation.* Finding the solution(s) to simple equations can be illustrated by the balance scale model. This model, which is used in the following two examples, can be helpful for illustrating the concepts of equality and equations to elementary school students. The steps for determining the number of balls needed to balance one box on the balance scale are similar to the steps for finding the value of x in the equation.

Here are the steps for solving $3x + 5 = 14$ and the corresponding steps on the balance scale:

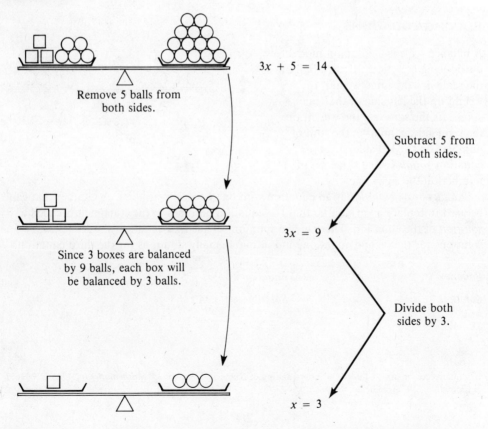

Remove 5 balls from both sides.

Since 3 boxes are balanced by 9 balls, each box will be balanced by 3 balls.

$3x + 5 = 14$

Subtract 5 from both sides.

$3x = 9$

Divide both sides by 3.

$x = 3$

Check　If each box on the first balance scale is replaced by 3 balls, the scale will balance with 14 balls on both sides. Similarly, if x is replaced by 3 in $3x + 5 = 14$, the equation becomes a true statement.

In the equation $7x + 2 = 3x + 10$ the variable occurs on both sides of the equation. The first step in the solution is to eliminate the variable from one side of the equation. This is done in the first step of the following illustration:

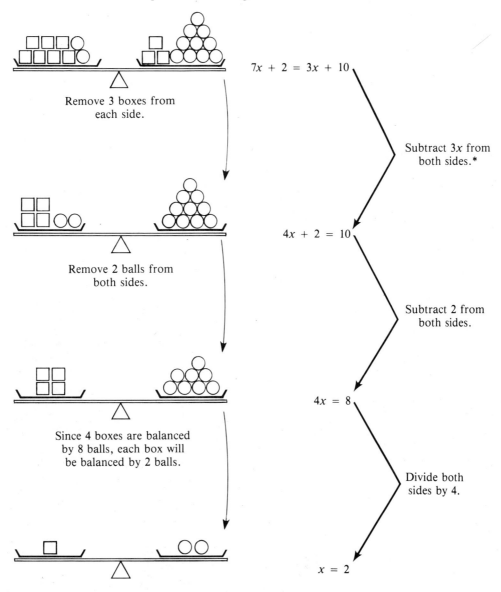

Remove 3 boxes from each side.

$7x + 2 = 3x + 10$

Subtract $3x$ from both sides.*

Remove 2 balls from both sides.

$4x + 2 = 10$

Subtract 2 from both sides.

Since 4 boxes are balanced by 8 balls, each box will be balanced by 2 balls.

$4x = 8$

Divide both sides by 4.

$x = 2$

*$7x - 3x = (7 - 3)x$ because of the distributive property and $(7 - 3)x = 4x$.

Check   If each box on the first scale is replaced by 2 balls, the scale will balance with 16 balls on each side. Replacing x by 2 in $7x + 2 = 3x + 10$ shows that 2 is a solution for this equation.

When using the balance scale, the same amount must be put on or removed from each side to maintain a balance. Similarly, with an equation, the same operation must be performed on each side to maintain an equality. In other words, *whatever is done to one side of an equation must be done to the other side.* Specifically, here are five methods of changing an equation:

1. Add the same number to both sides.

2. Subtract the same number from both sides.

3. Multiply both sides by the same nonzero number.

4. Divide both sides by the same nonzero number.

5. Take the square root of both sides if they are not negative.

6. Use known facts, such as the number properties (pages 448 and 449), to replace any term by an equal term.

Changing an equation by one or more of the preceding methods produces an *equivalent equation,* that is, another equation with the same solutions.

In the next two examples we will determine the steps for solving an equation by first listing the operations on x in the order they are performed. Then we will reverse this order and perform the inverse of each operation.

"I think you should be more explicit here in step two."

Here are the operations in the order they are performed on x in $9(x/4 - 3) = 108$: *step 1, divide* by 4; *step 2, subtract* 3; and *step 3, multiply* by 9. The result is 108. The equation is solved by performing the inverse of these operations in reverse order:

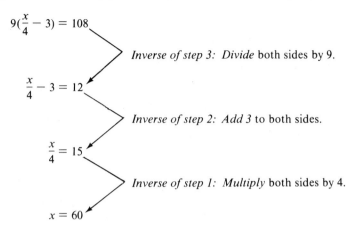

$9(\frac{x}{4} - 3) = 108$

Inverse of step 3: Divide both sides by 9.

$\frac{x}{4} - 3 = 12$

Inverse of step 2: Add 3 to both sides.

$\frac{x}{4} = 15$

Inverse of step 1: Multiply both sides by 4.

$x = 60$

Check Replace x by 60: $9(60/4 - 3) = 9(15 - 3) = 9(12) = 108$.

Let's use this method to solve another equation. Here are the operations in the order they are performed on x in $4x^2 + 17 = 161$: *step 1, square* the number; *step 2, multiply* by 4; and *step 3, add* 17. The result is 161. The equation is solved by performing the inverse of these operations in reverse order:

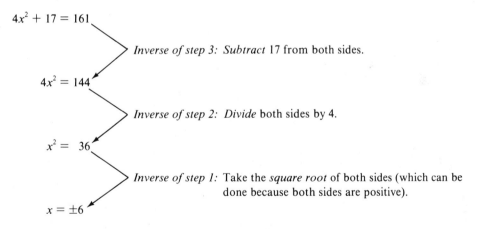

$4x^2 + 17 = 161$

Inverse of step 3: Subtract 17 from both sides.

$4x^2 = 144$

Inverse of step 2: Divide both sides by 4.

$x^2 = 36$

Inverse of step 1: Take the *square root* of both sides (which can be done because both sides are positive).

$x = \pm 6$

Check When x is replaced by either 6 or $^-6$, the equation is true:

$$4(6)^2 + 17 = 4(36) + 17 = 144 + 17 = 161$$
$$4(^-6)^2 + 17 = 4(36) + 17 = 144 + 17 + 161$$

When the variable occurs more than once, either on the same side of the equation or on opposite sides, the first step in solving the equation is to *collect terms* so that the variable occurs only once.

In the next solution the first three equations are replaced by the fourth equation in which the variable occurs only once. The second and third equations are obtained by using number properties:

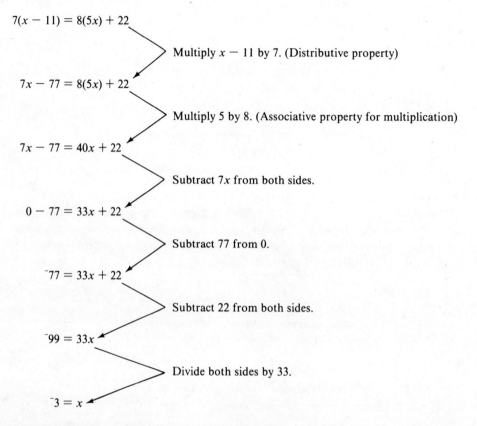

$$7(x - 11) = 8(5x) + 22$$

Multiply $x - 11$ by 7. (Distributive property)

$$7x - 77 = 8(5x) + 22$$

Multiply 5 by 8. (Associative property for multiplication)

$$7x - 77 = 40x + 22$$

Subtract $7x$ from both sides.

$$0 - 77 = 33x + 22$$

Subtract 77 from 0.

$$^-77 = 33x + 22$$

Subtract 22 from both sides.

$$^-99 = 33x$$

Divide both sides by 33.

$$^-3 = x$$

Check When x is replaced by $^-3$, the equation is true:

$$7(^-3 - 11) = 8(5)(^-3) + 22 = ^-98$$

Sometimes the variable in an equation will occur in the denominator of a fraction. This is common in problems involving ratios. In the first step of the next solution, both sides of the equation are multiplied by x to eliminate x from the denominator:

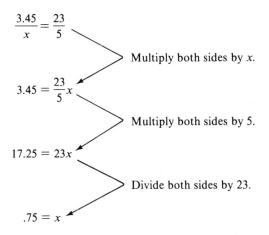

$$\frac{3.45}{x} = \frac{23}{5}$$

Multiply both sides by x.

$$3.45 = \frac{23}{5}x$$

Multiply both sides by 5.

$$17.25 = 23x$$

Divide both sides by 23.

$$.75 = x$$

Check   When x is replaced by .75, the equation is true:

$$\frac{3.45}{.75} = \frac{23}{5} = 4.6$$

IDENTITIES

We have been solving equations for which there is only one solution or a restricted set of solutions. Sometimes, however, an equation is true for all replacements of the variable(s). In this case the equation is called an *identity*. The following two mathematical statements are identities for all real numbers. That is, regardless of which real numbers are used in place of *a, b,* and *c,* both equations are true.

Multiplication is distributive over addition.   Addition is associative.

$$a(b + c) = ab + ac \qquad\qquad (a + b) + c = a + (b + c)$$

The distributive property and the associative property for addition are two of the real number properties. All of the equations for the real number properties (page 449) are identities. Here is another identity that is true for all real numbers that are used in place of the variables *a* and *b*:

$$(a + b)^2 = a^2 + 2ab + b^2$$

This statement says, the square of the sum of two numbers $(a + b)^2$ is equal to the sum of the squares of the two numbers, $a^2 + b^2$, plus twice their product $2ab$. Since the area of a square is found by multiplying the length of the side of the square by itself, this identity can be illustrated by a square whose sides have length $a + b$. The product $(a + b)$ times $(a + b)$ is the area of the square. On the other

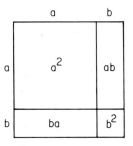

hand the square is divided into four pieces, and the sum of the areas of these pieces is $a^2 + ba + ab + b^2$. Since this sum equals $a^2 + 2ab + b^2$, we see that

$$(a + b)^2 = a^2 + 2ab + b^2$$

PROBLEM SOLVING

Algebra is a powerful problem-solving strategy. Many of the problems and exercises in Section 1.3, which were solved by drawing pictures, guessing and checking, and listing possibilities, can be solved by elementary algebra. The following three problems are from that section.

PROBLEM (page 45) A woman weighs 70 pounds plus half her weight. How much does she weigh?

Solution Let w represent the woman's weight. The given information is expressed by $w = 70 + \frac{1}{2}w$. Here are the steps for solving this equation:

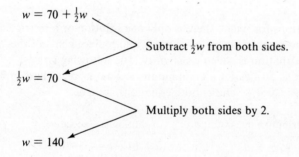

$w = 70 + \frac{1}{2}w$

Subtract $\frac{1}{2}w$ from both sides.

$\frac{1}{2}w = 70$

Multiply both sides by 2.

$w = 140$

Check The woman's weight is 140 pounds. Does this check in the first equation?

PROBLEM (page 36) John and Harry earned the same amount of money, although one worked 6 days more than the other. If John earns $12 a day and Harry earns $20 a day, how many days did each person work?

Solution It is helpful to list the given information in a table. Since John earned less money per day, he must have worked more days. Therefore, if x represents the number of days Harry worked, then $x + 6$ is the number of days John worked.

	Amount earned each day	Number of days worked	Total amount earned
Harry	$20	x	$20x$
John	$12	$x + 6$	$12(x + 6)$

The fact that the total amount earned by each person is the same is expressed by $20x = 12(x + 6)$. Here are the steps for solving this equation:

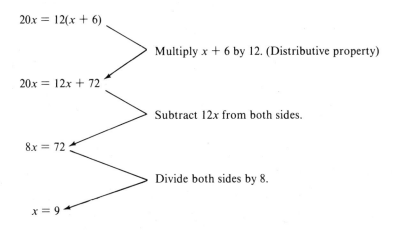

$$20x = 12(x + 6)$$

Multiply $x + 6$ by 12. (Distributive property)

$$20x = 12x + 72$$

Subtract $12x$ from both sides.

$$8x = 72$$

Divide both sides by 8.

$$x = 9$$

Check   Harry worked 9 days, which means that John worked 15 days. Check these answers to see if each person earned the same amount of money.

PROBLEM   (page 43) In driving from town *A* to town *D*, you pass through town *B* and then through town *C*. It is 10 times farther from *A* to *B* than from *B* to *C* and 10 times farther from *B* to *C* than from *C* to *D*. If it is 1332 miles from *A* to *D*, how far is it from *A* to *B*?

Solution   A diagram helps to visualize the relative distances between the towns. If x represents the distance from *C* to *D*, then $10x$ is the distance from *B* to *C* and $10(10x)$ is the distance from *A* to *B*:

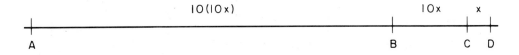

The given information is expressed by $10(10x) + 10x + x = 1332$. This equation is solved by the following steps:

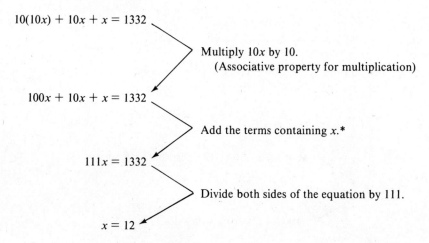

$10(10x) + 10x + x = 1332$

Multiply $10x$ by 10.
(Associative property for multiplication)

$100x + 10x + x = 1332$

Add the terms containing x.*

$111x = 1332$

Divide both sides of the equation by 111.

$x = 12$

Check   The distance from town C to town D is 12 miles. Use this distance to obtain the distances from B to C and A to B, and check to see if the sum of these distances is 1332 miles.

ALGEBRAIC PROOFS

When a statement involves an infinite set of numbers, it is impossible to prove it is true by trying all of the numbers. Such statements must be proved by using algebra and deductive reasoning. For example, on page 7 we showed several cases in which the sum of three consecutive whole numbers is divisible by 3. This result can now be proved with elementary algebra.

> THEOREM   The sum of any three consecutive whole numbers is divisible by 3.

> *Proof*   Let x be any whole number. Then $x + 1$ and $x + 2$ are the next two consecutive whole numbers. We must show that $x + (x + 1) + (x + 2)$ is divisible by 3. The commutative and associative properties for addition allow us to rewrite $x + (x + 1) + (x + 2)$ as $3x + 3$. Then by using the distributive property, $3x + 3$ can be written as $3(x + 1)$.

$$x + (x + 1) + (x + 2) = 3x + 3 = 3(x + 1)$$

Since $3(x + 1)$ divided by 3 is $x + 1$, and $x + 1$ is a whole number (why?), this shows that 3 divides the sum of any three consecutive whole numbers. Furthermore, it shows that the result of dividing by 3 is the middle number, $x + 1$. For example, $36 + 37 + 38$ divided by 3 is 37.

*$100x + 10x + x = (100 + 10 + 1)x$ because of the generalized distributive property, and $(100 + 10 + 1)x = 111x$.

Many simple number tricks and so-called magic formulas can be analyzed by elementary algebra. Select a number and perform the following operations:

Add 4 to your number; multiply the result by 6; subtract 9; divide by 3; add 13; divide by 2; and then subtract the number you started with.

If you performed these operations correctly, your final answer is 9, regardless of the number you started with.

Proof Let N represent an arbitrary number. Here are the preceding operations:

Add 4	$N + 4$	
Multiply by 6	$6(N + 4) = 6N + 24$	Distributive property of multiplication
Subtract 9	$(6N + 24) - 9 = 6N + 15$	Associative property of addition (why?)
Divide by 3	$(6N + 15)/3 = 2N + 5$	Distributive property $\left(\text{multiply by } \frac{1}{3}\right)$
Add 13	$(2N + 5) + 13 = 2N + 18$	Associative property of addition
Divide by 2	$(2N + 18)/2 = N + 9$	Distributive property $\left(\text{multiply by } \frac{1}{2}\right)$
Subtract N	$(N + 9) - N = 9$	Associative and commutative properties of addition

These equations show that it doesn't matter what number N represents. In the final step N is subtracted and the end result is always 9. Since the number properties hold for all real numbers, N can be a positive, a negative, a rational, or an irrational number.

The calendar patterns and relationships which were arrived at by inductive reasoning (page 9) can now be proved by algebra and deductive reasoning. Here is one of these relationships.

```
┌─────────────────────────────────┐
│         Nov. 1985               │
│  Sun Mon Tue Wed Thu  Fri  Sat  │
│                          1    2 │
│   3    4    5  6    7    8    9 │
│  10   11   12 13   14   15   16 │
│  17   18   19 20   21   22   23 │
│  24   25   26 27   28   29   30 │
└─────────────────────────────────┘
```

THEOREM   The sum of any 3 by 3 array of dates from a calendar is always 9 times the middle number.

Proof   Let x be the number in the upper left corner of the array. Then the 3 by 3 array is

x	$x + 1$	$x + 2$
$x + 7$	$x + 8$	$x + 9$
$x + 14$	$x + 15$	$x + 16$

The commutative and associative properties allow us to write the sum as $9x + 72$, and by the distributive property, $9x + 72$ can be written as $9(x + 8)$:

$$x + (x + 1) + (x + 2) + (x + 7) + (x + 8) + (x + 9) + (x + 14) + (x + 15) + (x + 16) =$$
$$9x + 72 = 9(x + 8)$$

Since $x + 8$ is the middle number of the 3 by 3 array of dates, these equations show that the sum of numbers is 9 times the middle number of the array.

COMPUTER APPLICATIONS

Some algebra problems can be solved by programming the computer to try numbers systematically for the variables, especially when the solutions are whole numbers.

> The square of Kathy's age plus John's age is 62 and the square of John's age plus Kathy's age is 176. What are their ages?

If K represents Kathy's age and J represents John's age, we must satisfy these two conditions:

$$K^2 + J = 62 \quad \text{and} \quad J^2 + K = 176$$

We know their ages are less than 62. Why? Therefore, we only need to try numbers for K and J which are less than 62. This is done with two loops in the following program. First K is equal to 1, and the loop from lines 20 to 40 is used 62 times with J taking on the whole numbers from 1 to 62. Then K is set equal to 2, and once again J takes on the whole numbers from 1 to 62. In this manner combinations for K and J are tried until the condition in line 30 is satisfied.*

Program 8.1A

```
10   FOR K = 1 TO 62
20   FOR J = 1 TO 62
30   IF K * K + J = 62 AND J * J + K = 176 THEN   GOTO 60
40   NEXT J
50   NEXT K
60   PRINT "KATHY IS "K" YEARS OLD AND JOHN IS "J" YEARS OLD. "
70   END

RUN

KATHY IS 7 YEARS OLD AND JOHN IS 13 YEARS OLD.
```

*Note: $K * K$ is used in line 30 rather than $K \wedge 2$ because the exponentiation operation ($\wedge$) sometimes causes round-off errors.

The printout for this program shows Kathy's and John's ages. However, there may be other solutions since the computer stopped when it found two values for *K* and *J* which satisfied the equations. To check for a different pair of numbers, we can add the condition that $K \neq 7$. *Not equal to* ($\neq$) is represented in BASIC by using both inequality symbols as shown in the following line:

```
30   IF K * K + J = 62 AND J * J + K = 176 AND K < > 7 THEN
        GOTO 60
```

Here's another way to revise line 30 to find all the solutions for Kathy's and John's ages:

```
30   IF K * K + J = 62 AND J * J + K = 176 THEN   PRINT "KATHY
        IS "K" YEARS OLD AND JOHN IS "J" YEARS OLD."
```

If this line 30 is used, line 60 can be replaced by

```
60   PRINT "THE COMPUTER HAS FINISHED LOOKING FOR SOLUTIONS."
```

Revise the preceding program to see if there are more solutions.

SUPPLEMENT (Activity Book)

Activity Set 8.1 Geometric Models for Algebraic Expressions
Just For Fun: Algebraic Skills Game

Exercise Set 8.1: Applications and Skills

1. This puzzle was posed by one of America's greatest puzzle experts, Sam Loyd (1841–1911).* What is the weight of the top? (See left pan of upper balance.)

   a. Use the first balance scale to solve this problem by guessing (and checking) the number of marbles required to balance each block.

   ★ b. Explain how the information from the second balance scale can be used with the first balance scale to solve this problem.

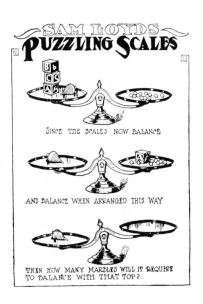

Mathematical Puzzles of Sam Loyd (New York: Dover 1959), p. 101.

2. In research conducted at the University of Massachusetts, Peter Rosnick found that $37\frac{1}{3}$ percent of a group of 150 entering engineering students were unable to write the correct equation for the following problem:*

   Write an equation using variables S and P to represent the following statement: "At this university there are six times as many students as professors." Use S for the number of students and P for the number of professors.

   a. What is the correct equation?

★ b. The most common error was $6S = P$. Give an explanation for this.

3. The following questions are part of an Algebra I test that was given to 3000 high school students in England to test their knowledge of variables. Perform the operations and answer the questions.

   a. Add 4 to $3n$.

★ b. Multiply $n + 5$ by 4.

   c. Which expression is the largest? $n + 1, n - 3, n + 4, n - 7$

★ d. (True or False) $2n$ is always greater than $n + 2$.

   e. If $e + f = 8$, then $e + f + g =$

   f. If $n - 246 = 762$, then $n - 247 =$

★ g. Is $L + M + N = L + P + N$ always, sometimes, or never true?

4. Determine the number of balls that will balance each box. Draw balance scales to show the steps in your solution. Write the equation that is represented by each balance scale and show the steps in solving this equation.

   a.   ★   b.

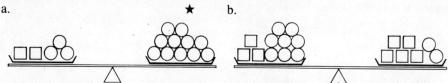

5. For each equation below there is a list of operations that were performed on the variable. Determine the inverse of each operation and list them in reverse order. Solve each equation by performing the inverse operations on both sides of the equation.

   a. $9(3x - 4) = 126$ Multiply by 3; subtract 4; multiply by 9

★ b. $(z/5 + 19)/2 = 37$ Divide by 5; add 19; divide by 2

   c. $\sqrt{(4.6y + 6)/3} = 5$ Multiply by 4.6; add 6; divide by 3; take the positive square root

*Peter Rosnick, "Some Misconceptions Concerning the Concept of a Variable," *The Mathematics Teacher,* 74 No. 6 (September 1981), 418–420.

6. In each of the following equations the variable occurs more than once. Solve each equation by first collecting terms so that the variable occurs only once.

   a. $x + \frac{1}{7}x = 19$

   ★ b. $5(2x - 3) = 7x$

   c. $3(5y) + 7y = 14y + 24$

   ★ d. $5b - (b + 4)/2 = 47 + b/2$

   e. $18 - x/24 = 8(2x - 94)$

7. Each of the following exercises are from Section 1.3 on problem solving. Use an equation with one variable to solve each problem.

   a. *Section 1.3, page 34* (*Hint:* Draw a diagram and let x represent the distance from town A to town B.)

   ★ b. *Exercise Set 1.3, #5* (*Hint:* Let c represent the number of checks and determine the value of c for which both plans cost the same.)

   c. *Exercise Set 1.3, #9* (*Hint:* Let p represent the number of postcards and complete the table.)

	Number	Total Cost
Postcards	P	
Letters		

   ★ d. *Exercise Set 1.3, #15e* (*Hint:* The total cost of the bottle and the cider is 86 cents.)

8. Give an algebraic proof of each statement:

   a. The sum of any three consecutive dates from a column of a calendar is three times the middle number.

   ★ b. The sum of five consecutive whole numbers is divisible by five.

   c. If the following operations are performed on any number, the result will be the number 3:

   Add 221 to the number; multiply by 2652; subtract 1326; divide by 663; subtract 870; divide by 4; and subtract the original number.

   ```
   May  1985

   Sun Mon Tue Wed Thu Fri Sat
                 1   2   3   4
    5   6   7  (8)  9  10  11
   12  13  14 (15) 16  17  18
   19  20  21 (22) 23  24  25
   26  27  28  29  30
   ```

9. The distance between rowlocks is called the span of a rowboat. This distance is the basis for computing the proper length of an oar. If the distance between rowlocks is s inches, the length of the oar in feet should be $25(s/2 + 2)/84$.

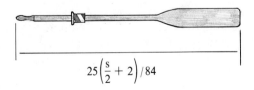

$25\left(\dfrac{s}{2} + 2\right)/84$

a. Determine the proper length of an oar to the nearest tenth of a foot, if the span is 42 inches.

★   b. If the length of the oar is 8 feet, determine the proper span of the boat to the nearest inch.

10. Suppose the price of an object is p, the tax is t, and there is a discount d. Should the discount be taken and then the tax computed on the remaining amount, or should the tax be added to the price and then the discount taken?

   a. What is the cost to the customer for both of the preceding methods, if $p = \$24$, $d = 20$ percent, and $t = 5$ percent.

   b. The cost for the first method (first discount, then tax) is represented by $(p - dp) + t(p - dp)$. The cost for the second method (first tax, then discount) is represented by $(p + tp) - d(p + tp)$. Show that these two algebraic expressions are equal.*

★ 11. *Birthdate Trick:* Ask a person to write the number of the month of his or her birth and perform the following operations: multiply by 5; add 6; multiply by 4; add 9; multiply by 5; and add the number of the day of birth. When 165 is subtracted from this number, the result will be a number that contains the day and month of birth. Try it. *Analysis:* Let D and M equal the day and month, respectively. The preceding formula is represented by the following algebraic expression. Prove that this expression is equal to $100\,M + D$.

$$5[4(5\,M + 6) + 9] + D - 165$$

12. *Lucky Birthdate:* Add the numbers of the day, the month, and the year you were born: subtract 23 times the day; add 21 times the month; add your age on December 31, 1991. If this result is divisible by 11, then you were born on a lucky day. *Analysis:* Let D, M, and Y equal the day, month, and year of birth, respectively. The preceding formula is represented by the following algebraic expression. Prove that this expression is divisible by 11. (*Note:* $1991 - Y$ is your age on December 31, 1991.)

$$(D + M + Y) - 23\,D + 21\,M + (1991 - Y)$$

Exercise Set 8.1: Problem Solving

1. *Number Array Trick:* The teacher asks the class to select a 4 by 4 square array of numbers from the 10 by 10 number chart (consecutive whole numbers from 1 to 100) and use only those numbers for the following four-step process:

---

*R. M. Knaus and C. G. Knaus, "An Application of an Algebraic Proof," *The Mathematics Teacher,* 72 No. 5 (May 1979), 44–45.

1. Circle any number and cross out the remaining numbers in its row and column.

2. Circle another non-crossed-out number and cross out the remaining numbers in its row and column.

3. Repeat step 2 until there are four circled numbers.

4. Add the four circled numbers.

For any sum that the teacher receives from a student, the teacher is able to predict the number in the upper left corner of the student's square by subtracting 66 from the sum and dividing the result by 4. Show why this formula works.

a. *Understanding the Problem:*
Let's follow through the steps for the 4 by 4 array shown here. The first circled number is 46, and the remaining numbers in its row and column have been crossed out. The next circled number is 38. Continue the four-step process. Does the teacher's formula produce the number in the upper left corner of this 4 by 4 array?

1	2	3	4	5	6	7	8	9	10
11	12	13	14	15	16	17	18	19	20
21	22	23	24	25	26	27	28	29	30
31	32	33	34	35	~~36~~	37	(38)	39	40
41	42	43	44	45	(46)	~~47~~	~~48~~	~~49~~	50
51	52	53	54	55	~~56~~	57	58	59	60
61	62	63	64	65	~~66~~	67	68	(69)	70
71	72	73	74	75	76	77	78	79	80
81	82	83	84	85	86	87	88	89	90
91	92	93	94	95	96	97	98	99	100

10 By 10 Number Chart

★ b. *Devising a Plan:* This problem can be solved by algebra. If we represent the number in the upper left corner by x, the remaining numbers can be represented in terms of x. Complete the next two rows of the 4 by 4 array of algebraic expressions.

x	$x + 1$	$x + 2$	$x + 3$
$x + 10$	$x + 11$	$x + 12$	$x + 13$

c. *Carrying out the Plan:* Carry out the four-step process on the 4 by 4 array in part **b.** Use the results to show that the teacher's formula works.

★ d. *Looking Back:* One variation to this number trick is to change the size of the array that the student selects. For example, what formula would the teacher use if a 3 by 3 array is selected from the 10 by 10 number chart? (*Hint:* Use a 3 by 3 algebraic array.) Another variation is changing the number chart. Suppose that a 3 by 3 array is selected from a calendar. What is the formula in this case?

2. *A Weighty Dilemma*:* Use the
information from the first two
balance beams to determine the
number of nails that are needed to
balance one cube.

a. Solve this problem without
algebra.

b. Solve this problem with algebra.
(*Hint:* Let *c* equal the weight of a
cube, *b* the weight of a bolt, and
n the weight of a nail.)

How many nails
will balance the cube?

3. *Identical Temperatures:* A Celsius thermometer and a Fahrenheit thermometer are
both placed in a solution simultaneously. The temperature is the same number on
both scales. What is the temperature? (*Hint:* Use a formula from page 278.)

★ 4. *Octogenarian Race:* Two octogenarians agree to a 12-kilometer race under the
following conditions: one person is to run half the distance and walk half the
distance; and the second person is to run half the time and walk the other half of the
time. If they both run at 6 kilometers per hour and walk at 3 kilometers per hour,
which person will win the race?

5. *An Old Card Trick:* Use an ordinary deck of 52 cards for this trick. Turn a card
face up, and then place cards face up on top of it by counting from the numerical
value of the card to 13. (Regard the jack as 11, queen as 12, and king as 13.) Repeat
this process to form several piles until no more piles can be formed. Keep the extra
cards. Then have a spectator turn the piles over and move them around. Now ask
someone to remove all but three of the piles and give the removed cards to you.
Then by turning over the top cards on any two of the three piles, you can tell the top
card on the third pile. This can be done because the sum of the three top cards plus
10 equals the number of cards in your hand. Try this trick a few times. Use algebra

*Balance beam art from The Arithmetic Teacher, *19 No. 6 (October 1972), 460–461.*

to show why it works. (*Hint:* Let x, y, and z represent the numerical values of the top cards on the three piles. Then the numbers of cards in each of these piles are 14 − x, 14 − y, and 14 − z.)

Exercise Set 8.1: Computers

1. Solve these age problems by writing a computer program that systematically tries numbers for the variables. (*Note:* You may need to use products rather than exponentiation.)

   a. The square of Bill's age is 15 years less than the cube of Mary's age. What are their ages?

   b. The square of Sonya's age plus Nike's age is twice the square of Nike's age plus Sonya's age. What are their ages?

   c. There are two solutions to part **b.** Revise the program to find another solution.

2. The sum of a whole number and its square and its cube is 20,439. Write a program to find this whole number. How do we know that this number is less than 30?

3. A monument with three different size cubes sits on a square plaza. Both the monument and the plaza were constructed from the same number of cubes. The length of the plaza is equal to the sum of the lengths of the cubes. What is the length of the plaza? [*Hint:* Write a computer program with three loops to solve $X^3 + Y^3 + Z^3 = (X + Y + Z)^2$.]

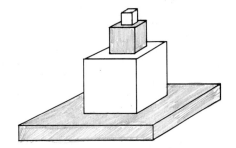

4. There is an interesting five-digit number that if 1 is placed to the right of its units digit, is three times greater than if 1 is placed on the left of its ten thousands digit. If N is the five-digit number, then N satisfies this equation:

$$10(N) + 1 = 3(100{,}000 + N)$$

   a. Solve this equation for N. Check your result.

   b. The number in part **a** was discovered by the program below. The computer found a solution for the equation in line 30 with $A = 1$. Revise this program for $A > 1$ to find another interesting five-digit number N. (*Hint:* N is between 85,000 and 86,000.)

```
10   FOR A = 1 TO 9
20   FOR N = 10000 TO 99999
30   IF 10 * N + A = 3 * (A * 100000 + N) THEN   GOTO 80
40   NEXT N
50   NEXT A
60   PRINT "THERE ARE NO SOLUTIONS. "
70   GOTO 90
80   PRINT "A = "A" AND N = "N
90   END
```

The Daniel Webster Hoan Memorial Bridge in Milwaukee, Wisconson
Courtesy, American Institute of Steel Construction

8.2
EQUATIONS AND
GRAPHS

Today's technology requires a wide range of curves. The designing of bridges, roads, and buildings is only one area of the many applications. Each of the curves in these photographs can be represented by equations and graphs using coordinate geometry. In this section we will look at the equations and graphs of some familiar curves.

Network of roads outside Empire State
Plaza, Albany, New York

Kresge Auditorium, Massachusetts
Institute of Technology

RECTANGULAR COORDINATES

There are many types of coordinate systems, each with its own method of locating points with respect to some frame of reference. Points on the earth's surface are located by a global coordinate system whose frame of reference is the equator and the zero meridian (page 233). Without a frame of reference the location of a point gives little or no information, as the fellow in this cartoon has just discovered.

One frame of reference for points in a plane, such as those on this sheet of paper, is a pair of perpendicular lines called the *x axis* and the *y axis*. Every point in the plane, including those on the axes, can be located by an ordered pair of numbers. The first number of the pair (x, y) is called the *x coordinate* and tells the distance from the *y* axis. Positive numbers are used for the distances to the right of the *y* axis and negative numbers for distances to the left of the *y* axis. The second number of the pair (x, y) is called the *y coordinate* and tells the distance

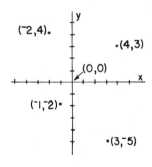

from the x axis. Positive numbers are used above the x axis and negative numbers below. The coordinates of several points are given in the preceding figure. The intersection of the x and y axes is called the *origin* and has coordinates (0, 0).

This method of locating points is called the *rectangular coordinate system,* or *Cartesian coordinate system.* The name "Cartesian" is in honor of Rene Descartes (1596–1650), the French mathematician and philosopher who first used this system of coordinates for representing geometric figures.

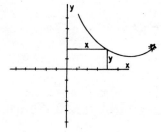

The story is told that the idea of coordinates in geometry came to Descartes while he lay in bed and watched a fly crawling on the ceiling. Whether the story is true or not, it is useful in illustrating the relationship between curves and their equations. Each position of the fly can be given by two distances from the edges of the ceiling where the walls and ceiling meet. Descartes discovered that these distances can be related by an equation. That is, each point on the curve has coordinates that are solutions to an equation, and conversely, every two numbers x and y that are solutions to the equation correspond to a point on the curve. This discovery makes it possible to study geometric figures by using equations and algebra. The link between geometry and algebra is one of the greatest mathematical achievements of all times.

GRAPHS AND THEIR EQUATIONS

Descartes was primarily interested in beginning with a line or curve and finding its equation. The line shown here has been drawn through the points (0, 0) and (1, 2). What is the relationship between each x coordinate and its corresponding y coordinate? The equation for this line is

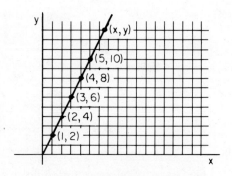

$$y = 2x$$

The circle shown here was drawn with a compass using (0, 0) as the center and a radius of 5 units. The coordinates of a few points on the circle have been labeled. The sum of the squares of the x and y coordinates is 25 for each of these points. Here are some examples:

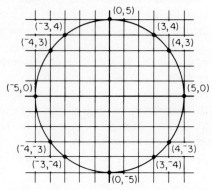

484

$$3^2 + 4^2 = 25 \qquad (^-4)^2 + 3^2 = 25 \qquad 4^2 + (^-3)^2 = 25$$

The equation for this circle is

$$x^2 + y^2 = 25$$

EQUATIONS AND THEIR GRAPHS

During the same period that Descartes was finding the equations of known curves, Pierre Fermat was working independently on techniques for graphing equations. The most straightforward method of graphing equations is to select some values for x and compute the corresponding values for y. The table below contains a few values of x and the corresponding values of y which satisfy

$$y = x^2$$

Notice that each value of y is positive or zero. Also, for a given value of x and its opposite, the values of y are equal. Because of these conditions the graph opens upward and is symmetric to the y axis. This kind of curve is a *parabola*.

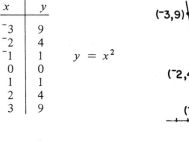

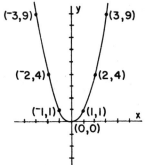

A point at which a curve crosses the y axis is called a *y intercept*. The x coordinate of such a point is always zero. Similarly, a point at which a curve crosses the x axis is called an *x intercept* and the y coordinate is zero. The x and y intercepts are used for graphing the next curve.

The table at the top of the next page contains the coordinates of the four intercepts on the graph whose equation is

$$\frac{x^2}{9} + \frac{y^2}{4} = 1$$

This is the equation of an *ellipse*. Try the coordinates of these points in the equation.

x	y
0	2
0	⁻2
3	0
⁻3	0

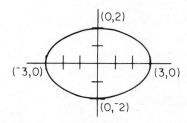

Not all curves have x intercepts and y intercepts. In the next graph the curve does not cross either the x axis or the y axis. Several values of x and y are shown in the table below for the equation

$$y = \frac{1}{x}$$

Notice that x cannot equal zero since division by zero is not allowed. Also, there is no value of x for which y is zero. Therefore, the graph for this equation does not cross the x axis or the y axis. This type of curve is a *hyperbola*.

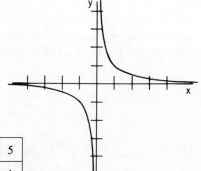

x	⁻5	⁻4	⁻3	⁻2	⁻1	1	2	3	4	5
y	$\frac{\text{-}1}{5}$	$\frac{\text{-}1}{4}$	$\frac{\text{-}1}{3}$	$\frac{\text{-}1}{2}$	⁻1	1	$\frac{1}{2}$	$\frac{1}{3}$	$\frac{1}{4}$	$\frac{1}{5}$

In these next few pages we will look more closely at the equations and graphs of lines, parabolas, ellipses, and hyperbolas.

STRAIGHT LINES

Highway engineers measure the slope or steepness of a road by comparing each 100 feet of horizontal distance with the corresponding vertical rise. The Federal Highway Administration recommends a maximum of 12 feet vertically for each 100 feet of horizontal distance. Many secondary roads and streets are much steeper. Filbert Street in San Francisco has a vertical rise of approximately 1 foot for each 3 feet of horizontal distance. By comparison, the east and west walls of the Daytona International Speedway have a vertical rise of 3 feet for each 5 feet of horizontal distance. These banks enable a car to turn at the ends of the speedway while maintaining speeds of 180 to 200 miles per hour.

East wall of the Daytona International Speedway with a slope of 31°

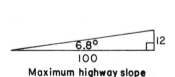

Maximum highway slope

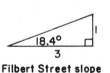

Filbert Street slope

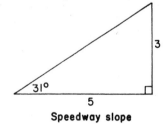

Speedway slope

The slope of a line on a rectangular coordinate system is measured by horizontal and vertical distances. The horizontal distance, which is the difference between the *x* coordinates of two points on a line, is called the *run*. The vertical distance is the difference between the *y* coordinates of the two points and is called the *rise*. The ratio of these distances, rise/run, is the *slope* of the line. In this example, the points (¯2, 2) and (1, 4) were used to

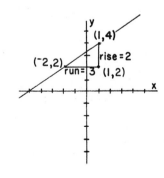

find the slope of 2/3. Using any two points on the same line, you will always get a slope of 2/3.

To distinguish between lines such as *L* and *K,* which are inclined in opposite directions, lines running from lower left to upper right have a *positive slope* and those running from upper left to lower right have a *negative slope:*

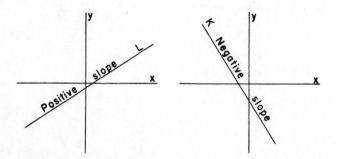

Lines that are parallel to the *x* axis have a slope of 0. For example, the line through (2, ⁻3) and (5, ⁻3) has a rise of 0 and a run of 3 for these two points. Therefore, its slope is 0/3, or 0. Lines that are perpendicular to the *x* axis, such as the line through (⁻2, 1) and (⁻2, 3), have a slope that is undefined. In this example the rise is 2 and the run is 0, but 2/0 is undefined. Between these two extremes the slopes of lines can be any positive or negative real numbers.

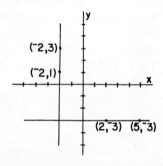

Notice the similarity in the following three graphs and their equations. The lines all have a slope of 2 and the coefficient of *x* in each equation is 2. Furthermore, the *y* intercepts are 0, 1, and 5. Can you see how these numbers are obtained from the equations?

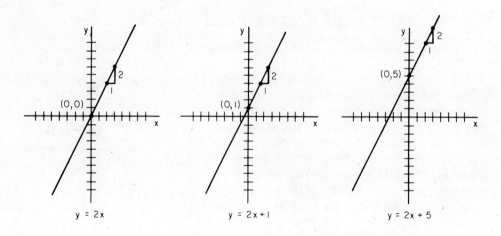

$$y = 2x \qquad y = 2x+1 \qquad y = 2x+5$$

Every straight line, except those that are perpendicular to the x axis, has an equation of the form

$$y = mx + b$$

where m and b are any real numbers. The m in this equation is the slope of the line, and b is the y coordinate of the point where the line intercepts the y axis. In the preceding three equations the y intercepts are 0, 1, and 5. When $b = 0$, the equation becomes $y = mx$, which corresponds to a line through the origin with a slope of m.

The equation $y = mx + b$ is called the *slope-intercept form* of a line. When the equation of a line is not in this form, it can be rewritten. For example, $3x + 4y = 7$ is the equation of a line that has a slope of $\frac{-3}{4}$ and a y intercept of $\frac{7}{4}$:

$$3x + 4y = 7$$
$$4y = {}^{-}3x + 7$$
$$y = \tfrac{-3}{4}x + \tfrac{7}{4}$$

When the slope of a line and the y intercept are known, the equation can be written immediately. This information is often given in applications. For example, a *rate,* such as miles per hour or cost per unit, can be described by a linear equation in which the *slope of the line* is the *rate.* Suppose it costs \$3 per hour to rent a lawn mower; it will cost \$6 for 2 hours, \$9 for 3 hours, etc. If x denotes the number of hours and y the total cost, this information is described by

$$y = 3x$$

Now if there is an initial fee in addition to the hourly rate, say \$5, then the equation becomes

$$y = 3x + 5$$

In general, the rate is the slope of a straight line and the initial cost is the y intercept.

PARABOLAS

The paths of balls, bullets, or other objects which are thrown or shot into the air in a nonvertical direction are parabolas. The distance an object travels varies with the angle it is aimed at, its *angle of elevation.* An object will travel its greatest distance away (horizontal distance) for a 45° angle of elevation. Using this condition, which waterspout on this fountain (see photo) appears to have an angle of elevation that is closest to 45°? While parabolas have many different shapes, as shown by these paths of water, each parabola satisfies the following definition.

Definition: A *parabola* is the set of all points in a plane that are the same distance from a fixed point, called the *focus,* as from a fixed line, called the *directrix.* (The distance from a point to a line is the perpendicular distance.)

In both of the following figures the focus is labeled *F* and the directrix *D.* The line passing through *F* and perpendicular to line *D* is the *axis* of the parabola. Notice how the distances for the points *S, T,* and *R* on these curves satisfy the conditions of the definition. Select another point on these curves and use a piece of paper to mark off its distances to the focus and directrix. Compare these distances.

Fountain at Swirbul Library,
Adelphi University

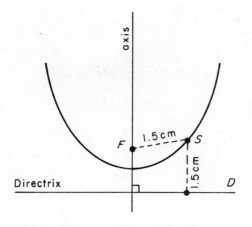

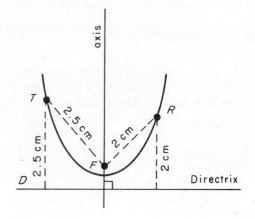

CHAPTER 8   ALGEBRA AND FUNCTIONS

The equation of the parabola on this coordinate system is

$$y = \frac{x^2}{8}$$

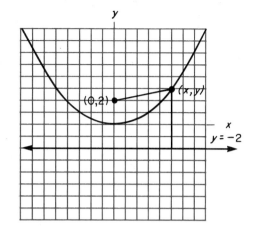

Any pair of numbers that satisfy this equation are the coordinates of points which are on this graph. As examples, $\left(2, \frac{1}{2}\right)$, $\left(-2, \frac{1}{2}\right)$, $(4, 2)$ and $(-4, 2)$ can all be used in this equation and you will find the corresponding points on the graph. Conversely, for every point on the graph, its coordinates will satisfy the equation $y = x^2/8$. The focus for this parabola is at $(0, 2)$, and the directrix is the line parallel to the x axis and two units below it ($y = -2$). Select a point on the graph and use a piece of paper to mark off its distances to the focus and directrix. Compare these measurements.

Parabolas have a very important reflection property that distinguishes them from other curves. Select any point P on a parabola and draw the tangent T to this point. (Intuitively, a tangent to a parabola is like a tangent to a circle—it touches the curve at one point but does not pass through it.) Lines from P to the focus and from P parallel to the axis of the parabola form equal angles (A and B) with the tangent.

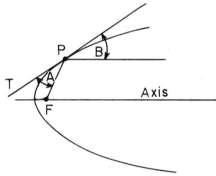

This property has applications in the *paraboloid,* a three-dimensional parabolic-shaped surface. These surfaces are used in searchlights, automobile headlights, and other reflectors. A light at the focus is reflected off the surface in rays which are parallel to the axis. This reflecting principle is used in reverse in radio telescopes and radar antennas. In radio telescopes weak radio waves from space arrive in nearly parallel rays. These rays are reflected off the parabolic-shaped receiver and concentrated at the focus. Radar antennas use the reflecting principle for

Radio telescope,
Millstone Hill Radar Observatory

both receiving and transmitting radar signals. The radar antenna pictured on page 491 was used to transmit signals to the moon and then moments later to receive the echoes which bounced back.

ELLIPSES

For 2000 years scientists thought that the planets moved in circular paths and at constant speeds. Then in 1609, the German astronomer Johannes Kepler announced his first two laws: (1) the planets move about the sun in elliptical paths; and (2) the planets do not move at a constant speed. The earth, for example, moves in an elliptical path about the sun as shown in the accompanying diagram. If it

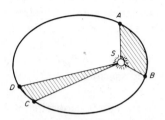

moves from *A* to *B* in 1 month, then it will also take 1 month to move from *C* to *D*, provided the areas of regions *ASB* and *CSD* are equal.

The location of the sun for the elliptical path of a planet is called the *focus*. Every ellipse has two focuses (foci).

Definition: An ellipse is the set of points in a plane such that the sum of the distances from each point to two fixed points, called *focuses,* is a constant.

The focuses for the ellipse shown here are labeled F_1 and F_2. For any point *P* which is chosen on this ellipse, the sum of the distances PF_1 and PF_2 will equal 5 centimeters. Select another point on this curve and mark off its distances to F_1 and F_2 on a piece of paper. Compare the total length to the 5-centimeter line segment below the ellipse.

The equation of the ellipse on the top of the following page is

$$9x^2 + 25y^2 = 225$$

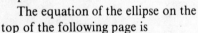

Every pair of numbers which satisfy this equation are the coordinates of a point on this curve, and conversely. As examples, the coordinates of points *A, B, C, D* are (0, 3), (⁻5, 0), (0, ⁻3), and (5, 0), and each satisfies the equation of the ellipse. The focuses are at ($^-$4, 0) and (4, 0). The sum of the distances from point G to F_1 and F_2 is 10 units. Select any point on the ellipse and use a piece of paper to mark off the sum of its distances to F_1 and F_2. Compare this length to 10 units on this coordinate system.

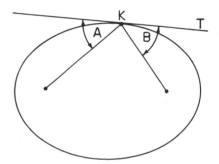

Every ellipse has a reflection property that involves its focuses. Select any point (K) on an ellipse and draw the tangent (T) to the ellipse at this point. Lines from the focuses to K form equal angles (A and B) with the tangent. This property has some interesting applications. Consider an elliptical-shaped pool table with a pocket at one focus. A ball shot from the other focus to any point on the ellipse will rebound into the pocket.

HYPERBOLAS

Hyperbolas are not seen as frequently in everyday life as parabolas and ellipses. One of the few times we see a hyperbola is by observing the shadow on a wall that is cast from a lamp with a cylindrical or a conical shade. The shadow above the lamp forms one branch of a hyperbola, and the shadow below forms the other branch. Hyperbolas also occur as the paths of comets. Comets that stay in the solar system follow elliptical paths. Those which enter the solar system and then leave again follow parabolic or hyperbolic paths.

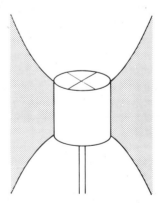

Every hyperbola has two focuses and two separate branches which satisfy the following definition.

Definition: A *hyperbola* is the set of all points in the plane such that the difference between the distances from any point to two fixed points, called *focuses,* is a constant.

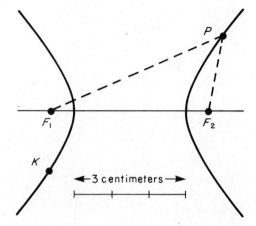

The focuses for the hyperbola in the accompanying diagram are labeled F_1 and F_2. The difference between the distances PF_1 and PF_2 is 3 centimeters. Try computing a similar difference for the point K. For any point on the hyperbola, use a piece of paper to mark off two distances to the focuses. Compare the difference between these distances to the length of the 3-centimeter line segment below the hyperbola.

The equation of the hyperbola on this coordinate system is

$$16x^2 - 9y^2 = 144$$

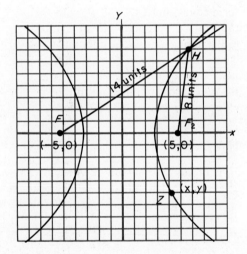

Every point on this curve has coordinates that satisfy this equation, and conversely. For example, the points whose coordinates are ($^-3$, 0) and (3, 0) are on the curve, and it is easy to see that their coordinates satisfy the equation of this hyperbola. The difference between the distances from point H to F_1 and F_2 is $14 - 8$, or 6 units. Select another point and mark off its distances to F_1 and F_2. The difference should equal 6 units on this coordinate system.

COMPUTER APPLICATIONS

The graph of an equation in X and Y can be approximated by selecting values for X and computing the corresponding values for Y. The curve shown here was obtained by selecting values for X which vary from $^-3$ to 3 in increments of .5 and computing the corresponding values of *Y*. This is carried out by the program below. The values of *Y* are rounded off by line 40 to one decimal place. Which four of the ordered pairs in the printout are not shown on the graph of the equation?

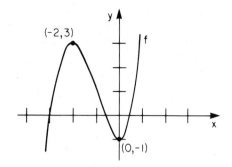

Program 8.2A

```
10   PRINT "THIS PROGRAM PRINTS ORDERED PAIRS (X,Y) FOR
        Y = X ^ 3 + 3 * X ^ 2 - 1."
20   FOR X =  - 3 TO 3 STEP .5
30   LET M = X ^ 3 + 3 * X ^ 2 - 1
40   LET Y =  INT (10 * M + .5) / 10
50   PRINT "("X", "Y") ";
60   NEXT X
70   END

     RUN

THIS PROGRAM PRINTS ORDERED PAIRS (X,Y) FOR Y = X^3 + 3*X^2 - 1.
(-3, -1) (-2.5, 2.1) (-2, 3) (-1.5, 2.4) (-1, 1) (-.5, -.4)
(0, 1) (.5, -.1) (1, 3) (1.5, 9.1) (2, 19) (2.5, 33.4) (3, 53)
```

Every equation of the form $Y = AX^2 + BY + C$, with $A \neq 0$, is the equation of a parabola. The next program prints ordered pairs that are solutions to this equation for given values of *A*, *B*, and *C*. Try this program for an equation of this form.

Program 8.2B

```
10   PRINT "THIS PROGRAM PRINTS SOLUTIONS FOR Y = A*X^2 + B*X
        + C. TYPE NUMBERS FOR A, B, AND C."
20   INPUT A,B,C
30   FOR X =  - 5 TO 5
40   LET M = A * X ^ 2 + B * X + C
50   LET Y =  INT (10 * M + .5) / 10
60   PRINT "("X", "Y") ";
70   NEXT X
80   END
```

SUPPLEMENT (Activity Book)

Activity Set 8.2 Conic Sections (Drawings by string and paper-folding)
Just for Fun: Coordinate Games

The paths of hot lava being shot into the air by this explosion of Mount Etna form mathematical curves called parabolas. Two thousand years before Galileo made his important discoveries about the parabolic paths of projectiles, the Greeks had studied this curve as well as the ellipse and hyperbola.

Exercise Set 8.2: Applications and Skills

1. The equation for the Rose of Grandi is

$$(x^2 + y^2)^3 = 72x^2y^2$$

Show that the coordinates of the five points that are labeled on this curve are solutions for the equation.

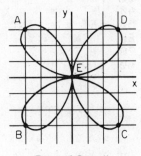

Rose of Grandi

2. Find some ordered pairs that satisfy these equations by letting x equal the integers from ⁻3 to 3. Plot these ordered pairs on a rectangular coordinate system and sketch the curves. Which of these curves are symmetric to the y axis?

a. $y = x^2 - 3$ ★

b. $y = \dfrac{x^3}{5}$

c. $y = \dfrac{1 - x^2}{2}$

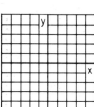
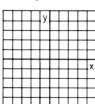
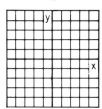

3. Here are six equations of lines. What is the slope and y intercept of each line? (*Hint:* The equation may need to be rewritten in the slope-intercept form.)

a. $x + y = 4$ ★

b. $y = 5x + 3$

c. $2x + 3y - 6 = 0$

d. $y + 3 = x$ ★

e. $y = \dfrac{x}{3}$

f. $8x - y = 4$

4. Here are the graphs of three lines. What is the slope of each line?

a. ★   b.   c.

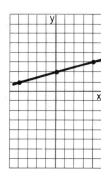
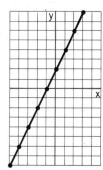

5. The graphs of three lines and their equations are shown at the top of the following page.

★   a. What is the slope of each line?

★   b. Draw another line on the first coordinate system whose slope is greater than the given line. Is there any limit to how large the slope of a line can become?

c. Draw a line on the third coordinate system whose slope is less than the slope of the given line. Is there any limit to how small the slope of a line can become? (*Hint:* The slope of a line can be negative.)

L_1 $y = 10x$ L_2 $y = x$ L_3 $y = \frac{1}{2}x$

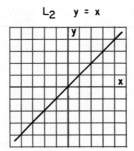
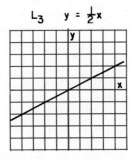

6. Leaky Boat Rentals charges 1 dollar per hour for renting a canoe. If you are a member of their club, there is no initial fee. Nonmembers who are state residents pay an initial fee of 2 dollars, and out-of-state people pay an initial fee of 5 dollars. The graphs of these rates are the lines shown in the accompanying figure.

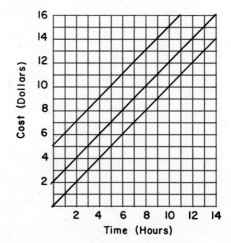

   a. Label the graphs that correspond to members; state residents who are nonmembers; and out-of-state residents. Are the slopes of these lines equal?

★ b. How much more will it cost an out-of-state resident than a club member to rent a canoe for 8 hours? Label the portion of the graph which corresponds to this difference.

   c. Answer the question in part **b** for 11 hours.

7. For each of the following word problems write an equation for the cost and then find one solution for the equation.

   a. A new push-button phone costs $36 to be installed and $11 each month. Let n be the number of months and c the total cost. Write an equation for the cost of using the phone for n months. What is c when n is 7?

★ b. A car rental company charges an initial fee of $50 plus 15 cents per mile. Let n be the number of miles and c the total cost. Write an equation for the cost of renting the car for n miles (not including gas). What is c when $n = 860$?

   c. A racquetball club charges $15 per month plus $6 for each hour of court time. Let n be the number of hours and c the total cost. Write an equation for the cost of playing racquetball for n hours. What is c when $n = 14$?

8. The solar furnace pictured here was constructed from the curved mirror of an old army searchlight. Instead of reflecting light outward from a central bulb, this mirror now reflects the sun's rays inward to a focal point, where temperatures reach as high as 3500°C. This effect accounts for the use of the word "focus," which in Latin means a hearth or burning-place.

★ a. What is the shape of the surface of the searchlight's reflector?

b. Explain why this surface enables a searchlight to be converted to a solar furnace. (*Note:* The sun's rays are nearly parallel when they reach the earth.)

Solar furnace, constructed from a searchlight

9. In 1594 Galileo discovered the laws of falling objects through his experiments at the Leaning Tower of Pisa in Italy. One of his discoveries was the formula for the distance in feet (d) traveled by an object falling from rest in a given number of seconds (t): $d = 16t^2$

a. Complete the table and plot the six points:

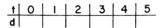

t	0	1	2	3	4	5
d						

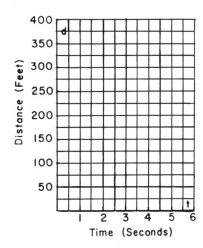

Time (Seconds)

★ b. Connect the points with a smooth curve. What would this curve look like if t were allowed to be negative as well as positive? What is the name of this curve?

10. The coordinates are given for three points on each ellipse. Try these coordinates in the equations in order to match each curve with its equation.

★   a. $\dfrac{x^2}{16} + \dfrac{y^2}{9} = 1$ b. $\dfrac{x^2}{4} + \dfrac{y^2}{25} = 1$ c. $\dfrac{x^2}{36} + y^2 = 1$

I.
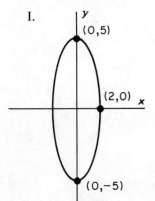

II.
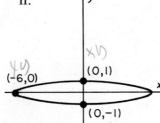

III.
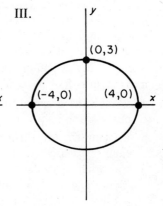

11. By knowing only its *x* and *y intercepts,* it is possible to sketch roughly an ellipse that is centered at the origin. The intercepts are the four points where the ellipse intersects the coordinate axes. The coordinates of these points will have either *x* or *y* equal to 0. Find the missing coordinates for these points and write them in the tables. Plot these points and sketch each ellipse.

★   a. $\dfrac{x^2}{25} + \dfrac{y^2}{16} = 1$ b. $\dfrac{x^2}{9} + \dfrac{y^2}{36} = 1$

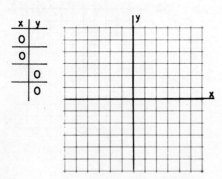

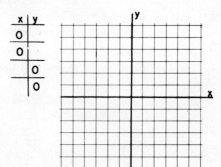

12. The coordinates are given for two points on each hyperbola. Try these coordinates in the equations in order to match each curve with its equation.

a. $\dfrac{y^2}{16} - \dfrac{x^2}{9} = 1$ b. $xy = 1$ c. $\dfrac{x^2}{4} - \dfrac{y^2}{9} = 1$

I.
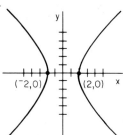

II.
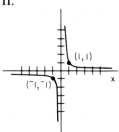

III.
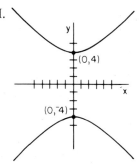

13. Use the definitions of an ellipse and a hyperbola to determine the distance d in each figure.

a.
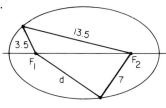

★ b.
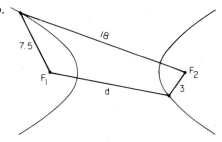

Exercise Set 8.2: Problem Solving

1. *Slopes on Geoboards:* A rectangular geoboard has rows and columns of pegs or nails. Line segments can be formed by stretching rubber bands between the pegs. The line segment on this geoboard has a run of 5 and a rise of 4. Its slope is $\frac{4}{5}$. How many line segments with different nonnegative slopes can be formed on this geoboard?

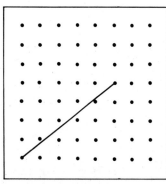

8 by 8 Geoboard

★ a. *Understanding the Problem:*
First notice that it is only necessary to consider the line segments
whose left endpoints are at the lower left corner
of the geoboard. Why? Explain why we do
not want to count line segments that are
parallel to the left edge of the geoboard.

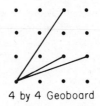

4 by 4 Geoboard

b. *Devising a Plan:* The slope of each line
segment can be determined from the coordinates of its right endpoint. What is the slope
of the line shown here? Explain why it is only
necessary to count the points whose coordinates are relatively prime.

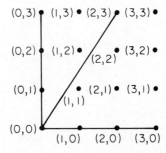

c. *Carrying Out the Plan:* Label the coordinates
of each point on the 8 by 8 geoboard and
count the points whose coordinates are
relatively prime. [*Note:* The coordinates of (1,
0) are relatively prime, but those of (2, 0), (3,
0), etc., are not relatively prime. Why?]

★ d. *Looking Back:* For any size geoboard the number of line segments with different
nonnegative slopes is even. Here are examples of the first four geoboards.
Explain why this number is always even. [*Hint:* Consider points (*a, b*) and (*b, a*).]

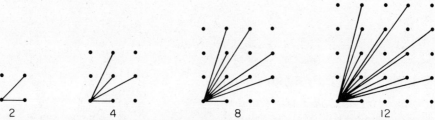

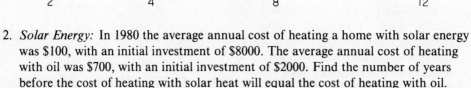

2   4   8   12

2. *Solar Energy:* In 1980 the average annual cost of heating a home with solar energy
was $100, with an initial investment of $8000. The average annual cost of heating
with oil was $700, with an initial investment of $2000. Find the number of years
before the cost of heating with solar heat will equal the cost of heating with oil.

a. *Understanding the Problem:* Let's look at the total costs for the first few years.
The first year of heating with oil costs $2700 and heating with solar heat costs
$8100. What is the cost of each system for the first 3 years?

b. *Devising a Plan:* One approach to solving this problem is to write equations for
the cost of oil heat and solar heat and graph these equations. The equation for
heating with oil is $y = 700x + 2000$, where y is the total cost for the first x years.
Use the variables x and y to write an equation for heating with solar heat.

c. *Carrying Out the Plan:* The graph of the equation for heating with oil is shown here. Graph the equation for heating with solar heat. Use these graphs to determine the number of years before the costs of heating with oil and solar heat are equal. What is this cost?

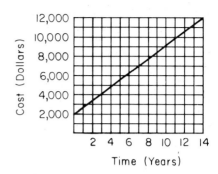

d. *Looking Back:* Another approach to this problem is to list the total year-by-year costs of using oil and solar heat.

Year	1	2	3	4	5	6	7	8	9	10	11	12
Oil	2700	3400	4100	4800	5500	6200	6900	7600	8300	9000	9700	10,400
Solar	8100	8200	8300	8400	8500	8600	8700	8800	8900	9000	9100	9200

How much more does it cost to heat with oil than with solar heat for the first 12 years? Mark the portion of the graph that shows this difference.

Exercise Set 8.2: Computers

1. Write a program to print some of the ordered pairs (X, Y) that are solutions to

$$Y = MX + B$$

Let X vary from -5 to 5. Try your program for $Y = 4X + 5$. What is the printout?

★ 2. What is the printout for Program 8.2B (page 495) for the following equation. Sketch the graph of this equation.

$$Y = 2X^2 - 5X - 6$$

3. Write a program to print some of the ordered pairs $(X,\ Y)$ that are solutions to

$$Y = X^4 - 3X^3 - 4X^2 + 12X$$

Let X vary from -3 to 4 in steps of .5. Sketch the graph of this equation.

★ 4. What two numbers with a sum of 20 have the maximum product? Write a program to solve this problem. [*Hint:* If X is one number, then the other is $20 - X$ and their product is $X(20 - X)$.]

Nine-person star with five people making approaches for slots, over California

8.3
FUNCTIONS AND SEQUENCES

FUNCTIONS

The distance a sky diver falls is related to the time that has elapsed during the jump. By the end of the first second the sky diver has fallen 16 feet, and after 2 seconds the distance is 62 feet. The distances for the first 30 seconds are shown in the table at the right. This table matches each time from 1 to 30 seconds with a unique (one and only one) distance. Since the distance fallen depends on time, distance is said to be a *function of* time.

> **Definition:** A *function* is two sets and an assignment of each element in the first set to a unique element of the second set.

Distance Fallen in Free-Fall
Stable Spread Position

Seconds	Distance	Seconds	Distance
1	16	16	2179
2	62	17	2353
3	138	18	2527
4	242	19	2701
5	366	20	2875
6	504	21	3049
7	652	22	3223
8	808	23	3397
9	971	24	3571
10	1138	25	3745
11	1309	26	3919
12	1483	27	4093
13	1657	28	4267
14	1831	29	4441
15	2005	30	4615

In the sky diving example each element in the first set, the set of whole numbers from 1 to 30, is assigned to one and only one element in the second set, the set of distances. This table of times and distances defines a function from one set of numbers to another.

Examples of Functions   Here are some examples of functions from the earlier chapters of this text. For each function there are two sets and a rule for assigning the element of the first set to those of the second set. One arrow is missing in each example. Draw this arrow to connect the element of the first set to the appropriate element in the second set.

Example 1   For each triangle there is one and only one area.

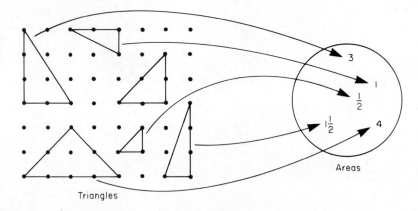

Triangles   Areas

Example 2   For each Fahrenheit temperature there is one and only one Celsius temperature.

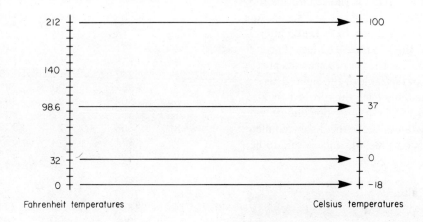

Fahrenheit temperatures   Celsius temperatures

Example 3 For each pair of numbers there is one and only one sum.

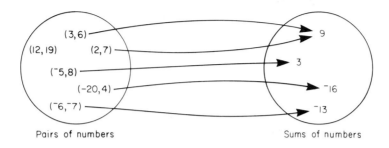

Pairs of numbers   Sums of numbers

Example 4 For each polygon there is one and only one perimeter.

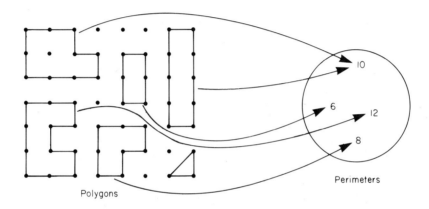

Polygons   Perimeters

Example 5 For every whole number greater than 1 there is a unique set of prime factors.

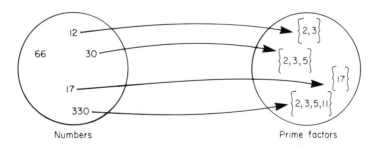

Numbers   Prime factors

The first set of elements in a function is called the *domain* and the second set is called the *range* (or *codomain*). The elements in the domain and range are not always numbers, as shown by the preceding examples. In Example 1 the elements in the domain are triangles; in Example 4 the elements in the domain are polygons; and in Example 5 the elements in the range are sets of numbers.

Not all assignments between sets are functions. Suppose we assign each positive number to its positive and negative square roots. Then each number corresponds to two numbers, and this violates one of the conditions of a function because each number is not assigned to one and only one number.

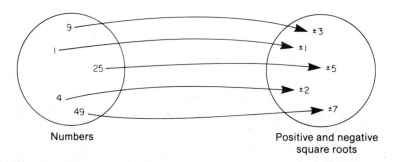

Numbers

Positive and negative square roots

FUNCTIONS FROM EQUATIONS

For every function there are two sets and some rule for assigning the elements in the domain to the elements in the range. In the preceding examples we assigned triangles to their areas; polygons to their perimeters; numbers to the prime factors; etc.

An equation is often used as the rule for assigning one set of numbers to another. The equation

$$y = 3x + 2$$

assigns each value of x with the resulting value of $3x + 2$. Several values of x and the corresponding values of $3x + 2$ are shown in this table. The replacements for x are the domain elements, and the resulting values of y are the range elements. Since each value of y depends on a value of x, y is called the *dependent variable* and x is called the *independent variable*.

Domain x	Range y
0	2
1	5
2	8
3	11
4	14
5	17
6	20
7	23

$$y = 3x + 2$$

The domain of the function defined by $y = 3x + 2$ is all real numbers, unless otherwise stated, since any real number may be used for x. Sometimes, however, when a function is defined by an equation, there will be some numbers that cannot be elements of the domain. Here are three examples in which the domain is restricted. In the first two examples x should not be replaced by a number that will make the denominator zero. In the third example x should not be a negative number, since the square root of a negative number is not a real number.

$$y = \frac{1}{x^2 - 1}$$

Domain: All real numbers except 1 and ⁻1.

$$y = \frac{1}{x - 3}$$

Domain: All real numbers except 3.

$$y = \sqrt{x}$$

Domain: All nonnegative real numbers.

CHAPTER 8 ALGEBRA AND FUNCTIONS

GRAPHS OF FUNCTIONS

Graphs are the most common method of visualizing functions. This graph shows the height of a ball as a function of time. The function is defined by

$$d = 128t - 16t^2$$

As time varies from 0 to 8 seconds, the ball rises to a height of 256 feet and then falls. Since distance is a function of time, the independent variable is t and the dependent variable is d. The domain of this function is all real numbers t such that $0 \leq t \leq 8$. What is the range of this function?

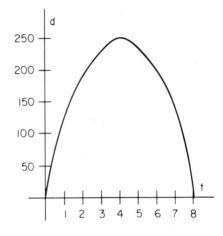

Vertical Line Test   Not all equations define a function. Once you know the graph of an equation, there is an easy test to determine if the equation defines a function. Simply draw vertical lines (lines perpendicular to the x axis), and if a line intersects the curve more than once, the curve is not the graph of a function. Here are three equations that do not define functions:

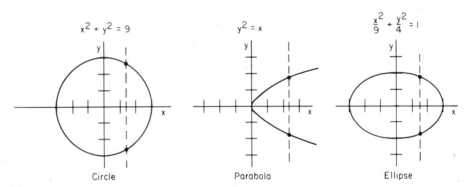

Circle   Parabola   Ellipse

For each of these equations the dotted line intersects its graph in two points. This means that for a value of x there are two corresponding values of y. However, for a function each value of x should correspond to one and only one value of y.

LINEAR FUNCTIONS

The distance to an approaching thunderstorm can be determined by counting the seconds between a flash of lightning and the resulting sound of thunder. For every 3 seconds, sound travels approximately 1 kilometer. If you can count up to 6 seconds before hearing the thunder, the storm is approximately 2 kilometers away. Distance in this example is a function of time. A few times and the corresponding distances are shown in the following table. The equation for this function is $y = \frac{1}{3}x$ and its graph is a straight line.

Time	Distance
3	1
6	2
9	3
12	4

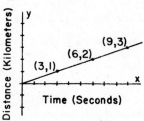

Sound travels faster in water than in air. In water it travels about 1.5 kilometers per second. In 2 seconds it travels 3 kilometers; in 3 seconds it travels 4.5 kilometers; etc. This is another example in which distance is a function of time. The graph of this function is shown at the right and its equation is

$$y = 1.5x$$

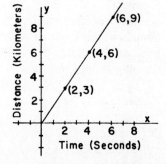

Notice that the slope of this line is greater than the slope of the line for the speed of sound in air because the speed of sound in water is greater.

Both of the preceding equations, $y = \frac{1}{3}x$ and $y = 1.5x$, are special cases of the slope-intercept form of a straight line:

$$y = mx + b$$

where m is the slope of the line and b is the y intercept. Since any vertical line will intersect the graph of $y = mx + b$ in only one point, any such equation defines a function. These functions are called *linear functions*. The only straight line that is not the graph of a function is a line perpendicular to the x axis. Use the vertical line test to show that $x = 3$ is not a function.

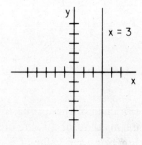

EXPONENTIAL FUNCTIONS

The graphs of growth rates for sizes, weights, populations, etc., of living things have similar shapes. If growth can be measured in the early and late stages of the life of an organ (heart, liver, brain, etc.), organism (plant, animal, insect, etc.), or population (bacteria, animal, insect, etc.), then the graph of the growth rate has three phases: the *lag phase,* the *exponential phase,* and the *stationary phase.* These phases are shown on the accompanying graph of a 5-day experiment in the growth of mold. This is the typical *sigmoid curve,* or *S-shaped curve,* of growth. Two more illustrations of S-shaped curves are contained in the graphs that follow:

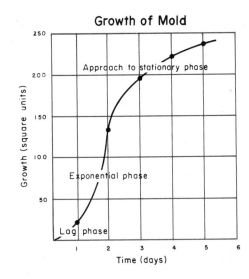

Growth of Mold

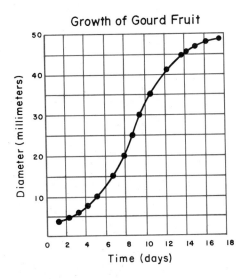

Growth of Gourd Fruit

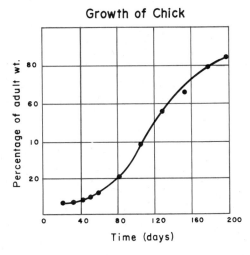

Growth of Chick

Theoretically, during the exponential phase of growth, 1 cell produces 2 cells, then 4, 8, 16, and so on, in a geometric sequence. This is represented by the exponential equation $y = 2^x$. The function this equation defines is called an *exponential function.* When x takes on the values 0, 1, 2, 3, 4, . . . , from the domain, y equals 1, 2, 4, 8, 16, . . . , in the range. When x is negative, the

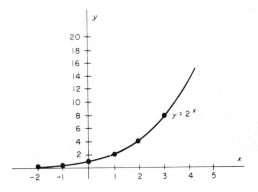

values of y are less than 1. For example, if $x = {}^-3$, $y = 2^{-3} = \frac{1}{2^3} = \frac{1}{8}$. The following table has a few values of x and the corresponding values of y.

x	$^-5$	$^-4$	$^-3$	$^-2$	$^-1$	0	1	2	3	4	5
y	$\frac{1}{32}$	$\frac{1}{16}$	$\frac{1}{8}$	$\frac{1}{4}$	$\frac{1}{2}$	1	2	4	8	16	32

The domain of this function is all real numbers, and the range is all positive real numbers. The value of 2^x for any x (fraction or irrational number) can be approximated by connecting the points shown in the graph on the preceding page to form a smooth curve.

In general, the equation

$$y = k^x$$

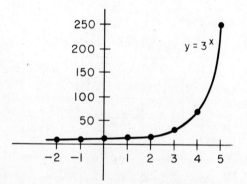

where k is any constant positive real number, defines an exponential function with base k. The domain of these functions is all real numbers, and the range is all positive real numbers. The graph of the exponential function with base 3 is shown at the right. A few integer values of x and the corresponding values of y are shown in this table:

x	$^-5$	$^-4$	$^-3$	$^-2$	$^-1$	0	1	2	3	4	5
y	$\frac{1}{243}$	$\frac{1}{81}$	$\frac{1}{27}$	$\frac{1}{9}$	$\frac{1}{3}$	1	3	9	27	81	243

There is one number in particular for the base k in the equation $y = k^x$, which occurs so often in the physical sciences that it is called the "natural base" and denoted by the letter e. This is an irrational number, which approximated to 5 decimal places is 2.71828. Since e is greater than 2 and less than 3, the graph of $y = e^x$ is between the graphs of $y = 2^x$ and $y = 3^x$. The equation $y = e^x$ is so important, its function is called the *exponential function*.

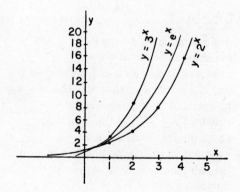

SEQUENCES AND FINITE DIFFERENCES

The five chords in this circle divide the circle into 16 regions. What is the maximum number of regions that can be formed by 50 chords in a circle?

To solve this problem let's begin by simplifying the question for smaller numbers of chords. Here are the maximum numbers of regions that can be formed in circles having 1, 2, 3, 4, and 5 chords (see figures):

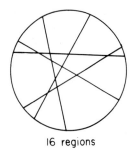

16 regions

2   4   7   11   16

Use inductive reasoning to predict the next number in this sequence.

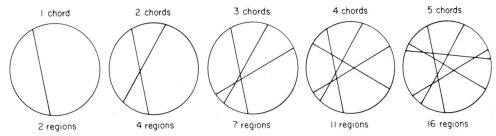

1 chord	2 chords	3 chords	4 chords	5 chords
2 regions	4 regions	7 regions	11 regions	16 regions

Finite Differences To make a conjecture about the number of regions for six chords, we can compute the differences between successive pairs of numbers in the original sequence. This produces a new sequence called *first differences.* By computing the differences again we get the *second differences,* which in this example are all 1's. If we assume that the next number in this column is a 1, we can work our way back from the column of second differences to the original sequence by filling in the boxes at the bottom of each column of numbers. The next number after 5 is 6, and, therefore, the next number after 16 is 22. Try drawing a circle with 6 chords that has 22 regions.

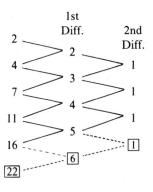

This approach to finding a next number in a sequence is called the *method of finite differences.* In general, the differences are computed until a sequence of constant terms is obtained.

A sequence of numbers is a special type of function whose domain is the set of positive whole numbers and whose range is the terms of the sequence. The sequence from the preceding page is shown here with 1 assigned to the first term of the sequence; 2 assigned to the second term; 3 assigned to the third term; etc. We would like to predict the fiftieth term without going through all the terms in between.

The method of finite differences that was used to find a next term in this sequence can also be used to find an equation for the function. When the second differences are all constant, as in this example, the equation for the function has the form

$$y = ax^2 + bx + c$$

The following tables contain a few values from the domain ($x = 1, 2, 3$, etc.) and the corresponding values from the range of this function.

Domain	Range
1	→ 2
2	→ 4
3	→ 7
4	→ 11
5	→ 16
6	→ 22
⋮	
50	→ ?

Domain	Range	1st Diff.	2nd Diff.
1	②		
		②	
2	4		①
		3	
3	7		1
		4	
4	11		1
		5	
5	16		
⋮	⋮	⋮	

Domain x	Range $ax^2 + bx + c$	1st Diff.	2nd Diff.
1	$a + b + c$		
		$3a + b$	
2	$4a + 2b + c$		2a
		$5a + b$	
3	$9a + 3b + c$		$2a$
		$7a + b$	
4	$16a + 4b + c$		$2a$
		$9a + b$	
5	$25a + 5b + c$		
⋮	⋮	⋮	

The circled algebraic terms in the table on the right correspond to the circled numbers in the table on the left. That is,

$$2a = 1 \qquad 3a + b = 2 \qquad a + b + c = 2$$

Solving these equations for a, b, and c,

$$2a = 1 \qquad\qquad 3\left(\tfrac{1}{2}\right) + b = 2 \qquad\qquad \tfrac{1}{2} + \tfrac{1}{2} + c = 2$$

$$a = \tfrac{1}{2} \qquad\qquad \tfrac{3}{2} + b = 2 \qquad\qquad 1 + c = 2$$

$$b = \tfrac{1}{2} \qquad\qquad c = 1$$

and substituting the values into the equation $ax^2 + bx + c$, we get

$$y = \frac{x^2}{2} + \frac{x}{2} + 1$$

as the equation of the function. Check this equation by letting $x = 1, 2, 3, 4$, and 5 to see if you get the first few terms of the sequence 2, 4, 7, 11, 16, ... In particular, when $x = 50$, we get 1276 as the number of regions for 50 chords in a circle.*

$$\frac{50^2}{2} + \frac{50}{2} + 1 = \frac{2500}{2} + \frac{50}{2} + 1 = 1276$$

Pyramid of Cannon Balls   Let's solve another problem by using the method of finite differences. One method of stacking cannonballs is to form a pyramid with a square base. The figure shown here has a 6 by 6 square base, and the pyramid has a total of 91 cannonballs. How many cannonballs would there be in a pyramid with a 30 by 30 base?

91 cannon balls

We can simplify this problem by looking at pyramids with smaller bases. The number of cannonballs in each layer of a pile is a square number. This observation enables us to get the number of cannonballs for the first few piles. For example, in the fourth pile there are $1 + 4 + 9 + 16 = 30$ cannonballs.

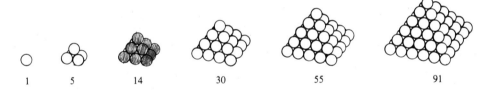

| 1 | 5 | 14 | 30 | 55 | 91 |

Let's see if the method of finite differences can be used to obtain the number of cannonballs in the seventh pile. The process of computing differences does not produce constants until the column of third differences.

Note: We are using inductive reasoning to draw this conclusion, since we have assumed that the numbers in the column of second differences on page 514 will continue to be 1's.

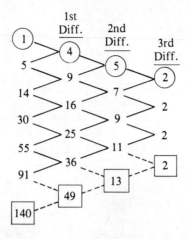

We can now work our way back from the column of third differences to the original sequence. The next number in this sequence is 140. Show that this is the number of cannonballs in the seventh pile by computing $1 + 4 + 9 + 16 + 25 + 36 + 49$.

The first few numbers in the cannonball sequence are shown in the range of the function at the right. If the numbers in the column of third differences in the above table continue to be equal, this function has an equation of the form

$$y = ax^3 + bx^2 + cx + d$$

Domain		Range
1	$\longrightarrow$	1
2	$\longrightarrow$	5
3	$\longrightarrow$	14
4	$\longrightarrow$	30
5	$\longrightarrow$	55
6	$\longrightarrow$	91
.		.
.		.
.		.
30	$\longrightarrow$	?

Let's use the method of finite differences to find this equation. The following table contains the first few positive whole numbers from the domain of this function and the corresponding values from the range:

Domain	Range			
x	$ax^3 + bx^2 + cx + d$	1st Diff.	2nd Diff.	3rd Diff.

1	$\boxed{a + \ b + \ c + d}$			
		$7a + \ 3b + c$		
2	$8a + \ 4b + 2c + d$		$\boxed{12a + 2b}$	
		$19a + \ 5b + c$		$\boxed{6a}$
3	$27a + \ 9b + 3c + d$		$18a + 2b$	
		$37a + \ 7b + c$		$6a$
4	$64a + 16b + 4c + d$		$24a + 2b$	
		$61a + \ 9b + c$		$6a$
5	$125a + 25b + 5c + d$		$30a + 2b$	
		$91a + 11b + c$		
6	$216a + 36b + 6c + d$			

Setting the four circled algebraic terms in this table equal to the corresponding circled numbers at the top of page 516, we get the following equations:

$$6a = 2 \qquad 12a + 2b = 5 \qquad 7a + 3b + c = 4 \qquad a + b + c + d = 1$$

Solving these equations for a, b, c, and d,

$$6a = 2 \qquad 12\left(\tfrac{1}{3}\right) + 2b = 5 \qquad 7\left(\tfrac{1}{3}\right) + 3\left(\tfrac{1}{3}\right) + c = 4 \qquad \tfrac{1}{3} + \tfrac{1}{2} + \tfrac{1}{6} + d = 1$$

$$a = \tfrac{1}{3} \qquad 4 + 2b = 5 \qquad 3\tfrac{5}{6} + c = 4 \qquad 1 + d = 1$$

$$2b = 1 \qquad c = \tfrac{1}{6} \qquad d = 0$$

$$b = \tfrac{1}{2}$$

and substituting the values into the equation $y = ax^3 + bx^2 + cx + d$, we get

$$y = \frac{1}{3}x^3 + \frac{1}{2}x^2 + \frac{1}{6}x$$

Letting $x = 30$, we see that the number of cannonballs in the thirtieth pyramid is 9455:*

$$\frac{30^3}{3} + \frac{30^2}{2} + \frac{30}{6} = 9000 + 450 + 5 = 9455$$

*We are using inductive reasoning because we assumed that the numbers in the column of third differences on page 516 will continue to be 2's.

Pattern of Blocks   Find the pattern in these blocks and predict the next number in this sequence:

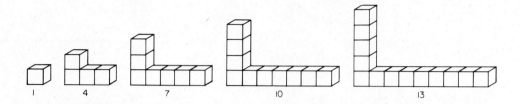

This is an arithmetic sequence (see page 6) with a common difference of 3. The next number in this sequence is 16. What is the 500th number in this sequence? To answer this question we will find the equation for the function whose range is this sequence of numbers. Since the first differences are all constant, this function has a linear equation of the form

$$y = ax + b$$

Domain		Range
1	$\longrightarrow$	1
2	$\longrightarrow$	4
3	$\longrightarrow$	7
4	$\longrightarrow$	10
5	$\longrightarrow$	13
6	$\longrightarrow$	16
.		
.		
.		
500	$\longrightarrow$	?

To find the equation for this function we will use the method of finite differences.

Domain	Range	1st Diff.
1	①	
		③
2	4	
		3
3	7	
		3
4	10	
		3
5	13	
.	.	
.	.	
.	.	

Domain	Range	1st Diff.
x	ax + b	
1	(a + b)	
		ⓐ
2	2a + b	
		a
3	3a + b	
		a
4	4a + b	
		a
5	5a + b	
.	.	
.	.	
.	.	

Setting the two circled algebraic terms in the table equal to the corresponding numbers in the table on the left, we get

$$a = 3 \qquad a + b = 1$$

The solutions to these equations are $a = 3$ and $b = {}^-2$. Therefore, the equation for the function is $y = 3x - 2$, and the 500th term in the sequence is 1498:

$$3(500) - 2 = 1500 - 2 = 1498$$

The method of finite differences will not always produce a sequence of constants. For example, the table at the right contains some of the values for the exponential function whose equation is

$$y = 2^x$$

See what happens when the method of finite differences is used on the y values in this table.

x	y
1	2
2	4
3	8
4	16
5	32
6	64
7	128

COMPUTER APPLICATIONS

The routine arithmetic that occurs in the method of finite differences can be carried out by the computer. Let's look at a program that does this. Line 30 computes the first differences: *G, H, I, J, K.* If these differences are equal, line 50 sends the computer to line 130. If these differences are not equal, line 60 computes the second differences. What happens if these differences are equal? What happens if they are not equal? The printout for the sequence 3, 11, 33, 75, 143, 243 shows that the third differences are equal.

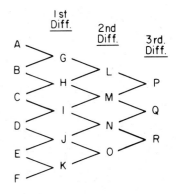

Program 8.3A

```
10   INPUT A, B, C, D, E, F
20   PRINT "THE ORIGINAL SEQUENCE IS "A", "B", "C", "D", "E", "F"."
30   LET G = B - A:H = C - B:I = D - C:J = E - D:K = F - E
40   PRINT "THE FIRST DIFFERENCES ARE "G", "H", "I", "J", "K"."
50   IF G = H AND H = I AND I = J AND J = K THEN  GOTO 130
60   LET L = H - G:M = I - H:N = J - I:O = K - J
70   PRINT "THE SECOND DIFFERENCES ARE "L", "M", "N", "O"."
```

```
80    IF L = M AND M = N AND N = O THEN   GOTO 130
90    LET P = M - L:Q = N - M:R = O - N
100   PRINT "THE THIRD DIFFERENCES ARE "P", "Q", "R"."
110   IF P = Q AND Q = R THEN   GOTO 130
120   GOTO 140
130   PRINT "THESE DIFFERENCES ARE EQUAL."
140   END

RUN

?3, 11, 33, 75, 143, 243
THE ORIGINAL SEQUENCE IS 3, 11, 33, 75, 143, 243.
THE FIRST DIFFERENCES ARE 8, 22, 42, 68, 100.
THE SECOND DIFFERENCES ARE 14, 20, 26, 32.
THE THIRD DIFFERENCES ARE 6, 6, 6.
THESE DIFFERENCES ARE EQUAL.
```

Now let's revise this program so that if the differences are equal, it will print out the next number in the original sequence. If the first differences are equal ($G = H = I = J = K$), then adding K to F will produce the next number in the original sequence (see line 130). If the second differences are equal ($L = M = N = O$), then adding $O + K$ to F will produce the next number in the original sequence (see line 134). Why? What number must be added to F if the differences are not equal until the sequence of third differences? Try this program to obtain the next number in the sequence: 3, 11, 33, 75, 143, 243:

Program 8.3B

```
10    INPUT A, B, C, D, E, F
20    PRINT "THE ORIGINAL SEQUENCE IS "A", "B", "C", "D", "E", "F"."
30    LET G = B - A:H = C - B:I = D - C:J = E - D:K = F - E
40    PRINT "THE FIRST DIFFERENCES ARE "G", "H", "I", "J", "K"."
50    IF G = H AND H = I AND I = J AND J = K THEN   GOTO 130
60    LET L = H - G:M = I - H:N = J - I:O = K - J
70    PRINT "THE SECOND DIFFERENCES ARE "L", "M", "N", "O"."
80    IF L = M AND M = N AND N = O THEN   GOTO 134
90    LET P = M - L:Q = N - M:R = O - N
100   PRINT "THE THIRD DIFFERENCES ARE "P", "Q", "R"."
110   IF P = Q AND Q = R THEN   GOTO 138
120   GOTO 140
130   PRINT "THESE DIFFERENCES ARE EQUAL. THE NEXT NUMBER IN THE
      ORIGINAL SEQUENCE IS "K + F"."
132   GOTO 140
134   PRINT "THESE DIFFERENCES ARE EQUAL. THE NEXT NUMBER IN THE
      ORIGINAL SEQUENCE IS "O + K + F"."
136   GOTO 140
138   PRINT "THESE DIFFERENCES ARE EQUAL. THE NEXT NUMBER IN THE
      ORIGINAL SEQUENCE IS "R + O + K + F"."
140   END
```

Activity Set 8.3 Patterns and Finite Differences
Just For Fun: Curves from Line Designs

Exercise Set 8.3: Applications and Skills

1. Experiments with rats at the University of London have tested the conjecture that the motivation level for learning a task is a function of the difficulty level of the task.*

   a. What information does the graph below show about the level of motivation and the difficulty level of the task?

   ★ b. What type of function has this graph?

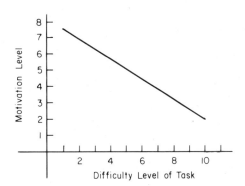

2. Electrical impulses that accompany the beat of the heart are recorded by an electrocardiograph. The electrocardiograph (ECG) measures electrical changes in millivolts (1/1000 of a volt). The following graph shows the changes in millivolts (mV) as a function of time for a measure of a normal heartbeat.

   ★ a. How much time was required for this graph if each small space on the horizontal axis represents 0.04 second?

*P. L. Broadhurst, "Emotionality and the Yerkes-Dodson Law," *Journal of Experimental Psychology,* 54 (1957), 345–352.

★ b. The tall rectangular part of the graph was caused by a 10-millivolt signal from the ECG machine. This is called a calibration pulse. What was the length of the time for this signal?

c. This graph shows 10 heartbeats, or pulses. Approximately how much time is there between each pulse (from the end of one pulse to the end of the next pulse)? At this rate how many pulses will there be per minute?

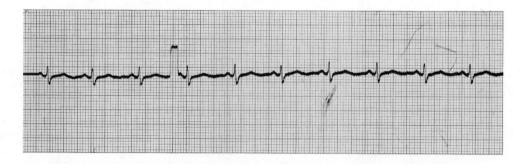

3. The cost of first-class postage is a function of weight. In 1984 the first ounce cost 20 cents and each additional ounce or fraction thereof cost 17 cents. For example, 1.4 ounces cost 37 cents and 2.7 ounces cost 54 cents.

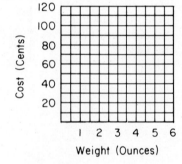

★ a. Find the costs for these weights:

   3.2 ounces   4.2 ounces   4.8 ounces

★ b. This postal rate holds for weights less than 13 ounces. Therefore, the domain of this function is all positive numbers less than 13. There are 13 numbers in the range. List these numbers.

c. Graph this function for weights less than or equal to 4 ounces.

4. Use the functions in Examples 1 through 5 on pages 506 and 507 to answer the following questions:
   a. A function is called *many-to-one* if there are two or more numbers in the domain assigned to the same number in the range. Which functions are many-to-one?

★ b. A function is called *one-to-one* if for each element in the range there is only one element in the domain assigned to it. Which functions are one-to-one?

c. (True or False) All linear functions are one-to-one.

5. Use the vertical line test to determine which of the following curves are not the graphs of functions:

a.   ★ b.   c.   ★ d.

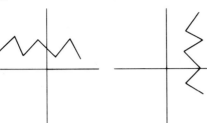 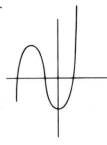

★ e.   f.   g.   h.

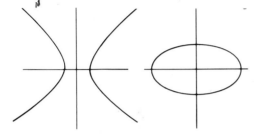 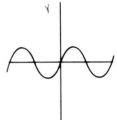

6. In 1984, Great Britain's pound was worth about $1.50 in U.S. currency. That is, 1£ = $1.50, 2£ = $3.00, etc. This is a linear function whose graph is shown here.

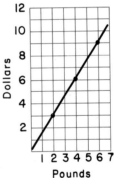

★ a. Use this graph to find the approximate value in dollars of these amounts in British currency:

£1.75   and   £2.50

★ b. Use the graph to find the approximate value in pounds of these amounts in U.S. currency:

$5   and $8

c. Letting P represent the independent variable (number of English pounds) and D represent the dependent variable (number of U.S. dollars), write an equation for this graph.

7. The *Escherichia coli* is a bacterium which under ideal conditions doubles every 20 minutes. Beginning with a single bacterium, there will be 8 at the end of the first hour. Compute the number of these bacteria for each of the hours from 2 to 6 and graph the results. Connect these points with a smooth curve. What is the equation of this curve?

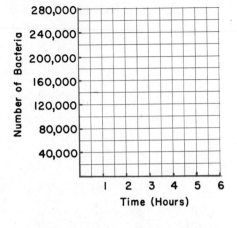

8. Graph the following coordinates, where the first number represents time in weeks and the second number represents weight in grams of a corn plant. Connect these points with an S-shaped curve. Mark the approximate locations of the lag, exponential, and stationary phases of this curve.

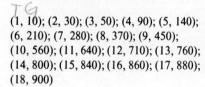

(1, 10); (2, 30); (3, 50); (4, 90); (5, 140); (6, 210); (7, 280); (8, 370); (9, 450); (10, 560); (11, 640); (12, 710); (13, 760); (14, 800); (15, 840); (16, 860); (17, 880); (18, 900)

★   a. How many weeks was this plant in its exponential growth phase?

b. How much weight did it increase during the exponential phase?

9. This graph shows the population growth in the United States since 1660:

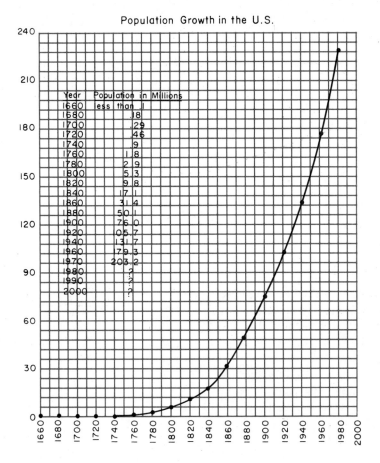

Population Growth in the U.S.

Year	Population in Millions
1660	less than 1
1680	.18
1700	.29
1720	.46
1740	.9
1760	1.8
1780	2.9
1800	5.3
1820	9.8
1840	17.1
1860	31.4
1880	50.1
1900	76.0
1920	105.7
1940	131.7
1960	179.3
1970	203.2
1980	?
1990	?
2000	?

★ a. Assuming that we continue in the exponential phase of this curve, predict the population in the year 2000.

b. Predict the population in the year 2050 if the exponential phase continues that long.

c. If the 1980s mark the beginning of the stationary phase, estimate the upper limit of population for this country.

10. The Environmental Protection Agency (EPA) has published the following results concerning fuel economy and safety of heavy-duty trucks:

★ a. In a crash, the force of a truck (its kinetic energy) is a function of the truck's mass and its speed. This force is computed by multiplying the mass by a speed factor. Use this table to find an equation for the speed factor of the force y as a function of the truck's speed x.

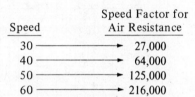

Speed	Speed Factor for Force
30	900
40	1600
50	2500
60	3600

★ b. When the speed is doubled from 30 to 60 miles per hour (mph), how many times greater is the force of the truck?

c. The air resistance to a moving truck is also a function of its mass times a speed factor, but in this case the speed factor is greater than that for force. Use this table to find an equation for the speed factor of the air resistance y as a function of speed x.

Speed	Speed Factor for Air Resistance
30	27,000
40	64,000
50	125,000
60	216,000

d. As the speed is doubled from 30 to 60 mph, how many times greater is the air resistance to the truck?

11. In the early 1970s a group of MIT scientists created a computer model acclaimed as able to forecast the world population growth. This World Model takes into account such factors as pollution, food supplies, and natural resources.*

★ a. What is this graph's domain? What part of this domain represents prediction of the future?

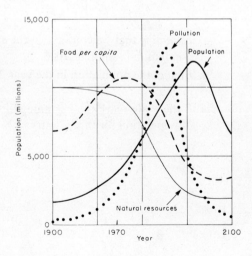

*Donnella H. Meadows et al., *Limits to Growth* (New York: Universe Press, 1972).

b. In 1900 all but one of these curves were increasing. Which curves are increasing and which are decreasing in the year 2000?

★ c. What are the approximate intervals of years during the exponential phases when the population and pollution graphs are increasing? What is happening to the food supply during these intervals?

12. *Calculator Exercise:* The exponential equation $y = e^x$ and variations of this equation occur frequently in analyzing rates of growth and decay. On a calculator that has a key for the exponential function, e^x can be found by two steps. The second step here shows e^2 to 9 decimal places.

Steps	Displays
1. Enter 2	2.
2. $\boxed{e^x}$	$\boxed{7.389056099}$

★ a. The number e is irrational. Use your calculator to find the first few decimal places in e by evaluating e^1 (e to the first power).

★ b. For negative exponents, e^x is less than 1. What are the first few decimal places in e^{-3}? (*Hint:* $e^{-3} = 1/e^3$.)

c. Find e^x for each of the values of x that are marked on the x axis. Round off these numbers to the nearest tenth. Plot the pairs (x, y) and connect these points to form the graph of $y = e^x$.

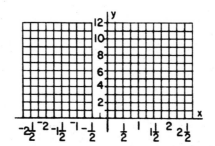

Exercise Set 8.3: Problem Solving

1. *Squares Problem:* How many squares can be formed on a chessboard by using the lines on the board?

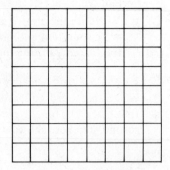

★ a. *Understanding the Problem:* There are sixty-four 1 by 1 squares and one large 8 by 8 square (the board). How many 2 by 2 squares can be formed using just the top two rows of squares?

b. *Devising a Plan:* Simplify the problem by determining the number of squares for each of the following squares. (Three have been done for you.) The number of squares is a function of the size of the large square. Use inductive reasoning and the method of finite differences to obtain the equation for this function.

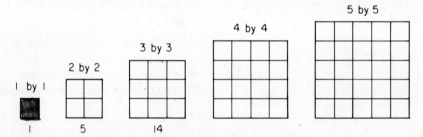

c. *Carrying Out the Plan:* Use the method of finite differences to determine the number of squares that can be formed in a 6 by 6 square. Use the equation to compute the total number of squares on the chessboard.

d. *Looking Back:* Check your equation by using it to compute the number of squares for the figures in part **b.** How many squares can be formed in a 100 by 100 square?

2. Use the method of finite differences to solve these problems:

a. A polygon with 100 sides has how many diagonals?

b. For five points on a circle there are 10 chords that have these points as endpoints. How many chords are there for 100 points on a circle?

★ c. This pyramid of cannonballs has a triangular base. Each side of the base has 6 cannon-balls, and the total number of balls in the pyramid is 56. How many cannonballs are in a triangular pyramid with 100 balls on each side of the base? (*Hint:* The number of cannonballs in each layer of the pyramid is a triangular number.)

Triangular Pyramid
of Cannonballs

Exercise Set 8.3: Computers

1. What is the printout for Program 8.3B (page 520) if the first six numbers of this sequence are typed for the input?

$$4, 15, 48, 115, 228, 399$$

★ 2. What are the eighth, ninth, and tenth numbers in the sequence in Exercise 1. Explain how Program 8.3B can be used to find these numbers.

3. Program 8.3B (page 520) requires that the first six numbers of a sequence be typed for input. Revise this program so that it can be used when only the first five numbers of a sequence are known.

4. Write a computer program to print the ordered pairs (X, Y) that are solutions to $Y = 2^X$ with X increasing from $^-3$ to 3 in steps of .5. Round off each value of Y to the nearest tenth.
   a. Use this program to sketch the graph of $Y = 2^X$.
   b. Revise this program to print the solutions for $Y = 2^{-X}$.
★ c. What can be said about the curves of the equations in parts **a** and **b** if they are graphed on the same coordinate system?

★ 5. The natural base e occurs so often in exponential functions that computers are programmed to compute the values of e^X. This function is usually written in the form Y = EXP(X). Write a program that prints out pairs (X, Y) for Y = EXP(X) with X varying from $^-3$ to 3 in steps of .5. Round off the values of Y to one decimal place.

9

Motions in Geometry

Mathematics is an obscure field, an abstruse science, complicated and exact; yet so many have attained perfection in it that we might conclude almost anyone who seriously applied himself would achieve a measure of success.

CICERO

"We're here to fix the copier."

9.1
CONGRUENCE
MOTIONS

There is an old belief that everyone has a "double"—someone who looks exactly like him or her, somewhere in the world. Each man in this cartoon has two doubles, which seems appropriate for a team of copier repairers. Copy machines have the ability to reproduce quickly words and figures that have the same *size* and *shape* as the original. Such figures are said to be *congruent*. Intuitively, we say that two plane figures are *congruent* if one can be *moved* onto the other so that they coincide. The idea of motion or movement is one of the more important concepts in mathematics. The particular motions that are associated with congruence will be studied in this section.

MAPPINGS

If triangle *ABC* is traced on paper and flipped over, it can be placed on triangle *RST* so that the points of each triangle coincide. The correspondence of point *A* with *R*, *B* with *S*, and *C* with *T* is indicated by

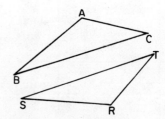

$$A \leftrightarrow R \qquad B \leftrightarrow S \qquad C \leftrightarrow T$$

By placing triangle *ABC* onto triangle *RST*, each point on the first triangle corresponds to exactly one point on the second triangle. This one-to-one correspondence of points is a special type of function. In Section 8.3 there are functions of numbers in which each number in one set is assigned to one and only one number in a second set. Similarly, there are functions of points for which each point in one set is assigned to a unique point in a second set. In geometry, functions are called *mappings*, or *transformations*. The mapping of triangle ABC to triangle RST is denoted by $\triangle ABC \rightarrow \triangle RST$. This notation indicates that the following sides and angles are matched with each other:

Corresponding Sides	Corresponding Angles
$\overline{AB} \leftrightarrow \overline{RS}$	$\angle ABC \leftrightarrow \angle RST$
$\overline{BC} \leftrightarrow \overline{ST}$	$\angle BCA \leftrightarrow \angle STR$
$\overline{AC} \leftrightarrow \overline{RT}$	$\angle CAB \leftrightarrow \angle TRS$

These pairs of sides and angles are called *corresponding sides* and *corresponding angles* for the mapping $\triangle ABC \rightarrow \triangle RST$. For each point in triangle ABC, the point it is *mapped to* is called its *image*. This terminology will be used to examine three mappings and their corresponding motions.

TRANSLATIONS

A translation is a special kind of mapping that can be described by a sliding motion. Each point is moved the same distance and in the same direction. The translation on the right maps A to A', B to B', C to C', $\overline{BC}$ to $\overline{B'C'}$, and pentagon K to pentagon K'. This translation is completely determined by the point A and its image A'. That is, given any point X we can find its image X' by moving in the *same direction* as from A to A' and the *same distance* as AA'.

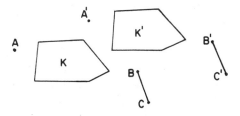

Translations occur for space figures as well as for plane figures. Just as in the case of two-dimensional figures, a translation in three dimensions is described by a sliding motion of every point in the same direction and for the same distance. This photo shows a sliding motion of the earth's crust, which geologists call a block fault. The arrow points to one side of the fault along which the earth's crust has been displaced.

Fault line showing displaced rock

REFLECTIONS

A reflection about a line is a mapping that can be described by folding. If this page is folded about line L, each point will coincide with its image. E will be mapped to E', F to F', $\overline{EF}$ to $\overline{E'F'}$, and figure M to figure M'. Since point S is on L, it does not move for this mapping. S and all other points on L are called *fixed points* for the reflection about L.

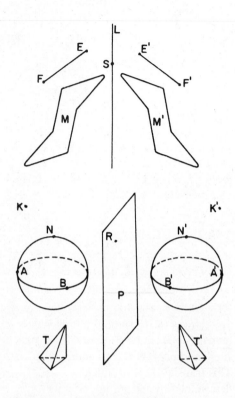

Reflections in space take place about planes. Each point to the left of plane P a unique image on the right side of P. The sphere is mapped to the sphere, point K to K', tetrahedron T to tetrahedron T', etc. Point R and all other points on the plane are *fixed points* of the mapping. That is, each point on the plane is its own image. Reflections in space can be illustrated by mirrors. If plane P is replaced by a mirror, so that the figures to the left of the mirror are reflected, their images will appear to be in the positions of the figures on the right side of the mirror.

Surprisingly clear reflections can be created by mirror images from pools. Pick out some points on the building shown in this photo and their images.

Model of the New Delhi United States Embassy, 1959

In the previous mappings about line *L* and plane *P*, each point and its image are on lines that are perpendicular to the line or plane of reflection. For example, $\overline{EE'}$ is perpendicular to *L*, and $\overline{NN'}$ is perpendicular to *P*. Furthermore, each point is the same distance from the line or plane as is its image. These two conditions hold for all reflections.

ROTATIONS

The third type of mapping is a rotation. To illustrate a 90° rotation about *O*, place a piece of paper on this page and trace $\overline{FG}$ and quadrilateral *ABCD*. Hold a pencil at point *O* and rotate the paper 90° in a clockwise direction. (A 90° rotation can be determined by placing the edges of the

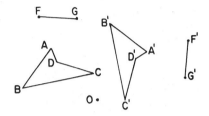

paper parallel to the edges of this page.) Each of the points you trace will coincide with its image after this rotation. Quadrilateral *ABCD* is mapped to quadrilateral *A'B'C'D'*, and $\overline{FG}$ is mapped to $\overline{F'G'}$. The only fixed point for this mapping is point *O*.

Space figures are rotated about lines. If the sphere shown here is rotated 90° about the vertical axis through *N* and *S*, *H* will be mapped to *H'* and *B* to *B'*. Each point will be mapped to a new location except for the points *N* and *S*, which remain fixed.

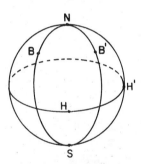

The restaurant and observation deck at the top of the 60-story Space Needle in Seattle, Washington, rotates once every 60 minutes. Each point on this moving structure traces out a circular path during one complete revolution. These moving points are constantly changing their locations and being mapped to each other. For example, consider the locations of points at 10:00 A.M. and again at 10:15 A.M. During this 15-minute interval, each point rotates 90° and finishes in a position that was previously occupied by another point of the structure.

Space Needle, Seattle, Washington

COMPOSITION OF MAPPINGS

The wood engraving by M. C. Escher combines translations and reflections. The white swan W is mapped onto the black swan B by a translation followed by a reflection. This mapping can be carried out by tracing swan W and its center line on a piece of paper. Then slide the paper diagonally to swan B so that the two center lines coincide. The swan that was traced can now be made to coincide with swan B by a reflection about the center line. A translation followed by a reflection is called a *glide reflection*.

Swans, wood engraving by M. C. Escher

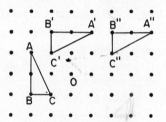

When one mapping is followed by another, it is called a *composition* or *product* of mappings. Any combination of translations, reflections, or rotations can be used. In the figure shown here, a 90° clockwise rotation about point *O* is followed by a translation of each point three spaces to the right. Triangle *ABC* is mapped to triangle *A'B'C'* by the rotation, and then the translation maps triangle *A'B'C'* to triangle *A"B"C"*. The composition of the rotation and translation is the mapping that takes triangle *ABC* to triangle *A"B"C"*.

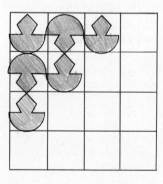

Two-dimensional patterns such as those on wallpaper and tiled floors are systematically created by translations, reflections, rotations, and compositions of these mappings. This is accomplished by beginning with a basic figure, such as the mushroom-shaped figure in the upper left square of this grid. To obtain the position of the figure in each square of the rows, the mushroom will be rotated 180° about the center of the edge of an adjacent square. To go down the columns from square to square, each mushroom will be reflected about the common edge of the square.

The Moors of Spain used combinations of mappings to generate two-dimensional patterns. In commenting about their work, Escher said, "What a pity that Islam did not permit them to make 'graven images.' They always restricted themselves, in their massed tiles, to designs of an abstract geometrical type."* The following patterns from

*M. C. Escher, *The Graphic Works of M. C. Escher* (New York: Random House, 1967).

the Alhambra were sketched by Escher. The first pattern was obtained by merely translating the basic figure (shown above the pattern) across the rows and down the columns. In the second pattern the basic figure is rotated 90° and then translated both down and to the right to the adjacent figures. The third pattern is generated by the same transformations as used in the rows and columns of the mushroom pattern, described in the preceding paragraph.

Alhambra drawings by M. C. Escher

The composition or product of mappings is similar to an operation on numbers. For example, the product of two integers is always another integer (closure property), and the composition of two mappings is always another mapping. The composition of some pairs of mappings is quite easy to determine. A 30° rotation followed by a 45° rotation can be replaced by a 75° rotation.

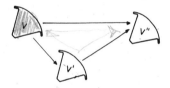

Two translations can always be replaced by one translation, as illustrated by the figure shown above. The first translation maps V to V' and the second maps V' to V''. The result of these two translations can be accomplished by the single translation which maps V to V''.

The situation for reflections is more interesting as shown by the figure on the right. For a reflection about line M, figure H is mapped to H'. Then H' is mapped to H'' by a reflection about line N. These two reflections can be replaced by a single rotation about point O which maps H to H''.

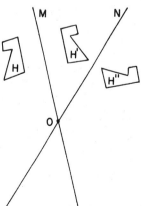

Compositions of different types of mappings, such as a rotation followed by a translation, or a reflection followed by a rotation, can also be replaced by a single mapping. It can be proved that the composition of any two of the three mappings (rotation, translation, or reflection) is a rotation, translation, reflection, or glide reflection.

CONGRUENCE

Translations, reflections, and rotations all have something very important in common. If A and B are any two points and A' and B' are their respective images, then the distance between A and B is the same as the distance between A' and B'. That is, for these mappings the lengths of line segments are the same as the lengths of their images.

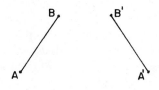

Such mappings are called *distance preserving* or *isometric*. What this means intuitively is quite simple: As figures are rotated, translated, or reflected, their size and shape do not change.

Up to this point we have thought of two figures as being congruent if they have the same size and shape or if one can be made to coincide with the other. For example, figure W is congruent to figure Z because W can be placed on Z so that they coincide. Thinking of congruence in terms of coinciding figures is fine for an introductory notion, but this concept can be defined more explicitly in terms of mappings. In this example figure W is congruent to figure Z because W can be mapped to Z by a glide reflection (a translation to the right and a reflection about line L).

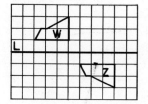

The need for a more careful definition of congruence becomes evident when we consider congruence in three dimensions. For example, we need a way of defining congruence for these two kitchen grinders, but it does not make sense to say that they coincide.

A suitable definition of congruence for both plane and space figures can be given in terms of mappings.

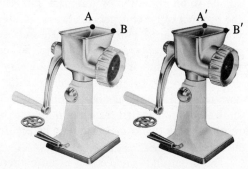

> **Definition:** Two geometric figures are congruent if and only if there exists a mapping (translation, reflection, rotation, or glide reflection) of one figure onto the other.

This definition says that for each of these four mappings a figure is congruent to its image. Conversely, if two figures are congruent, one can always be mapped to the other by one of these mappings. This definition gives us a way of viewing congruence of both plane and space figures. The plane figures W and Z shown above are congruent because there is a glide reflection mapping one figure to the other. The two grinders are congruent because there is a translation that maps one to the other. Each point on the left grinder is mapped to a corresponding point on the right grinder. The distances between any two points on the left grinder, such as points *A* and *B*, and their images, *A'* and *B'*, are equal. Defining congruence in terms of distance-preserving mappings is the mathematical way of saying that two objects have the same size and shape.

MAPPING FIGURES ONTO THEMSELVES

Michael Holt and Zoltan Dienes in their book *Let's Play Math* describe the following scheme for coloring pictures of a house.* Cut out a square and color the corners with four different colors. Both the front and back sides of each corner should have the same color. Place the square on a piece of paper and draw a frame around it. At each corner of the frame write (or draw pictures for) the words "wall," "roof," "door," and "window." This is called the Rainbow Toy. The different positions in which the square can be placed

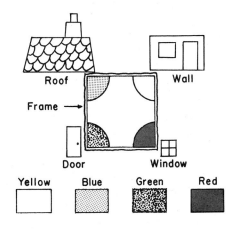

on the frame determine different arrangements of colors for the wall, roof, door, and window of the house. For the position of the square that is shown here, we get the colors for house 1. Color schemes for houses 2, 3, and 4 are obtained by rotating the square into three different positions. By flipping the square over, there are four more positions for the color schemes for houses 5 through 8.

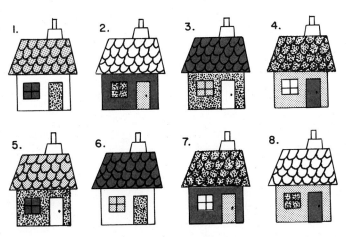

*M. Holt and Z. Dienes, *Let's Play Math* (New York: Walker and Company, 1973), pp. 88–94.

The Rainbow Toy is an elementary way of illustrating the eight different mappings of a square back onto itself. Four of these are rotations of 90°, 180°, 270°, and 360°. The other four mappings are reflections. The reflections about D_1 (the diagonal from upper left to lower right) interchanges corners b and d but leaves corners c and a in the same location. There is also a reflection in diagonal D_2 (the diagonal from upper right to lower left) and reflections about the horizontal and vertical lines, H and V.

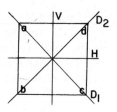

If we cut out a square and draw a frame about it, as for the Rainbow Toy, some interesting observations can be made about the composition of the eight mappings. Let's label the corners of the square and the frame by a, b, c, and d. The position shown here, with a in the a corner of the frame, b in the b corner, etc., will be called the Initial Position of the square. Now, if the square is rotated 90° clockwise and then reflected about the horizontal axis, b and d will end up in the same corner in which they started, but c and a will be interchanged.

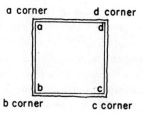

a corner   d corner

b corner   c corner

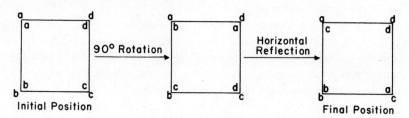

We can get the same Final Position by beginning with the Initial Position and reflecting about D_2 (the diagonal from upper right to lower left). That is, the composition of a 90° rotation followed by a horizontal reflection is the same as a reflection about D_2.

Something surprising happens when we reverse the order of these two mappings. The following figures show that by first carrying out the horizontal reflection and then a 90° rotation, we do not get the same Final Position as we did in the previous example:

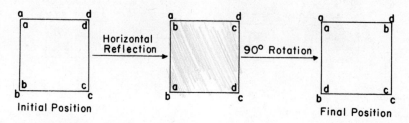

540

This time c and a are in the same position in which they started, but b and d have changed places. The composition of these two mappings is the same as a reflection about D_1 (the diagonal from upper left to lower right). These examples show that the order in which mappings are carried out is significant. In other words, the composition of mappings is not commutative.

COMPUTER APPLICATIONS

The turtle's position at any point on the screen can be described by x and y coordinates. On the screen shown here the x coordinates vary from ⁻140 to 140 and the y coordinates vary from ⁻120 to 120.

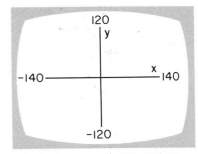

Cartesian coordinate system

The Logo command SETXY can be used to send the turtle to any point (X, Y) on the screen. For example, typing

SETXY ⁻40 65

will move the turtle to the point (⁻40, 65) and leave its heading unchanged.* If the y coordinate is negative, it must be enclosed in parentheses. For example, to send the turtle to (70, ⁻30) we type

SETXY 70 (⁻30)

In the following paragraphs we will define procedures for three transformations, and the command SETXY will be used in two of them.

The figures below were obtained from the procedures RIGHTPLANE and LEFTPLANE. They are mirror images of each other. Either procedure can be obtained from the other by interchanging the RT and LT commands. Here are the commands for RIGHTPLANE:

```
TO RIGHTPLANE
 RT 120 FD 50 RT 150 FD 86.6 RT 90
 FD 36 RT 90 FD 10 RT 90 FD 11
 LT 90 FD 33.3 LT 90
END
```

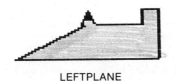

LEFTPLANE

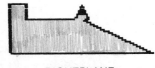

RIGHTPLANE

*In Apple Logo this command is SETPOS (for SETPOSITION). It must be typed with square brackets. For example, SETPOS [⁻40 65].

Translations The translation below maps point (⁻40, ⁻10) to its image, (65, 60). Each point on RIGHTPLANE in the lower left of the screen is moved in the same direction and through the same distance to its image point in the upper right of the screen. The procedure for this translation is called TRANSRIGHTPLANE. It has four variables for the coordinates of two points. The first two variables are for moving the turtle to point (*A, B*) before RIGHTPLANE is drawn. The second two variables are the coordinates of the image of (*A, B*). The figures for this translation were obtained by typing TRANSRIGHTPLANE ⁻40 (⁻10) 65 60.*

```
TO TRANSRIGHTPLANE :A :B :C :D
  HIDETURTLE
  PENUP SETXY :A :B PENDOWN
  RIGHTPLANE
  PENUP SETXY :C :D PENDOWN
  RIGHTPLANE
END
```

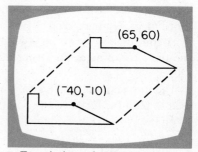

Translation of RIGHTPLANE

Rotations The rotation below rotates point A 120° to its image A′. Each point on RIGHTPLANE at the top of the screen is rotated through the same angle to its image in the lower right of the screen. The procedure for this rotation is called ROTATERIGHTPLANE. It has two variables. The first variable is for moving the turtle to (*O, A*) before the figure for RIGHTPLANE is drawn. The second variable is for the number of degrees in the rotation. The figures for this rotation were obtained by typing ROTATERIGHTPLANE 70 120.

```
TO ROTATERIGHTPLANE :A :ANGLE
  PENUP FD :A PENDOWN
  RIGHTPLANE
  PENUP BK :A RT :ANGLE FD :A PENDOWN
  RIGHTPLANE
  PENUP HOME
END
```

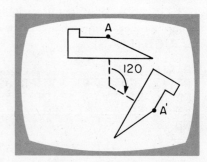

Rotation of RIGHTPLANE

Reflections The following reflection maps each point on RIGHTPLANE about the north-south line to its image point on LEFTPLANE. This procedure is called REFLECTRIGHTPLANE. It has two variables for the coordinates of points. First, the

Note: For Apple Logo the command SETXY :A :B must be replaced by SETPOS SE :A :B.

turtle is moved to the point (A, B) to draw RIGHTPLANE. Then the turtle is moved to $(^-A, B)$ to draw the image of RIGHTPLANE. The figures for this reflection were obtained by typing REFLECTRIGHTPLANE 60 60.

```
TO REFLECTRIGHTPLANE :A :B
  HIDETURTLE
  PENUP SETXY :A :B PENDOWN
  RIGHTPLANE
  PENUP SETXY - :A :B PENDOWN
  LEFTPLANE
END
```

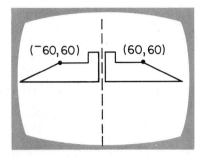

Reflection of RIGHTPLANE

SUPPLEMENT (Activity Book)

Activity Set 9.1 Translations, Rotations, and Reflections (Illustrations of mappings and their compositions by geometric models and circular geoboards)

Exercise Set 9.1: Applications and Skills

1. This silhouette of lines and angles was produced by joining six photographs of a construction staging, side by side.

   ★ a. Are these six photographs congruent?

   b. Are the six photographs of the staging congruent?

   ★ c. What mapping is suggested by this picture?

2. The design on this nineteenth century quilt also occurs in the fourteenth century Moorish palace Alhambra. Beginning with the white figure in the upper left corner, the top row can be generated by a sequence of 180° rotations.

   a. Locate the centers of rotation for these mappings.

   ★ b. Beginning with the figure in the upper left corner, what mapping can be carried out to generate the left column of figures?

Arabic lattice patchwork quilt, ca. 1850
Collections of Greenfield Village and the
Henry Ford Museum, Dearborn, Michigan

3. The importance of triangles to architecture is due to a basic mathematical fact, which is not true for polygons with more than three sides. The following questions will help you to see why triangles are common in construction:

   a. Construct two noncongruent quadrilaterals whose sides are congruent to the following four line segments:

   Construction of Lundolm Field House, University of New Hampshire

   ★ b. How many noncongruent quadrilaterals can be constructed whose sides are congruent to the four line segments in part **a**? (*Hint:* Imagine changing the shape of a quadrilateral formed by linkages; see page 204.)

   c. Construct a triangle using segments that are congruent to these line segments:

   ★ d. Can you construct another triangle from the line segments in part **c** which is not congruent to your first triangle?

   e. A triangle is the basic figure for strengthening buildings, bridges, and other structures because triangles hold their shape better than polygons with four or more sides. Use your results from parts **a** through **d** to explain why this is true.

4. For the translation that takes A to A', sketch the image of the hexagon.

   ★ a. The line through point D and its image is parallel to $\overleftrightarrow{AA'}$. Is this true for every point on the hexagon and its image?

   b. If E' and G' are the respective images of E and G, how does the length $E'G'$ compare with the length EG?

   ★ c. Compare the area of the hexagon with the area of its image.

   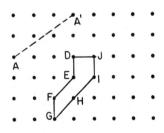

5. Sketch the image of the pentagon for a reflection about line L. draw reflection

★   a. If R' is the image of R, what is the measure of the angles formed by the intersection of $\overleftrightarrow{RR'}$ and L? $90°$

   b. If U' is the image of U, how does the distance from U to L compare with the distance from U' to L?

★   c. What are the fixed points for this mapping?

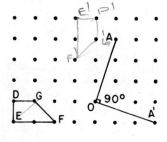

6. Sketch the image of the quadrilateral for the 90° rotation about O which takes A to A'. (Trace the figure onto a piece of paper and rotate.)

★   a. If E' is the image of E, what is the measure of $\angle EOE'$?

   b. If G' is the image of G, how does the length of EG compare with the length of $E'G'$?

   c. Are there any fixed points for this mapping?

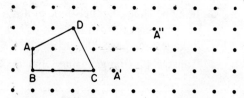

7. Map quadrilateral $ABCD$ to quadrilateral $A'B'C'D'$ by the translation that moves point A to A'. Then map quadrilateral $A'B'C'D'$ to quadrilateral $A''B''C''D''$ by the translation that takes A' to A''.

★   a. The translation that maps A to A' can be described by "over 4 and down 1." The translation from A' to A'' is "over 2 and up 2." Describe the translation that maps A to A''.

   b. What is the image of quadrilateral $ABCD$ for the composition of the following two mappings: the translation which takes A to A'; followed by the translation which takes A' to A?

8. Reflect pentagon $RSTUV$ about line M and then reflect its image about line N.

★   a. What single mapping (rotation, translation, or reflection) is equal to the composition of these two reflections?

   b. Let R' be the image of R for the reflection about M, and R'' be

the image of R' for the reflection about N. Compare the distance from R to R″
with the distance from line M to line N. What relationship do you find? Will this
hold for other points and their images?

9. The hexagons in the accompanying
   sketch can be mapped to each
   other by compositions of
   reflections about lines M and N.

   a. Sketch the image of hexagon F
      for a reflection about line M.
      The reflection of this image
      about line N should coincide
      with hexagon G.

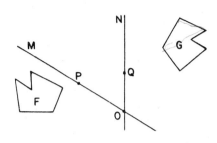

   ★ b. There is a rotation about point O that will map figure F to figure G. How is the
         number of degrees in this rotation related to ∠POQ?

10. Complete the patterns for the grids
    by carrying out the mappings on
    the basic figure.

   ★ a. Rows: Rotate 180°.
         Columns: Reflect about edge of
         square.

      b. Rows: Reflect about edge of
         square.
         Columns: Reflect about edge of
         square.

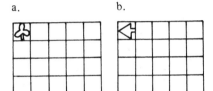

a.   b.

★ 11. Mark off a square region that can be used as a
      basic figure to generate this wallpaper pattern.
      Describe the mappings for obtaining the rows
      and columns.

12. A translation maps P with coordinates (⁻2, 1) to
    P' with coordinates (3, 2). Graph these points
    and draw an arrow from P to P' to indicate the
    length and direction of the translation.

   a. Sketch the image of △ABC for this
      translation.

   ★ b. If A', B', and C', are the images of A, B, and
         C, what are their coordinates?

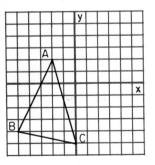

13. Draw the image of quadrilateral *DEFG* for a reflection about the *x* axis. Label the images of these vertices as *D'*, *E'*, *F'*, and *G'*.

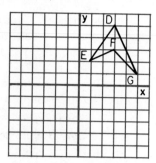

a. What are the coordinates of *D'*, *E'*, *F'*, and *G'*?

b. Reflect quadrilateral *D'E'F'G'* about the *y* axis and label its vertices by *D"*, *E"*, *F"*, and *G"*. What are the coordinates of the vertices of this figure?

c. What single rotation will map quadrilateral *DEFG* to quadrilateral *D"E"F"G"*?

14. Mappings are sometimes given by describing what will happen to the coordinates of each point. For example, the mapping $(x, y) \rightarrow (x + 2, y - 3)$ is a translation that maps each point two units to the right and down three units. To find the image of a particular point, such as $(1, 7)$, substitute these values for *x* and *y* into $(x + 2, y - 3)$. In this case, $(1, 7)$ maps to $(3, 4)$. Find the image of each figure for the following mapping. Label each mapping as a translation, rotation, reflection, or glide reflection. — composition of translation & reflection

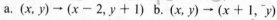

★ a. $(x, y) \rightarrow (x - 2, y + 1)$ b. $(x, y) \rightarrow (x + 1, {}^{-}y)$ c. $(x, y) \rightarrow ({}^{-}x, {}^{-}y)$

$(2, -2, 0+1)(0,1)$ glide reflection

$(5-2, 0+1)$
$(3, 1)$

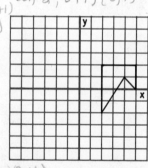
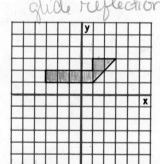
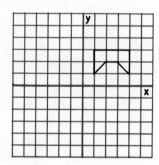

$(5-2, 2+1)$
$3, 3$

Exercise Set 9.1: Problem Solving

1. *Isolations:* How many different ways can five consecutive whole numbers be placed in a row so that no two consecutive whole numbers are next to each other?

a. *Understanding the Problem:* Here is one solution for the first five consecutive whole numbers. If one arrangement can be obtained from another by a reflection, such as shown here, then we will consider the two arrangements as the same. Find another solution.

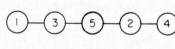

Five-in-a-row

b. *Devising a Plan:* This type of problem can be solved by making a systematic list. For example, you might begin by listing all the different arrangements with 1 in the first position; then 2 in the first position; etc. If you do this without checking for duplication due to reflections, what adjustment must be made to the total number of arrangements?

c. *Carrying Out the Plan:* Follow the system suggested above or one of your own to solve this problem.

d. *Looking Back:* There are several ways to extend this problem. For example, there are 45 different solutions for 6 consecutive whole numbers. You may want to see if you can find them. Another possibility is using different configurations. How many different solutions are there for the following figures? (*Reminder:* Two arrangements are the same if they can be obtained from each other by a reflection.)

★

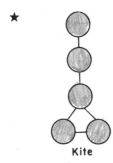

Kite

Orange Crate

2. *Center of Rotation:* Figure *A* maps onto figure *A'* by a rotation. Describe a method (other than guessing) of locating the center of this rotation. Check your method by rotating figure *A* onto figure *A'*. (*Hint:* The perpendicular bisector of a chord on a circle passes through the center of the circle.)

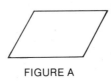

FIGURE A

FIGURE A'

3. *Freize Patterns:* An ornamental design that extends to the right and left (around rooms, buildings, pottery, etc.) is called a freize.

★ a. What transformations will map each of the following freize patterns onto itself? (Consider these patterns as extending in both directions.)

Container with cover, Attica, Greece, eighth century B.C. Courtesy Museum of Fine Arts, Boston.

Chinese ornament painted on porcelain

French Renaissance ornament from casket

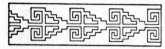

Masonry fret, temple at Mitla, Mexico

Indian painted lacquer work

b. Design a freize pattern that will be transformed onto itself by all of the following transformations: translation, 180° rotation, horizontal reflection, and vertical reflection.

4. *Isolation Puzzle*:* This configuration of eight circles is another isolation number puzzle. It is from *An Unexpected Hanging,* by Martin Gardner. If we agree that all solutions that can be obtained from each other by rotations and reflections of this diagram are the same, then there is only one solution. Find this unique solution.

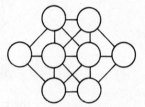

★ 5. *Different Viewpoints:* Which two of these figures are different views of the same die? The sum of the numbers on opposite faces of a die is 7.

Exercise Set 9.1: Computers

1. Sketch the figure for each procedure (see pages 542 and 543).
   a. TRANSRIGHTPLANE ⁻50 70 60 (⁻30)
   ★ b. ROTATERIGHTPLANE 80 270
   c. REFLECTRIGHTPLANE ⁻50 (⁻50)

---

*More isolation puzzles are described by D. T. Piele, "Isolations," *The Mathematics Teacher* **67** No. 8 (December 1974), 719–22.

2. Here are the commands for the procedure RIGHTVENT:

```
TO RIGHTVENT
 FD 100 RT 90 FD 80 RT 90 FD 50
 RT 90 FD 30 RT 90 FD 20 LT 90
 FD 30 LT 90 FD 70 RT 90 FD 20 RT 90
END
```

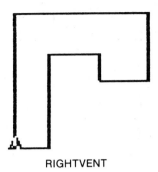

RIGHTVENT

Define the following procedures and sketch the figure for each one:
a. TRANSRIGHTVENT ⁻90 ⁻70 40 20
b. ROTATERIGHTVENT 20 270
★ c. REFLECTRIGHTVENT ⁻100 20

3. Suppose a given procedure is revised by interchanging FD (the FORWARD commands) with BK (the BACKWARD commands). Try this for a procedure that has both FD and BK commands. What type of transformation will map the figure that is obtained from the procedure onto the figure that is obtained from the revised procedure?

★ 4. Suppose a given procedure is revised by interchanging the FD with the BK commands and the RT with the LT commands. What type of transformation will map the figure that is obtained from the procedure onto the figure which is obtained from the revised procedure?

5. Sketch part of the figure for the command MANYRIGHTPLANE. What type of transformation does this procedure illustrate?

```
TO MANYRIGHTPLANE
 RIGHTPLANE
 PENUP RT 30 FD 35 LT 30 PENDOWN
 MANYRIGHTPLANE
END
```

Model of Fort Wayne, Indiana, in environmental wind tunnel

9.2
SIMILARITY MOTIONS

Our intuitive notion of similar figures is "same shape" with no special requirement on size. Such figures occur in a wide variety of ways. An indispensable application in the design of large objects is the use of models. Research in ship and plane design, for example, is routinely carried out by testing models in ship-model towing tanks and wind tunnels. The scale model of Fort Wayne, Indiana, shown in the above photo, is being tested in an environmental wind tunnel in the Fluid Dynamics and Diffusion Laboratory at Colorado State University.

Virtually all manufactured objects, whether large or small, are designed and constructed from scale drawings. A drawing or plan may be scaled down or scaled up, depending on the size of the original object. This picture shows a computer-on-a-chip and its scaled-up enlargement.

All of our familiar optical instruments involve similar figures. When you look through a magnifying glass or a microscope, you see an image that is similar to the original but scaled up. Telescopes, binoculars, cameras, and even our eyes all produce images that are similar to objects before them.

Enlargement of computer-on-a-chip, actual chip on finger

SHADOWS AND MAPPINGS

Similar figures can be created by lights and shadows. Hold a flat object perpendicular to a flashlight's rays, and the light will produce a shadow similar to that of the original figure. Because the light is from a small bulb and spreads out in the shape of a cone, the shadow is larger than the object. The sun, on the other hand, is so large that its rays are almost parallel as they strike the earth. If a flat object is placed perpendicular to the sun's rays and its shadow is cast onto a surface that is also perpendicular to the sun's rays, the shadow will be congruent to the original figure.

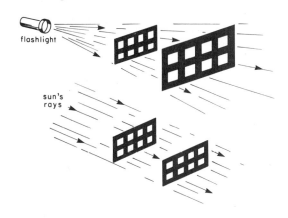

Light rays and shadows are like mappings and their images. For each point on the object there is a corresponding "shadow point" that is its image. The rays of light are like lines projecting from a central source to the object and then to its image. This type of mapping is illustrated here by quadrilateral *ABCD* and its image. Point *O*, which is taking the place of the light source, is called a *projection point,* and the mapping of one figure to the other is called a *similarity mapping.* Each point on quadrilateral *A'B'C'D'* is twice as far from point *O* as its corresponding point on quadrilateral *ABCD*. For instance, *OA'* is twice *OA*, *OB'* is twice *OB*, etc. Because of this relationship between points and their images, this mapping is said to have a *scale factor* of 2.

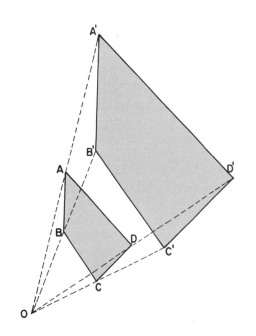

SCALE FACTORS

If the scale factor for a similarity mapping is greater than 1, the image is an enlargement of the original figure. The following mapping from figure *XZW* to figure *X'Z'W'* has a scale factor of 3. Each image point of figure *X'Z'W'* is 3 times further from *O* than its corresponding point on figure *XZW*. That is, *OX'* = 3 times *OX, OZ'* = 3 times *OZ*, etc.

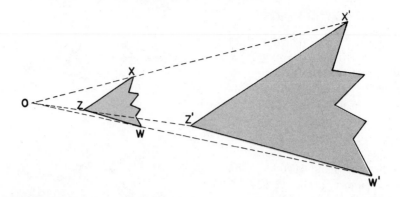

When the scale factor is less than 1, the image is a *reduction* of the original figure. In this similarity mapping, each of the distances from O to points A, B, C, and D has been multiplied by $\frac{1}{3}$ to get the image points A', B', C', and D'. That is, the larger figure has been reduced by a scale factor of $\frac{1}{3}$.

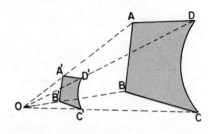

It is even possible for a scale factor to be negative. In this case the original figure and its image will be on opposite sides of the projection point, and the image will be "upside down" from the original figure. The larger flag shown here is projected through point O to the smaller flag by a scale factor of $\frac{-1}{2}$. In particular, A is mapped to A', and G is mapped to G'. As in the previous examples, the scale factor determines

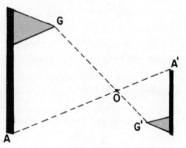

the size of the image. For a scale factor of $\frac{-1}{2}$, each image point is half as far from the projection point as is its corresponding point (and these points are on opposite sides of the projection point). For example, OG' is half of OG, and OA' is half of OA.

The lenses of our eyes and of cameras invert the images of scenes much like a similarity mapping with a negative scale factor. These lenses are the projection points, and the scene is produced upside down on the retinas of our eyes and the film of a camera.

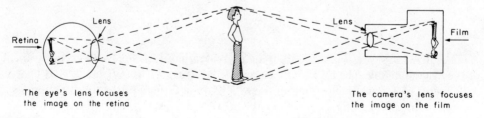

The eye's lens focuses the image on the retina

The camera's lens focuses the image on the film

SIMILARITY MAPPINGS

The pantograph is a mechanical device for enlarging and reducing figures which employs the principle of a similarity mapping. It can easily be constructed from four strips of wood that are hinged at points *A, B, C,* and *P* so that they move freely. Point *O* is the projection point and should be held fixed. As point *P* traces the original figure, a pencil at *P'* (its image) traces the enlargement. Since *P'* is twice as far from the projection point *O* as point *P,* the scale factor for this mapping is 2. In order to reduce a figure, the pencil is held at *P,* and *P'* is moved around the original figure. In this case the scale factor will be $\frac{1}{2}$. The pantograph can be changed for different scale factors by adjusting the settings at points *B* and *C.*

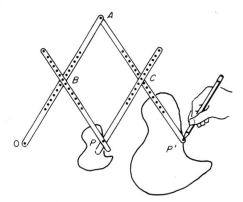

Pantograph, for reproducing similar figures

The ability of the pantograph to reproduce figures with the same shape is due to two important properties of similarity mappings. First, the sizes of angles do not change. Consider this similarity mapping of triangle *DEF.* Angle *EDF* is congruent to $\angle E'D'F'$, and $\angle DEF$ is congruent to $\angle D'E'F'$. In general, all angles are congruent to their images under a similarity mapping.

Second, the lengths of line segments all change by the same factor. Each side of triangle *D'E'F'* is twice as long as its corresponding side in triangle

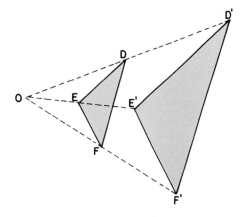

DEF. In general, if *k* is the scale factor of a similarity mapping, each line segment will have an image that is *k* times as long. Another way of stating this second condition is to say that similarity mappings preserve ratios. For instance, the ratio *DE/EF* is equal to *D'E'/E'F'*. This is true because both *D'E'* and *E'F'* are twice as long as *DE* and *EF.*

In our study of congruence the mappings we used (rotations, translations, and reflections) preserved size and shape. Similarity mappings, on the other hand, do not preserve size, but they do preserve shape. What we have been intuitively referring to as the "same shape" can now be made more precise by the following definition of similar figures.

Definition: Two geometric figures are similar if and only if there exists a similarity mapping of one figure onto the other.

This definition also holds for space figures. These two boxes are similar because the smaller one can be mapped onto the larger one by using projection point *O*, inside the small box, and a scale factor of 3.6. The larger box has a width, length, and height that are each 3.6 times the corresponding dimension of the smaller box. As with plane figures, a scale factor greater than 1 produces an enlargement, and a scale factor between 0 and 1 reduces the original figure.

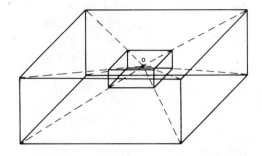

SIMILAR POLYGONS

There are many applications of similar figures for which it is inconvenient or even impossible to set up similarity mappings by using projection points. A solution to this problem is to produce similar figures from measurements of angles and distances. The construction of maps and charts is one example. The polygon on this chart connects five points and is similar to the large polygon over the water that connects the actual landmarks. These positions on the chart could have been plotted by measuring the five vertex angles and five distances between these islands. The following theorem shows that these measurements are all that is necessary to obtain similar polygons.

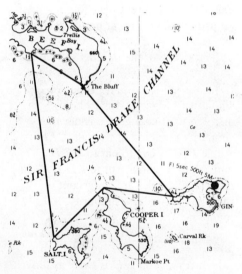

Virgin Islands, West Indies,
scale factor 139,000

Theorem: Two polygons are similar if there is a one-to-one correspondence between their vertices such that:

(1) Corresponding angles are congruent.
(2) Lengths of corresponding sides have the same ratio.

Both conditions (1) and (2) are usually necessary for two polygons to be similar. A square and a rectangle have congruent angles (all 90°), but their sides do not satisfy condition (2). Therefore, they are not similar. The rectangle and the parallelogram satisfy condition (2) because the sides of one are equal to the corresponding sides of the other, but the angles in the parallelogram are not congruent to those in the rectangle. Therefore, these two figures are not similar.

When the two polygons are triangles, they will be similar if either condition (1) or condition (2) of the preceding theorem is satisfied. Triangles *KLM* and *K'L'M'* satisfy condition (2) because the sides of the larger triangle are two times longer than the corresponding sides of the smaller triangle. Therefore, we can conclude that the two triangles are similar and that the vertex angles at *K, L,* and *M* are congruent to the corresponding angles at *K', L',* and *M'*.

On the other hand, if two triangles satisfy condition (1), they are similar. For example, these two triangles are similar because the angles of one are congruent to the angles of the other. In fact, because there are 180° in every triangle, it is necessary only to find two angles from one triangle that are congruent to two angles from the other. Why? When this condition is satisfied, we can conclude that the lengths of the corresponding sides have the same ratio.

These facts about similar triangles have many applications. In particular, they form the basis of indirect measurement.

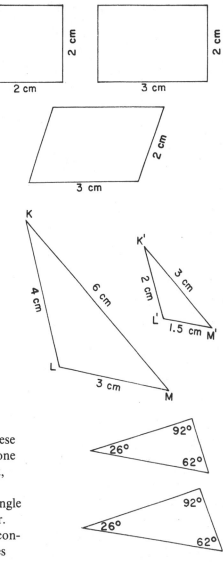

INDIRECT MEASUREMENT

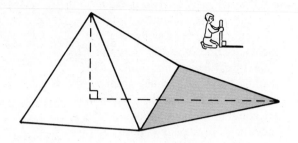

The Greek mathematician Thales (ca. 600 B.C.) used the shadow of a staff to measure the heights of the Egyptian pyramids. This is the earliest known example of determining the heights of objects by indirect measurement. By holding a stick vertically to the ground, the stick and its shadow form a small right triangle that is similar to a right triangle formed by the pyramid and its shadow.

Here is an example of how this method can be used to find the height of a tree. Both the stick and the tree in the next figure form a right angle with the ground. Furthermore, $\angle CAB$ and $\angle STR$ are congruent because they are formed by the sun's rays. With this information about the angles, we know that the two triangles are similar, with $\overline{AC}$ corresponding to $\overline{TS}$, $\overline{AB}$ to $\overline{TR}$, and $\overline{CB}$ to $\overline{SR}$. Therefore, the ratios of the corresponding sides are equal. In particular, $TR/AB = SR/CB$. Three of these distances, TR, AB, and CB, can be measured directly and are given in the diagram. The height of the tree, SR, can be computed from these ratios:

$$\frac{35}{200} = \frac{SR}{80}$$

This equation is satisfied for $SR = 14$. Therefore, the height of the tree is 14 meters.

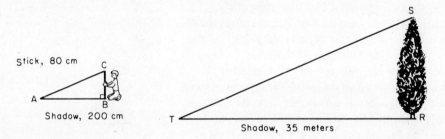

Stick, 80 cm

Shadow, 200 cm

Shadow, 35 meters

SCALE FACTORS— AREA AND VOLUME

The smaller of the two knives pictured on page 559 is a regular-size Scout knife, whose length is about equal to the width of the palm of your hand. The newspaper clipping says that the bigger knife is "three times larger" than the conventional Scout knife. Does this mean that the length of the larger knife is three times greater, or that its surface area, or volume is three times greater? Phrases such as "twice as large" or "three times bigger" can be misleading. They often refer to a comparison of linear dimensions, as in the case of these knives (compare their lengths). The "three" in this example refers to the scale factor. It means that the dimensions of length, width, and height of the big knife are three times greater than the corresponding dimensions of the smaller knife. As another example, when scientists say that a microscope enlarges an object 1000 times, they are referring to the linear dimensions and not an increase in area. On the other hand, when a realtor speaks of one plot of land as being twice as large as another, he or she is referring to the

Prepared for anything

What could be the world's largest Scout-type knife is ready for the world's largest potato. Wayne Goddard, a professional knife-maker who works at his home at 473 Durham St., Eugene, turned this one out for Dennis and Raymond Ellingsen, Eugene knife collectors. Completely functional, the knife is 24½ inches long when opened. It weighs 4¼ pounds and is three times larger than the conventional Scout knife.

area and not its length or width. If a farmer wants a silo that is twice as large, this refers to the volume and not the height or surface area of the silo. In the following paragraphs you will see the effect that scale factors have on surface area and volume.

Area These two rectangles are similar. The scale factor from the smaller to the larger is 3. That is, the length and width of the larger are 3 times greater than the length and width of the smaller. However, the area of the larger rectangle is 9 times greater. (The smaller rectangle has an area of 8 square units, and the larger rectangle has an area of 72 square units.) In general, if one figure is similar to another by a scale factor of k, where k is any positive number, then one will have an area which is k^2 times the area of the other.

The relationship between the scale factor and the surface area of a three-dimensional figure is the same as that for plane figures. Consider the two boxes shown here. The scale factor from the smaller box to the larger is 2. That is, the length, width, and height of the larger box are each two times greater than the corresponding dimensions of the smaller box. Let's compare the areas of the sides of these

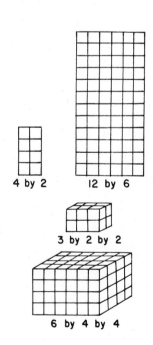

4 by 2   12 by 6

3 by 2 by 2

6 by 4 by 4

boxes. The front side of the smaller box has an area of 6, and the front side of the larger box has an area of 24. The larger area is 2^2 or 4 times greater than the smaller area. If a similar comparison is made between each face of the smaller box and the corresponding face of the larger box, you will see that the larger area is 4 times greater than the smaller area. In general, the surface areas of two similar figures are related by the *square of their scale factor*. If the scale factor is k, where k is any positive number, then one figure will have a surface area with is k^2 times the surface area of the other figure.

Volume   There is also a relationship between the volumes of two similar space figures. There are 12 cubes in the smaller box on the preceding page and 96 cubes in the larger one. The larger box has 8 (or 2^3) times more cubes. In general, the volumes of two similar figures are related by the *cubes of their scale factors*. If two figures are similar and their scale factor is any positive number k, the volume of one will be k^3 times the volume of the other.

We are now prepared to examine the relationships between the areas and volumes of the two knives shown in the news clipping. The scale factor from the smaller knife to the larger is 3. Therefore, the larger knife has a surface area that is 3^2 or 9 times greater than the surface area of the smaller knife. The volume of the larger knife is 3^3 or 27 times greater than the volume of the smaller knife.

Let's apply the relationships between scale factor and area and volume to another example. Pictured here is a nineteenth century scale model of a cookstove. This miniature stove was used by a traveling salesperson as a sample and has all the features of the life-size appliance. The scale factor from this miniature to the life-size stove is 5. Therefore, the larger stove has a surface area that is 5^2 or 25 times greater than that of the smaller stove. Since the surface area of the top of the smaller stove is 300 square centimeters,

Nineteenth century scale model of cookstove

the surface area of the top of the big stove is 25 × 300, or 7500 square centimeters. The increase in volume between the two stoves is determined by the cube of the scale factor. The volume of the life-size stove is 5^3 or 125 times greater than the volume of the miniature stove. The oven of the small stove has a volume of 1000 cubic centimeters. Therefore, the volume of the oven in the large stove is 125 × 1000, or 125,000 cubic centimeters.

SIZES AND SHAPES OF LIVING THINGS

The relationship between the surface area and volume of similar figures has some important and interesting applications in nature. For every type of animal there is a most convenient size and shape. A flea cannot have the shape of a hippopotamus, and an elephant cannot be smaller than a rabbit. One factor that governs the size and shape of a living thing is the ratio of its surface area to its volume. All warm-blooded animals at rest lose the same amount of heat for each unit area of skin. Small animals have too much surface area for their volumes, and so a major reason for eating is to keep warm. For example, 5000 mice weigh as much as a person, but their combined surface area and food consumption are about 17 times greater! At the other end of the scale, large animals tend to overheat because they have too little surface area for their volumes. This is one reason why you do not see elephants and other large animals with fur.

Let's take a closer look at the relationship between surface area and volume as the size of an object increases. This table contains the surface areas and volumes of four different-size cubes. As the sizes of the cubes increase, both the surface areas and the volumes increase, but the

Box Size	Surface Area	Volume	Area / Volume
2 by 2 by 2	24	8	3
3 by 3 by 3	54	27	2
4 by 4 by 4	96	64	1.5
10 by 10 by 10	600	1000	.6

volumes increase at a faster rate. One way of viewing this change is to form the ratio of surface area to volume. This ratio is 3 for a 2 by 2 by 2 cube, and as the dimensions of the cubes increase, the ratio decreases. For a 7 by 7 by 7 cube it is less than 1.

Another way to compare the change in surface area and volume as the size of a cube increases is by graphs. For a cube whose length, width, and height are x, its area is $6x^2$ and its volume is x^3. The graphs of $y = 6x^2$ and $y = x^3$ show that for $x < 6$ the area is greater than the volume; for $x = 6$ the area equals the volume; and for $x > 6$ the volume is greater than the area.

Areas and Volumes — Volume of cubes — Area of cubes

Dimensions of boxes

To relate these changes in surface area and volume to the problem of maintaining body temperatures, assume that the three cubes shown here are animals and that the ideal ratio between surface area and volume is 2. In this case the 2 by 2 by 2 animal has

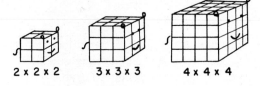

2 x 2 x 2   3 x 3 x 3   4 x 4 x 4

too much surface area (it would tend to be too cold) because its area-to-volume ratio is 3; the area-to-volume ratio for the 3 by 3 by 3 animal is 2, which is "just right"; and the 4 by 4 by 4 animal has too little surface area (it would tend to be too hot) because its ratio of area to volume is 1.5.

COMPUTER APPLICATIONS

The lengths of corresponding sides of two similar figures are proportional, and the corresponding angles are equal. Compare the following two procedures and their figures. Each side of the smaller figure is one-half as long as the corresponding side of the larger figure, and each angle of the smaller figure is equal to the corresponding angle of the larger figure.

```
TO FULLSCALE              TO HALFSCALE
  FD 60 LT 45               FD 30 LT 45
  FD 50 LT 100              FD 25 LT 100
  FD 150 LT 138             FD 75 LT 138
  FD 124.5 LT 77            FD 62.25 LT 77
END                       END
```

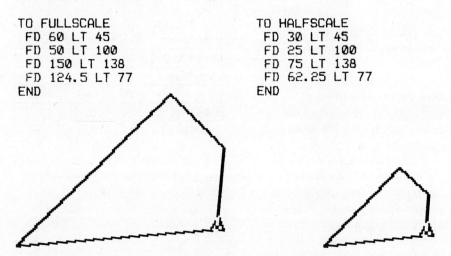

The similar figures shown next were drawn from the procedure SKYHOOK. The largest figure was produced by beginning with SIDE equal to 80. The next-to-last line of this procedure sends the computer back to the beginning with SIDE being half as big. Therefore, the scale factor from each figure to the next smaller figure is one-half. The conditional command stops the turtle when SIDE becomes less than 15.

```
TO SKYHOOK :SIDE
 HIDETURTLE
 FD :SIDE RT 90 FD :SIDE LT 45
 FD :SIDE/5 RT 90 FD :SIDE/3
 RT 45 FD :SIDE/5
 PENUP HOME PENDOWN
 IF :SIDE < 15 STOP
 SKYHOOK :SIDE/2
END
```

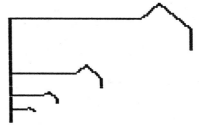

SKYHOOK 80

A scale factor greater than 1 causes an enlargement. The figures below were drawn from the procedure called HORSEHEAD. The smallest figure was drawn first by beginning with SIDE equal to 20. The scale factor in this procedure is 1.4. The conditional command stops the turtle when SIDE becomes greater than 60.

```
TO HORSEHEAD :SIDE
 FD :SIDE RT 65 FD :SIDE RT 60
 FD :SIDE RT 80 FD .3 * :SIDE
 RT 90 FD :SIDE/2 LT 100
 PENUP HOME PENDOWN
 IF :SIDE > 60 STOP
 HORSEHEAD 1.4 * :SIDE
END
```

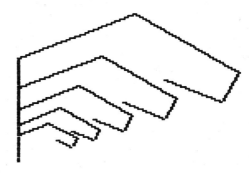

HORSEHEAD 20

SUPPLEMENT (Activity Book)

Activity Set 9.2 Devices for Indirect Measurement (Stadiascope, clinometer, hypsometer, transit, and plane table)

Exercise Set 9.2: Applications and Skills

1. Engineers use models of planes to gain information about wing and fuselage (central body) designs. The scale factor from this model of the Boeing B52 to the full-size plane is 100.

Boeing B52 being adjusted for wind tunnel test

★ a. The lift of an airplane depends on the surface area of its wings. How many times greater is the surface area of the B52 than the surface area of its model?

   b. The weight of a plane depends on its volume. How many times greater is the volume of the B52 than the volume of its model?

★ c. In actual flight the wings of the B52 flap up and down a distance of 6 meters. How much wing flap can be expected in a wind tunnel test of the model?

2. For each figure and the given scale factor, sketch a similar figure.

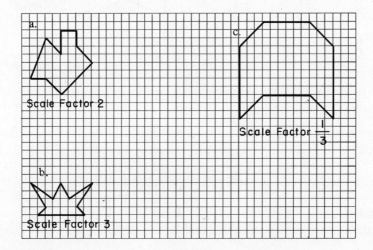

3. Use O as a projection point and find the images of triangle T for scale factors of 2, 3, and $\frac{1}{2}$. Look for a relationship between the coordinates of the vertices of T and the coordinates of their images. Triangle T has an area of 16 square units. What are the areas of its three images?

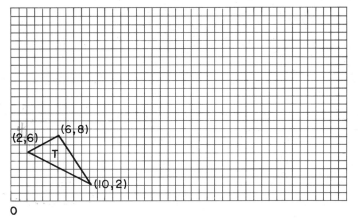

(2,6)   (6,8)

T

(10,2)

O

4. The similarity mappings in parts **a, b,** and **c** describe what will happen to each
point on the given figures by relating the coordinates of each point to the
coordinates of its image. The mapping in part **a** *doubles the coordinates* of each
point. For example, point (⁻1, 1) gets mapped to (⁻2, 2). Sketch the images of the
given figures. What is the scale factor for each mapping?

★ a. $(x, y) \rightarrow (2x, 2y)$ b. $(x, y) \rightarrow \left(\dfrac{x}{2}, \dfrac{y}{2}\right)$ c. $(x, y) \rightarrow (^-3x, ^-3y)$

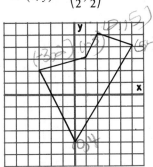

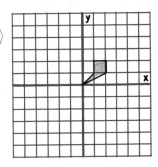

(⁻1,1)
$(2(⁻1), 2(1)) = (⁻2,2)$ (3,3)
                              $(2(3), 2(3)) = (6,6)$

5. A pinhole camera can easily be made from a box. When the pinhole is uncovered,
rays of light from the object strike a light-sensitive film. These rays of light travel in
straight lines from the object to the pinhole of the camera, like lines through a
projection point. Without a lens to gather light rays and increase their intensity, it
takes from 60 to 75 seconds for enough light to pass through the pinhole to record
the image.

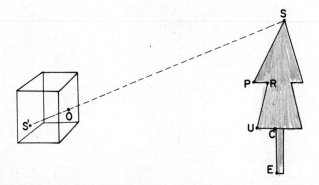

a. Draw lines from the letters of this tree to point O and label their images on the back wall (film) of the camera.

b. Sketch the complete image of the tree.

★ c. What is the scale factor for this projection?

6. Using projection points K and M and a scale factor of 2, sketch the two images of the pentagon.

★ a. Is each image similar to the original pentagon?

★ b. Are the two images congruent to each other?

c. Explain why the size of the image of a similarity mapping depends only on the scale factor and not on the location of the projection point.

7. *Diagonal Test:* There is an easy way to test for similar rectangles. If one rectangle is placed on the other with one of their right angles coinciding, as shown in these figures, then the rectangles are similar if their diagonals lie on the same line. For example, rectangle $AEFG$ is similar to rectangle $ABCD$, but the two rectangles in the lower figure are not similar.

★ a. Explain why rectangle $ABCD$ is similar to rectangle $AEFG$.

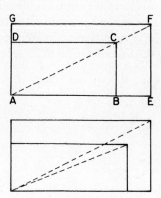

566

★   b. Use the diagonal test to find out which two of the following rectangles are similar:

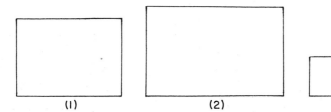

(1)   (2)   (3)   (4)

8. Similar figures can be obtained by reproducing a figure from one grid to another grid of a different size. This practice is common for enlarging patterns in quilting and sewing. The patchwork doll pattern shown here was enlarged by a scale factor of 3 to get the larger doll.

    Reproduce the figure from Grid A onto Grids B and C by copying a square at a time. (The figure will become larger on Grid B and smaller on Grid C.)

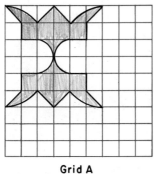
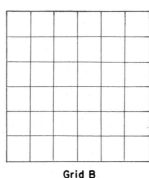
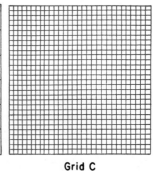

Grid A   Grid B   Grid C

★   a. What is the scale factor from Grid A to Grid B? Grid B to Grid C?

      b. The scale factor from Grid A to Grid C is $\frac{3}{10}$. How can this scale factor be obtained from the two scale factors in part **a**?

★   c. How many times greater is the area of the figure on Grid B than the area of the figure on Grid A?

★ 9. The larger box shown here has linear dimensions that are twice those of the smaller box. Doubling the linear dimensions produces a similar box with a scale factor of 2, and tripling the dimensions is equivalent to a scale factor of 3. For the scale factors in the following table, record the dimensions of the boxes which are similar to the small box. Compute the areas and volumes of these boxes.

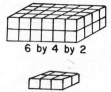

6 by 4 by 2

3 by 2 by 1

	Scale Factors				
	1	2	3	4	5
Dimensions	3 x 2 x 1	6 x 4 x 2			
Surface Area	22 cm²				
Volume	6 cm³				

10. This is not an example of trick photography. There are two average-size adults sitting in this chair, and there is room for several more. The scale factor from the small chair to the large chair is 8.

★ a. The small chair is 40 ×8 = 320 centimeters tall and its seat is 24 centimeters wide. What is the height of the large chair and the width of its seat? 8×24

b. How many times more paint will the large chair require than the small chair? 8²= 64

★ c. The small chair weighs about 1 kilogram. Both chairs are made of the same kind of wood. What is the weight of the large chair?

similarity mapping

11. In *Gulliver's Travels,* by Jonathan Swift, Gulliver went to the kingdom of Lilliput where he found that he was 12 times taller than the average Lilliputian.

   a. The Lilliputians computed Gulliver's surface area to make him a suit of clothes. About how many times more material would be needed for his suit than one of theirs?

★ b. They computed Gulliver's volume to determine how much he should be fed. About how many times more food would he require than a Lilliputian?

   c. Tiny people like the Lilliputians can exist only in fairy tales because the ratio of their volume to surface area would not enable them to maintain the proper body temperature. Would they be too warm or too cold?

12. *Similarity Stretcher:* Loop or knot two rubber bands together and hold one end fixed at point *O*. Stretch the bands so that as the knot at point *P* traces one figure, a pencil at point *P′* traces an enlargement. Use this method to enlarge the map of the United States. Assume the distance from *O* to *P′* is 2.3 times the distance from *O* to *P*. How many times greater will the distance from Denver to Kansas City be on the enlarged map than on the smaller map?

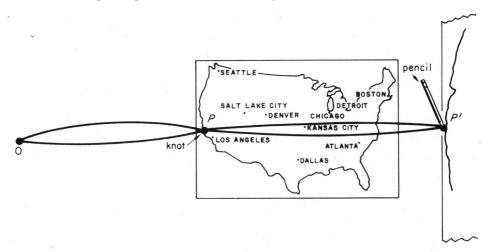

Exercise Set 9.2: Problem Solving

1. *Copy Machines:* The Xerox 7000 in Mr. Gary's printshop has five switches to determine the size of a reproduction. If switch 1 is used, the reproduction is congruent to the original. Switches 2, 3, 4, and 5 reduce the original size by scale factors of .85, .76, .65, and .58, respectively. Using combinations of three or fewer switches, what scale factors can Mr. Gary obtain on this copier?

★   a. *Understanding the Problem:* These copies of hexagons were obtained from the Xerox 7000. The length of the sides of hexagon 2 is .85 times the length of the sides of hexagon 1. The smallest hexagon was obtained by using switch 2 to obtain a reproduction, and then using this reproduction and switch 3. What is the scale factor from hexagon 1 to this hexagon?

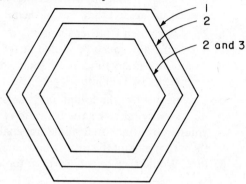

b. *Devising a Plan:* We could begin by listing the 10 different ways that switches 2, 3, 4, and 5 can be paired (a switch can be paired with itself). What are they? Why isn't switch 1 included in these pairs?

c. *Carrying Out the Plan:* List the scale factors that can be obtained by using three or fewer switches. Round off each scale factor to the nearest hundredth.

d. *Looking Back:* Mr. Gary also has a copier that enlarges the size of a figure by a scale factor of 2. Using this machine once and the Xerox 7000 either once or twice, what additional scale factors can be obtained?

★ 2. *Book Page Production:* The pages of a book come from the printing press as a pile of large flat rectangular sheets of paper. Each sheet is fed through a series of rollers and folded in half several times. A sheet that has been folded once is called a *folio.* Half of a folio is a *quarto,* and half of a quarto is an *octavo.* Each fold is perpendicular to the previous fold. Which of these rectangles (original, folio, quarto, octavo) are similar? Explain why. If this folding pattern is continued, which of the resulting rectangles will be similar?

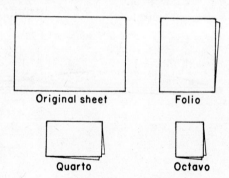

3. *Similar Figures:* These two figures are similar. Each dimension of the larger figure is twice the corresponding dimension of the smaller figure. If this doubling process is continued, how many blocks will there be in the tenth figure?

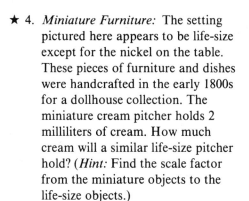

★ 4. *Miniature Furniture:* The setting pictured here appears to be life-size except for the nickel on the table. These pieces of furniture and dishes were handcrafted in the early 1800s for a dollhouse collection. The miniature cream pitcher holds 2 milliliters of cream. How much cream will a similar life-size pitcher hold? (*Hint:* Find the scale factor from the miniature objects to the life-size objects.)

5. *Paper Chase Game:* Player 1 and Player 2 each choose beginning points near the center of a grid. For the example shown here these points are *A* and *B.* Each player will draw a connected path of line segments by drawing only one line segment on each turn. The line segments may be drawn in a horizontal, vertical, or diagonal direction. Play begins with Player 1 drawing a line segment from *A.* Then Player 2 draws a line segment from point *B,* which is twice as long and in the

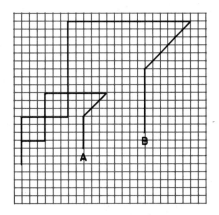

same direction. Play continues in this manner with both players beginning at the ends of their previous line segments and Player 2 always drawing a segment which is twice as long as Player 1's and in the same direction.

The object of the game is for Player 1 to move so that when he or she finishes a turn, Player 2 finishes his or her turn at the same point. When the game is over, the players' paths will be similar figures. Describe a strategy for Player 1 that will end the game in one move. (*Hint:* The point at which the game ends has a very special relationship to the starting points.)

Exercise Set 9.2: Computers

1. List the commands for drawing a figure that is similar to the figure FULLSCALE on page 562. The scale factor from the figure for FULLSCALE to this new figure should be .8. Sketch the figure that will be obtained from these commands.

2. Here is a revision of the procedure called POLYGONS, with two new lines to make several copies of similar polygons. The variable, NUMBER, is for the number of sides in a polygon. These hexagons were produced by typing ROWOFPOLYGONS 6 30.

```
TO POLYGONS :NUMBER :SIDE
  REPEAT :NUMBER [FD :SIDE RT
  360/:NUMBER]
  IF :SIDE < 10 STOP
  PENUP RT 90 FD 2.5 * :SIDE LT
  90 PENDOWN
  POLYGONS :NUMBER .6 * :SIDE
END
```

```
TO ROWOFPOLYGONS :NUMBER :SIDE
  HIDETURTLE PENUP SETXY - 140 0 PENDOWN
  POLYGONS :NUMBER :SIDE
END
```

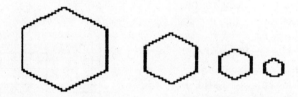

a. What type of polygons, and how many, will be produced by typing ROWOFPOLYGONS 5 40?

★ b. What type of polygons, and how many, will be produced by typing ROWOFPOLYGONS 4 50?

3. Take the turtle on a walk and return it to its starting position. Record each turn and move of its path. Use these turns and moves to define a procedure called WALK. Then write a procedure for a similar path that is one-half as long and call it HALFWALK. (*Hint:* To complete the last leg of the trip that returns the turtle to its starting position, use HOME.)

4. Define the procedure BIGRIGHTPLANE, whose dimensions are twice the dimensions of RIGHTPLANE (see page 541). Sketch the figures that are produced by the following commands. Are these figures similar?

        HIDETURTLE
        BIGRIGHTPLANE
        PENUP FD 40 PENDOWN
        RIGHTPLANE

5. Define a procedure for drawing the following figures. Then define a procedure for drawing a larger similar figure with a scale factor of 2.5.

★ a. Scalene triangle (Name these procedures TRIANGLE and ENLARGE.TRIANGLE.)

b. Nonconvex quadrilateral (Name these procedures QUADRILATERAL and ENLARGE.QUADRILATERAL.)

Anamorphic painting of H.M.S. *Victory* by James Steere

9.3
TOPOLOGICAL
MOTIONS

Topology has been called "the mathematics of distortion." It is a special
kind of geometry that involves
analyzing figures and surfaces when
they are twisted, bent, stretched,
shrunk, or, in general, distorted from
one shape to another. These descriptions are all examples of topological
motions. A figure may be changed so
much by a topological motion that it
bears little resemblance to its original
shape. The above painting of the
H.M.S. *Victory* by James Steere
is a topological distortion called

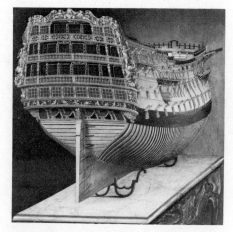

H.M.S. *Victory* by Joseph Marshall

574

anamorphic art. The reflection of this painting in the cylindrical mirror that has been placed at the center of the picture brings the picture into clear perspective. It shows the stern of the ship from the painting by Joseph Marshall. You may wonder how mathematicians find anything left to study when figures can be so drastically changed. In the following paragraphs we will consider some of the properties of figures which remain unchanged by topological motions.

TOPOLOGICAL MAPPINGS

Topology is sometimes referred to as rubber-sheet geometry because topological mappings in the plane can be performed as if the plane were a rubber sheet. As the rubber sheet in these figures is stretched, the face is distorted into several shapes. In each picture the original circle has changed shape, but it is still a simple closed curve.

Similar distortions can take place in space by thinking of an object as made of an elastic material. In the following sequence a donut-shaped object, called a *torus*, has been deformed in three steps into the shape of a cup. Since topologists view each of these as being "the same" in the sense that they have the same topological properties, a topologist is sometimes described as a person who doesn't know the difference between a donut and a cup of coffee.

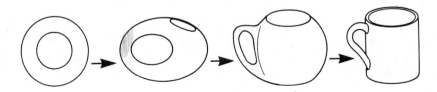

These stretching and bending motions from one object to another are like mappings, with each point having a corresponding image point. The definition of a topological mapping for both plane and space figures contains two conditions.

Definition: A mapping from one set of points to another is a *topological mapping* if:

(1) There is a one-to-one correspondence between the two sets.
(2) The mapping is continuous.

If there is a topological mapping from one set to another, the two sets are said to be *topologically equivalent*.

The basic idea of continuity, in condition (2) of the definition, is that points "close together" cannot be cut or torn apart and separated by the mapping. As examples, it is not permissible to deform a curve by cutting it or to deform a region by punching a hole in it. In this figure the simple closed curve C is not topologically equivalent to curve C', and region R is not topologically equivalent to R'.

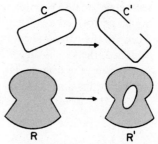

Nontopological Mappings

There is, however, one exception to this restriction on cutting. Cutting is permissible as long as the points along the cut are matched back together again. For example, the solder wire in the form of a simple closed curve in Figure A has been cut at point K; knotted in Figure B; and resoldered at point K in Figure C. The curves formed by the solder in Figures A and C are topologically equivalent.

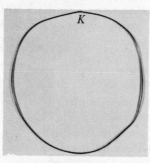

Figure A

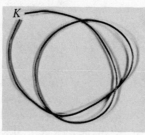

Figure B

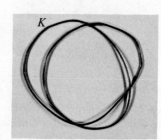

Figure C

All simple closed curves are topologically equivalent. This includes circles, squares, triangles, and irregular simple closed curves such as the one shown on the right. All line segments and simple nonclosed curves such as the following figures are topologically equivalent. If we think of these figures as

rubber string, each could be bent or stretched into the shape of another without cutting or separating points.

In three dimensions spheres (balls), cubes (boxes), ellipsoids (footballs), and cylinders (cans) are all topologically equivalent. If we think of these as being made of clay, any one of them could be deformed into the shape of another without punching holes or separating points.

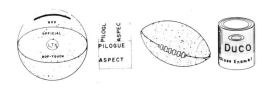

TOPOLOGICAL PROPERTIES

In Section 9.1 we saw that the size and shape of an object remain the same for congruence mappings. To state this another way, these mappings preserve the distance between points and their images. In Section 9.2, we saw that an object changes its size but not its shape for a similarity mapping. In this case, the mapping preserves measures of angles, and ratios of the lengths of line segments, but does not preserve distances between points. Now, with topological mappings both the size and the shape can vary greatly between an object and its image. In other words, topological mappings do not preserve distances between points, measures of angles, or ratios of the lengths of line segments. In spite of this freedom there are properties of figures which are preserved by topological mappings. One of these is the number of sides and edges of a surface.

A sheet of paper has two sides and one edge, in the sense that an ant crawling on one side must cross over an edge to get to the other side. Deform the sheet by crumpling it in your hands, and it still has two sides and one edge. The two-sidedness and one-edgedness are properties that are preserved under a topological mapping.

A cylindrical band of paper has two sides and two edges. A surface must be crossed to get from edge to edge, and an edge must be crossed to get from surface to surface.

In the nineteenth century the German mathematician Ferdinand Moebius (1790–1868) made a surprising discovery by putting a half-twist in a strip of paper and fastening the edges. The resulting surface has only one edge and one side! This means

Flat sheet of paper

Two sides and one edge

Cylindrical band

Two sides and two edges

that an ant crawling on this surface can reach any other spot without crawling over an edge. This surface is called a Moebius band or Moebius strip. While the Moebius band is only a slight variation of a cylindrical band, these two figures cannot be topologically equivalent because of their different number of sides.

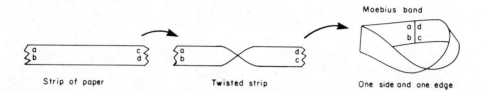

Strip of paper   Twisted strip   One side and one edge

The Moebius band has fascinated people for years and has been a frequent subject in literature and in art. It is pictured here in Escher's 1963 woodcut, *Moebius Strip II*. In at least one case the Moebius band has been put to practical use. The Sandia Corporation has invented a Moebius band nonreactive resistor that has been patented by the U.S. Atomic Energy Commission. This resistor has several desirable electrical properties that ordinary bands do not have. The following sketch is from this patent:

Metal Moebius band electrical resistor

Moebius Strip II, 1963
woodcut by M. C. Escher

Another property that is preserved by topological mappings can be roughly described by speaking of the number of "holes" in an object. For example, the four objects pictured next are all topologically different. Mathematicians describe their differences by referring to the genus of a surface. The *genus* is the maximum number of circular cuts (simple closed curves) that can be made without dividing the surface into two separate pieces. For example, two cuts can be made (as shown) on the two-holed surface without separating its surface into disconnected pieces. Similarly, the torus can be cut once without separating its surface into disjoint pieces. Any circular cut on the surface of a sphere will divide its surface into two pieces. The numbers under these figures are their *genus numbers*.

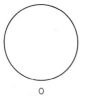  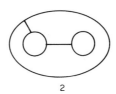 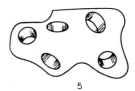

0   1   2   5

NETWORKS

In the eighteenth century in the town of Königsberg, Germany, a favorite pastime was walking along the Pregel River and crossing the town's seven bridges. During this period a natural question arose: Is it possible to take a walk and cross each bridge only once? This question was solved by the Swiss mathematician Leonhard Euler. His solution was the beginning of network theory, which is an important branch of topology.

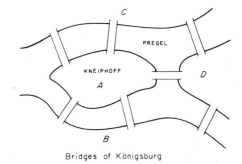

Bridges of Königsburg

Euler represented the four land areas of Königsberg (*A, B, C,* and *D* in the previous figure) by four points, and the seven bridges by seven lines joining these points. For example, the island of Kneiphoff has five bridges, and in the diagram at the right there are five lines from point *A*. The three lines from point *D* represent three bridges, etc. This kind of a diagram is called a *network*. Notice that Euler was concerned not with the size and shape of the bridges and land regions but rather with how the bridges were connected.

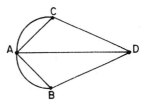

A *network* is a collection of points, called *vertices,* and a collection of lines connecting these points, called *arcs.* A network is *traversible* if you can trace each arc exactly once by beginning at some point and not lifting your pencil from the paper. The problem of crossing each bridge exactly once reduces to one of traversing the network for these bridges.

Euler made a remarkable discovery about networks which depends on the number of odd vertices. In the Königsberg network there are an odd number of arcs at point *A* and so *A* is called an *odd vertex.* If the number of arcs meeting at a point is even, it is called an *even vertex.* Euler found that the only traversible networks are those that have either no odd vertices or exactly two odd vertices! Since the Königsberg network has four odd vertices, it is not traversible.

Here are some examples of traversible and nontraversible networks:

| Traversible | Traversible | Nontraversible | Nontraversible |
| (2 odd vertices) | (No odd vertices) | (4 odd vertices) | (6 odd vertices) |

FOUR-COLOR PROBLEM

If you try to color this map of eight states with only three colors so that no two states with a common boundary have the same color, you will find it cannot be done. It can, however, be colored with four colors.

For the past 125 years mathematicians have been trying to prove that four colors are enough to color any map no matter how many regions it contains or how they are situated. The only condition that must be satisfied is that regions with a common boundary must be colored differently. It is agreed that if two regions meet only at a point, then they do not share a common boundary. Since no one was able to produce a map that required more than four colors, it was conjectured that four colors were sufficient to color any map. This simple-appearing conjecture became one of the most famous unsolved problems in mathematics.

Then on July 22, 1976, two University of Illinois mathematicians, Kenneth Appel and Wolfgang Haken, announced they had proved that four colors are all that is necessary to color any map. The problem was solved by representing map regions as points and boundaries as arcs connecting these points. Using this network approach they were able to reduce the problem to the examination of 1936 basic map forms. All other maps are topologically equivalent to these basic forms or else they differ in some insignificant way. Then they fed their map forms into the computer, and after 1200 hours the computer determined that each map could be colored with four colors or less.

COMPUTER APPLICATIONS

Here are four closed curves that were produced by line segments. Curves (a) and (b) are topologically equivalent and curves (c) and (d) are topologically equivalent.

CHAPTER 9 MOTIONS IN GEOMETRY

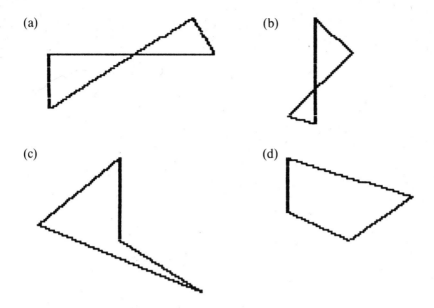

(a)

(b)

(c)

(d)

These curves were drawn by procedure CC4 (*C*losed *C*urves with *4* sides). This procedure uses two new commands: RANDOM and MAKE. Before we examine procedures for drawing arbitrary closed curves, let's look at some examples that use these new commands.

RANDOM This command can be used to select random whole numbers from 0 to any number. In the following example, RANDOM 100 randomly selects whole numbers as small as 0 and as large as 99; and RANDOM 360 selects whole numbers from 0 to 359. The figure shown here was produced by the following commands:

```
FD RANDOM 100
RT RANDOM 360
FD RANDOM 100
```

MAKE The command MAKE is another way of creating or making a Logo variable. In the following example, MAKE sets the variable SIDE equal to 50.* This command together with SQUARE (page 88) instructs the turtle to draw a square whose sides

*MAKE is similar to LET in BASIC. For example, LET X = 50.

have length 50. The quote symbol is used in front of SIDE because SIDE is the name of the variable. The colon symbol : is used with SIDE to represent the number that SIDE is equal to.

```
MAKE "SIDE 50
SQUARE :SIDE
```

Now we are ready to consider procedure CC4. The first three uses of MAKE and RANDOM in this procedure set X, Y, and Z equal to random distances. Notice that 10 is added to the number produced by RANDOM 100 to ensure that the distance will be greater than zero. The next two uses of MAKE and RANDOM set S and T equal to random numbers of degrees for the turns. Why isn't 10 added to RANDOM 360? Procedure CC4 produced the figures on page 581 and the quadrilateral shown here.

```
TO CC4
 HIDETURTLE
 MAKE "X 10 + RANDOM 100
 MAKE "Y 10 + RANDOM 100
 MAKE "Z 10 + RANDOM 100
 MAKE "S RANDOM 360
 MAKE "T RANDOM 360
 FD :X RT :S FD :Y RT :T FD :Z HOME
END
```

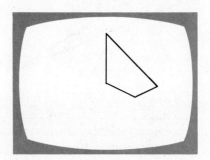

The next procedure uses recursion to draw a simple closed curve having a variable number of line segments. Each time the computer reaches the next-to-last line of the procedure, it is sent back to the beginning. The number of line segments, which is typed in at the beginning of the procedure, is decreased each time through the procedure until it equals 1. When :NUMBER = 1 the conditional command sends the turtle home and stops the program. (See the footnote on page 253 regarding conditional commands in Apple logo.)

```
TO C :NUMBER
 HIDETURTLE
 MAKE "X 10 + RANDOM 50
 MAKE "T RANDOM 360
 FD :X RT :T
 MAKE "NUMBER :NUMBER -1
 IF :NUMBER = 1 HOME STOP
 C :NUMBER
END
```

The following closed curves were drawn by the above procedure. Curves (a) and (b) were obtained by typing C 5 , and curves (c) and (d) by typing C 6.

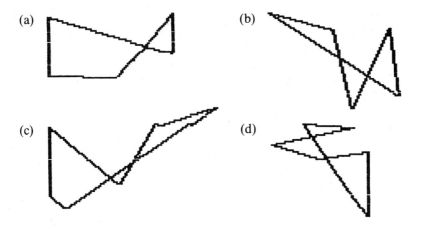

(a)

(b)

(c)

(d)

SUPPLEMENT (Activity Book)

Activity Set 9.3 Topological Entertainment (Moebius bands, topological tricks, Gale Game, and game of "Sprouts")

Exercise Set 9.3: Applications and Skills

1. Each figure in parts **a** through **g** is topologically equivalent to the sphere in Figure 1, or the torus in Figure 2, or the two-holed object in Figure 3. Match the correct number with each one.

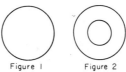

Figure 1   Figure 2   Figure 3

★ a. figure 3   b. figure 2 ★ c. Fig 1   d. Fig 3

★ e. Fig 1   f. Fig 1   g. Fig 1

2. According to the Swiss psychologist Jean Piaget, children's first discoveries of geometry are topological, and by the age of 3 they readily distinguish between open and closed figures.* Assuming that a child "thinks topologically," match up four pairs of figures that he or she would consider to be "the same."

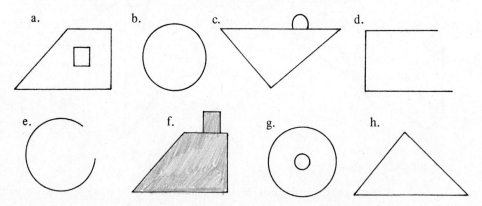

3. Two of these networks are traversible. For each traversible network, show a beginning point and an ending point, and mark the route with arrows.

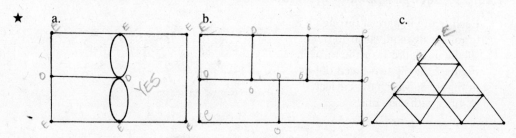

★ 4. Each of these networks has two odd vertices and is traversible. Show a beginning point and an ending point for each. What will always be true about the beginning and ending points of networks with exactly two odd vertices?

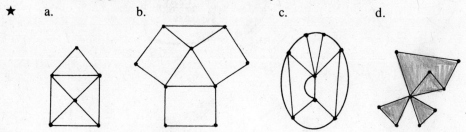

*J. Piaget, "How Children Form Mathematical Concepts," *Scientific American,* **189** No. 5 (1953), 74–79.

5. The vertices and edges of polyhedra are three-dimensional networks.

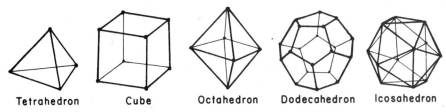

Tetrahedron   Cube   Octahedron   Dodecahedron   Icosahedron

★   a. Which one of the five regular polyhedra is traversible?

★   b. What is the least number of diagonals that need to be drawn on the faces of the cube before the network will be traversible?

★ 6. Many years after Euler proved that it was impossible to take a walk in which each of the seven bridges of Königsberg is crossed over exactly once, an eighth bridge was built. Sketch a network with four vertex points for the land areas *A*, *B*, *C*, and *D*, and eight arcs for the bridges. Is this network traversible?

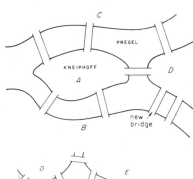

7. This is a sketch of 2 islands (*A, B*), 4 land regions (*C, D, E, F*), and 15 bridges. Is it possible to plan a walk in which each bridge is crossed exactly once? Explain the reason for your answer.

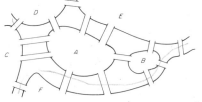

8. The curves shown next are simple closed curves. The Jordan curve theorem (page 203) says that if a point inside a simple closed curve is connected to a point outside the curve, the connecting arc will intersect the curve. In these examples, *B* is outside. Can you draw an arc from *A* to *B* that does not intersect the curve? In other words, is *A* inside or outside?

a.

b.

★   c. Draw line segments $\overline{AB}$ for the curves in parts **a** and **b**. Count the number of times that $\overline{AB}$ intersects each curve. How can this number be used to tell when a point is inside or outside a simple closed curve? Check your answer on a few curves.

9. This simple closed curve is from the NCTM publication *Puzzles and Graphs* by John Fujii. Are the pairs of points that are given in parts **a, b, c,** and **d** on the same side of the curve? (*Hint:* Connect pairs of points by a line segment and count the number of intersections of the line segment and the curve.)

★   a. *A, B*   b. *B, C*   c. *D, C*   d. *B, D*

10. To color maps it is desirable for regions with a common boundary to have a different color.

★   a. What is the minimum number of colors required to color any map, regardless of the number of regions?

       b. What is the minimum number of colors required for this map?

11. Anamorphic art is a special type of topological mapping. One example of this art is the *slant picture,* which is designed to be viewed at an angle.* The technique for drawing such pictures is to use grids as we did for producing similar figures in Exercise 8, page 567. However, this time the grid is distorted along one axis, such as shown in Figure 2. Finish copying the design in Figure 1 onto the corresponding regions of the trapezoidal grid in Figure 2. The completed figure should look normal if viewed by placing your eye to the right of this grid and a little above the surface of this page.

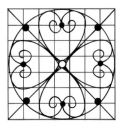

Figure 1

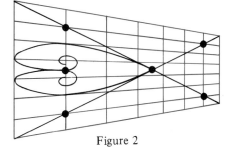

Figure 2

Exercise Set 9.3: Problem Solving

1. *Museum Tours:* A tour guide is planning a tour of a museum. The guide would like to have the tour pass through each door of the museum exactly once. For what type of floor plans is this possible?

   a. *Understanding the Problem:* The tour may begin at any point inside or outside of the museum and end at any point. Show that a tour can begin inside or outside for the floor diagrams shown here.

   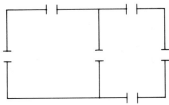

   b. *Devising a Plan:* One approach is to draw floor diagrams with different numbers of rooms and doors and try planning tours. By counting the numbers of rooms, doors, and doors per room, you may discover the necessary conditions. Draw a floor diagram for which a tour cannot be conducted by passing through each door exactly once.

   c. *Carrying Out the Plan:* Use the above approach or one of your own to solve this problem.

---

*For more details on slant drawings as well as techniques for sketching cylindrical and conical anamorphic pictures, see M. Gardner, "The Curious Magic of Anamorphic Art," *Scientific American,* **232** No. 1 (1975), 110–114.

★ d. *Looking Back:* Suppose the tour is to begin outside the museum and end outside. What are the conditions (number of doors, etc.) for a tour to pass through each door of the museum exactly once?

e. *Looking Back Again:* Perhaps it occurred to you that this type of problem can be solved by using networks. Form a network by connecting the five inside points and point *P* with arcs that run through each of the doors for the following floor diagram. Why is this network nontraversible? What does this tell you about a tour that passes through each door exactly once?

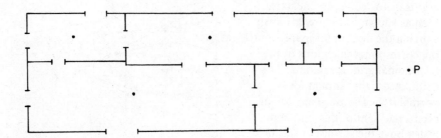

2. *Continuous Arcs:* Figure A contains four polygons that are connected by common sides. A type of puzzle that is similar to planning the museum walk in Exercise 1, requires that a continuous arc be drawn which passes through each side of each polygon exactly once. The solution in this example is shown in Figure B.

Figure A   Figure B

Under what conditions can a continuous arc be drawn that will pass through each side of each polygon exactly once? Experiment with the following figures or draw some of your own. (*Hint:* Note the number of polygons with an odd number of sides.)

★ a.   b.   c.   d.

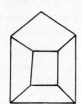

★ 3. *Coloring Problem:* Some maps require two different colors in order for regions with a common boundary to be colored differently, and others require three or four colors. What are the conditions under which maps can be colored with just two colors? Try the maps shown below or draw a few of your own. (*Hint:* Count the number of even and odd vertices that are not on the outer boundary of the map.)

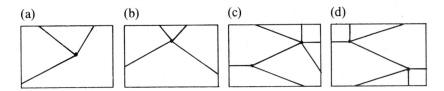

(a)   (b)   (c)   (d)

★ 4. *Matchstick Problem:** The three matchstick patterns shown here are the only topologically different patterns that can be formed with three matches, subject to the condition that the matches meet only at their endpoints with no overlapping. (The line segment with three matches in a row is topologically equivalent to the Z-shaped pattern.) How many topologically distinct patterns can be formed with four matches, subject to the above conditions? How can this problem be extended?

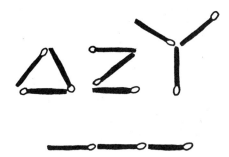

Exercise Set 9.3: Computers

1. Which curves on page 583 are topologically equivalent?

★ 2. It is possible to obtain a closed curve by typing C 6 (page 582), which is not topologically equivalent to any of the curves on pages 581 or 583. Sketch such a curve.

3. Suppose the third command in the procedure C (page 582) was changed to

MAKE "T 10 + RANDOM 80

   a. Explain why all the figures produced by C 3 and C 4 would be topologically equivalent. What type of curves would they be?

★ b. Show why a curve produced by C 5 or C 6 might not be topologically equivalent to those in part **a**.

*M. Gardner, *The Unexpected Hanging* (New York: Simon and Schuster, 1969), pp. 79–81.

4. Use the procedure SPOKES to answer the questions below:

```
TO SPOKES :NUMBER
   HIDETURTLE
   MAKE "X 40 + RANDOM 80
   MAKE "T RANDOM 360
   RT :T FD :X BK :X
   MAKE "NUMBER :NUMBER - 1
   IF :NUMBER = 0 STOP
   SPOKES :NUMBER
END
```

   a. What type of figure is produced by SPOKES 20? Sketch this figure.

★ b. Will all of the figures produced by SPOKES 20 be topologically equivalent?

5. The subprocedure at the right is used in the procedure below. Sketch a figure that might be obtained from FLOWERS 10.

```
TO FLOWER
   FD 20
   LT 45
   REPEAT 4 [FD 10 RT 90]
   RT 45
   BK 20
END
```

```
TO FLOWERS :NUMBER
   MAKE "X ( - 120 + RANDOM 240 )
   MAKE "Y ( - 80 + RANDOM 160 )
   PENUP SETXY :X :Y PENDOWN
   FLOWER
   MAKE "NUMBER :NUMBER -1
   IF :NUMBER = 0 STOP
   FLOWERS :NUMBER
END
```

Probability and Statistics

"Statistical thinking will one day be as necessary for efficient citizenship as the ability to read and write."

H.G. WELLS

10.1
PROBABILITY

Probability, a relatively new branch of mathematics, was started in Italy and in France during the sixteenth and seventeenth centuries to provide strategies for gambling games. The Italian mathematician Jerome Cardan made the first notable contribution to probability in *The Book on Games of Chance*. In this book Cardan analyzed the probability of obtaining a sum of 7 on a roll of two dice. Cardan also included tips on how to cheat and how to detect cheating. In the seventeenth century the mathematicians Blaise Pascal and Pierre Fermat worked together to develop what is considered to be the first theory of probability. Once again, it was questions about dice games that provided the initial incentive.

From these beginnings, probability has evolved to where it has applications in all walks of life. Life insurance companies use probability to estimate how long a person is likely to live. Each different age has a different probability, called *life expectancy*. Scientists use rules of probability for interpreting statistics and estimating values for experimental data. The outcomes of athletic events are predicted in terms of odds, which is another way of stating a probability. If the odds are high, the probability is close to 1, and if the odds are low, the probability is close to 0.

THEORETICAL PROBABILITY (COMPUTING PROBABILITY IN ADVANCE)

Before the theory of probability was developed, the chance of an event happening was determined solely by experience. Pascal and Fermat, on the other hand, were interested in determining the probabilities of events without relying on experiments. For example, on the roll of an ordinary die any one of six faces has an equally likely chance of turning up. Since there are six possible outcomes, the *probability* of rolling any particular number, such as 5, is $\frac{1}{6}$. This is written as

$$P(5) = \frac{1}{6}$$

and is read as, "the probability of 5 equals $\frac{1}{6}$." Similarly, the probability of rolling either a 3 or a 5 is $\frac{2}{6}$ because two out of six faces have the desired outcomes.

$$P(3 \text{ or } 5) = \frac{2}{6}$$

In general, when each of the outcomes is equally likely, we have the following definition:

Definition: Probability of event $= \dfrac{\text{Number of favorable outcomes}}{\text{Total number of outcomes}}$

Because this probability is computed theoretically and does not rely on experiments, it is called *theoretical probability*. Since the number of favorable outcomes is always less than or equal to the total number of outcomes, this definition shows that the probability of an event is some fraction from 0 to 1. If there are no favorable outcomes, such as rolling a die and getting an 8, the probability is 0. If every outcome is a favorable outcome, the probability is 1. For example, the probability of rolling a die and getting a number less than 7 is 1.

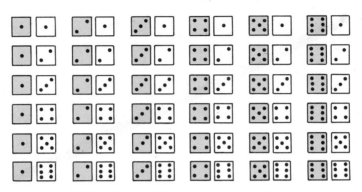

Thirty-six possible outcomes for two dice

When two dice are rolled, there are 36 equally likely outcomes. Find the pairs of dice in this array whose numbers have a sum of 7. You will see something special about their locations. Since six of these pairs have a sum of 7, the probability of rolling a 7 is $\frac{6}{36}$, or $\frac{1}{6}$.

$$P(\text{sum of } 7) = \frac{1}{6}$$

Locate the dice whose numbers have a sum of 3. What is the probability of rolling a sum of 3?

Many questions in probability can be answered by listing all possible outcomes and then determining the number of favorable outcomes. To consider another illustration of this approach suppose that the four playing cards facing down are numbered 2, 3, 4, and 5. If two cards are selected, what is the probability of getting the cards with 2 and 3? In all, there are six different ways that pairs of cards can be selected (see next page). These six pairs of cards represent equally likely outcomes. Since only one pair, namely, the cards with 2 and 3, is a favorable outcome, the probability of selecting a 2 and a 3 is $\frac{1}{6}$.

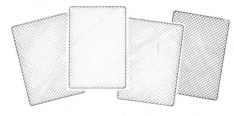

EMPIRICAL PROBABILITY
(COMPUTING PROBABILITY
FROM EXPERIENCE)

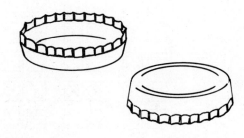

When a probability is determined from experience or by observing the results of experiments, it is called *empirical probability*. Suppose we were to flip a bottle cap instead of a coin. Because of its construction we cannot assume it is equally likely to land right side up as upside down. If we flip the bottle cap a large number of times and find that it lands right side up 20 percent of the time, we would conclude that the probability of landing upside down is greater than the probability of landing right side up. If many repeated experiments yield similar results, we will conclude from experience that the probability of landing right side up is about 20/100, or 1/5.

The empirical approach to probability is the only means of determining probability in many fields. Insurance companies measure the risks against which people are buying insurance in order to set premiums. A person's age and life expectancy are important factors. To compute the probability that a person 20 years old will live to be 65 years old, insurance companies gather birth and death records of large numbers of people and compile mortality tables. One such table indicates that for every 10 million births, 6,800,531 will live to age 65 (see table on page 602). Thus, the probability that a newborn baby will live to be 65 is 6,800,531/10,000,000 or about .68 (68 percent). According to the table, 9,664,994 people will live to age 20. Therefore, the probability of a baby living to age 20 is about .966, or 96.6 percent. We can compare the numbers from the table for two different ages to determine the probability of living from one age to another. The probability that a person of age 20 will live to be 65 is 6,800,531/9,664,994 or approximately .7 (70 percent).

ODDS

Racetracks state probabilities in terms of odds. Suppose that the odds against Blue Boy winning are 4 to 1. This means that for every dollar you bet on Blue Boy, the racetrack management will match it with $4. For each win, you receive the money you bet plus the money matched by the racetrack. That is, for $1 you receive $5, for $2 you receive $10, etc. With 4-to-1 odds, the racetrack management expects Blue Boy to lose four out of every five races. Thus, the probability of Blue Boy losing the race is $\frac{4}{5}$ and the probability of his winning is $\frac{1}{5}$.

This example suggests a procedure for converting odds to probabilities. If the *odds in favor* of an event are m to n, then the probability of the event is $m/(m + n)$. In this case, the *odds against* this event will be n to m, and the probability of the event not happening will be $n/(m + n)$. These relationships will be illustrated in the following examples.

"BLUE BOY SEEMS TO BE HOLDING BACK A BIT."

In a deck of 52 cards there are four aces. The probability of drawing an ace is 4/52:

$$\frac{\textbf{Number of favorable outcomes}}{\textbf{Total number of outcomes}} = \frac{4}{52} \ \text{ or } \ \frac{1}{13}$$

The probability of not drawing an ace is 48/52:

$$\frac{\textbf{Number of unfavorable outcomes}}{\textbf{Total number of outcomes}} = \frac{48}{52} \ \text{ or } \ \frac{12}{13}$$

The odds of drawing an ace are 1 to 12.

Number of favorable outcomes to *Number of unfavorable outcomes* = 4 to 48 (1 to 12)

The odds of not drawing an ace are 12 to 1.

Number of unfavorable outcomes to *Number of favorable outcomes* = 48 to 4 (12 to 1)

These examples show the close relationship between odds and probability. Both are just different ways of presenting the same information. This relationship can be clarified further by a diagram. The bar in this sketch has 13 equal parts, 12 to represent unfavorable outcomes and 1 representing a favorable outcome. The *probability* of an event is the ratio of the number of favorable outcomes to the total number of outcomes. This is illustrated in

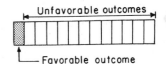

Unfavorable outcomes

Favorable outcome

the diagram by comparing the number of shaded parts to the total number of parts, 1 to 13 or $\frac{1}{13}$. The *odds* of an event is a ratio between the number of favorable outcomes and the number of unfavorable outcomes. This is illustrated by comparing the number of shaded parts of the bar to the number of unshaded parts, 1 to 12.

MATHEMATICAL EXPECTATION

To evaluate the fairness of a game we must consider the prize or monetary gain as well as the probability of winning. The probability of winning may be fairly small, but if the prize is large enough the game may be a good risk. We know, for example, that the probability of rolling a sum of 7 on two dice is $\frac{1}{6}$. This means that on the average a sum of 7 will come up once on every six rolls. If it costs $1 every time you roll the dice and the payoff is $6 for rolling a 7, then over a period of time you could expect to break even. The probability of winning multiplied by the value of the prize is called the *mathematical expectation*. In this game the mathematical expectation is 1 dollar: $\frac{1}{6} \times 6 = 1$. Since this is also the cost of playing each game, this game is considered to be *fair*.

The mathematical expectation for roulette shows that this game favors the "house." A roulette wheel has 38 compartments: Two are numbered 0 and 00 and are colored green; the remaining ones are numbered 1 through 36, and half are red and half are black. One way of playing this game is to bet on the red or black colors. The probability of a red is 18/38 or 9/19 (number of red compartments/total number of compartments). If you bet $1 on the red and win, you are paid $2. This game has a mathematical expectation of 9/19 $\times$ $2, or approximately 95¢. Since this is less than a $1 bet, the game is unfair.

Sometimes there is more than one category for winning. Several states, for example, have sweepstakes and for each ticket there are several categories of prizes. One type of ticket sold in New Hampshire has two five-digit numbers. The amount you win depends on which digits of the two numbers match. For example, if the two digits in the ten thousands places are equal, you win $2000. If the two digits in the thousands places are equal, you win $20, etc., as shown on the lottery ticket. Anyone who wins the $5-, $2-, or $1-ticket prizes is eligible for the $100,000 grand prize. If the probabilities of winning

$100,000, \$2000, \$20, \$5, \$2 or one ticket are $1/100{,}000$, $1/20{,}000$, $1/200$, $1/25$, $1/10$, and $1/5$ respectively, then the mathematical expectation is

$$\frac{1}{1{,}000{,}000}\,(\$100{,}000) + \frac{1}{20{,}000}\,(\$2000) + \frac{1}{200}\,(\$20) + \frac{1}{25}\,(\$5) + \frac{1}{10}\,(\$2) + \frac{1}{5}\,(\$1) = 90\text{¢}$$

MONTE CARLO METHOD

Sometimes it is difficult to compute a theoretical probability and impractical to determine an empirical probability by experimenting with concrete objects.

BIRTH-MONTH PROBLEM   What is the probability that in a group of five people chosen at random, at least two will have a birthday in the same month?

To find an empirical probability by polling the birth-months of a large group of people would be time-consuming. Another approach is to simulate the mathematics in this problem with a model. For example, we can use a device for randomly generating the numbers from 1 to 12, for the 12 months, and count the number of times there are two or more equal numbers in each group of five. The repeated use of a model involving a random generator is called the *Monte Carlo method*.

The following groups of random numbers were obtained from a spinner. There are 20 groups and 12 have the same number occurring two or more times. Therefore, the probability for this simulation of the birth-month problem is $12/20 = .6$.

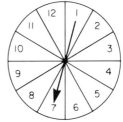

9	3	9	9	11	7	2	5	4	4	6	9	9	11	12	1	1	10	8	5
1	2	6	12	10	3	9	10	9	1	9	5	9	11	11	7	11	12	2	6
2	7	5	9	2	5	11	4	9	7	11	12	5	8	6	12	3	2	12	8
2	6	3	11	2	7	10	11	7	9	7	2	6	3	12	2	12	5	1	7
7	6	10	9	7	3	8	2	3	12	9	12	4	8	7	6	9	1	4	12

The theoretical basis for the Monte Carlo method is called the *law of large numbers.* This states that the more times the simulation is carried out, the closer the probability,

Number of favorable outcomes
———————————————————
Total number of outcomes

is to the theoretical probability.*

COMPUTER APPLICATIONS

The computer can be programmed to simulate random outcomes such as those obtained from dice, coins, spinners, etc. The following sequence of numbers was obtained from a program that simulates the toss of a die. Notice that 6 occurs once in the first three "tosses," once in the second three "tosses," and once in the third three "tosses."

1 3 6 | 1 5 6 | 6 2 4 | 1 1 2 | 3 5 5 | 6 6 3 | 6 1 4 | 5 1 6 | 6 4 5 | 5 6 2 | 3 6

PROBLEM   What is the probability of obtaining at least one 6 in three tosses of a die?

Let's obtain a simulated probability by using the computer and the Monte Carlo method. The computer simulates random outcomes by using the random number function, RND(X). Different values for X produce varying results depending on the type of computer. On some computers if X is a whole number greater than 1, RND(X) will produce a random whole number from 1 to X; and if X = 1, RND(1) will produce a random decimal between 0 and 1.† The following program uses RND(1) to generate 20 random decimals:

Program 10.1A

```
10  FOR N = 1 TO 20
20  PRINT  RND (1)" ";
30  NEXT N
40  END
```

```
RUN
.418370941 .0104457493 .724081156 .688485832 .783594547
.785313342 .282624159 .234071481 .0926492609 .556071849
.678107428 .315772133 .74983173 .831719192 .664750396
.359356579 .301872403 .882338336 .737358009 .690074356
```

---

*Kenneth J. Travers, "Using Monte Carlo Methods to Teach Probability and Statistics," *Teaching Statistics and Probability, 1981 Yearbook* (Reston, Virginia: The National Council of Teachers of Mathematics, 1981), pp. 203–219.

†*Note:* Some versions of BASIC may require RND(0) or RND(X) to produce random decimals between 0 and 1.

The random number function can be combined with the integer function to produce random whole numbers. The next program generates 50 random whole numbers between 0 and 25. A decimal is obtained from the RND function and multiplying it by 25 produces a number between 0 and 24. The INT function strips off the decimal part of the number leaving a whole number which is greater than or equal to zero and less than 25.

Program 10.1B

```
10   FOR N = 1 TO 50
20   PRINT  INT (25 *  RND (1))" ";
30   NEXT N
40   END
```

```
RUN
```

```
9 8 20 10 15 14 23 12 13 24 23 16 12 1 6 14 17 3 21 14 3 16 5 1 14
14 20 8 15 1 17 13 9 13 2 22 12 9 12 5 0 22 10 3 16 16 21 13 13 15
```

Now let's look at a program that simulates 60 tosses of a die. The integer and random number functions in line 20 produce a whole number that is greater than or equal to 0 and less than or equal to 5. By adding 1 we obtain a whole number from 1 to 6.

Program 10.1C

```
10   FOR N = 1 TO 60
20   PRINT  INT (6 *  RND (1)) + 1" ";
30   NEXT N
40   END
```

```
RUN
```

```
4 6 6 1 6 4 3 2 3 3 5 3 5 2 1 5 3 1 2 5
5 6 2 6 2 1 6 1 2 2 3 3 3 6 5 1 4 1 6
1 1 5 1 4 3 5 2 2 3 4 2 4 4 4 1 5 1 1 3
```

Group the numbers in this printout by 3's and count the number of groups with one or more 6's. What is the simulated probability of obtaining at least one 6 in three tosses of a die? The theoretical probability is approximately .42.

SUPPLEMENT (Activity Book)

Activity Set 10.1 Probability Experiments
Just for Fun: Buffon's Needle Problem

"The greatest advantage in gambling comes in not gambling at all." JEROME CARDAN

Exercise Set 10.1: Applications and Skills

1. Out of 36 possible ways two dice can come up, 6 ways produce a sum of 7. Compute the probability of rolling each of the other sums. (You may want to use the array of dice on page 593.)

Sum	2	3	4	5	6	7	8	9	10	11	12
Probability						$\frac{6}{36}$					

★   a. What is the probability of a sum greater than or equal to 8?

★   b. What is the probability of a sum greater than 4 and less than 8?

2. The dice game called "craps" is played by the following rules. If the player rolling the dice gets a sum of 7 or 11, he or she wins. If this player rolls a sum of 2, 3, or 12, he or she loses. If the first sum rolled is a 4, 5, 6, 8, 9, or 10, the player continues rolling the dice. The player now wins if he or she can obtain the first sum rolled before rolling a 7.

a. What is the probability of rolling a 7 or an 11? (See table in Exercise 1.)

★   b. What is the probability of losing on the first roll?

c. Suppose a player rolls an 8 on the first turn. Which is the better probability on the second turn: rolling a 7 or an 8?

3. A chip is to be drawn from a box containing the following numbers of colored chips: 8 orange, 5 green, 3 purple, and 2 red. Determine the probabilities of selecting the following chips:

a. A purple chip.

★   b. A green or purple chip.

c. A chip that is not orange.

★ 4. The five regular polyhedra, called Platonic solids, are pictured next. These are the only polyhedra that can be used for fair dice. The regular polyhedron with 20 faces (see upper right corner) was used as a die by the Egyptians more than 2000 years

600

ago. The faces of these five polyhedra are labeled with consecutive whole numbers beginning with 1. For example, the tetrahedron has numbers 1, 2, 3, 4; and the icosahedron has numbers from 1 to 20. Determine the probabilities of rolling the numbers in the following table for each of these "dice."

tetrahedron
(4 faces)

cube
(6 faces)

octahedron
(8 faces)

dodecahedron
(12 faces)

icosahedron
(20 faces)

	Tetrahedron	Cube	Octahedron	Dodeca-hedron	Icosahedron
★ a. A number less than 3	2/4	2/6	2/8	2/12	2/20
★ b. An even number					
★ c. The number 2					

5. A deck of 52 cards contains 13 cards in each of four suits: spades, hearts, clubs, and diamonds. The clubs and spades are black, and the diamonds and hearts are red. Each suit has three face cards: jack, queen, and king. Compute the probabilities and odds of drawing each of the following cards:

		Prob.	Odds
★	a. An ace		
	b. A face card		
★	c. A diamond		
	d. A black face card		

6. When the probability that an event will happen is added to the probability that the event will not happen, the sum is 1. This fact and your answers in Exercise 5 will be helpful in computing the following probabilities:

★ a. Not drawing an ace.
★ b. Not drawing a face card.
★ c. Drawing a club, heart, or spade.
★ d. Not drawing a black face card.

7. For each of the given events and probabilities, determine the odds in favor of each event.

   ★ a. The probability of living to age 65 is 7/10.

   b. The probability of selecting a person with type O blood is 3/5.

   ★ c. The probability of winning a certain raffle is 1/500.

   d. The probability of rain on Monday is 80 percent.

8. Gambling syndicates predict the outcomes of sporting events in terms of odds. Convert each of the following odds to probabilities:

   ★ a. The odds on the Packers over the Rams are 4 to 3.

   b. The odds on the University of Michigan defeating Ohio State are 7 to 5.

   c. The odds on the Yankees winning the pennant are 10 to 3.

9. The accompanying mortality table is based on the lives and deaths of policyholders in several large insurance companies. Use this information to answer the following questions:

   ★ a. What is the empirical probability that a child will live to be 1 year old?

   b. What is the empirical probability of a person living to age 50?

   ★ c. What is the empirical probability that a person of age 34 will live to age 65?

   d. What is the empirical probability that a 60-year-old person will live to be 65?

   ★ e. If an insurance company has 7000 policyholders at age 28, how many death claims must they be prepared to pay before these people reach age 29?

Age	Number Living	Number Dying	Age	Number Living	Number Dying
0 . . .	10,000,000	70,800	20 . . .	9,664,994	17,300
1 . . .	9,929,200	17,475	21 . . .	9,647,694	17,655
2 . . .	9,911,725	15,066	22 . . .	9,630,039	17,912
3 . . .	9,896,659	14,449	23 . . .	9,612,127	18,167
4 . . .	9,882,210	13,835	24 . . .	9,593,960	18,324
5 . . .	9,868,375	13,322	25 . . .	9,575,636	18,481
6 . . .	9,855,053	12,812	26 . . .	9,557,155	18,732
7 . . .	9,842,241	12,401	27 . . .	9,538,423	18,981
8 . . .	9,829,840	12,091	28 . . .	9,519,442	19,324
9 . . .	9,817,749	11,879	29 . . .	9,500,118	19,760
10 . . .	9,805,870	11,865	30 . . .	9,480,358	20,193
11 . . .	9,794,005	12,047	31 . . .	9,460,165	20,718
12 . . .	9,781,958	12,325	32 . . .	9,439,447	21,239
13 . . .	9,769,633	12,896	33 . . .	9,418,208	21,850
14 . . .	9,756,737	13,562	34 . . .	9,396,358	22,551
15 . . .	9,743,175	14,225	35 . . .	9,373,807	23,528
16 . . .	9,728,950	14,983	36 . . .	9,350,279	24,685
17 . . .	9,713,967	15,737	37 . . .	9,325,594	26,112
18 . . .	9,698,230	16,390	38 . . .	9,299,482	27,991
19 . . .	9,681,840	16,846	39 . . .	9,271,491	30,132

CHAPTER 10   PROBABILITY AND STATISTICS

Age	Number Living	Number Dying	Age	Number Living	Number Dying
40 ...	9,241,359	32,622	55 ...	8,331,317	108,307
41 ...	9,208,737	35,362	56 ...	8,223,010	116,849
42 ...	9,173,375	38,253	57 ...	8,106,161	125,970
43 ...	9,135,122	41,382	58 ...	7,980,191	135,663
44 ...	9,093,740	44,741	59 ...	7,844,528	145,830
45 ...	9,048,999	48,412	60 ...	7,698,698	156,592
46 ...	9,000,587	52,473	61 ...	7,542,106	167,736
47 ...	8,948,114	56,910	62 ...	7,374,370	179,271
48 ...	8,891,204	61,794	63 ...	7,195,099	191,174
49 ...	8,829,410	67,104	64 ...	7,003,925	203,394
50 ...	8,762,306	72,902	65 ...	6,800,531	215,917
51 ...	8,689,404	79,160	66 ...	6,584,614	228,749
52 ...	8,610,244	85,758	67 ...	6,355,865	241,777
53 ...	8,524,486	92,832	68 ...	6,114,088	254,835
54 ...	8,431,654	100,337	69 ...	5,859,253	267,241

10. The odds posted by racetracks are the "odds against" a particular horse, dog, or cat entered in the event. For each of these bets and odds, determine how much money the bettor receives back on a win.

	a. ★ b.		c.
Bet	$2	$6	$10
Odds	3 to 2	4 to 1	7 to 5

5?

11. Suppose that there are three red, four blue, and five green chips in a bag and that you win by selecting either a red or a green chip. If you get a red chip you win $3 and a green chip pays $2.

★ a. What is the mathematical expectation?

b. If it costs $1.50 to play this game, would you expect to win or to lose in the long run?

12. One type of bet that can be made in roulette is to place a chip on a single number. If the ball lands in the compartment with your number, the "house" pays you 36 chips.

★ a. What is the probability that the ball will land on 13?

b. If each chip is worth $1, what is the mathematical expectation in this game?

★ c. Is the mathematical expectation for playing a color (as computed on page 596) greater than, less than, or equal to that for playing a particular number?

13. Use the Monte Carlo method to obtain a simulated probability for each of the following questions. Describe your model and the number of experiments to obtain the probability.

a. What is the probability that in a family of three boys and two girls the three boys were born in succession?

★ b. On a certain quiz show people are required to guess which of three identical looking envelopes contains a $1000 bill. What is the probability that exactly four out of eight people will guess the correct envelope?

Exercise Set 10.1: Problem Solving

1. *Craps with Dodecahedra:* Redesign the dice game "craps" (see page 600, Exercise 1) to be played with two dodecahedra dice.

Dodecahedra Dice

★   a. *Understanding the Problem:* Each dodecahedron die has whole numbers from 1 to 12 on its faces. How many different sums are possible by rolling these two dice?

     b. *Devising a Plan:* In the regular game of craps the probability of winning on the first toss is $\frac{2}{9}$ (rolling a 7 or 11) and the probability of losing on the first toss is $\frac{1}{9}$ (rolling a 2, 3, or 12). One plan for designing the new game is to keep these probabilities of winning and losing the same. To determine new sets of numbers for winning and losing it will be helpful to know how many times each sum can be obtained. Complete the following table:

Sums	2	3	4	5	6	7	8	9	10	11	12	13	14	15	16	17	18	19	20	21	22	23	24
No. of Sums	1	2																					

     c. *Carrying Out the Plan:* What new numbers can be included with 7 and 11 to produce a winning probability of $\frac{2}{9}$? What new number can be included with 2, 3, and 12 to produce a losing probability of $\frac{1}{9}$? Suppose a player does not win or lose on the first toss. Suggest a method for finishing the game.

★   d. *Looking Back:* Suppose the dice for this game are octahedra. How many different sums are possible? What is the probability of rolling a sum of 7 or 11 on these dice?

★ 2. *New Cubical Dice:* Design a numbering system for the faces of two cubes so that when they are rolled, the sum can be any whole number from 1 to 12 and each sum has the same probability of occurring. Solve a similar problem for the other four regular polyhedral dice (see page 600, Exercise 4).

3. Galileo was once asked the following question by an Italian noble: When three dice are tossed, why does the total 10 show more often than the total 9? What is the probability that the sum from three dice will be 9? 10? (*Hint:* Galileo answered the question by making a table of the 216 equally likely outcomes when rolling three dice.)

★ 4. There are five prize amounts under the six rectangles on the following lottery ticket. If the same prize amount appears in three separate rectangles, the ticket owner wins that prize. The five prizes are $5000, $25, $5, $2 and 1 ticket (worth $1). Are these tickets printed so that each prize has a $\frac{1}{5}$ chance of occuring in each rectangle? (*Hint:* Use the Monte Carlo method to obtain a simulated probability of winning a prize, and then compute the mathematical expectation.

Exercise Set 10.1: Computers

1. Use a computer program and the Monte Carlo method to find the probability of obtaining at least one 6 in four tosses of a die. Explain how you obtain your answer.

★ 2. Write a program that will simulate outcomes for the spinner on page 597. Use this program to print out 100 numbers. Determine a simulated probability that at least two people in a group of five will have a birthday in the same month.

3. There are six digits under the six squares on this sweepstakes ticket. If the same digit occurs five times, you win $10,000, and so forth, as shown on the ticket. Only the digits 1, 2, 3, 4, 5, 6, 7, 8, and 9 are used. Write a computer program to select randomly six of these nine digits. Use the Monte Carlo method and this program to determine the following probabilities: (Note: These probabilities are higher than the actual probabilities for the sweepstakes tickets.)

   a. 2 of a kind

   ★ b. 3 of a kind

4. Write a program for playing the "red" or "black" on a roulette wheel. (*Hint:* Generate random numbers from 1 to 38 for the 38 compartments.)

Photo from NASA's Synchronous Meteorological Satellite-1

10.2
COMPOUND PROBABILITIES

Meteorologists use computers and probability to analyze weather patterns. In recent years meteorological satellites have improved the chances of accurate weather forecasting. NASA's Synchronous Meteorological Satellite-1 returned the above photograph of the Western Hemisphere on May 28, 1974. Four storms can be seen across the top of the picture from western Canada to the Atlantic Ocean. The small-scale structure of cumulus clouds over Florida and the Caribbean Sea allows meteorologists to infer wind speed and direction. Weather forecasts are usually stated in terms of probability: "There is a 20 percent chance of rain." The probability we are given is determined by a combination of probabilities. For example, there may be one probability of a cold front approaching from the north and a different probability for a storm approaching from the west. The probability of two or more events happening is called a *compound probability*.

Many of the questions that led to the development of probability contained two or more events. For example, if one die is to be rolled twice,

Second Toss

	·	··	·.·	::	:·:	:::
·	1,1	1,2	1,3	1,4	1,5	1,6
·.	2,1	(2,2)	2,3	(2,4)	2,5	(2,6)
·.·	3,1	3,2	3,3	3,4	3,5	3,6
::	4,1	(4,2)	4,3	(4,4)	4,5	(4,6)
:·:	5,1	5,2	5,3	5,4	5,5	5,6
:::	6,1	(6,2)	6,3	(6,4)	6,5	(6,6)

First Toss

what is the chance of getting an even number on each roll? The table on page 606 shows that there are 36 possible outcomes for these two events and that 9 of these contain pairs of even numbers (circled in table). Therefore, the probability of both numbers being even is 9/36 or 1/4. Find all the pairs of numbers that have at least one 3. What is the probability of rolling a die twice and getting at least one 3?

INDEPENDENT EVENTS

In the preceding examples, the first toss of the die had no effect on the outcome of the second toss. Two or more events that have no influence on each other are called *independent events*. The probability of several independent events all happening can be computed by the following formula.

> **Theorem:** The probability of independent events is the product of the separate probabilities of each event.

Let's apply this formula to one of the preceding examples. The probability of rolling an even number on a die is 1/2. Therefore, the probability of an even number on both tosses of the die is $1/2 \times 1/2 = 1/4$.

Here is another example of independent events. What is the probability of drawing a 4 from the five cards shown here and also rolling a 5 on the die? The probability of drawing a 4 is 3/5 (3 of the 5 cards have 4's), and the probability of rolling a 5 is 1/6. By the formula for the probability of independent events, the probability of drawing a 4 and rolling a 5 is $3/5 \times 1/6 = 3/30$ or 1/10. We can see that this probability is correct by listing all possible outcomes (as shown in the table) and then counting those which consist of a 4-card and a toss of 5 (circled pairs).

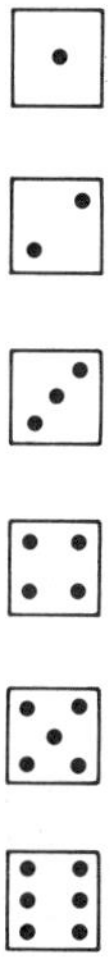

The formula we are using shows that the probability of two events occurring in succession is much poorer than the probability of either event. Consider the chance of drawing two aces in succession from a deck of 52 cards. The probability of selecting an ace on the first draw is 4/52 or 1/13. If the card selected is then replaced, the probability of drawing an ace the second time is also 1/13. These draws are independent of each other, so the probability of selecting two aces is $1/13 \times 1/13$, or 1/169.

	4D	4H	4S	2C	7D
1	1,4D	1,4H	1,4S	1,2C	1,7D
2	2,4D	2,4H	2,4S	2,2C	2,7D
3	3,4D	3,4H	3,4S	3,2C	3,7D
4	4,4D	4,4H	4,4S	4,2C	4,7D
5	(5,4D)	(5,4H)	(5,4S)	5,2C	5,7D
6	6,4D	6,4H	6,4S	6,2C	6,7D

DEPENDENT EVENTS

Suppose that the first card selected in the previous example is not replaced for the second drawing. Then the chance of drawing an ace on the second draw depends on the outcome of the first. If the first card is an ace, then only three of the remaining 51 cards are aces. In this case the probability of an ace on the second draw is 3/51. When one event affects the outcome of another, they are called *dependent events*. As with independent events, the probability of two dependent events can be computed by multiplying the probabilities of each event. In this example, the probability of drawing two aces is $1/13 \times 3/51 = 3/663$ or $1/221$. As you would expect, this probability is smaller than getting two aces when the first card is replaced (1/169).

In most card games the different outcomes are dependent events. What is the probability of being dealt two hearts from the following simplified deck of cards: 9 of hearts, 8 of hearts, 2 of hearts, 9 of clubs, and 5 of spades? The probability of a heart on the first card is 3/5, and if the first card is a heart, the chance that the second card will be a heart is 2/4. The probability of getting two hearts is $3/5 \times 2/4 = 6/20$ or $3/10$. This can be verified by comparing the pairs of cards with both hearts (circled pairs) to the total number of pairs in the table.

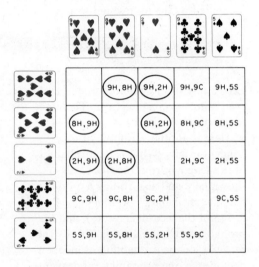

	9H	8H	2H	9C	5S
9H		(9H,8H)	(9H,2H)	9H,9C	9H,5S
8H	(8H,9H)		(8H,2H)	8H,9C	8H,5S
2H	(2H,9H)	(2H,8H)		2H,9C	2H,5S
9C	9C,9H	9C,8H	9C,2H		9C,5S
5S	5S,9H	5S,8H	5S,2H	5S,9C	

In most familiar card games there is such a variety of combinations that one rarely gets the same hand twice. The total number of different five-card poker hands is 2,598,960, and there are 635,013,559,600 different bridge hands. The chance of getting any particular hand involves dependent events. Consider, for example, the chance of receiving a flush in poker (all five cards of the same suit). The first card dealt could be any suit. Suppose it was a club. The second card must also be a club and the probability of getting it would be 12/51. The probabilities that the third, fourth, and fifth cards are clubs are 11/50, 10/49, and 9/48, respectively. Thus, the probability of getting five clubs (or five cards of any one suit) is

Poker hand—a flush

$$\frac{12}{51} \times \frac{11}{50} \times \frac{10}{49} \times \frac{9}{48} = \frac{11,880}{5,997,600} \quad \text{or about} \quad \frac{1}{505}$$

BINOMIAL PROBABILITY

According to birth records there are more boys born than girls. However, since each sex is approximately 50 percent of the total, it is customary to assume that the probability of a girl is $1/2$ and that for a boy is $1/2$. If we also assume that having two or more children are independent events, then the probability of two girls is $1/2 \times 1/2 = 1/4$. Similarly, the probability of having three boys is $1/2 \times 1/2 \times 1/2 = 1/8$. The probability of several repeated independent events in which there are two outcomes for each event is called a *binomial probability*.

The Harrison family from Tennessee had 13 boys. The probability of this happening is

$$\frac{1}{2} \times \frac{1}{2} \times \frac{1}{2} \times \frac{1}{2} \times \frac{1}{2} \times \frac{1}{2} \times \frac{1}{2} \times \frac{1}{2} \times \frac{1}{2} \times \frac{1}{2} \times \frac{1}{2} \times \frac{1}{2} \times \frac{1}{2} = \frac{1}{8192}$$

The chance of 14 consecutive boys is $1/16,384$. Since the probability of this happening is very small, you might expect that if a couple had 13 boys, the next child would likely be a girl. However, since each event is independent, the probability of another boy is still $1/2$.

Here is another example that is contrary to our intuition. It seems reasonable to expect that if a couple is planning to have 4 children, the probability of having 2 boys and 2 girls is $1/2$. To determine the correct probability, consider the combinations of 4 children which are listed in the right-hand column of this tree diagram. The various paths of this diagram show that there are 16 combinations and that only 6 of these paths have 2 boys and 2 girls. Therefore, the probability of 2 boys and 2 girls is $6/16$ (or $3/8$) rather than $1/2$.

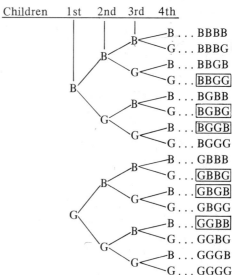

COMPLEMENTARY PROBABILITIES

The probability that an event will happen and the probability that it will not happen are called *complementary probabilities*. The sum of two complementary probabilities is always 1. The probability of not getting a sum of 7 on a roll of two dice is $5/6$, because the probability of getting a 7 is $1/6$. The $5/6$ and $1/6$ are complementary probabilities.

There is a particular type of problem in which the use of complementary probabilities is especially convenient. Consider the following example:

What is the probability of getting at least one 6 in four rolls of a die?

One way to handle this problem is to find the probability of not getting a 6 on any roll. There is a 5/6 chance of not getting a 6 each time the die is rolled, and so the probability of no 6's in four rolls is

$$\frac{5}{6} \times \frac{5}{6} \times \frac{5}{6} \times \frac{5}{6} = \frac{625}{1296} \quad \text{or approximately 48 percent}$$

The complement to this probability, 52 percent, is the probability of getting at least one 6.

In this example, finding the probability of rolling at least one 6 without using complementary probability is much more difficult. This is because of the many ways of succeeding. Rolling a 6 four times in succession or getting two 6's in four rolls, etc., are all ways of satisfying the condition of "at least one 6."

The Birthday Problem   This famous problem has a solution that is unexpected and difficult to believe.

What is the smallest arbitrary group of people for which there is better than a 50 percent chance that at least two of them will have a birthday on the same day of the year?

Surprisingly, the answer is only 23 people. In fact, for 30 people the probability of two birthdays on the same day is 70 percent, and for 50 people it is about 97 percent. This graph shows that for more than 50 people we can be almost certain of two birthdays on the same day.

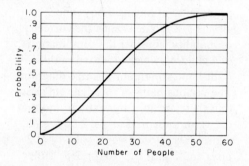

The Birthday Problem can be solved by complementary probabilities. That is, we will determine the probability that all 23 people have birthdays on different days, and then subtract this probability from 1. To begin with, consider the problem for just two people. It doesn't matter what date the first person's birthday is on; there is a probability of 364/365 that the second person's birthday *will not* be on the same day. When a third person joins this group the probability that his or her birth date will differ from those of the other two people is 363/365. Therefore, the probability that three people will not share a birthdate is 364/365 × 363/365.

Similarly, the probability that all 23 people will have different birthdays is

$$\frac{364}{365} \times \frac{363}{365} \times \frac{362}{365} \times \cdots \times \frac{344}{365} \times \frac{343}{365} \quad \text{or approximately 49 percent}$$

Therefore, the probability that there will be two or more more birthdays on the same day, for a group of 23 people, is $100\% - 49\% = 51\%$. Make a prediction next time you're in a group of 23 or more people. The odds will be in your favor that there will be at least two birthdays on the same day.

COMPUTER APPLICATIONS

There are eight possible outcomes when a coin is tossed three times in succession:

HHH   HHT   HTH   THH   TTT   TTH   THT   HTT

The mathematician Walter Penny discovered that no matter which of these combinations is selected, it is always possible to select a second combination that will have a better chance of occurring in a random sequence of heads and tails. For example, HTT is more likely to occur before TTH. The following sequence of heads and tails was obtained from a program that simulates the toss of a coin. Which combination occurs first in this sequence, HTT or TTH?

HHHTTHTHTH

The following table contains the eight possible outcomes, and above each one is a dominating outcome that is more likely to occur in any random sequence. Select two of the outcomes from a column of this table and repeatedly flip a coin to see which occurs first.

Dominating Outcomes	THH	THH	HHT	TTH	HTT	HTT	TTH	HHT
Eight Outcomes	HHH	HHT	HTH	THH	TTT	TTH	THT	HTT

Let's use the Monte Carlo method to determine how much more likely one outcome is than another. First, we need a computer program that will simulate the tossing of a coin. Since the computer can be programmed to distinguish between odd and even numbers, it is customary to assign "heads" and "tails" to these types of numbers. Line 20 of Program 10.2A produces a random whole number, possibly as large as 9999, and line 30 determines if this number is even or odd. This program will print a sequence of 10 H's and T's. Will "T" or "H" be printed if $Z = 346$ in line 30?

Program 10.2A

```
10   FOR N = 1 TO 10
20   LET Z =   INT (10000 *   RND (1))
30   IF Z / 2 =   INT (Z / 2) THEN   GOTO 70
40   PRINT "T ";
50   NEXT N
60   GOTO 90
70   PRINT "H ";
80   GOTO 50
90   END
```

Here are 10 sequences of H's and T's for 10 different runs of this program. Since HTT occurs before TTH in 8 of these sequences, we obtain a simulated probability of .8 that HTT will occur before TTH. (*Note:* A greater number of printouts should be used to give a more reliable probability.)

```
H T H T T T H T T T
H H H T T T H H H T
H T T T T H H H H H
T T H T H T T T T H
H H H T T H H T T H
H T H H T H T T H T
T H T T T T T T T H
H H H H T T H H H T
T T T T H T T H H H
H T T H T T T H T T
```

SUPPLEMENT (Activity Book)

Activity Set 10.2 Compound Probability Experiments
Just for Fun: Trick Dice

If you had it all to do over, would you change anything? "Yes, I wish I had played the black instead of the red at Cannes and Monte Carlo."

WINSTON CHURCHILL

Exercise Set 10.2: Applications and Skills

1. Find the following roulette wheel probabilities. (There are 18 red, 18 black, and 2 green compartments.) P(once on red) $\frac{18}{38} = \frac{9}{19}$

★ a. Winning twice in a row on red. P(twice on red) $\frac{9}{19} \times \frac{9}{19} = \frac{81}{368}$

b. Losing twice in a row if you bet on red each time. (*Hint:* You lose if the ball lands in a green or a black compartment.) P(lose) = $\frac{20}{38}$ twice = $\frac{20}{38} \times \frac{20}{38}$

★ c. Winning three times in a row on black.

d. Winning 26 times in a row on black. Indicate your answer, but don't multiply it out. While the chance of black coming up 26 times is very small, it has happened before as related by Darrell Huff in *How to Take a Chance.** On August 18, 1913, at a casino in Monte Carlo, black came up 26 times in a row on a roulette wheel. After about the fifteenth occurrence of black, there was a near-panic rush to bet on red. People kept doubling their bets in the belief that, after black came up the twentieth time there was not a chance in a million of another repeat. The casino came out ahead by several million francs.

★ 2. If you flipped a fair coin 9 times and got 9 heads, what is the probability of a head on the next toss? $\frac{1}{2}$ independent events

*D. Huff, *How to Take a Chance* (New York: W.W. Norton, 1959), pp. 28–29.

612   CHAPTER 10   PROBABILITY AND STATISTICS

3. Classify the following events as *dependent* or *independent* and compute their probabilities:

   $P(\text{3 heads}) = \frac{1}{2} \times \frac{1}{2} \times \frac{1}{2}$
   $= \frac{1}{8}$

   ★ a. Tossing a coin three times and getting three heads in a row.

   ★ b. Drawing 2 aces from a complete deck of 52 playing cards. (The first card selected is not replaced.)

   c. Rolling two dice and getting a sum of seven twice in succession.

   d. Selecting two green balls from a bag of five green and three red. (The first ball selected is not replaced.)

4. Alice and Bill make one payment each week, and it is determined by the "debits spinner." Assume each outcome on the spinner is equally likely.

   "O.K., Alice, spin the wheel and let's see who gets paid this week."

   $p(\text{fuel}) \; \frac{1}{12}$

   ★ a. What is the probability of making a fuel payment 2 weeks in a row? $P(2 \text{ fuel}) = \frac{1}{12} \times \frac{1}{12} \left(\frac{1}{144}\right)$

   b. What is the probability they will not make an electricity payment this week?

   ★ c. If they don't make an electricity payment within the next 3 weeks, their lights will be shut off. What is the probability they will lose their lights?

5. The paths of this tree diagram show the various combinations of three children. There are three ways of having two girls (BGG, GBG, and GGB); there is one way of having no girls, that is, all three children are boys; etc.

   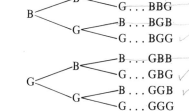

   ```
   1st    2nd    3rd
                  B . . . BBB
              B
                  G . . . BBG
          B
                  B . . . BGB
              G
                  G . . . BGG ✓
                  B . . . GBB
              B
                  G . . . GBG ✓
          G
                  B . . . GGB ✓
              G
                  G . . . GGG
   ```

   a. Complete the following table with the numbers of combinations for having 0, 1, 2, or 3 girls in a family of three children.

   ★ b. What is the probability of having one girl and two boys? $\frac{3}{8}$

   c. If a family has two girls, what is the probability that the next child will be a boy? $\frac{3}{8}$

Number of Girls	0	1	2	3
Combinations	1		3	

6. A couple is planning to have four children.

   a. Determine the numbers of combinations for which there will be 0, 1, 2, 3, or 4 girls. (*Hint:* See the tree diagram on page 609.)

Number of Girls	0	1	2	3	4
Combinations	1	4	6	4	1

★ b. The numbers in Pascal's triangle (page 6) can be used to find the numbers of combinations in part **a** of Exercises 5 and 6. Explain how.

```
        1
       1 1
      1 2 1
     1 3 3 1
    1 4 6 4 1
   1 5 10 10 5 1
```

   c. Use your observations in part **b** to determine the numbers of combinations for having 0, 1, 2, 3, 4, or 5 girls in a family of five children. Write them in the table.

Number of Girls	0	1	2	3	4	5
Combinations						

★ d. What is the probability of having one girl and four boys in a family of five children?

7. A bureau drawer contains 10 black socks and 10 brown socks. Assuming the socks are to be selected in the dark, find the probabilities of the following events:

dependent

★ a. Selecting two black socks. $P(1 \ black) = \frac{10}{20}$ or $\frac{1}{20}$ $P(2 \ black) = \frac{1}{2} \times \frac{9}{19} = \frac{9}{38}$

   b. Selecting a black sock and then a brown sock.

★ c. Selecting three socks and having two of the same color.

$\left(\frac{1}{2}\right)^{10} \rightarrow$ 8. On a test of 10 true or false questions what is the probability of getting them all correct by guessing? *independent events* $P(getting \ \#1 \ correct) \ \frac{1}{2}$

$\frac{1}{1024}$
$= .00098$

$19 \ 2 \big/ \frac{1}{8} \ 3$

9. In an experiment designed to test estimates of probability, people were given the three alternatives in parts **a** and **b**. In each case which response seems most preferable? Check your answers by determining the probabilities.

★ a. 1) Drawing once for a winning ticket in a box of 10.

2) Drawing for a winning ticket in a box of five, two different times, and winning both times.

3) The chances are the same in both cases.

b. 1) Drawing once for a winning ticket in a box of 10.

2) Drawing 2 times for a single winning ticket in a box of 20. (*Hint:* There are 190 different combinations of 2 tickets and 19 winning pairs.)

3) The chances are the same in both cases.

10. Determine the probabilities of the following events. (*Hint:* Use complementary probabilities.)

P(not getting 7)
$= \frac{36}{36} = \frac{5}{6}$

★ a. Getting a sum of 7 at least once on four rolls of a pair of dice. *4 rolls* *P(sum of 7)*

b. Getting at least one 6 on four rolls of one die.

★ c. Getting at least one sum of 7 or 11 on three rolls of a pair of dice.

11. Solve each of the following problems by using: (1) the Monte Carlo method, and (2) complementary probability. Describe your model for the Monte Carlo method and the number of experiments to obtain the simulated probability.

a. A system with three components fails if one or more components fail. The probability that any given component fails is 1/10. What is the probability that the system fails?

★ b. A manufacturer of bubble gum puts a 5¢ coupon in one out of every five packages of gum. What is the probability of obtaining at least one of these coupons in four packages of gum?

12. Mr. and Mrs. Petritz of Butte, Montana, have five children who were all born on April 15. In the following questions, assume that it is equally likely that a child will be born on any of the 365 days of the year:

a. After the first Petritz child was born, what was the probability the second child would be born on April 15?

★ b. The third and fourth Petritz children are twins. What was the probability that the second and third children would be born on April 15?

c. If a couple has a child on April 15, what is the probability that their next three children will be born on April 15, if there are no multiple births?

13. The typical slot machine has three wheels that operate independently of one another. Each wheel has six different kinds of symbols that occur various numbers of times, as shown in the chart. If any one of the winning combinations appears in a row, the player wins money according to the payoff assigned to each combination. Find the probabilities of the following events:

	Wheel I	Wheel II	Wheel III
Cherries	7	7	0
Oranges	3	6	7
Lemons	3	0	4
Plums	5	1	5
Bells	1	3	3
Bars	1	3	1
	20	20	20

★ a. A bar on wheel I. $P(Bar) = \frac{1}{20}$

b. A bar on all three wheels. $P(3\ bars)\ \frac{1}{20} \cdot \frac{3}{20} \cdot \frac{1}{20} = \frac{3}{8000}$

★ c. Bells on wheels I and II and a bar on wheel III.

d. Plums on wheels I and II and a bar on wheel III. $\frac{5}{20} \cdot \frac{1}{20} \cdot \frac{1}{20} = \frac{5}{8000}$

Exercise Set 10.2: Problem Solving

1. Two players have invented a game. A bowl is filled with an equal number of white and red marbles. One player, called the *holder,* holds the bowl while the other player, called the *drawer,* is blind-folded and selects two marbles. The drawer wins if both marbles have the same color; otherwise, she loses. Is this a fair game?

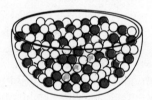

Bowl of marbles

★ a. *Understanding the Problem:* The two marbles selected will either both be white, both be red, or the colors will be different. The drawer feels that she has a $\frac{2}{3}$ chance of winning since there are three outcomes and two are favorable. Is this true?

b. *Devising a Plan:* One approach is to try some numbers (both large and small) for different numbers of marbles and compute compound probabilities. Drawing two marbles is the same as drawing one and then without replacing it, drawing a second. Are these events independent or dependent?

c. *Carrying Out the Plan:* Who has the best chance of winning this game? What happens to the probability if greater numbers of marbles are used?

★ d. *Looking Back:* Suppose the game continues with the drawer selecting two marbles at a time until there are no marbles left. Does this game favor the drawer or the holder? (*Hint:* Try some experiments and determine an empirical probability.)

★ 2. There are three containers. Container I has one black marble and one white marble, container II has one black and two white, and container III has one black and one white. You select at random from container I and put that marble in container II; then select at random from container II and put that marble in container III. What is the probability of now drawing a black marble from container III?

3. In Sweden a motorist was accused of overparking in a restricted time zone. A police officer testified that this particular parked car was seen with the tire valves pointing to 1 o'clock and to 6 o'clock. When the officer returned later (after the  allowed parking time had expired), this same car was there with its valves pointing in the same directions—so a ticket was written. The motorist claimed that the car had been driven away from that spot during the elapsed time, and upon returning later the tire valves coincidentally happened to come to rest in the same positions as before. The driver was acquitted, but the judge remarked that if the position of the tire valves of all four wheels had been recorded and found to point in the same direction, the coincidence claim would be rejected as too improbable.*

Using only the 12 hour-hand positions, what is the probability that two given tire valves of a car would return to the same position? What is the probability of all four tire valves returning to the same position? (Assume that due to variations in tire sizes and turning corners, the tires will not turn the same amount.)

Exercise Set 10.2: Computers

1. It is necessary to type RUN for Program 10.2A on page 611 to obtain each sequence of 10 H's and T's. Revise this program so that 20 sequences of 10 H's and T's will be printed each time the program is run.

✆ ★ 2. Use the revised program in Exercise 1 and the Monte Carlo method to determine a simulated probability for the occurrence of each outcome in the top row of the table before the outcome below it.

Dominating Outcomes	THH	THH	HHT	TTH	HTT	HTT	TTH	HHT
Eight Outcomes	HHH	HHT	HTH	THH	TTT	TTH	THT	HTT
Simulated Probabilities								

✆ 3. What are the chances of getting three or more consecutive heads in six tosses of a coin? 50/50? Revise Program 10.2A on page 611 to determine a simulated probability for this event.

*See F. Mosteller et al., *Statistics: A Guide to the Unknown* (New York: Holden-Day, 1972), p. 102.

4. A gambler once asked the French mathematician Blaise Pascal why he had been losing consistently when he bet even money that a double six would show at least once in 24 rolls of two dice.

   a. Write a computer program to simulate the toss of two dice 24 times.

   b. Use the Monte Carlo method to determine a simulated probability of obtaining at least one double six in 24 rolls of two dice.

   c. This is one of the gambling questions that led to the development of probability. Pascal solved this problem with theoretical probability. Find the probability that at least one double six will occur in 24 rolls of two dice. (*Hint:* Use complementary probability. The probability of not obtaining a double six on one roll is 35/36.)

"Hello? Beasts of the Field? This is Lou, over in Birds of the Air. Anything funny going on at your end?"

10.3 DESCRIPTIVE AND INFERENTIAL STATISTICS

The employees of the Animated Animal Company keep records of the number of toys each machine produces and the number of breakdowns. The table shows that machine III outproduced machines I and II, and machine II had the most problems. Numerical information such as this is often called

	M	T	W	Th	F	Break-downs
Machine I	165	158	98	125	260	13
Machine II	117	82	46	6	30	24
Machine III	182	243	196	305	261	4

Weekly record of birds produced

statistics. More technically, statistics is the branch of mathematics that deals with the analysis of numerical data. There are two main branches: descriptive statistics and inferential statistics. The processes of describing and interpreting data is called *descriptive statistics.* The average number of birds produced weekly by machine I is an example of descriptive statistics. On the other hand, the processes for using data to predict future outcomes is called *inferential statistics*. For example, sampling is an important part of inferential statistics. If the manager of Birds of the Air randomly selects 100 birds off the assembly line and finds that 3 are defective, he can estimate that the number of bad birds in a batch of 5000 will be 150.

Statistics had its beginning in the seventeenth century through the work of the English businessperson John Graunt. Graunt used a publication called "Bills of Mortality," which listed births, christenings, and deaths. Here are some of his conclusions: The number of male births exceeds the number of female births; there is a higher death rate in urban (as opposed to rural) areas; and more men die violent deaths than women. Graunt used this source to publish his book *Natural and Political Observations of Mortality*. In his work he summarized great amounts of data to make it understandable (descriptive statistics) and made conjectures about large populations based on small samples (inferential statistics).

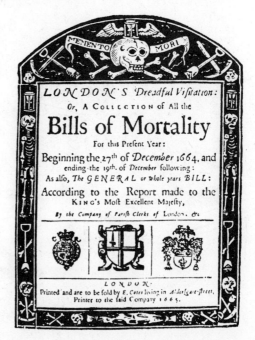

DESCRIPTIVE STATISTICS

The accompanying table contains the world heavyweight boxing champions from 1885 to 1984.* By a quick glance at these columns of numbers it is difficult to know which measurements occur most often and which occur least often. There are several ways of summarizing such lists of data. One method is to use the following frequency distribution. This table contains the heights of the boxing champions to the nearest inch and the number of men having each height. The heights of 6 feet, 6 feet 1 inch, and 6 feet 3 inches occurred most frequently. Heavyweight champions with heights below 5 feet 11 inches and above 6 feet 3 inches do not occur very often. The bar graph and frequency polygon are pictorial methods of illustrating frequency distributions. Notice how these graphs are high near the middle portion of the scale and decrease in height toward the ends of the scale.

*Source: B. R. Sugar et al. (eds.), *The Ring: 1981 Record Book and Boxing Encyclopedia* (New York: Atheneum, 1981), pp. 464–489.

Modern World Heavyweight Boxing Champions

		Height	Weight	Age*
1885–1892	JOHN L. SULLIVAN	5-10½	190	27
1892–1897	JAMES J. CORBETT	6-1	178	26
1897–1899	BOB FITZSIMMONS	6	150	34
1899–1905	JAMES J. JEFFRIES	6-2½	216	22
1905–1906	MARVIN HART	5-11	190	29
1906–1908	TOMMY BURNS	5-7	175	25
1908–1915	JACK JOHNSON	6-1	195	30
1915–1919	JESS WILLARD	6-6	225	34
1919–1926	JACK DEMPSEY	6-½	187	24
1926–1928	GENE TUNNEY	6-½	174	29
1930–1932	MAX SCHMELING	6-1	189	25
1932–1933	JACK SHARKEY	6	196	30
1933–1934	PRIMO CARNERA	6-5½	260	27
1934–1935	MAX BAER	6-2½	210	25
1935–1937	JAMES J. BRADDOCK	6-3	162	29
1937–1949	JOE LOUIS	6-1½	200	23
1949–1951	EZZARD CHARLES	6	182	28
1951–1952	JERSEY JOE WALCOTT	6	194	37
1952–1956	ROCKY MARCIANO	5-11	184	29
1956–1959	FLOYD PATTERSON	6	182	21
1959–1960	INGEMAR JOHANSSON	6-½	195	27
1960–1962	FLOYD PATTERSON			
1962–1964	SONNY LISTON	6-1	200	30
1964–1970	MUHAMMAD ALI	6-3	186	22
1970–1973	JOE FRAZIER	5-11½	205	26
1973–1974	GEORGE FOREMAN	6-3	220	25
1974–1978	MUHAMMAD ALI			
1978	LEON SPINKS	6-2	200	25
1978–1980	MUHAMMAD ALI			
1980–1984	LARRY HOLMES	6-4	215	31

*Age when championship was won.

Frequency Distribution

Height (Feet-Inches)	5–7	5–8	5–9	5–10	5–11	6–0	6–1	6–2	6–3	6–4	6–5	6–6
Frequency	1	0	0	0	3	6	7	2	5	1	0	2

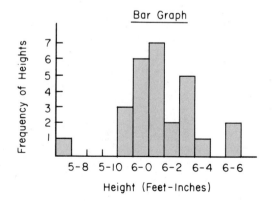

Bar Graph

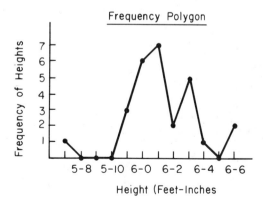

Frequency Polygon

Sometimes the data are spread over a wide range, and there may be only a few or even no numbers occurring with the same frequency. In this case it is convenient to group the data in intervals. The numbers of days for the gestation periods of animals have been grouped in 30-day intervals in the frequency distribution that follows. This table shows that two animals have a gestation period of less than 30 days and the most common gestation periods fall in intervals of 30 to 59 days, 60 to 89 days, and 90 to 119 days.

		Number of Days in Gestation Period	Frequency
Ass	365	0– 29	2
Baboon	187	30– 59	6
Badger	60	60– 89	5
Bat	50	90–119	5
Bear		120–149	1
Black	219	150–179	3
Grizzly	225	180–209	3
Polar	240	210–239	3
Beaver	122	240–269	4
Buffalo	278	270–299	2
Camel	406	300–329	0
Cat (domestic)	63	330–359	2
Chimpanzee	231	360–389	3
Chipmunk	31	390–419	1
Cow	284	420–449	1
Deer	201	450–479	0
Dog	61	480–509	1
Elk	250		
Fox	52		
Giraffe	425		
Goat (dom.)	151		
Goat (mtn.)	184		
Gorilla	257		
Guinea Pig	68		
Horse	330		
Kangaroo	42		
Leopard	98		
Lion	100		
Monkey	165		
Moose	240		
Mouse (meadow)	21		
Opossum	14–17		
Pig	112		
Puma	90		
Rabbit	37		
Rhinoceros	498		
Sea Lion	350		
Sheep	154		
Squirrel	44		
Tiger	105		
Whale	365		
Wolf	63		
Zebra	365		

MEASURES OF CENTRAL TENDENCY

Another approach to summarizing data is to represent all the values by one number, such as an average. In statistics there are three types of averages: the *mean,* the *median,* and the *mode.* These are called *measures of central tendency.* To illustrate these measures consider this table of major twentieth century earthquakes. Scanning the deaths column we see that the greatest life toll was 800,000 and the least was 21. Let's see how the three measures of central tendency can be used to "average" these extremes and to obtain "typical" values.

Major Earthquakes from 1946 to 1980

Date	Place	Deaths	Magnitude
1946 Dec. 21	Japan, Honshu	2000	8.4
1948 June 28	Japan, Fukui	5131	7.3
1949 Aug. 5	Ecuador, Pelileo	6000	6.8
1950 Aug. 15	India, Assam	1530	8.7
1953 Mar. 18	NW Turkey	1200	7.2
1956 June 10–17	N. Afghanistan	2000	7.7
1957 July 2	Northern Iran	2500	7.4
1957 Dec. 13	Western Iran	2000	7.1
1960 Feb. 29	Morocco, Agidir	12,000	5.8
1960 May 21–30	Southern Chile	5000	8.3
1962 Sept. 1	Northwestern Iran	12,230	7.1
1963 July 26	Yugoslavia, Skopje	1100	6.0
1964 Mar. 27	Alaska	114	8.5
1966 Aug. 19	Eastern Turkey	2520	6.9
1968 Aug. 31	Northeastern Iran	12,000	7.4
1970 Mar. 28	Western Turkey	1086	7.4
1970 May 31	Northern Peru	66,794	7.7
1971 Feb. 9	Cal., San Fernando Valley	65	6.5
1972 Apr. 10	Southern Iran	5057	6.9
1972 Dec. 23	Nicaragua	5000	6.2
1974 Dec. 28	Pakistan (9 towns)	5200	6.3
1975 Sept. 6	Turkey (Lice, etc.)	2312	6.8
1976 Feb. 4	Guatemala	22,778	7.5
1976 May 6	Northeast Italy	946	6.5
1976 June 26	New Guinea, Iran, Jaya	443	7.1
1976 July 28	China, Tangshan	800,000	8.2
1976 Aug. 17	Philippines, Mindanao	8000	7.8
1976 Nov. 24	Eastern Turkey	4000	7.9
1977 Mar. 4	Romania, Bucharest, etc.	1541	7.5
1977 Aug. 19	Indonesia	200	8.0
1977 Nov. 23	Northwestern Argentina	100	8.2
1978 June 19	Japan, Sendai	21	7.5
1978 Sept. 16	Northeast Iran	25,000	7.7
1979 Sept. 12	Indonesia	100	8.1
1979 Dec. 12	Colombia, Ecuador	800	7.9
1980 Oct. 10	Northwest Algeria	4500	7.3
1980 Nov. 23	Southern Italy	4800	7.2

The *mean* is what we often refer to as "the average." This is the sum of all values divided by the number of values. The sum of the number of deaths from the 37 quakes listed here is 1,026,068. The mean (average) is 1,026,068/37, or approximately 27732 deaths per quake.

The *median* is the middle value of a list when the numbers are listed in increasing order. Half of the values are greater than or equal to the median and half are less than or equal to it. The numbers of deaths in the 37 earthquakes are listed next in increasing order. The median or middle number in this list is 2500 (circled number). Eighteen numbers are less than 2500 and 18 numbers are greater.

21, 65, 100, 100, 114, 200, 443, 800, 946, 1086, 1100, 1200,

1530, 1541, 2000, 2000, 2000, 2312, (2500), 2520, 4000, 4500,

4800, 5000, 5000, 5057, 5131, 5200, 6000, 8000, 12,000,

12,000, 12,230, 22,778, 25,000, 66,794, 800,000

In an ordered list with an odd number of values, such as the earthquake list, the median is always the middle value. If there are an even number of values, the median is halfway between the two middle numbers. For example, consider the first four values in the earthquake list: 21, 65, 100, and 100. The median is 82.5, the number that is half way between 65 and 100.

The *mode* is the value that occurs most frequently. For the earthquakes it is the number 2000, which occurs in the list three times. If there are two different values that occur the same number of times, the data has two modes and is called *bimodal*.

The mean, median, and mode each have their advantages, depending on the type of data and information desired. Which do you feel is the best measure of central tendency for describing the earthquakes: the mean, 27,732; the median, 2500; or the mode, 2000?

Sometimes the mode and median are better descriptions than the mean. In this example of the salaries of a small company, the mean, $16,312.50, is misleading. It is more than twice the salary of most of the employees. In this case both the median, 6000, and the mode, 5000, are more representative of the majority of salaries.

One president	$100,000
One vice-president	60,000
One salesperson	20,000
One supervisor	11,000
One machine operator . . .	10,000
Five mill workers (each earning)	6,000
Six apprentice workers (each earning)	5,000

MEASURES OF DISPERSION

These two lists of numbers have the same mean (23), median (20), and mode (20). However, since list B is more spread out than list A, this example illustrates the need for ways of measuring dispersion. The *range* and the *standard deviation* are two such measures.

List A

30, 28, 26, 20, 20, 19, 18

List B

60, 50, 20, 20, 10, 1, 0

The *range* is the difference between the largest and smallest values. The range for list A is 12, but for list B it is 60.

The *standard deviation* measures the spread of values about the mean. It is computed by the following steps: (1) Determine the mean; (2) Find the difference between each value and the mean; (3) Square these differences; (4) Determine the mean of these squared differences; and (5) Compute the square root of this mean. These steps are used in the following tables to compute the standard deviations for lists A and B.

List A (mean = 23)

Values	Difference from Mean	Squares of Differences
30	7	49
28	5	25
26	3	9
20	3	9
20	3	9
19	4	16
18	5	25
	Total	142

Standard deviation $\approx \sqrt{20.3} \approx 4.5$

List B (mean = 23)

Values	Difference from Mean	Squares of Differences
60	37	1369
50	27	729
20	3	9
20	3	9
10	13	169
1	22	484
0	23	529
	Total	3298

Standard deviation $\approx \sqrt{471.1} \approx 21.7$

Let's review these steps by computing the standard deviation for list A: (1) The mean for list A is 23; (2) The differences between 23 and the numbers in the first column of the table are the numbers in the second column (the difference is computed so that the numbers in the second column are positive); (3) The numbers in the third column are the squares of the corresponding numbers in the second column; (4) The sum of the numbers in the third column is 142 and the mean of these numbers is 142 ÷ 7, which is approximately 20.3; (5) The square root of the mean, 20.3, is approximately 4.5. Carry out these five steps to obtain the numbers in the table for list B. The average of the numbers in the third column of this table is approximately 471.1. The square root of 471.1 gives a standard deviation of about 21.7 for list B.

The standard deviation for list A is about 4.5, compared to a standard deviation of approximately 21.7 for list B. The greater standard deviation is caused by the numbers being more dispersed or spread out about the mean.

Standard deviations determine intervals about the mean. For list A, 1 standard deviation above the mean is 23 + 4.5, or 27.5, and 1 standard deviation below the mean is 23 − 4.5, or 18.5. The interval within ±1 standard deviation of the mean is the interval from 18.5 to 27.5. The interval within ±2 standard deviations of the mean is from 14 to 32. What is the interval within ±1 standard deviation of the mean for list B?

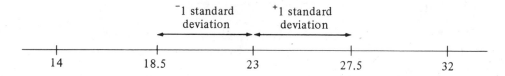

INFERENTIAL STATISTICS

Making predictions from samples is an important part of statistics. Because of the possibilities for errors, strict procedures must be followed in gathering a sample. The sample must be large enough and it must be a representative cross section of the whole.

The need for scientific sampling techniques was dramatically illustrated in the 1936 presidential election. The *Literary Digest,* which had been conducting surveys of elections since 1920, sent questionnaires to 10 million voters. Their sample was obtained from telephone directories and lists of automobile owners. Rather than sample voters from different income levels, they had selected people with above-average incomes. The *Digest* predicted that Alfred Landon would win, and instead the election was a landslide victory for Franklin D. Roosevelt. The *Digest* went out of publication the following year.

Topics of the day

LANDON, 1,293,669; ROOSEVELT, 972,897
Final Returns in The Digest's Poll of Ten Million Voters

Well, the great battle of the ballots in the Polls of ten million voters, scattered throughout the forty eight capital states of the Union, is now finished, and in the table below we record the figures received up to the hour of going to press.

These figures are exactly from more than op polled in our c

tran National Committee purchased *THE LITERARY DIGEST?*" And all types and variables, including: "Have the Jews purchased *THE LITERARY DIGEST?*" "Is the Pope of Rome a stockholder of *THE LITERARY DIGEST?*" And so it goes— all equally absurd and amusing. We could add more this list, and not all of t tions rent days

Random Sampling   Random sampling is probably the most important sampling procedure in statistical applications. To use this method two conditions must be satisfied. First, every member of a population must have the same chance of being selected; and second, the selection of one member should not influence the chances of any other member being selected.

Cluster Sampling In cluster sampling the population is divided into groups or clusters, and a certain number of these clusters are randomly selected. In one example a city council wanted to sample voter support for a plan to build a public swimming pool. The city had 316 blocks, each containing about 20 houses. The council used random numbers to select 15 of the blocks and then interviewed each household in these blocks. Why might this method of sampling produce a bias sample?

Stratified Sampling This method is often more reliable than cluster sampling. In stratified sampling the population is divided into groups called *strata.* In the swimming pool example the population might be divided into three strata: high-income, middle-income, and low-income families. Each group is then randomly sampled in proportion to its actual percentage. If 20 percent of the population is low-income, 70 percent middle-income, and 10 percent high-income, then 20 percent of the sample is taken from low-income families, etc.

"That's the worst set of opinions I've heard in my entire life."

The largest regular sampling procedure in the world is the survey of 47,000 households that is conducted monthly by the U.S. government's Bureau of Census. To ensure that these surveys will represent the civilian population, stratified sampling is used with several strata, such as urban, central city, rural nonfarm, and rural farm.

MONTE CARLO METHOD

In Section 10.1 we used the Monte Carlo method for determining probabilities. Here is a type of sampling problem that can also be analyzed by the Monte Carlo method:

PROBLEM   Each package of a certain brand of cereal contains one of seven cards about superheroes. A student in an elementary school class wants to know how many boxes of cereal he can expect to buy before getting the entire set.*

*Ann E. Watkins, "Monte Carlo Simulations: Probability the Easy Way," *Teaching Statistics and Probability, 1981 Yearbook* (Reston, Virginia: The National Council of Teachers of Mathematics, 1981), pp. 203–209.

Answering this question by purchasing boxes of cereal would be expensive. The elementary school class solved this problem by writing the names of the seven superheroes on slips of paper and randomly selecting them from a box. They continued until each name was drawn at least once and then recorded the number of "boxes purchased." They repeated this process 20 times. Here are the number of "boxes" for each trial:

Numbers of Boxes for Obtaining All Seven Superheroes

20	14	27	18	17	15	19
19	19	20	16	11	15	21
15	22	20	28	12	26	

The average (mean) number of boxes a student can expect to buy before getting the entire set of cards is the sum of these numbers divided by 20:

$$\frac{374}{20} = 18.7 \approx 19$$

The students could see that it might take more than 19 boxes to get all 7 cards if they were unlucky or fewer than 19 if they were lucky.

COMPUTER APPLICATIONS

The computer is an invaluable aid for analyzing statistical data. Let's begin with an elementary program for computing the mean of seven numbers.

Program 10.3A

```
10  PRINT "THIS PROGRAM COMPUTES THE MEAN OF 7 NUMBERS. TYPE
        EACH NUMBER SEPARATED BY A COMMA."
20  INPUT A, B, C, D, E, F, G
30  LET M = (A + B + C + D + E + F + G) / 7
40  PRINT "THE MEAN OF THESE 7 NUMBERS IS "M"."
50  END
```

For large sets of data such as those in the tables on pages 621 through 623 the INPUT command is not practical. There is another type of input command in BASIC. It is called READ. This command reads from a list of data that is entered into the computer when the operator types the program; otherwise it works in a manner similar to INPUT. The list of data for a program is contained in a data statement that begins with the word DATA. READ and DATA are used below to revise the program for computing the mean of seven numbers. Lines 30 through 70 form a loop. The first time through this loop the READ command sets A equal to 30, the first number in the DATA statement. The second time through the loop the READ command sets A equal to the next unused number from the DATA statement, which is 28. This process continues until A has been set equal to each of the seven numbers in line 20. What lines of the program compute the sum and mean of these numbers?

Program 10.3B

```
10   PRINT "THIS PROGRAM COMPUTES THE MEAN OF THE 7 NUMBERS
         LISTED BELOW."
20   DATA  30, 28, 26, 20, 20, 19, 18
30   FOR I = 1 TO 7
40   READ A
50   PRINT A" ";
60   LET T = T + A
70   NEXT I
80   LET M = T / 7
85   PRINT
90   PRINT "THE MEAN OF THESE 7 NUMBERS IS "M"."
100  END
```

Now let's make a few revisions in Program 10.3B to handle larger sets of data. The INPUT command in line 15 of the next program allows the operator to type the number of data when the program is run. The DATA statements for this program contain the numbers from the table of major earthquakes (page 623). Notice that each new line of data requires a new line number and DATA statement. Compare this program to the preceding one to determine what changes have been made.

Program 10.3C

```
10   PRINT "THIS PROGRAM COMPUTES THE MEAN OF THE NUMBERS LIST
         ED BELOW. TYPE THE NUMBER OF NUMBERS."
15   INPUT N
20   DATA 2000, 5131, 6000, 1530, 1200, 2000, 2500, 2000, 1200
         0, 5000, 12230
21   DATA 1100, 114, 2520, 12000, 1086, 66794, 65, 5057, 5000,
         5200, 2312
22   DATA 22778, 946, 443, 800000, 8000, 4000, 1541, 200, 100,
         21, 25000
23   DATA  100, 800, 4500, 4800
30   FOR I = 1 TO N
40   READ A
50   PRINT A" ";
60   LET T = T + A
70   NEXT I
80   LET M = T / N
85   PRINT
90   PRINT "THE MEAN OF THESE "N" NUMBERS IS "M"."
100  END
```

SUPPLEMENT (Activity Book)

Activity Set 10.3 Applications of Statistics (Averages, standard deviation, and sampling)

"It is truth very certain that, when it is not in one's power to determine what is true, we ought to follow what is more probable." RENÉ DESCARTES

Exercise Set 10.3: Applications and Skills

★ 1. Find the numbers of incomes from the list of per capita incomes that fall in the given intervals in the frequency distribution table. For example, there is only one income in the interval from 7000 to 7499. Use the frequencies from the table to complete the bar graph.

Frequency Distribution Table

Interval	Frequency
7000–7499	1
7500–7999	
8000–8499	
8500–8999	
9000–9499	
9500–9999	
10,000–10,499	
10,500–10,999	
11,000–11,499	
11,500–11,999	
12,000–12,499	
12,500–12,999	
13,000–13,499	
13,500–13,999	
14,000–14,499	

Per Capita Income by States*

State and region	Income	State and region	Income	State and region	Income
New England		**Plains**		**Southwest**	
Connecticut	12,995	Iowa	10,149	Arizona	9693
Maine	8655	Kansas	10,870	New Mexico	8654
Massachusetts	11,158	Minnesota	10,747	Oklahoma	10,210
New Hampshire	10,073	Missouri	9876	Texas	10,743
Rhode Island	10,466	Nebraska	10,296		
Vermont	8654	North Dakota	10,525	**Rocky Mountains**	
		South Dakota	8793	Colorado	11,142
Mideast				Idaho	8906
Delaware	11,279	**Southeast**		Montana	9676
Dist. of Columbia	12,050	Alabama	8200	Utah	8307
Maryland	11,534	Arkansas	8042	Wyoming	11,780
New Jersey	12,115	Florida	10,050		
New York	11,440	Georgia	8960	**Far West**	
Pennsylvania	10,373	Kentucky	8455	California	12,057
		Louisiana	8456	Nevada	11,633
Great Lakes		Mississippi	7256	Oregon	9991
Illinois	11,479	North Carolina	8679	Washington	11,266
Indiana	9,656	South Carolina	8050		
Michigan	11,009	Tennessee	8604	Alaska	14,190
Ohio	10,371	Virginia	10,445	Hawaii	11,096
Wisconsin	10,056	West Virginia	8334		

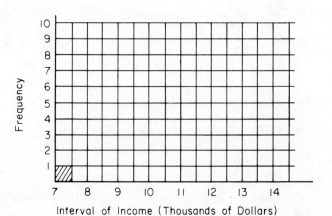

★ 2. Which interval in Exercise 1 has the most incomes (mode interval)? Which interval has the median income? The mean income for the United States in 1981 was $10,517. Does this fall within the intervals containing the median or the mode?

*U.S. Bureau of the Census, *Statistical Abstract of the United States: 1982–83,* 103d ed., (Washington, D.C.: 1982), p. 427.

3. Use the list at the right to count the number of
   states having each of the different gasoline
   taxes.* Sketch the frequency polygon for this
   data.

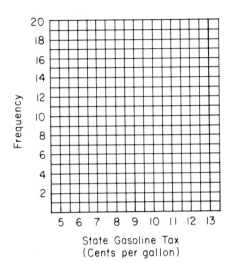

State Gasoline Tax
(Cents per gallon)

State	State Gas Tax (cents per gallon)
Alabama	11
Alaska	8
Arizona	8
Arkansas	9.5
California	7
Colorado	7
Connecticut	11
Delaware	9
Dist. of Col.	11
Florida	8
Georgia	7.5
Hawaii	8.5
Idaho	9.5
Illinois	7.5
Indiana	8.5
Iowa	10
Kansas	8
Kentucky	9
Louisiana	8
Maine	9
Maryland	9
Massachusetts	9.8
Michigan	11
Minnesota	11
Mississippi	9
Missouri	7
Montana	9
Nebraska	13.5
Nevada	6
New Hampshire	11
New Jersey	8
New Mexico	8
New York	8
North Carolina	9
North Dakota	8
Ohio	7
Oklahoma	6.5
Oregon	7
Pennsylvania	11
Rhode Island	10
South Carolina	11
South Dakota	12
Tennessee	7
Texas	5
Utah	9
Vermont	9
Virginia	11
Washington	12
West Virginia	10.5
Wisconsin	9
Wyoming	8

*U.S. Bureau of the Census, *Statistical Abstract of the United States: 1982–83,* 103d ed., (Washington, D.C.: 1982), p. 612.

4. Which of the "averages," mean, median, or mode, is the best number for describing the following instances?

★   a. The "average" size of hats sold in a store. mode

   b. The heights of players on a basketball team. mean

★   c. The "average" age of seven people in a family if six of them are under 40 and one is 96 years old. median

   d. The "average" size of bicycles (by tire size) sold by a bike shop. mode

5. According to this description from Norris and McWhirter's 10 best oddities, a woman golfer took 166 strokes for the 130-yard sixteenth hole in a Ladies Invitational Golf Tournament. Hypothetical scores for the remaining 17 holes are given in the table.

> *The Worst Woman Golfer.* "A woman player in the qualifying round of the Shawnee Invitational for Ladies at Shawnee-on-Delaware, Pa., in c. 1912, took 166 strokes for the 130-yard 16th hole. Her tee shot went into the Binniekill River and the ball floated. She put out in a boat with her exemplary, but statistically minded, husband at the oars. She eventually beached the ball 1½ mi. downstream, but was not yet out of the woods. She had to play through one on the home stretch."

Number of Hole	1	2	3	4	5	6	7	8	9	10	11	12	13	14	15	16	17	18
Score	4	5	4	5	4	6	4	5	5	4	7	3	7	5	5	166	7	6

★   a. What was her average (mean) score for the 18 holes?

   b. What were her median and mode scores?

★   c. Which of these "averages" is most appropriate for summarizing her performance?

   d. If it took 165 strokes to the green from 1.5 miles (7920 feet) downstream, what was the average (mean) length in feet of each return shot?

6. *Calculator Exercise:* Whenever anyone mentions the good old days, they were always "good." How good were they? Here are some surprising facts that were reported by the National Institute of Health Federal Credit Union.

★   a. Divide each 1925 time by the corresponding 1976 time to find out how many times longer you would have worked for each item in 1925. Compute your answer to two decimal places.

   b. Compute the average (mean) of your answers in part **a.** On the average, how many times longer would you have had to work for these items in 1925 compared to 1976?

THE GOOD OLD DAYS

To Buy	You Would Work in	
	1925	1976
New car	41½ wks	26½ wks
Year in college	31 wks	15 wks
Gas range	138 hrs	61½ hrs
Washing machine	120½ hrs	54 hrs
Sewing machine	101½ hrs	22 hrs
Woman's skirt	6½ hrs	3 hrs
Dozen oranges	3/4 hr	1/4 hr
Dozen eggs	53 min	13 min
Pound of coffee	48 min	20 min
Pound of butter	1 hr	15 min

7. The home run leaders in the National and American Leagues from 1960 to 1983 are given below. Find the mean, median, and mode for each list of home runs.

Home Run Leaders

Year	National League	H.R.	Year	American League	H.R.
1960	Ernie Banks, Chicago	41	1960	Mickey Mantle, New York	40
1961	Orlando Cepeda, San Francisco	46	1961	Roger Maris, New York	61
1962	Willie Mays, San Francisco	49	1962	Harmon Killebrew, Minnesota	48
1963	Hank Aaron, Milwaukee		1963	Harmon Killebrew, Minnesota	45
	Willie McCovey, San Francisco	44			
1964	Willie Mays, San Francisco	47	1964	Harmon Killebrew, Minnesota	49
1965	Willie Mays, San Francisco	52	1965	Tony Conigliaro, Boston	32
1966	Hank Aaron, Atlanta		1966	Frank Robinson, Baltimore	49
	Willie McCovey, San Francisco	44			
1967	Hank Aaron, Atlanta	39	1967	Carl Yastrzemski, Boston	
				Harmon Killebrew, Minnesota	44
1968	Willie McCovey, San Francisco	36	1968	Frank Howard, Washington	44
1969	Willie McCovey, San Francisco	45	1969	Harmon Killebrew, Minnesota	49
1970	Johnny Bench, Cincinnati	45	1970	Frank Howard, Washington	44
1971	Willie Stargell, Pittsburgh	48	1971	Bill Melton, Chicago	33
1972	Johnny Bench, Cincinnati	40	1972	Dick Allen, Chicago	37
1973	Willie Stargell, Pittsburgh	44	1973	Reggie Jackson, Oakland	32
1974	Mike Schmidt, Philadelphia	36	1974	Dick Allen, Chicago	32
1975	Mike Schmidt, Philadelphia	38	1975	George Scott, Milwaukee	
				Reggie Jackson, Oakland	36
1976	Mike Schmidt, Philadelphia	38	1976	Greg Nettles, New York	32
1977	George Foster, Cincinnati	52	1977	Jim Rice, Boston	39
1978	George Foster, Cincinnati	40	1978	Jim Rice, Boston	46
1979	Dave Kingman, Chicago	48	1979	Gorman Thomas, Milwaukee	45
1980	Mike Schmidt, Philadelphia	48	1980	Reggie Jackson, New York	
				Ben Oglivie, Milwaukee	41
1981	Mike Schmidt, Philadelphia	31	1981	Bobby Grich, California	
				Tony Ames, Oakland	
				Dwight Evans, Boston	
				Eddie Murry, Baltimore	22
1982	Dave Kingman, New York	37	1982	Gorman Thomas, Milwaukee	
				Reggie Jackson, California	39
1983	Mike Schmidt, Philadelphia	40	1983	Jim Rice, Boston	39

a. Which league's home run leaders have the greater mean (average) of home runs from 1960 to 1983?

   b. Which league has the greater median?

c. Which league has the greater total?

   d. In how many different years did the National League's home run leaders hit more home runs than the American League's?

   e. Which league's home run leaders have the better record? Support your conclusion.

8. The grades of eight students on a 10-point test were 1, 3, 5, 5, 7, 8, 9, and 10.

a. Compute the mean and record it above the table.

   b. Use this table to compute the standard deviation of these scores.

c. Another class took the same test and had the same mean. What can be said about the two sets of test scores if the second class had a standard deviation of 2?

Mean = $\boxed{46}$

Score	Difference between Mean and Score	Squares of Differences
1	5	25
3	3	9
5	1	1
5	(1)3	1
7	1	1
8	2	4
9	3	9
10	4	16
	Total	66

Standard Deviation = 2.87 $8\sqrt{66} = 8.25$
$8^2 = 64$

9. *Calculator Exercise:* This table contains the number of nuclear power reactors in 1981 for 19 countries and their gross capacity in megawatts (1 million watts).*

Country	Number of Reactors	Megawatt Capacity
United States	77	61,036
Argentina	1	357
Belgium	3	1,744
Canada	9	5,588
China: Taiwan	3	2,257
Germany, F.R. of	11	8,996
Great Britain	33	9,012
Finland	4	2,296
France	30	23,068
India	4	860
Italy	4	1,490
Japan	23	15,676
Korea, Rep. of	1	587
Netherlands	2	529
Pakistan	1	137
Spain	4	2,047
Sweden	9	6,710
Switzerland	4	2,034
Yugoslavia	1	664

*U.S. Bureau of the Census, *Statistical Abstract of the United States: 1982–83,* 103d ed., (Washington, D.C.: 1982), p. 589.

a. What is the average (mean) capacity in megawatts (to the nearest tenth) of the reactors in the United States? Answer this question for France, Japan, and Great Britain.

★ b. Which of the four countries in part **a** has the smallest average megawatt capacity per reactor?

c. What is the average (mean) number of reactors for these 19 countries? Explain why the mean is a misleading average in this example.

10. Here are the pulse rates of 55 people, taken at rest. The mean of these rates is 72, and the standard deviation is approximately 9.2.

```
51  56  56  57  57      61  62  62  62  63      64  65  65  65  66      67  67  68  68  69
69  70  70  70  70      70  70  72  73  73      74  74  74  74  75      75  76  76  76  77
78  79  79  80  80      80  81  82  84  84      86  86  89  91  92
```

a. Circle the pulse rates that are within 1 standard deviation above and 1 standard deviation below the mean. What percent of the total are these rates?

★ b. What percent of the rates are within 2 standard deviations above and 2 below the mean?

c. What percent of the rates are more than 2 standard deviations above or 2 standard deviations below the mean?

11. In the book *How to Take a Chance,** Darrell Huff relates the following study and its conclusion:

A large metropolitan police department made a check of the clothing worn by pedestrians killed in traffic at night. About four-fifths of the victims were wearing dark clothes and one-fifth light-colored garments. This study points up the rule that pedestrians are less likely to encounter traffic mishaps at night if they wear or carry something white.

Explain why this conclusion does not necessarily follow from the study.

12. Wildlife biologists have devised a sampling procedure for computing the numbers and types of fish in a lake. Sample catches are made by electronic shocking, and the fish are marked and returned to the lake. Sometime later, another sample is taken, and the ratio of marked fish to the number of fish in the sample is used to compute the total number of fish in the lake. In one survey 232 pickerel were caught and marked. About 2.5 months later a second sample of 329 was caught and 16 were found to be marked.

★ a. In the second sample, what is the ratio of the number of marked fish to the number of fish in the sample?

★ b. Assuming that the marked fish intermingled freely with the unmarked fish during the 2.5-month period, the ratio in part **a** should be equal to the ratio of 232 to the total pickerel population of the lake. Use a proportion to estimate the number of pickerel in the lake.

*D. Huff, *How to Take a Chance* (New York: W. W. Norton, 1959), p. 164.

13. Use the Monte Carlo method to solve the following problems. Describe your use of this method.

★   a. A manufacturer puts one of five different coloring markers in each box of crackerjacks. What is the average number of boxes that must be purchased to obtain all five different markers?

   b. A cloakroom attendant receives nine hats from nine men and gets the hats mixed up. If she returns the hats at random and all simultaneously, what is the average number of hats that will go to the correct owner?

Exercise Set 10.3: Problem Solving

1. *Guessing a Mean:* One student told his friend that adding all the numbers to find the mean was too much work. He said, "I simply guess a mean, subtract it from each number, and keep a running total of the (positive and negative) differences. Then I divide the total by the number of items and add the result to my guessed mean." Does this method work?

★   a. *Understanding the Problem:* Let's try this method with the six scores in the table. If we guess that the mean is 16, then the differences between these six scores and the mean are 0, ⁻2, 2, ⁻4, 6 and ⁻8. What is the mean of these six numbers? Is the sum of this mean and 16 equal to the mean of the six scores?

Scores	Differences from Guessed Mean
16	0
14	⁻2
18	2
12	⁻4
22	6
8	⁻8

   b. *Devising a Plan:* One approach is to try the student's method of computing a mean for several examples. Use this method to compute the mean of the six scores in part **a** by guessing a mean of 12. Even if this method produces the correct mean for several examples, we cannot be sure it will always work. Another approach is to use algebra and deductive reasoning.

   c. *Carrying Out the Plan:* Using the algebraic approach let G equal the guessed mean of the six scores, and write the differencess in this table. Continue using this approach with the student's method. What does this show? Explain why this is not a general proof that the student's method will always work.

Scores	Differences from Guessed Mean
16	$16 - G$
14	
18	
12	
22	
8	

★ d. *Looking Back:* This bar graph provides visual evidence for the student's method. The horizontal line represents the mean of 15. Suppose the guessed mean is too high, say 24. Then the differences between the 6 scores and 24 are all negative. What is the effect of adding the average of these negative differences to 24? Suppose the guessed mean is too low, say 12. Explain why the student's method increases the mean in this case. Suppose the guessed mean is 15. What is the sum of the differences between the 6 scores and 15?

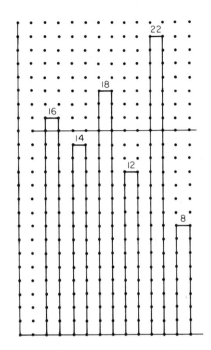

★ 2. At one time the state of New Hampshire had $1 lottery tickets, with a square on each ticket that covered a letter. There was a prize of $50 for collecting the letters N, E, and W and a prize of $10,000 for collecting H, A, M, P, S, H, I, R, and E. Did each of the 10 different letters in NEW HAMPSHIRE occur with the same frequency on these tickets? (*Hint:* Assume that there is a $\frac{1}{10}$ chance that any of these letters will occur on a ticket, and use the Monte Carlo method to determine the average number of tickets that would be needed to obtain the letters in HAMPSHIRE.)

Exercise Set 10.3: Computers

1. The following list of numbers contains the life spans of the 35 U.S. presidents from George Washington to Lyndon Johnson. Adapt Program 10.3C (page 629) to compute the average (mean) life span of these people.

   67, 90, 83, 85, 73, 80, 78, 79, 68, 71, 53, 65, 74, 64, 77, 56, 66, 63, 70, 49, 57, 71, 67, 58, 60, 72, 67, 57, 60, 90, 63, 88, 78, 46, 64

★ 2. Here is the beginning of a program to compute the standard deviation of the numbers in line 30 of the following program. Finish writing this program so that it prints the standard deviation rounded off to one decimal place. Test this program to see if you obtain a standard deviation of 4.5.

```
10   PRINT "THIS PROGRAM COMPUTES THE STANDARD DEVIATION OF THE
        NUMBERS LISTED BELOW. TYPE THE NUMBER OF THESE NUMBERS A
        ND THEIR MEAN. "
20   INPUT N, M
30   DATA  30, 28, 26, 20, 20, 19, 18
40   FOR I = 1 TO N
50   READ A
60   LET R = R + (A - M) ^ 2
70   NEXT I
```

3. Adapt the program from Exercise 2 to find the standard deviation of the 35 life spans from Exercise 1.

4. Solve the following problems by writing a computer program to simulate the mathematics. Use the model together with the Monte Carlo method.

   a. Pepe and Anna are playing a penny-tossing game. The player who can toss 10 heads in the fewest number of tosses wins the game. How many tosses on the average are required to obtain 10 heads?

   ★ b. A newly married couple would like to have a child of each sex. What is the average number of children in order for this to occur?

Alexander Hamilton

FEDERALIST

PAPERS

James Madison

10.4
DISTRIBUTIONS

Studies of frequency distributions have a wide range of application. In some cases such studies have been used to determine authorship. Consider the solution to the authorship controversy of the *Federalist Papers*.* This is a collection of 85 political papers that were written in the eighteenth century by Alexander Hamilton, John Jay, and James Madison. There is general agreement on the authorship of all but 12 of these papers. To determine who should be given credit for these papers frequency distributions have been compiled for certain filler words such as "by," "to," "of," "on," and several others. The graphs on the following page show the frequencies of usage of the words "by" and "to" from samples of Hamilton's and Madison's writings and the disputed papers. In both cases the graphs for the disputed papers (bottom graphs) are closer to the graphs for Madison than the graphs for Hamilton.

Let's look at the graphs for the rate of occurrence of "by." The top graph was obtained by counting all the words in each of Hamilton's 48 papers. It shows that his most frequent use of "by" was an average of 7 times (see horizontal axis) for every 1000 words (see tallest column of graph). This rate occurred in 18 out of his 48 (18/48) papers (see vertical axis). Madison's most frequent use of "by" was an average of 11 times for every 1000 words and this rate occurred in 16 out of his 50 papers (16/50). The bottom graph shows that the most frequent use of "by" in the 12 disputed papers was also 11 times per 1000. In general, the graphs for "by" show that the disputed papers more closely match Madison's papers than Hamilton's papers. The remaining

*F. Mosteller et al., *Statistics: A Guide to the Unknown* (New York: Holden-Day, 1972), pp. 164–75.

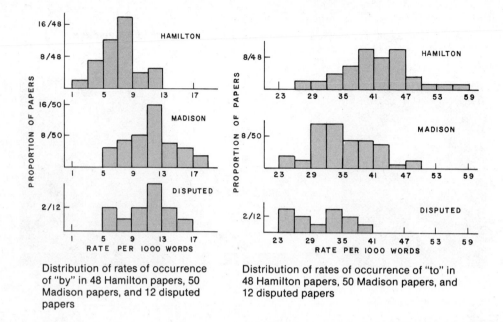

Distribution of rates of occurrence of "by" in 48 Hamilton papers, 50 Madison papers, and 12 disputed papers

Distribution of rates of occurrence of "to" in 48 Hamilton papers, 50 Madison papers, and 12 disputed papers

three graphs show a comparison of papers for the rate of occurrence of "to." Similar distributions for other filler words also support Madison as the author of the disputed papers.

RANDOMNESS AND UNIFORM DISTRIBUTIONS

Alexander Hamilton and James Madison tended to favor certain words over others. That is, knowingly or unknowingly, their selection of words was biased. Opposite to the notion of bias is that of *randomness.* When events occur at random, there is no way to predict their outcome. For example, when selecting a digit (0 through 9) at random, any one of them is equally likely. Yet, if 1000 digits are selected at random, we can predict that each digit will occur about 100 times. Another way of saying this is that the distribution of digits should be a uniform distribution. In a *uniform distribution* each event occurs about the same number of times.

Randomness is difficult to achieve. Repeated tosses of a coin may appear to be a random method of making "yes or no" type decisions, but imbalances in the coin's weight and tossing it to approximately the same height each time are two ways of causing a biased result. Similarly, dice and spinners produce fairly random results, but these also have slight biases due to their physical imperfections.

Randomness is so important in sampling techniques and experiments, and so difficult to attain without very special efforts, that tables and books of random digits are published. One of these sources contains a million random digits.* There are hundreds of ways that computer programs can be written to generate random digits. The following list of 250 digits is from such a program. The digits are printed in pairs and groups of ten for ease in reading and counting.

40 09 18 94 06	62 89 97 10 02	58 63 02 91 44	79 03 55 47 69	14 11 42 33 99
33 19 98 40 42	13 73 63 72 59	26 06 08 92 65	63 08 82 45 85	14 45 81 65 21
69 49 02 58 44	45 45 19 69 33	51 68 97 99 05	77 54 22 70 97	59 06 64 21 68
17 49 43 65 45	04 95 82 76 31	85 53 15 21 70	59 17 27 54 67	07 76 13 95 00
43 13 78 80 55	90 80 88 19 13	13 89 11 00 60	41 86 23 07 60	22 77 93 30 83

Let's consider an example of how a table of random digits can be used. Suppose that there are 650 different items that are numbered with whole numbers from 1 to 650 and that you would like to take a random sample of 65. One way to accomplish this is to start at any digit in the table and list consecutive groups of 3 digits until you have found 65 numbers between 1 and 650. The numerals 001, 002, etc., represent 1, 2, etc., and any triples of numbers from the table that are greater than 650 will be discarded. If this method is applied to the previous table, beginning with the first line, 400 is the first number. Then 918, 940, 662, 899, and 710 are discarded because they are greater than 650. The next acceptable number is 025 which represents 25. Continuing this process will produce a random sample.

NORMAL DISTRIBUTIONS

One of the most important distributions in statistics is called the *normal distribution*. This distribution has a symmetric bell-shaped curve, called the *normal curve*. The majority of values are clustered around the mean, with a gradual tapering off as the extremes are approached. The shape of these curves can vary somewhat as shown in the accompanying figures. The word "normal" is used to indicate that this type of curve is very common in nature. About 1833 the Belgian scientist L. A. J. Quetelet collected large amounts of data on human measurements: height, weight, length of limbs, intelligence, etc. He found that all measurements of mental and physical characteristics of human beings tended to be normally distributed. That is, the majority of people have measurements that are close to the mean (average), and moving away from the mean, the measurements occur less frequently. Quetelet was

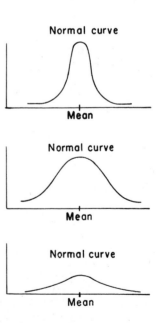

*Rand Corporation, *A Million Random Digits with 100,000 Normal Deviates*. (Glencoe, Illinois: Free Press, 1955).

convinced that Nature's creation of
people aims at the perfect person but
misses the mark and thus creates deviations on both sides of the ideal.

Normal distributions have the
following important properties: About
68 percent of the values are within 1
standard deviation above and below
the mean; about 96 percent are within 2
standard deviations of the mean; and
about 99.8 percent fall within 3
standard deviations of the mean. These
percentages hold for all normal distributions regardless of the mean or
size of the standard deviation.

The two normal curves shown next
are approximations to nearly normal
distributions of measurements. On the
left the distribution of college entrance

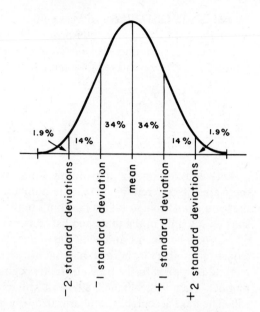

examination scores has a mean of 500 and a standard deviation of 100. We know from
this that 68 percent of the people taking these tests score between 400 and 600 and that
96 percent score between 300 and 700. The distribution on the right shows the frequency
of heights of 8585 men. The mean is approximately 67 inches or 5 feet 7 inches. A standard deviation of 3 implies that 68 percent of these men are between 64 inches and 70
inches in height. Two standard deviations below the mean is 61, and 2 standard deviations above the mean is 73. The normal distribution tells us that 96 percent of these
men have heights between 61 and 73 inches. We can also conclude that less than 2
percent of these men are taller than 73 inches (6 feet 1 inch).

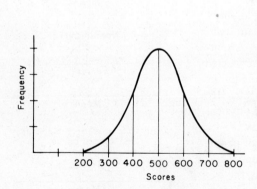

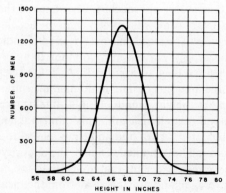

SKEW DISTRIBUTIONS

For a particular survey, this graph shows that about 97 families have no children, 110 families have one child, etc. The most common number of children per family is two. Graphs such as this, having the data piled up at one end of the scale and tapering off toward the other end, are called *skewed.* The direction of skewness is determined by the longer "tail" of the distribution. This graph is *skewed to the right* or *positively skewed.*

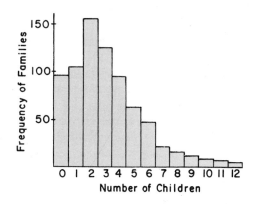

School test results sometimes produce skewed graphs, especially if the test is too difficult or too easy for the students. If a test that is designed for fifth graders is given to a class of second graders, the majority of students will get low scores. In this case the scores will pile up at the lower end of the scale and the distribution will be skewed to the right, as shown in graph a. If this test is given to a fifth grade class, the distribution of scores will be more or less normal as illustrated by graph b. Eighth graders taking this test will obtain high scores and very few of them will score poorly. In this case the distribution of scores will be skewed to the left, as shown in graph c.

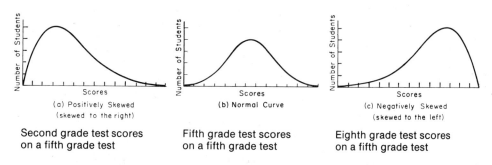

(a) Positively Skewed (skewed to the right)

Second grade test scores on a fifth grade test

(b) Normal Curve

Fifth grade test scores on a fifth grade test

(c) Negatively Skewed (skewed to the left)

Eighth grade test scores on a fifth grade test

PERCENTILES AND STANINES

For a given set of values we often wish to compare one value with the distribution of all values. This is especially important in analyzing test results. The mean is one common method of comparison. If the mean test score is 70 and a student has a score of 85, then we know the student has done better than "average." However, this information does not tell us how many students scored higher than 85 or whether or not 85 was the highest score on the test.

Percentiles One popular method of stating a person's relative performance on a test is to give the percentage of people who did not score as high. For example, a person who scores higher than 80 percent of the people taking a test is said to be at the 80th percentile. Percentiles range from a low of 1 to a high of 99. The 50th percentile is the *median,* and the 25th and 75th percentiles are called the *lower quartile* and the *upper quartile,* respectively.

It is customary on standardized tests to establish percentiles for large samples of people. When you take such a test your score is compared to the sample. A percentile score of 65 means that you did better than 65 percent of the sample group. The following table and chart show a student's performance on a differential aptitude test. There are nine categories listed in the table at the top of this form: verbal reasoning, numerical ability, VR + NA (verbal reasoning and numerical ability), abstract reasoning, etc. Each raw score under these categories is the number of questions that the student answered correctly. The student's percentile score is obtained by comparing these raw scores with the scores from a sample of thousands of other students.

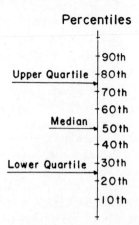

Percentiles

Upper Quartile — 80th
— 90th
— 70th
— 60th
Median → 50th
— 40th
Lower Quartile — 30th
— 20th
— 10th

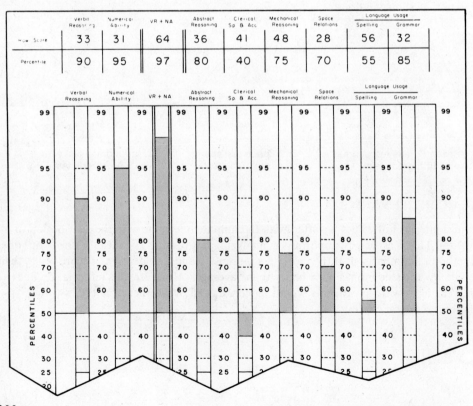

	Verbal Reasoning	Numerical Ability	VR + NA	Abstract Reasoning	Clerical Sp & Acc	Mechanical Reasoning	Space Relations	Language Usage	
								Spelling	Grammar
Raw Score	33	31	64	36	41	48	28	56	32
Percentile	90	95	97	80	40	75	70	55	85

The verbal reasoning score is at the 90th percentile. This means that the student scored higher than 90 percent of the students in the sample. This student scored at the 95th percentile in numerical ability, etc. The chart below this table (page 644) is a visual representation of these percentile scores. There is a horizontal line across the chart at the 50th percentile to make it easier to spot the scores above and below this level. We see that this student was below the 50th percentile in clerical speed and accuracy and just above the 50th percentile in spelling. Notice that the 75th and 25th percentiles are marked with dark horizontal line segments to show the upper quartile and the lower quartile. This student is in the upper quartile in six of these tests.

Stanines A stanine is a value on a 9-point scale of a normal distribution. The word "stanine" is a contraction of "standard nine." Rather than dividing the scale into 100 parts, as in the case of percentiles, there are only 9 subdivisions for stanines. The stanines are numbered from a low of 1 to a high of

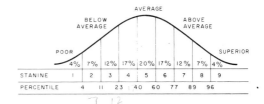

9, with 5 representing average performance. The percentage of values for each of the stanine categories is shown under this normal curve. The first stanine has the lowest 4 percent; the second stanine has the next 7 percent; etc.; to the 9th stanine which has the highest 4 percent. Thus, stanines are just subdivisions of the percentile range. If a person's test score is in the 8th stanine, we know this score is above the 88th percentile and below the 96th percentile. Stanines yield approximately the same information as percentiles, but there is the advantage of fewer subdivisions of the scale.

COMPUTER APPLICATIONS

An extensive amount of research has been devoted to developing techniques for generating sequences of random numbers. One test of randomness for an arbitrary sequence of digits is to count the number of times each digit occurs. In examining 1000 digits, for example, the percentage of 0's (or 1's, 2's, etc.) should be close to 10 percent, and for 10,000 digits the percentages for each digit should be even closer to 10 percent.

PROBLEM   How can a computer be programmed to check the percentages of occurrence for each type of number in a sequence?

In order for the computer to count the number of times a digit occurs in a sequence, each digit of the sequence must be set equal to a variable. The DIM (dimension) command in BASIC can be used to create a large number of variables, called an *array*. Here are some examples of how DIM is used. The number in parentheses tells the computer how many variables to create.

DIM A(15) creates 16 variables: A(0), A(1), A(2), . . . , A(15)
DIM B(100) creates 101 variables: B(0), B(1), B(2), . . . , B(100)

The numbers in parentheses for each variable are like subscripts: A_1, A_2, A_3, etc.

The following program generates 15 random digits and assigns each digit to a variable from $A(1)$ to $A(15)$. The printout shows these 15 variables and the random numbers.

Program 10.4A

```
10   DIM A(15)
20   FOR N = 1 TO 15
30   LET A(N) =   INT (10 *   RND (1))
40   PRINT "A("N") = "A(N)
50   NEXT N
60   END
```

```
RUN

A(1) = 2
A(2) = 1
A(3) = 6
A(4) = 1
A(5) = 1
A(6) = 6
A(7) = 3
A(8) = 6
A(9) = 5
A(10) = 7
A(11) = 7
A(12) = 3
A(13) = 6
A(14) = 2
A(15) = 7
```

The next program generates 100 random digits. Line 30 in this program assigns these numbers to the variables from $A(1)$ to $A(100)$. The values of these variables are printed by line 35. Lines 50, 60, and 70 instruct the computer to check each variable separately to determine if it is equal to 3; and if so, S is increased by 1. The printout contains 100 random digits. What percentage of these digits are 3's?

Program 10.4B

```
10   DIM A(100)
20   FOR N = 1 TO 100
30   LET A(N) =   INT (10 *   RND (1))
35   PRINT A(N)" ";
40   NEXT N
42   PRINT
50   FOR N = 1 TO 100
60   IF A(N) = 3 THEN   LET S = S + 1
70   NEXT N
80   PRINT "THE NUMBER OF 3'S IS "S"."
90   END
```

RUN

```
3 7 6 7 9 3 9 9 4 4 3 3 9 3 3 0 8 4 0 2 2 5 2 8 6
4 0 6 1 5 5 4 4 3 0 3 4 3 5 6 7 0 6 2 2 8 9 7 7 4
7 6 6 2 3 1 9 6 8 7 1 9 4 8 3 4 9 9 1 1 7 2 0 5 3
1 1 8 4 4 7 7 9 7 7 0 2 0 5 2 5 4 2 4 8 0 2 2 5 6
```

THE NUMBER OF 3'S IS 12.

Now with a few minor revisions the computer will print the frequency of each of the digits 0 through 9. Compare Program 10.4C with Program 10.4B. A FOR-NEXT command has been inserted in lines 44 and 85 to create a loop for the variable K. For each value of K from 0 to 9 the computer uses lines 50, 60, and 70 to determine which of the variables from $A(1)$ to $A(100)$ equal K. For example, when $K = 0$, the computer determines the number of variables that equal 0 and line 80 prints the result. Then line 85 sends the computer back to line 44 for another value of K. Why is S set equal to 0 for each new value of K? What percentage of the digits in the printout are 8's?

Program 10.4C

```
10   DIM A(100)
20   FOR N = 1 TO 100
30   LET A(N) =  INT (10 *  RND (1))
35   PRINT A(N)" ";
40   NEXT N
42   PRINT
44   FOR K = 0 TO 9
46   LET S = 0
50   FOR N = 1 TO 100
60   IF A(N) = K THEN  LET S = S + 1
70   NEXT N
80   PRINT "THE NUMBER OF "K"'S IS "S"."
85   NEXT K
90   END
```

RUN

```
4 5 9 5 8 9 1 7 2 0 4 0 8 5 6 4 3 9 5 9 3 5 7 2 2
0 3 8 3 9 3 8 4 6 7 7 4 7 6 6 2 3 1 9 6 8 7 1 9 4
8 3 4 9 9 1 1 7 2 3 7 6 7 9 3 9 9 4 4 3 3 9 3 3 0
8 4 0 2 2 5 2 8 0 5 3 1 1 8 4 4 7 7 9 7 7 0 2 0 5
THE NUMBER OF 0'S IS 8.
THE NUMBER OF 1'S IS 7.
THE NUMBER OF 2'S IS 9.
THE NUMBER OF 3'S IS 14.
THE NUMBER OF 4'S IS 12.
THE NUMBER OF 5'S IS 8.
THE NUMBER OF 6'S IS 6.
THE NUMBER OF 7'S IS 13.
THE NUMBER OF 8'S IS 9.
THE NUMBER OF 9'S IS 14.
```

Activity Set 10.4 Statistical Experiments (Uniform, nonuniform, and normal distributions)
Just for Fun: Cryptanalysis

Exercise Set 10.4: Applications and Skills

1. This article appeared in the *Washington Post* on November 27, 1965. The student recorded 17,950 coin flips and got 464 more heads than tails. He concluded that the U.S. Mint produces tail-heavy coins. For many repeated experiments of 17,950 tosses of a fair coin we can expect an approximately normal distribution with a mean of 8975 heads and a standard deviation of 67.

   ★ a. The area under a normal curve within ±1 standard deviation of the mean is 68 percent of the total area under the curve. Therefore, 68 percent of the time the number of heads should be between what two numbers?

   b. Numbers above 3 standard deviations from the mean will occur only 0.1 percent of the time. Therefore, 99.9 percent of the time the number of heads should be below what number?

   ★ c. Edward Kelsey got 9207 heads. Was he justified in concluding that the coins are tail-heavy?

> **Student Flips,
> Finds Penny
> Is Tail-Heavy**
>
> by Martin Weil
> Washington Post Staff Writer
>
> Edward J. Kelsey, 16, turned his dining room into a penny pitching parlor one day last spring, all in the name of science and statistics.
>
> In ten hours the Northwestern High School senior registered 17,950 coin flips and showed the world that you didn't get as many heads as tails. You get more.
>
> Edward got 464 more, enough to make him study the coins' balance and so discover that the United States Mint produces tail-heavy pennies.

2. Computers have calculated π to 100,000 places. Sketch a frequency polygon on the grid for the occurrence of the first 200 decimal digits in π (see listing that follows part **b**).

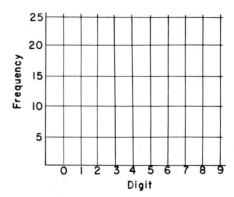

a. The digits of π are randomly distributed. This means that in an arbitrary sample of 200 digits, each digit would be expected to occur about 20 times. Which digits in the first 200 occur exactly 20 times?

★ b. Another condition for randomness is that each pair of digits should occur about 10 times in an arbitrary sample of 1000 digits, or about twice in a sample of 200. How many times do the following ordered pairs of digits occur in the first 200 digits of π: 23, 66, 74, 09?

$$\pi \approx \;3.14159\;\;26535\;\;89793\;\;23846\;\;26433\;\;83279\;\;50288\;\;41971\;\;69399\;\;37510$$
$$58209\;\;74944\;\;59230\;\;78164\;\;06286\;\;20899\;\;86280\;\;34825\;\;34211\;\;70679$$
$$82148\;\;08651\;\;32823\;\;06647\;\;09384\;\;46095\;\;50582\;\;23172\;\;53594\;\;08128$$
$$48111\;\;74502\;\;84102\;\;70193\;\;85211\;\;05559\;\;64462\;\;29489\;\;54930\;\;38196$$

3. The Montagnais-Naskapi, an Eastern Indian tribe, bake the shoulder blade of the caribou to help them make decisions concerning the well-being of their tribe. One bit of information they receive is determining the direction of the next hunt from the direction of the cracks that appear in the bone. This method of determining direction is a fairly random device that avoids human bias. It suggests that some practices in magic need to be reassessed.*

a. Explain how a table of random numbers can be used to determine random directions of 0° to 360° for hunting.

★ b. How can a table of random numbers be used to determine both random directions and random distances to be traveled for the hunt?

c. Use your method in part **b** and the list of random numbers in Exercise 4 to determine the direction and distance of your first hunt.

*O. K. Moore, "Divination—A New Perspective," *American Anthropologist,* **59** (1965), 121–128.

4. Computer programs generate random numbers for experiments and games. For example, a computer cannot flip coins and roll dice, but it can read numbers and simulate these activities.

★ a. Explain how to simulate the flipping of a coin (tossing heads or tails) by using the following numbers from a random number table. Use your method and record the first ten "coin tosses."

b. Devise a way to use these random numbers to simulate the rolling of a die. Use your method and record the first 10 "rolls of the die."

61 44 34 03 09	05 64 20 54 24	65 69 66 39 80	13 97 76 73 34
41 17 26 81 06	85 19 76 44 59	08 60 20 66 68	42 99 28 71 47
73 73 97 24 18	38 25 89 37 20		

5. Objects that are manufactured to certain specifications tend to vary slightly above and below their designed measurements. Answer the following questions by assuming the measurements are normally distributed:

★ a. A certain type of bulb has a mean life of 2400 hours with a standard deviation of 200 hours. What percentage of these bulbs can be expected to burn longer than 2600 hours?

b. A brand of crockpots has an average (mean) high temperature of 260° F and a standard deviation of 3° F. If the high temperature of these pots is below 254° F or above 266° F, they are considered defective. What percentage of these pots can be expected to be defective?

6. A certain university's Watts line can handle as many as 20 calls per minute. The average number of calls per minute during peak periods is 16 with a standard deviation of 4. What percent of the time will the Watts line be overloaded during peak periods? (Assume that the numbers of phone calls are normally distributed during peak periods.)

7. One method of grading that uses a normal curve gives students within 1 standard deviation above and below the mean a grade of C. Grades of A and B are given for intervals of 2 and 3 standard deviations above the mean, and grades of D and F are given for intervals of 2 and 3 standard deviations below the mean. Answer the following questions for a test that was given to 50 students. The mean score was 78 and the standard deviation was 6.

★ a. How many students received a C?
b. How many students received a grade below C?
★ c. How many students received an A?

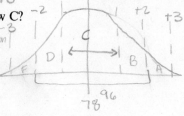

650   CHAPTER 10   PROBABILITY AND STATISTICS

8. For each of the following distributions, would its curve tend to be skewed to the right, normal, or skewed to the left?

★   a. Distances that a fifth grade class of boys can throw a football. $\frown$ *normal*

     b. Weights of newborn babies.

★   c. Numbers of people whose cars were built in the following years: 1970, 1971, . . . , 1983, 1984, 1985.

     d. Test scores of fifth graders taking a pretest on decimals at the beginning of the school year.

★   e. Shoe sizes of a college class of nurses.

9. Graphs of distribution are sometimes intended to be misleading. By using two different scales on the vertical axes of the following graphs, the distribution of sales for the 5-year period shown in this table will be skewed in one case and uniform in the other. Illustrate these sales with bar graphs. How do these two graphs differ in the impression that they give to the reader? Which graph best illustrates the true sales increase? Explain why.

Year	Sales
1973	$191,000,000
1974	191,500,000
1975	193,000,000
1976	195,000,000
1977	198,000,000
1978	200,500,000

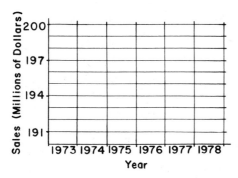

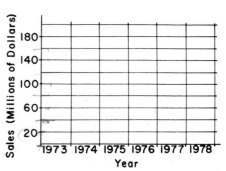

10. This table and the corresponding bar graph contain the diameters to the nearest whole inch of 100 trees. The mean diameter is 12 inches and the standard deviation is approximately 2 inches. Answer the following questions to see how close this graph is to a normal distribution:

★ a. What percentage of the diameters are within 1 standard deviation above and below the mean? 70 trees 70/100 = 70%

★ b. What percentage of the diameters are within 2 standard deviations of the mean?

Frequency Table of Diameter Measures of 100 Trees of the Same Species

Diameters (inches)	Number of trees
7	2
8	5
9	8
10	10
11	13
12	26
13	12
14	9
15	8
16	4
17	3
Total	100

s.d. = 2

s.d = 2(4)

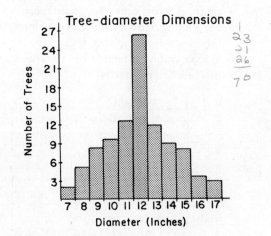

Tree-diameter Dimensions

Number of Trees (vertical axis: 3, 6, 9, 12, 15, 18, 21, 24, 27)

Diameter (Inches) (horizontal axis: 7 8 9 10 11 12 13 14 15 16 17)

11. The subtests shown on this score form are only a few of those on the Stanford Achievement Test, 1982 edition. The top row of numbers shows the number of questions answered correctly out of the total number of questions for each subtest. The second and third rows of numbers show the national and local percentiles and stanine scores.

a. Use the percentile scores to find the eight missing stanine scores.

★ b. All but one of the local percentile scores are lower than the corresponding national scores on this form. Explain what information this gives about the national and local test results.

Stanford ACHIEVEMENT TEST	SCORE TYPE	MATH COMP	READING COMP	VOCAB-ULARY	MATH APPL	SPELLING	LANGUAGE
GR 4 NORMS GR 4.8	RS/NO POSS	29/ 44	57/ 60	32/ 36	32/ 40	37/ 40	40/ 53
LEVEL INTER 1 FORM E	NAT'L PR-S	54 - 5	96 - 9	90 -	77 -	86 -	63 -
STUDENT NO 400000044	LOCAL PR-S	68 - 6	78 - 7	69 -	62 -	83 -	45 -
OTHER INFO 4	GRADE EQUIV	5.5	PHS	8.4	6.4	8.9	5.4
AGE 9- 6 TEST DATE 5/10/83							

★ 12. A score greater than or equal to the 23d percentile and less than or equal to the 39th is in the 4th stanine. For each stanine give the lower and upper percentiles.

Stanine	1	2	3	4	5	6	7	8	9
Lower Percentile				23					
Upper Percentile				39					

Exercise Set 10.4: Problem Solving

1. The device in the accompanying photo is called a probability machine and was described by Sir Francis Galton in 1889. There are 10 horizontal rows of pegs in the top half of this device. As a ball falls through the opening at the top center, it strikes the center peg in the top row and has an equal chance of going right or left. At each lower row the ball hits a peg and in each case has a 50/50 chance of falling right or left. The balls collect in 11 compartments in the lower half of this device. When a large number of balls are dropped, the distribution of the balls will be approximately normal, with the greatest number in the center compartment and the numbers decreasing as the compartments become further from the center. What is the probability of a ball falling into the center compartment of the probability machine in this photo?

★ a. *Understanding the Problem:* The probability of a ball falling into the center compartment can be computed by determining the number of ways a ball can fall into each of the compartments. Let's look at a simplified probability machine. The one shown here has only four rows of pegs. The path of one ball that has fallen into compartment B is marked by arrows. Determine the number of ways the balls can fall into each of the compartments A, B, C, D, and E. (*Hint:* The sum of these numbers is 16.)

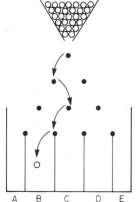

b. *Devising a Plan:* Since there are so many possibilities to consider for 10 rows of pegs, a natural approach is to solve this problem for simplified cases and hope that a pattern or some insight to a more general solution occurs. Solve this problem for three rows of pegs.

c. *Carrying Out the Plan:* Perhaps you noticed in solving parts **a** and **b** that the number of ways a ball can hit the pegs of a given row can be determined by the number of ways the ball can hit the pegs of the previous row. For example, the number under each peg shown here is the number of ways a ball can reach the peg. Continue this pattern to answer the original question.

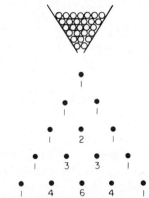

★ d. *Looking Back:* Did you recognize the triangular pattern of numbers in part **c**? This is Pascal's triangle (see page 6 and Exercise Set 1.1). If the probability machine in the photo has a total of 5120 balls, approximately how many balls will there be in the center compartment?

2. Cryptology is the science of making and breaking secret codes. One of the oldest recorded cryptograms dating from the fifteenth century B.C. has an unusual story surrounding it. A messenger journeyed from Persia to the house of Aristagoras, Greece, with a cryptogram tattooed on his scalp. When his head was shaved and the message was decoded, it was an order to Aristagoras from his father-in-law to start a revolt.

"I forgot the message!"

The frequencies of the letters that occur in our words is an important factor in enabling cryptographers to decode messages. This approach is used in parts **a**, **b**, and **c** to decode the following statement by the nineteenth century mathematician Pierre Laplace:

CA CW GZJHGBHYRZ AOHA H WVCZQVZ MOCVO YZFHQ MCAO AOZ
VLQWCXZGHACLQ LK FHJZW LK VOHQVZ WOLERX YZ ZRZSHAZX AL AOZ
GHQB LK AOZ JLWA CJILGAHQA WEYPZVAW LK OEJHQ BQLMRZXFZ.

★ a. Make a frequency distribution for the number of times each letter occurs. The four most often used letters in the English language are e, t, a, and o, in that order. This is also the order in which these letters are used in this "message." Substitute these letters into the four most frequently used letters of this code.

b. The letters h, n, i and s are four more that occur with high frequency in our language. Substitute these letters into the fifth, sixth, seventh, and eighth most frequently used letters of the code. (*Note:* The letters that represent i and s both occur the same number of times in this message.)

c. Identify the short words. With these new letters you should be able to finish the decoding.

Exercise Set 10.4: Computers

1. The printout for Program 10.4C (page 647) contains the frequency of each of the digits from 0 to 9 that occur in the 100 random digits.

a. What is the percentage of occurrence for each of these digits?

★ b. Revise this program to print out 1000 random digits and to determine the frequency of each digit.

↺ c. Run the program in part **b.** Determine the percentage of occurrence for each digit.

↺ d. Are the percentages in part **c** closer to 10 percent than the corresponding percentages in part **a**?

2. By assigning variables to each randomly generated number, these numbers can be used in a program several times. The following program generates 100 random digits and then uses this set of numbers twice; first to compute the mean, and then to compute the standard deviation.

```
10   DIM A(100)
20   FOR N = 1 TO 100
30   LET A(N) =  INT (10 *  RND (1))
35   PRINT A(N)" ";
40   NEXT N
42   PRINT
50   FOR N = 1 TO 100
60   LET T = T + A(N)
70   NEXT N
80   LET M = T / 100
90   PRINT "THE MEAN OF THESE 100 DIGITS IS "M"."
100  FOR N = 1 TO 100
110  LET R = R + (A(N) - M) ^ 2
120  NEXT N
130  LET S = (R / 100) ^ .5
140  PRINT "THE STANDARD DEVIATION OF THESE 100 DIGITS IS "S"."
150  END
```

a. List the lines that produce the 100 random digits.

★ b. List the lines that compute the mean.

c. List the lines that compute the standard deviation.

☛★ d. Try this program. Why will the mean of these numbers be close to 4.5?

e. Revise this program by using DATA and READ to compute the mean and standard deviation of the following numbers: 30, 28, 26, 20, 20, 19, 18. (*Hint:* The following are the first few lines:)

```
5   DATA  30, 28, 26, 20, 20, 19, 18
10   DIM A(7)
20   FOR N = 1 TO 7
30   READ A(N)
35   PRINT A(N)" ";
```

★ 3. Write a program to simulate 500 rolls of a pair of dice and determine the number of times each sum occurs. [*Hint:* Obtain the outcome for each die by INT(6 ∗ RND(1)) + 1. Use DIM A(500) and set each A(N) equal to a random sum.]

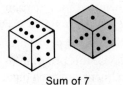

Sum of 7

4. Write a program that will simulate 500 rolls of a pair of dice and determine the number of times each sum occurs. Program the computer to print the mean and standard deviation of these sums. (*Hint:* See the programs in Exercises 2 and 3.)

☛ a. Run this program. What is your mean and standard deviation?

☛ b. What percentage of these sums are within 1 standard deviation above and below the mean?

c. What percentage of these sums are within 2 standard deviations of the mean?

d. Are these 500 sums close to a normal distribution?

Answers to Selected Exercises

EXERCISE SET 1.1

1. The sum divided by 3 is the middle number.   **c.** *Hint:* How many times greater is the sum than the middle number of the array?

2. Each sum is one less than a Fibonacci number. This Fibonacci number is two Fibonacci numbers beyond the last number in the sum.

5. 1, 2, 4, 8, 16, . . .   **a.** Geometric sequence

7. **a.** $22\frac{1}{2}$, 27, $31\frac{1}{2}$ $\left(\text{add } 4\frac{1}{2}\right)$ **c.** 8, 4, 0 (subtract 4)

8. **c.** 35, 51, 70   9. **a.** 5 lines produce 16 regions.

10. **a.** $4^2 + 5^2 + 20^2 = 21^2$ 11. Try a two-digit number.

12.
$$6 = 1 + 5 = 2 + 4 = 3 + 3$$
$$7 = 1 + 6 = 2 + 5 = 3 + 4$$
$$8 = 1 + 7 = 2 + 6 = 3 + 5 = 4 + 4$$
$$9 = 1 + 8 = 2 + 7 = 3 + 6 = 4 + 5$$
$$10 = 1 + 9 = 2 + 8 = 3 + 7 = 4 + 6 = 5 + 5$$

In general, the number of sums is the whole number of times 2 will divide into the number.

14. .4 .7 1. 1.6 2.8 5.2 10.   **a.** 2.8 astronomical units

15. **a.** 75   **b.** 75   It does not matter which of three Wednesday dates are chosen.

EXERCISE SET 1.2

1. **b.** Yes

2. **b.** If the knight lands on a black square then the knight began on a white square.
   d. If the Democrats win the election, they will take California.

3. **b.** If a person is an employee in the Tripak Company then he or she must retire by age 65.
   d. If a person is a pilot then he or she must have a physical examination every 6 months.

4. **b.** If the cards do not need to be dealt again then there is an opening bid.

5. **b.** Smith is guilty if and only if Jones is innocent.

   d. If there are negotiations then the damaged equipment will be repaired. If the damaged equipment is repaired then there will be negotiations.

6. **b.** Invalid.   **d.** Invalid

7. **b.** The patient's production of red blood cells will be slowed down. Law of detachment.

8. **b.** If we don't pay attention to the government then it will not improve.

   d. Statement **c** is stronger than statement **b**.

9. **b.** Invalid

10. **b.**

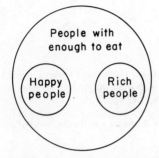

12. **b.** *Hint:* The broker and appraiser are both men.

13. **b.** Invalid.

EXERCISE SET 1.3

1. **a.** 4 feet, 2 feet

2. **d.** If each person can carry a 4-day supply of food and water, the leader can complete a 7-day trip across the desert with the aid of 6 people. (*Hint:* Some people will have to meet the returning people with food to help them back to the starting point.) Note: only 5 helpers are needed if food can be left at points in the desert.

3. **d.** 39 and 93

4. **c.** The counterfeit coin can be found from among 6 coins by beginning with 3 coins on each side of the balance scale. Explain how.

5. **d.** 19 checks.   6. **d.**

7. **d.** No.   9. 9 postcards   11. $340.

14. Number the links of the circular chain with the numbers from 1 to 12. Explain how 4 pieces of chain with 3 links each can be obtained by cutting links #1, #5 and #9.

15. **d.** The other coin is a nickel.   **f.** No dirt.
   i. The whites of the egg are not yellow.

EXERCISE SET 2.1: Applications and Skills

2. a. Column 2   **c.** These numbers are just before and just after 10 and 20.

3. a. Cardinal   **b.** Ordinal and naming use

4. a. Well-defined   **c.** Not well-defined

5. b. {0, 1, 2, 3, 5, 7, 8}   **d.** {1, 3, 4, 5, 6, 7}

6. a. *SW* and *SB*, *W* and *L*   **c.** *SW* and *SB*, *SB* and *W*, *SW* and *L*, *SB* and *L*
   e. *lwt, lwr, lwh*

7. a. *lbt, lbr, lbh, sbt, sbr, lwt, lwr, lwh, swt, swr*

8. a. *lbt, lbr, lbh*

9. b.

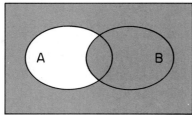

d. B′ ∪ A or B′ ∪ (A ∩ B)

10. a. No   **c.** Yes   **11. a.** d

13. b. An infinite number. (All the odd numbered rooms became vacant except room 1.)

EXERCISE SET 2.1: Problem Solving

1. b. $y = 3$

2. b. $7 + 8 = 15$ but the sum of the numbers for these regions of the diagram should be 13.
   d. 23

3. b. 55 men are married and have a telephone.   **d.** 10

EXERCISE SET 2.2: Applications and Skills

1. a. 1241   **c.**

Thousands	● ✗	M
Hundreds	● ●	D C
Tens	● ● ● ● ● ● ●	L X
Units	● ● ●	V I

2. a. Last hands in the second and fourth columns

    c. First and third columns; second and fourth columns

4. a. Four hands and two   **b.** Hand of hands, two hands and two

5. (8) Three on the other hand; (16) One on the other foot; (25) Hand on the next man.

7. b. ⋂⋂⋂⋂||||||||| XLVIII   •• / •••

8. b. DCIII   **9.** □ □ ○ ○ ○ ○ – – – – – – – ||||

12. a. Five million, four hundred thirty-eight thousand, one hundred forty-six

    c. Eight hundred sixteen billion, four hundred forty-seven million, two hundred ten thousand, three hundred sixty-one

13. b. 43,670,000

14. a.

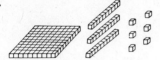

15. a.

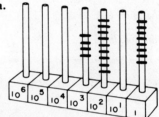

16. a. $10^{23} - 1 = 99{,}999{,}999{,}999{,}999{,}999{,}999{,}999$

17. a. Add 800,000

18. b. $42 = 32 + 8 + 2$

EXERCISE SET 2.2: Problem Solving

  1. d. The total number of grains of wheat for the first 63 squares is $2^{63} - 1$. Why is this less than the number of grains of wheat for the 64th square?

  2. c. $246 \times \$20 = \$4920$

EXERCISE SET 2.3: Logo

  2. b. _____
_____
_____
_____
_____
_____
_____
_____

  3. TO GRID
     PARALLELS
     PENUP HOME PENDOWN
     RT 90 PARALLELS
     END

4. a. TO V
        PENUP RT 90 FD 7.5 LT 90
        PENDOWN LT 20 FD 32 BK 32
        RT 40 FD 32 BK 32
        PENUP RT 70 FD 17.5
        LT 90 PENDOWN
        END

8. TO MANYKITE
    HIDETURTLE
    REPEAT 10 [KITE RT 36]
    END

6. TO SPOKE :NUMBER
    HIDETURTLE
    REPEAT :NUMBER [FD 80 BK 80 RT 360/:NUMBER]
    END

EXERCISE SET 2.3: BASIC

1. b. 3 + 9 IS EQUAL TO 12.
        THE AVERAGE OF 3 AND 9 IS 6.

2. b. THIS PROGRAM COMPUTES THE MONTHLY PROFIT FOR THE
        SALE OF THREE TYPES OF HEATERS. TYPE THE NUMBER OF TYPE
        I, TYPE II, AND TYPE III HEATERS.
        ? 250, 135, 190
        THE MONTHLY PROFIT IS $11543.75.

    c. ? 200, 150
        THE TOTAL COST FOR 200 STUDENTS BUYING HOT LUNCH TICKETS
        AND 150 STUDENTS BUYING COLD LUNCH TICKETS IS $510.

    e. ? 30
        THE COST IS NOT ABOVE $500.

3. b. 1, 3, 5, 7, 9, 11, 13, 15, 17, 19

4. 10 INPUT D
    20 LET W = .25 * D
    30 PRINT "THE NUMBER OF POUNDS LOST IN "D" DAYS IS "W"."
    40 END

6. 10 PRINT "TYPE THE NUMBER OF $8 TICKETS."
    20 INPUT X
    30 PRINT "TYPE THE NUMBER OF $11 TICKETS."
    40 INPUT Y
    50 PRINT "TYPE THE NUMBER OF $15 TICKETS."
    60 INPUT Z
    70 PRINT "THE TOTAL AMOUNT OF MONEY RECEIVED IS $"8 * X +
        11 * Y + 15 * Z"."
    80 END

```
9.  10    INPUT N
    20    IF N < = 25 THEN  GOTO 60
    30    IF N < = 50 THEN  GOTO 80
    40    LET C = .06 * N
    50    GOTO 90
    60    LET C = .10 * N
    70    GOTO 90
    80    LET C = .08 * N
    90    PRINT "THE COST OF COPYING "N" SHEETS IS $"C"."
    100   END

    ?60
    THE COST OF COPYING 60 SHEETS IS $3.6.
```

EXERCISE SET 2.3: Problem Solving

1. b. REPEAT 5 [SQUARE 10 PENUP RT 90 FD 15 LT 90 PENDOWN]
       PENUP HOME BK 15 PENDOWN

2. c.
```
   10    PRINT "THIS PROGRAM PRINTS THE FIRST 11 NUMBERS OF AN
             ARITHMETIC SEQUENCE. TYPE THE FIRST NUMBER AND THE
             COMMON DIFFERENCE. SEPARATE THESE NUMBERS BY A
             COMMA. "
   20    INPUT F,D
   30    FOR N = 0 TO 10
   40    LET X = F + N * D
   50    PRINT X
   60    NEXT N
   70    END
```

EXERCISE SET 3.1: Applications and Skills

1. a. Units wheel and hundreds wheel   c. Yes

2. a. The 10 was not regrouped ("1" was not carried) to the tens column.
   c. The "14" from 6 + 8 and the "12" from 5 + 7 were recorded side by side.

3. a. 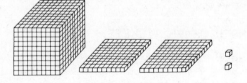   4. a.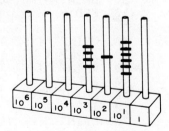

5. b.

6. a. and **d.**   **a.** 4   **b.** 4   **c.** 4   **d.** 5

7. a.   482
                 6731
               + 2064
               ─────
                 8̷1̷77
                   92

8. a. Associative   **c.** Commutative

9. b. 220   **d.** 410   **10. b.** $2411

12. There is no missing money. The desk clerk has $25, the bellboy has $2, and each man has $1. The confusion arises when the $27 (which includes the $2 "tip") is added to the $2. The situation is similar to the sum of numbers on the Hillsville sign in #2.

EXERCISE SET 3.1: Problem Solving

1. d. For 2 regions there are 4 sums. For 3 regions there are 10 sums.

2. b. Yes   **3. a.**   7 [7] 2 [9]
                              + [5] 4 [6] 2
                              ──────────
                               1 3 1 9 1

4. c.   4510
                   10
               + 2786
               ─────
                 7306

There are other answers to **4. c.**

EXERCISE SET 3.1: Computers

2. b. 1   **3. b.** 79 or 97

EXERCISE SET 3.2: Applications and Skills

1. a. Computed 6 − 4
  b. After regrouping a 10 to the units column, the "5" was not changed to a "4"
  c. The numbers were added.   **d.** Computed 9 − 7 = 2 and 4 − 3 = 1

2. a.

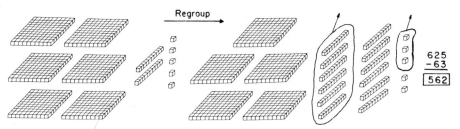

625
−63
───
562

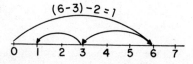

Regroup →

$$\begin{array}{r} 6245 \\ -2873 \\ \hline \boxed{3372} \end{array}$$

4. a.

(6-3)-2 = 1

6. b. No.

7. b. Change 4 to 14 in the tens column and change 3 to 4 in the hundreds column.

8. b. Add 4 to 3, add 2 to 2, add 5 to 0, and subtract 1 from 8.

11. b. $700 - 300 = 400$

12. The tens digit is always 9, and the sum of the units and hundreds digits is 9.

EXERCISE SET 3.2: Problem Solving

1. b. Subtract a single digit number so that the remaining number is 1.

2.

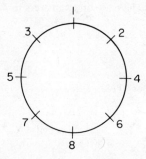

3.

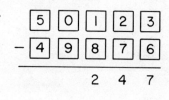

$$\begin{array}{r} 5\ 0\ 1\ 2\ 3 \\ -\ 4\ 9\ 8\ 7\ 6 \\ \hline 2\ 4\ 7 \end{array}$$

EXERCISE SET 3.2: Computers

1. b. Yes   **d.** Yes

3. If the difference between the units and hundreds digits is 1, two steps are needed. If the difference between the units and hundreds digits is 2, six steps are needed.

4. b. The special number for three-digit numbers is 495.

EXERCISE SET 3.3: Applications and Skills

1. b.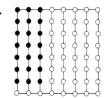

f. 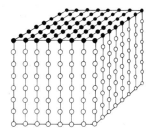 The bulbs in the top face of the cube are all lighted.

3. a.

5. b. The 2 and 1 in the tens column were added; or the 2 in the tens column was ignored.
   d. The 1, 3 and 2 in the tens column were added; or $2 \times 3 = 6$ plus 1 is 7.

6. a.

```
    1   0   1
2  /8  /0  /5
   /   /   /
   2   0   2
0  /4  /0  /0
   /   /   /
   4   0   3
9  /2  /0  /5
   9   3   5
```

9. a. Commutative property for multiplication
   c. Distributive property

10. b. $35 \times 19 = 35 \times (20 - 1)$
$$= 700 - 35$$
$$= 665$$

7. b.

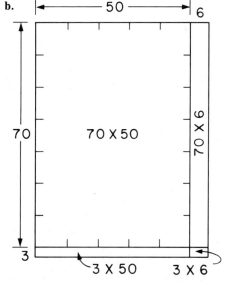

11. b. The numbers 12 and 14 are represented by 2 raised fingers on one hand and 4 raised fingers on the other hand. The product, 12×14, is equal to 100; plus the sum of the raised fingers, $2 + 4$, times 10; plus the product of the raised fingers. In general, the products of numbers from 10 to 15 are computed by beginning with 100 and using only the raised fingers.

12. **a.**

$$7\,\overline{\big|\,7\,\big|\,0\,\big|\,2\,\big|}$$
$$\big|\,2\,\big|\,0\,\big|\,4\,\big|$$
$$2\quad2\quad4$$

EXERCISE SET 3.3: Problem Solving

1. **b.** 5

2. **a.** This method of computing the product of two 2-digit numbers can be used when the tens digits are equal and the sum of the units digits is 10.

5. **a.** $\begin{array}{r} 51 \\ \times 61 \\ \hline \end{array}$

EXERCISE SET 3.3: Computers

1. **a.** All whole numbers from 1 to 28.

2. **b.** 39 is the only number in the chain.

4. **b.** $157 \times 75 = 11775$

```
10   FOR C = 1 TO 9
20   FOR D = 0 TO 9
30   FOR E = 1 TO 9
40   LET P = (C * 100 + D * 10 + E) * (E * 10 + D)
50   IF P = (C * 10000 + C * 1000 + E * 100 + E * 10 + D)
     THEN GOTO 100
60   NEXT E
70   NEXT D
80   NEXT C
90   GOTO 110
100  PRINT C * 100 + D * 10 + E" * "E * 10 + D" = "P"."
110  END
```

EXERCISE SET 3.4: Applications and Skills

1. **b.**

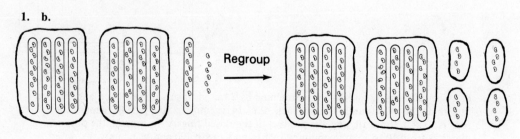

2. **a.** The number of groups of markers on each column is a digit in the quotient. For example, there are 4 groups of two markers on the 10^3 column, and the digit in the thousands column of the algorithm is 4.

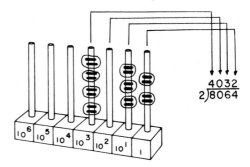

$$\begin{array}{r} 4032 \\ 2\overline{)8064} \end{array}$$

3. **a.** $15 \div 3 = 5$

4. **b.** The "6" and "8" have been written in the wrong places.
   d. The remainder 5 is greater than the divisor 4. The first number in the quotient should be 3.

5. **b.** Division is not commutative.

6. **a.** The sum of the numbers that correspond to the binary numbers 1, 2, and 8, equals 1232. Therefore, the quotient $1232 \div 112$ equals $1 + 2 + 8$, or 11.

7. **a.** $5^{14} \times 5^{20} = 5^{34}$ $10^{32} \div 10^{15} = 10^{17}$

8. **a.** 10^{19} **c.** 10^{13}

9. **a., b.** The steps in part **c** may produce the correct answer, depending on the type of calculator.

10. **a.** Remainder of 28   12. **a.** 10^7 **b.** 10^{53}

EXERCISE SET 3.4: Problem Solving

1. **b.** # occurs in squares 6, 12, 18, etc.   **d.** 2

2. Cut off $\frac{1}{4}$ of the end of the cake.

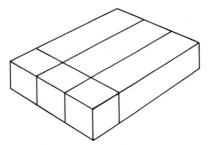

3. $1 = (4 + 4) \div (4 + 4)$ $3 = (4 \times 4 - 4) \div 4$
   $5 = (4 \times 4 + 4) \div 4$ $10 = (44 - 4) \div 4$

4. **a.** $([(22 - 19) + 2] \times 14) \div 10$

5. $354,393 \div 39$

EXERCISE SET 3.4: Computers

1. **c.** Rows 2, 4, 6 and 8

2. **b.**
```
10  PRINT "THIS PROGRAM PRINTS THE FIRST EIGHT POWERS OF
        A NUMBER. TYPE THE NUMBER."
20  INPUT X
30  FOR N = 1 TO 8
40  PRINT X ^ N" ";
50  NEXT N
60  END
```

3. **b.** 27,000

4. **b.** Yes, there are 32 such numbers. Here are the first two: 10,000 and 12,996.

EXERCISE SET 3.5: Applications and Skills

2. 47, 59, 73, 89, 107, 127, 149, 173, 199, 227, 257, $\boxed{289}$

4. **a.** Prime numbers   **c.** Square numbers have an odd number of factors.

5. **a.** Prime numbers

6. 2, 3, 5, 7, 11, 13, 17, and 19   **7.** **b** and **d**   **8.** **a.** $924 = 2 \times 2 \times 3 \times 7 \times 11$

9. 1,000,000,000 has a unique factorization containing only 2s and 5s. Since there is no other factorization, 7 is not a factor.

10. **a.** 3 yellow rods or 15 white rods   **c.** 1, 2, 3, 6, 7, 14, 21, and 42

11. **a.** $n = 4, n = 6$

12. **a.** $21 = 3 + 7 + 11$; $27 = 3 + 11 + 13$; $31 = 7 + 11 + 13$ **c.** True

EXERCISE SET 3.5: Problem Solving

1. **b.** Lockers 1, 4 and 9   **d.** Twice.

3. *Hint:* Determine the number of times 2 and 5 occur as factors.

4. **b.** Let $N = 2 \times 3 \times \ldots \times 1{,}000{,}001$. Consider $N + 2$, $N + 3$, etc.

EXERCISE SET 3.5: Computers

2. **b.** 5663 is not a prime.   **d.** 8317 is a prime.

```
10   PRINT "TYPE A NUMBER WHICH IS GREATER THAN 100 AND
     LESS THAN 10000. "
20   INPUT K
30   FOR X = 2 TO 100
40   IF K / X =  INT (K / X) THEN  GOTO 80
50   NEXT X
60   PRINT K" IS A PRIME. "
70   GOTO 90
80   PRINT K" IS NOT A PRIME. "
90   END
```

3. **b.** Tartaglia's conjecture does not hold. $2^N - 1$ is not prime for $N = 8, 9, 10, 11$ and 12.

4. **b.** $2^N - 1$ is prime for $N = 13, 17$ and 19.

EXERCISE SET 3.6: Applications and Skills

1. **b.** If the highest floor which must be delivered to has an odd number, deliver to all the odd numbered floors first. Then walk down one flight of stairs and use the elevator for the even numbered floors. A similar plan can be used if the highest floor which must be delivered to has an even number.

2. **a.** 648 seconds or 10.8 minutes   **c.** 32,400 seconds or 9 hours.

3. **a.** White rods, green rods, or dark green rods   **c.** Relatively prime

4. **a.** 7 purple rods and 4 black rods

6. **a.** Not necessarily   **7.** 512,112 and 4,328,104,292 are divisible by 4.   **8. d.** Yes

10. **a.**

g.c.f. $(198,165) = 3 \times 11 = 33$

11. **a.**

l.c.m. $(30,42) = 2 \times 3 \times 5 \times 7 = 210$

12. **a.**

$$\begin{array}{r} 64796 \\ \times\ 1560 \\ \hline 101181760 \end{array}$$

$$\begin{array}{r} \underline{\text{Excess}} \\ 5 \\ \underline{\times\ 3} \\ 15 \end{array}$$

The nines excess for 101181760 is 7, and for 15 it is 6. This indicates an error.

d. The remainder is 7.

EXERCISE SET 3.6: Problem Solving

1. **b.** (1) 7 tiles   (3) 4 tiles   (5) 6 tiles   (7) 12 tiles
   d. The four small rectangles are copies of each other.

2. *Hint:* Find the least common multiple of 7, 6, 5, 4, 3 and 2.

EXERCISE SET 3.6: Computers

2. **b.** Abundant   **d.** Abundant

4.
```
10   PRINT "THIS PROGRAM FINDS THE L.C.M. OF TWO NUMBERS.
        TYPE TWO WHOLE NUMBERS, SEPARATED BY A COMMA."
20   INPUT A, B
22   FOR N = B TO 1 STEP  - 1
24   IF B / N =  INT (B / N) AND A / N =  INT (A / N) THEN
        GOTO 30
26   NEXT N
30   PRINT "THE L.C.M. OF "A" AND "B" IS "(A * B) / N"."
40   END
```


EXERCISE SET 4.1: Applications and Skills

1. **b.** 120°   2. **b.** 60°

3. **a.** Four new line segments and a total of 10 line segments
   b. Five more line segments and a total of 15 line segments

4. **a.** 105   5. **a.** 0, 1, 3, 4, 5, 6

6. The number of regions plus the number of vertices minus 1 equals the number of line segments.

8. **a.** Angles *G, I, J*   **c.** Angles *B, E*

9. **a.**

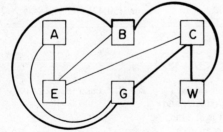

EXERCISE SET 4.1: Problem Solving

1. **b.** Pentagon, 5.   Hexagon, 9.   Heptagon, 14.

2. Five lines intersect in 10 points and six lines intersect in 15 points.

5.

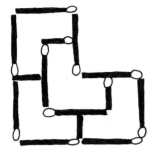

EXERCISE SET 4.1: Computers

3.
```
TO PARALLELOGRAM
   RT 35
   FD 40 RT 55
   FD 70
   RT 125
   FD 40
   HOME
END
```

5.
```
TO EQUITRIANGLE :SIDE
   RT 30 REPEAT 3 [FD :SIDE
   RT 120]
END
```

EXERCISE SET 4.2: Applications and Skills

1. **a.** Hexagons; no   **2.** **a.** 72°   **b.** Hexagon 60°, octagon 45°

5. **a.** Approximately 51.4°   **6.** **a.** Pentagon

7. **a.** False   **c.** False   **e.** True

8. **c.** No. Each vertex angle has a measure of 135° and 135 is not a factor of 360.

10. **a, c, e**

EXERCISE SET 4.2: Problem Solving

1. **b.** Pentagon, 540°. Hexagon, 720°. Heptagon, 900°.   **d.** Yes

3. For five motorcycles (a regular pentagon) the angle between the mirrors is 72°.

4. b.

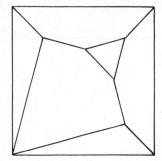

EXERCISE SET 4.2: Computers

1. b.
```
TO OCTAGON :SIDE
   REPEAT 8 [FD :SIDE RT 45]
END
```

2.
```
TO POLYGON :NUMBER :SIDE
   REPEAT :NUMBER [FD :SIDE RT 360 / :NUMBER]
END
```

4. a.
```
TO PETAL :SIZE
   ARC :SIZE
   RT 90
   ARC :SIZE
   RT 90
END
```

EXERCISE SET 4.3: Applications and Skills

2. a. No   **b.** Angle *a*   *Angles b* and *c* are right angles.

3. *CFGH, AEFH, ABCF, ACDH*

6. a. 1 through 7   **c.** 26   **7. a.** Conic   **b.** Plane   **c.** Cylindrical

8. Tokyo (35°N,140°E)
San Francisco (38°N,120°W)
Melbourne (38°S,145°E)
Glasgow (56°N,4°W)
Capetown (35°S,20°E)

9. a. (20°S,60°E)   **10.** Buenos Aires

11. a. September 15: (32.5°N,48°W)

EXERCISE SET 4.3: Problem Solving

1. b. (3) 12 vertices, 18 edges, 8 faces   **d.** No.
      (5) 16 vertices, 24 edges, 10 faces

2.

	Vertices	Faces	Edges
Tetrahedron	4	4	6
Cube	8	6	12
Octahedron	6	8	12
Dodecahedron	20	12	30
Icosahedron	12	20	30

4. *Hints:* Use a disc to obtain a right circular cone. Dip a right circular cone (or right circular cylinder) into a liquid at an angle.

5. White; one location is the North Pole. For other locations start 15 miles north of a 15-mile circle around the South Pole.

EXERCISE SET 4.3: Computers

2.
```
TO SQUAREPRISM :HEIGHT
   HIDETURTLE
   RT 30 FD 40 PRISMEDGE :HEIGHT
   RT 120 FD 40 PRISMEDGE :HEIGHT
   RT 210 FD 40 PRISMEDGE :HEIGHT
   RT 300 FD 40 PRISMEDGE :HEIGHT
   FD :HEIGHT SQUARE
END
```

3.
```
TO OCTAPRISM :SIDE :HEIGHT
   HIDETURTLE
   RT 30 FD :SIDE PRISMEDGE :HEIGHT
   RT 75 FD :SIDE PRISMEDGE :HEIGHT
   RT 120 FD :SIDE PRISMEDGE :HEIGHT
   RT 165 FD :SIDE PRISMEDGE :HEIGHT
   RT 210 FD :SIDE PRISMEDGE :HEIGHT
   RT 255 FD :SIDE PRISMEDGE :HEIGHT
   RT 300 FD :SIDE PRISMEDGE :HEIGHT
   RT 345 FD :SIDE PRISMEDGE :HEIGHT
   FD :HEIGHT RT 30
   OCTAGON :SIDE
END
```

EXERCISE SET 4.4: Applications and Skills

1. c. Rectangular windows   **2. a.** Eight

3. a. Two lines of reflection and two rotational symmetries
  c. Six lines of reflection and six rotational symmetries

4. b, g **a.** 4 **b.** 2 **c.** 5 **d.** 1 **e.** 2 **f.** 7 **g.** 3 **5.** a, c, e, f

6. *Hint:* There are four letters with two lines of reflection and three letters with two rotational symmetries but no lines of reflection.

7. a. A, H, I, M, O, T, U, V, W, X, Y Each of these letters has a vertical line of reflection.

8. a. Two, (180°, 360°)   **c.** Three, (120°, 240°, 360°)   **e.** Four, (90°, 180°, 270°, 360°)

9. a.   **10. a.**   **11. b.**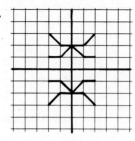

13. a. Cylinder, prism, sphere, cube
  c. Cone (infinite), cylinder (infinite), prism (3), sphere (infinite), cube (4), pyramid (4)

14. a. Eight rotational symmetries and no lines of reflection

15. a. Sixteen   **c.** Six

EXERCISE SET 4.4: Problem Solving

1. b. A figure with two lines of symmetry has two rotational symmetries.

2. a. 13   **b.** 9

EXERCISE SET 4.4: Computers

2. a.
```
TO EXERCISE#2.A
   REPEAT 10 [FLAG RT 36]
END
```

3. b.
```
TO EXERCISE#3.B
   REPEAT 10 [HOOK RT 36]
END
```

EXERCISE SET 5.1: Applications and Skills

1. a. 92   **c.** 60   **3.** *Across* 1. 165   4. 3150   5. 120
                                          8. 92   9. 7920   10. 5550

4. Potatoes, 4.536 kg; motor oil, 946 ml

5. a. From 111.5 kg to 112.5 kg   **c.** From 48.25 cm to 48.35 cm

6. a. 11.6 kg   **c.** $63.36   **e.** 320 days; 5¢ per day

7. a. 12ℓ   **8. a.** 7.2

9. a. 56   **c.** $108.48   **10.** 2

11. b. Length 158 m, breadth 26 m, height 16 m   **12. a.** 57.8°C   **c.** ⁻14.8°F

EXERCISE SET 5.1: Problem Solving

1. b. 12 cm.   24 cm.   **d.** Yes

2. *Hint:* How long before the trains collide?

4. *Hint:* Look at the sketch. Where is page 1 of book 1? Where is page 300 of book 4?

EXERCISE SET 5.1: Computers

1. b. 36   **e.** 9.8

2. b. THIS PROGRAM CONVERTS MILES TO KILOMETERS. TYPE THE
NUMBER OF MILES.
? 412
412 MILES IS APPROXIMATELY EQUAL TO 663.3 KILOMETERS.

c. TYPE A FAHRENHEIT TEMPERATURE.
? 98.6
TO THE NEAREST WHOLE NUMBER 98.6 DEGREES FAHRENHEIT
EQUALS 37 DEGREES CELSIUS.

3. a. Longer by 8.5 meters   **c.** Shorter by 2.6 meters.

EXERCISE SET 5.2: Applications and Skills

1. a. Approximately 1088 m^2 **b.** 106 m by 40 m; Area $=$ 4240 m^2

2. b. Dimensions (or perimeter) and area; the sum of the dimensions of the 3 by 5 rectangle, $3 + 5$, is equal to the sum of the dimensions of the 4 by 4 square, $4 + 4$.

3. a. Type A   **c.** $8.28   **e.** Approximately 1428.6 m^2

4. a. 1740 cm^2 **c.** 1696.25 cm^2 **5. a.** 1600 cm^2

6. a. 1375 mm^2 or 13.75 cm^2 **c.** 870 mm^2 or 8.7 cm^2

7. a. 982 m^2 (to the nearest square meter)   **c.** 111 m   **e.** 3220 m^2

8. a. 90,675 cm^2 **9. a.** 56 cm^2

10. a. $20,790   **c.** $1081.08 for the house in part **a**

11. The sectors of the circle should cover approximately 3.1 squares. This approximation will improve if the circle is cut into smaller sectors. Theoretically, the area of the circle is equal to π times the area of the square or approximately 3.14 squares.

12. **a.** 152 mm^2 **c.** Approximately 22%

EXERCISE SET 5.2: Problem Solving

1. **b.** The shape depends on the placement of the inner square. Here are
   two possibilities.

   e. The area of the small circle is one-half the area of the large circle.

2. 4π meters   **4. b.** Too large.

EXERCISE SET 5.2: Computers

1. **a.**

WIDTH	LENGTH	AREA
20	56	1120
21	54	1134
22	52	1144
23	50	1150
24	48	1152
25	46	1150
26	44	1144
27	42	1134
28	40	1120
29	38	1102
30	36	1080

```
]
```

2. **a.**
```
10  PRINT "TYPE A NUMBER FOR THE PERIMETER OF A SQUARE."
20  INPUT P
30  LET L = P / 4
35  LET S = L ^ 2
40  LET R = P / (2 * 3.14)
45  LET C = 3.14 * R ^ 2
50  PRINT "FOR A PERIMETER OF "P" THE AREA OF THE SQUARE
    IS "S" AND THE AREA OF THE CIRCLE IS "C"."
60  END
```

EXERCISE SET 5.3: Applications and Skills

1. **b.** Roof of the tower   **f.** Hands of the clock   **j.** Edges of the roof of the tower

2. a. 946 cm^3

3. a. 166,779 cm^3

4. a. 21,000 Btu units   **c.** Type B

5. a. 37.5

6. a. They counted the surface area that can be seen.   **7. a.** 2.68 m   **c.** No

8. b. 1570 kg

10. a. About 25 times greater   **11. a.** Two times greater   **c.** 2^{20}

12. a. 72 cm^3 **c.** 26.25 cm^3 **e.** 9.42 cm^3

EXERCISE SET 5.3: Problem Solving

1. a. The width is 18 cm and the length is 38 cm.
   d. Try 5 by 5 and 7 by 7 squares.

2. Try some numbers for W and L and compute the volumes.

4. Use the formula for the volume of a cone for the ends of the football, and the formula for the volume of a cylinder for the central section.

EXERCISE SET 5.3: Computers

1. b. 300 cubic units   **2. b.** 4206 m^2 **3. b.** 25 hours

EXERCISE SET 6.1: Applications and Skills

1. a. More

3. a. $\dfrac{25}{6} = 4\dfrac{1}{6}$ $\dfrac{17}{5} = 3\dfrac{2}{5}$ **b.** $2\dfrac{1}{5} = \dfrac{11}{5}$ $2\dfrac{5}{6} = \dfrac{17}{6}$

   c. $\dfrac{5}{3} = \dfrac{40}{24}$ $\dfrac{5}{6} = \dfrac{20}{24}$ **d.** $\dfrac{4}{12} = \dfrac{3}{9}$

4. c. $\dfrac{3}{15} = \dfrac{6}{30}$ $\dfrac{5}{6} = \dfrac{25}{30}$

5. a. $\dfrac{7}{10} = \dfrac{14}{20}$

6. a. Dark green rod   **c.** Brown rod

7. a. Dow Ch: $44.38   **b.** RCA: up 12.5 cents

8. a. Alaska Airlines   **c.** Less than

10. b. $\dfrac{2}{5} < \dfrac{4}{6}$ because $\dfrac{2}{5} < \dfrac{1}{2}$ and $\dfrac{4}{6} > \dfrac{1}{2}$.

11. c. 4   **f.** 0   **12. b.** $\dfrac{7}{12}$

EXERCISE SET 6.1: Problem Solving

1. **b.** Every moment 2 are born; every moment $1\frac{1}{2}$ are born.

2. *Hint:* The total number of plates must have 2, 3 and 4 as factors.

3. **b.** Cauliflower $\frac{2}{3}$ and celery $\frac{1}{2}$.

4. **c.** $\frac{4}{6} < \frac{5}{7}$ $\frac{5}{7} < \frac{6}{8}$ $\frac{6}{8} < \frac{7}{9}$, etc.

EXERCISE SET 6.1: Computers

1. **b.** Until fractions with denominators which are less than or equal to 10.

2. **b.** Yes.

4.
```
10   PRINT "TYPE THE NUMERATOR AND DENOMINATOR OF A FRAC
     TION, SEPARATED BY A COMMA."
20   INPUT A,B
22   FOR N = B TO 2 STEP  - 1
24   IF B / N =  INT (B / N) AND A / N =  INT (A / N) THEN
     GOTO 50
26   NEXT N
30   PRINT A"/"B" IS IN LOWEST TERMS. "
40   GOTO 60
50   PRINT A"/"B" IS EQUAL TO "A / N"/"B / N" IN LOWEST
     TERMS. "
60   END
```

EXERCISE SET 6.2: Applications and Skills

1. **b.**

$$\frac{5}{6} - \frac{1}{3} = \frac{3}{6}$$

c.

$$\frac{1}{12} \qquad \frac{1}{3} \times \frac{1}{4} = \frac{1}{12}$$

2. **b.**

$$\begin{array}{r}
\frac{4}{5} = \frac{32}{40} \\
+\ 1\frac{3}{8} = 1\frac{15}{40} \\
\hline
1\frac{47}{40} = 2\frac{7}{40}
\end{array}$$

3. **b.**

$$\begin{array}{r}
1\frac{1}{8} = 1\frac{5}{40} = \frac{45}{40} \\
-\ \frac{7}{10} = \frac{28}{40} = \frac{28}{40} \\
\hline
\frac{17}{40}
\end{array}$$

4. b. $\frac{13}{24}$ **k.** $2\frac{1}{8}$ **n.** $6\frac{4}{9}$ **p.** $20\frac{1}{2}$

5. a. Added numerators and added denominators.

   e. Subtracted $\frac{1}{4}$ from $\frac{3}{4}$.

   g. Added the same number to the numerator and denominator.

6. a. $3\frac{5}{8}$ **c.** Drew National $1\frac{1}{4}$ MEM Company $\frac{5}{8}$ Old Town $1\frac{3}{8}$

7. a. 5.4 meters

8. a. Commutative property for addition   **c.** Associative property for addition
   e. Distributive property

10. b. 14   **d.** 14   **11. b.** 44   **d.** $64\frac{1}{2}$

13. a. D, E, G, A, B

EXERCISE SET 6.2: Problem Solving

1. b. 16, 18, 20, 22, 24   **d.** The middle jar contains $13\frac{1}{7}$ bars.

2. c. No.

4. c. $\frac{1}{4} - \frac{1}{5} = \frac{1}{4} \times \frac{1}{5}$, $\frac{1}{5} - \frac{1}{6} = \frac{1}{5} \times \frac{1}{6}$ etc.

EXERCISE SET 6.2: Computers

1. b. .999511719   **d.** 1

2.
```
10   PRINT "HOW MANY FRACTIONS DO YOU WISH TO ADD? TYPE A
        WHOLE NUMBER. "
20   INPUT N
30   FOR X = 2 TO N + 1
40   LET S = S + (1 / X)
50   NEXT X
60   PRINT "THE SUM OF THE FIRST "N" FRACTIONS IS "S"."
70   END
```

   d. Sum of 100 fractions equals 4.19727851. Sum of 1000 fractions equals 6.48646987.

3. b. $\frac{1}{12} + \frac{1}{51} + \frac{1}{68} = \frac{2}{17}$ $\frac{1}{20} + \frac{1}{124} + \frac{1}{155} = \frac{2}{31}$

EXERCISE SET 6.3: Applications and Skills

1. **a.** $^-29°C$

3. **a.** $6 + {}^-5 = 1$

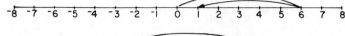

   e. $4 - 7 = {}^-3$

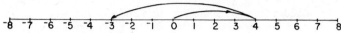

4. **a.** $^-6 \times 2 = {}^-12, \qquad {}^-20 \div 4 = {}^-5, \qquad \frac{1}{3} \times {}^-12 = {}^-4$

   b. $4 + {}^-14 = {}^-10, \qquad 6 - {}^-4 = 10 \qquad {}^-6 \times 2 = {}^-12, \qquad {}^-15 \div 5 = {}^-3$
   c. $^-3 + 1 = {}^-2$
   d. The negative of $\frac{7}{8}$ is $\frac{{}^-7}{8}$ and the reciprocal of $\frac{7}{8}$ is $\frac{8}{7}$.

5. **b.** $^-300$ **d.** $^-100$ 6. **b.** 1979

7. **a.** Commutative property for addition   **c.** Associative property for multiplication

9. **a.** The reciprocals of nonzero numbers in the interval from $^-1$ to 1 will be outside this interval.

11. **a.** 17 days and 17 hours

EXERCISE SET 6.3: Problem Solving

1. **b.** 1, 2, 3, and 4 must be neighbor numbers to obtain 10.

2. *Hint:* 1, 2, 3, 4, and 5 must be neighbor numbers.

3. All the even numbers.

5. **b.** Can 100 be obtained by inserting other numbers of plus and minus signs? What other numbers from 1 to 100 can be obtained?

EXERCISE SET 6.3: Computers

1. **c.** 1, 3, 6, 12

2. **b.**
```
10  PRINT "TYPE A FAHRENHEIT TEMPERATURE. "
20  INPUT F
30  LET C = 5 * (F - 32) / 9
40  LET R =  INT (10 * C + .5) / 10
50  PRINT "TO THE NEAREST TENTH OF A DEGREE "F" DEGREES
        FAHRENHEIT EQUALS "R" DEGREES CELSIUS. "
60  END
```

3. **b.** The sum of the first 10 fractions is .263455988.

EXERCISE SET 7.1: Applications and Skills

1. **a.** 9,193,000,000

3. **b.** 25 parts out of 100 is equal to 250 parts out of 1000.
   d. 3 parts out of 10 is greater than 125 parts out of 1000.

5. **a.** ⁻11.9° Celsius   **c.** ⁻13.6° Celsius

7. **a.** Three hundred sixty and two thousand, eight hundred sixty-six ten-thousandths
   e. Three hundred forty-seven and ninety-six hundredths

8. **a.** $\dfrac{837}{1000}$ **c.** $\dfrac{64}{99}$ 9. **a.** .44   **c.** 3.72

11. **a.** .4266   **c.** .43365   **e.** .436   12. **a.** .41$\overline{6}$

13. M. Sweeney 6.47 feet   W. G. George 4.21 minutes

EXERCISE SET 7.1: Problem Solving

1. **b.** Find the decimal for each of these fractions: $\frac{6}{9}, \frac{8}{12}, \frac{2}{3}, \frac{4}{6}$
   d. The only prime factors of **b** are 2 or 5.

2. **b.** *Hint:* Try some fractions whose denominators do not have 2 or 5 as factors.

EXERCISE SET 7.1: Computers

2. The length of the period is the number of 9's in the denominator.

3. **a.** Each sum is a multiple of 9.

EXERCISE SET 7.2: Applications and Skills

1. **b.** $325.   **d.** $2925.

2. **c.** Use a decimal square for .1 and divide the shaded part into 10 equal parts.
   e. Use a decimal square for .75 and mark off 15 groups of shaded parts each.

3. **a.**

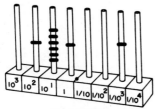

161.001

4. a.

$$\begin{array}{r} 1 \\ 4.821 \\ + 61.73 \\ \hline 66.551 \end{array} \qquad \frac{8}{10} + \frac{7}{10} = \frac{15}{10} = \frac{10}{10} + \frac{5}{10} = 1 + \frac{5}{10}$$

d.

$$\begin{array}{r} 5 \\ 6\cancel{0}.43 \\ - 41.72 \\ \hline 24.71 \end{array} \qquad 6 = 5 + 1 = 5 + \frac{10}{10}$$

6. b. The 6 in the hundredths column was not subtracted.
   d. The remainder of 2 was recorded in the quotient.

7. b. $480 - 170 = 310$ (actual, 312.7)   **d.** $4 \div .8 = 5$ (actual ≈ 5.27)

8. a. .8874106   **9. a.** $25.96   **c.** $14.74   **e.** $172.92

10. a. $1210.90   **c.** $3.76   **e.** $158.55

11. a. 852.6 kilowatt-hours   **c.** $34.10 **12. a.** 4 minutes and 11.76 seconds

EXERCISE SET 7.2: Problem Solving

1. a. 82 U.S. dollars; 60 Canadian dollars.   **d.** No

4. *Hint:* There are 2^{25} pieces of paper and 5280 feet in a mile.

EXERCISE SET 7.2: Computers

2. b.
```
TYPE THE AMOUNT OF THE FINANCE CHARGE.
? 300
THIS PROGRAM COMPUTES THAT PART OF $300 WHICH MUST BE
PAID AFTER THE FIRST K MONTHS. TYPE A NUMBER FOR K.
? 3
THE AMOUNT OF FINANCE CHARGE TO BE PAID AFTER 3 MONTHS
IS $89.47.
```

3. a. Yes   **c.** 4

EXERCISE SET 7.3: Applications and Skills

2. b. 2,250,000

3.
Mercury	36,002,000	A.U. = .39
Venus	6.7273×10^7	A.U. = .72
Earth	93,003,000	A.U. = 1.0

4. a. 3.2×10^{-2} **b.** .000000003048
635,000,000,000

5. a. 3.9952×10^4 kph   **c.** 2.551176×10^{18} km

6. The distributive property   **a.** $178.08   **7. a.** $50.54

8. a. Large size   **9. a.** 1 to 45   **10. a.** Approximately .2%   **c.** 27.168 tons

11. a. $1060   **c.** 12 years   **12. b.** 18

13. Alabama   30.5 to 1   **14. a.** New Jersey
      Florida   27.3 to 1
      Hawaii   23.4 to 1

EXERCISE SET 7.3: Problem Solving

1. d. No.   **2.** *Hint:* A 20% saving followed by a 30% saving reduces the original fuel requirement by 44%.

EXERCISE SET 7.3: Computers

1. b. $19,725,189

3.
```
10   PRINT "THIS PROGRAM COMPUTES THE AMOUNT A WHICH
     RESULTS FROM AN INVESTMENT P OVER N YEARS WITH AN IN
     TEREST RATE R COMPOUNDED DAILY. TYPE NUMBERS FOR P,
     N, AND R."
20   INPUT P,N,R
30   LET A = P * (1 + R / 365) ^ (365 * N)
40   LET M =  INT (100 * A + .5) / 100
50   PRINT "$"P" COMPOUNDED DAILY FOR "N" YEAR(S) BECOMES
      $"M"."
60   END
```

4. b. $5287.85.

EXERCISE SET 7.4: Applications and Skills

1. $\frac{1}{8}$, rational number and real number; $\sqrt{3}$, real number.

3. a. $3\sqrt{5}$ **c.** $2\sqrt{15}$ **4. a.** 10   **c.** 14   **5. a.** 5.7   **c.** 4.1

7. a. $\dfrac{3\sqrt{7}}{7}$ **c.** $\sqrt{5}$

8. a. 1   **b.** 1   **9. a.** No; $3 - 5 = {}^-2$ **d.** Yes

10. a. Associative property for multiplication   **c.** Commutative property for addition

11. b. Yes   **c.** $\sqrt{7.5} \approx 2.7$ Yes

12. a. 3.140845   **c.** 3.1415929   **e.** 3.1416

EXERCISE SET 7.4: Problem Solving

1. **a.** The inner square has an area of 2.
    d. The areas of the first five squares are 4, 2, 1, $\frac{1}{2}$, $\frac{1}{4}$, and $\frac{1}{8}$.

3. A line segment of length $\sqrt{5}$ can be constructed as the hypotenuse of a right triangle whose legs have lengths of 1 and 2.

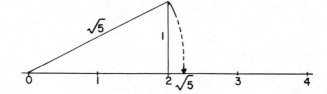

EXERCISE SET 7.4: Computers

1. **b.** 987 and 610.   3. **b.** 376.7 million miles in 258.9 days.

4. **b.** So that the satellite stays in the same position relative to the earth.

EXERCISE SET 8.1: Applications and Skills

1. **b.** Replace the top in the first scale by 1 block and 8 marbles.

2. **b.** The fact that for every six students there is one professor is directly (but incorrectly) translated as $6S = P$.

3. **b.** $4n + 20$ **d.** False   **g.** Sometimes true.

4. **b.** $3x + 8 = 5x + 2$
    $$8 = 2x + 2$$
    $$6 = 2x$$
    $$3 = x$$

5. **b.** $(z/5 + 19)/2 = 37$
$z/5 + 19 = 74$	Multiply by 2
$z/5 = 55$	Subtract 19
$z = 275$	Multiply by 5

6. **b.** $5(2x - 3) = 7x$
    $$10x - 15 = 7x$$
    $$3x = 15$$
    $$x = 5$$

    d. $5b - (b + 4)/2 = 47 + b/2$
    $$10b - (b + 4) = 94 + b$$
    $$9b - 4 = 94 + b$$
    $$8b = 98$$
    $$b = 12\frac{1}{4}$$

7. **b.** $200 + 15c = 300 + 8c$, $c = 14\frac{2}{7}$
    So, the new plan is cheaper for 15 or more checks per month.
    d. Let b equal the cost of the bottle and $b + 60$ equal the cost of the cider. Solve $b + (b + 60) = 86$.

8. **b.** Let the five numbers be x, $x + 1$, $x + 2$, $x + 3$, and $x + 4$. Add these terms and divide by 5.

9. b. 50 inches.

11. *Hint:* Begin with the inner parentheses and multiply 4 times 5m + 6.

EXERCISE SET 8.1: Problem Solving

1. b.

x + 20	x + 21	x + 22	x + 23
x + 30	x + 31	x + 32	x + 33

  d. To obtain the formula for a 3 by 3 array taken from a 10 by 10 number chart, solve this equation for *x*: 3x + 33 = s, where s = sum.

4. Let T equal the total time for the person who will run half the time. Solve the following equation to determine this person's total time:

$$6\left(\frac{T}{2}\right) + 3\left(\frac{T}{2}\right) = 12.$$

EXERCISE SET 8.1: Computers

1. b. S = 5 and N = 3, if the given information is interpreted as in line 30.

```
10   FOR S = 1 TO 80
20   FOR N = 1 TO 80
30   IF S * S + N = 2 * (N * N + S) THEN  GOTO 60
40   NEXT N
50   NEXT S
60   PRINT "SONYA IS "S" YEARS OLD AND NIKE IS "N" YEARS
     OLD. "
70   END
```

3. The length of the plaza is 6 units.

4. b. A = 2 and N = 85714

EXERCISE SET 8.2: Applications and Skills

2. b.

x	⁻3	⁻2	⁻1	0	1	2	3
y	⁻5.4	⁻1.6	⁻.2	0	.2	1.6	5.4

3. b. Slope 5 and y intercept 3   **e.** Slope $\frac{1}{3}$ and y intercept 0

4. b. $\frac{1}{4}$ **5. a.** Slope of L_1 is 10. Slope of L_3 is $\frac{1}{2}$.   **b.** No.

6. b. $5.   **7. b.** c = .15n + 50   **8. a.** Paraboloid   **9. b.** Parabola

10. a. III   **11. a.**

x	0	0	5	⁻5
y	4	⁻4	0	0

13. b. 13.5

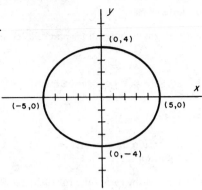

EXERCISE SET 8.2: Problem Solving

1. a. The slope is not defined.   **d.** *Hint:* Look for symmetry.

EXERCISE SET 8.2: Computers

2.
```
THIS PROGRAM PRINTS SOLUTIONS FOR Y = A*X^2 + B*X + C.
TYPE NUMBERS FOR A, B, AND C.
? 2, -5, -6
(-5, 69) (-4, 46) (-3, 27) (-2, 12) (-1, 1) (0, -6)
(1, -9) (2, -8) (3, -3) (4, 6) (5, 19)
```

4. 10 and 10

```
10  PRINT "1ST NO.","2ND NO.","PRODUCT"
15  PRINT
20  FOR X = 0 TO 20 STEP .5
30  PRINT X,20 - X,X * (20 - X)
40  NEXT X
50  END
```

EXERCISE SET 8.3: Applications and Skills

1. b. Linear function   **2. a.** 6.92 seconds   **b.** .12 seconds

3. a. 3.2 ounces cost 71 cents.   **b.** 20, 37, 54, 71, 88, etc.

4. b. Example 2 and Example 5.

5. b. Not a function.   **d.** Function.   **e.** Not a function.

6. a. £1.75 = $2.63   **b.** $5 = £3.33

8. a. 7 weeks (6th week to 13th week)   **9. a.** 280 million

10. a. Speed factor for force equals the square of the speed.　　**b.** 4

11. a. This graph's domain is the years from 1900 to 2100. The years from 1970 to 2100 represent prediction of the future.
　　c. Population: 1950 to 2050; pollution: 1960 to 2025; food supply is decreasing.

12. a. 2.71828183　　　　　　　　　　**b.** .049787068

EXERCISE SET 8.3: Problem Solving

1. a. 7　　　**2.** *Hint:* Use the method of finite differences on the numbers 1, 4, 10, 20, 35 and 56 to find a formula for the number of cannon balls.

EXERCISE SET 8.3: Computers

2. 963, 1380 and 1903

4. c. They are symmetric about the y axis.

5.
```
10  PRINT "THIS PROGRAM PRINTS ORDERED PAIRS (X,Y) FOR Y =
      EXP(X)."
20  FOR X =  - 3 TO 3 STEP .5
30  LET K =  EXP (X)
40  LET Y =  INT (10 * K + .5) / 10
50  PRINT "("X",  "Y") ";
60  NEXT X
70  END
```

EXERCISE SET 9.1: Applications and Skills

1. a. No.　　**c.** Translation　　　　**2. b.** Reflections or 180° rotations

3. b. Infinite number　　**d.** No　　　**4. a.** Yes　　**c.** Their areas are equal

5. a. 90°　　**c.** All points on line L　　**6. a.** 90°　　**7. a.** Over 6 and up 1

8. a. Translation　　**9. b.** It has twice as many degrees as $\angle POQ$.

10. a.

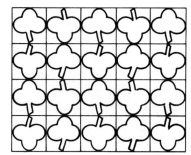

11. Reflect across the rows and down the columns.

12. **b.** $A'(3,3)$ $B'(0,^-3)$ $C'(5,^-4)$

14. **a.** Translation

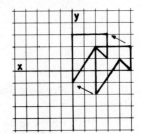

EXERCISE SET 9.1: Problem Solving

1. **d.**

3. **a.** Chinese ornament: Translation or glide reflection. Masonry fret: translation or horizontal reflection

5. *Hint:* The second die is one of the views. Make a model to find the other view.

EXERCISE SET 9.1: Computers

1. **b.**

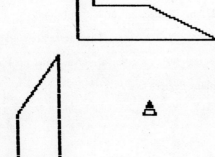

2. **c.**
```
TO REFLECTRIGHTVENT :A :B
    HIDETURTLE
    PENUP SETXY :A :B
     PENDOWN
    RIGHTVENT
    PENUP SETXY -:A :B
     PENDOWN
    LEFTVENT
END
```

4. A reflection about the x axis, if the turtle starts at home.

688   ANSWERS TO SELECTED EXERCISES

EXERCISE SET 9.2: Applications and Skills

1. a. 10,000   **c.** 6 cm

5. c. ¯18   **6. a.** Yes   **b.** Yes

7. a. $\triangle ABC$ is similar to $\triangle AEF$. Therefore, $\dfrac{AB}{AE} = \dfrac{BC}{EF} = \dfrac{AC}{AF}$. Furthermore, $\triangle ACD$ is similar to $\triangle AFG$. Therefore, $\dfrac{AD}{AG} = \dfrac{DC}{GF} = \dfrac{AC}{AF}$. All angles in the rectangles equal 90°.

b. Rectangle (2) is similar to rectangle (4).

8. a. Grid A to Grid B: scale factor $= \dfrac{3}{2}$.
Grid B to Grid C: scale factor $= \dfrac{1}{5}$.

c. $\dfrac{9}{4}$ or $2\dfrac{1}{4}$

9.

1	2	3
3 x 2 x 1	6 x 4 x 2	9 x 6 x 3
22 cm^2	88 cm^2	198 cm^2
6 cm^3	48 cm^3	162 cm^3

4. a.

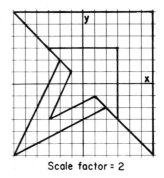

Scale factor = 2

10. a. Height: 320 cm; width: 192 cm
c. 512 kg

11. b. 1728

EXERCISE SET 9.2: Problem Solving

1. a. Approximately .65.

2. The original sheet is similar to the quarto; the folio is similar to the octavo.

4. The scale factor is 1 to 5.

EXERCISE SET 9.2: Computers

2. b. Five squares

5. a.
```
TO ENLARGE.TRIANGLE
   RT 40 FD 2.5 * 30 RT 110 FD 2.5 * 50 HOME
END

TO TRIANGLE
   RT 40 FD 30 RT 110 FD 50 HOME
END
```

EXERCISE SET 9.3: Applications and Skills

1. **a.** Figure 3 **c.** Figure 1 **e.** Figure 1 **3.** **a.**

4. **a.**

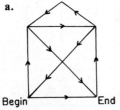 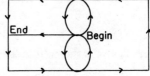

One odd vertex will be a beginning point and the other odd vertex will be an ending point.

5. **a.** Octahedron **b.** Three, connecting six different vertices.

6. Yes

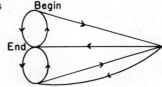

8. **c.** The point is inside if the line crosses the simple closed curve an odd number of times. Points of tangency should not be counted.

9. **a.** No **10.** **a.** 4

EXERCISE SET 9.3: Problem Solving

1. **d.** All the rooms must have an even number of doors.

2. **a.**

3. If there are one or more odd vertices, at least 3 colors will be needed.

4.

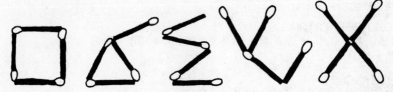

EXERCISE SET 9.3: Computers

2.

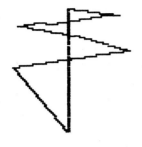

3. b. This curve was produced by C 5.
It is not a simple closed curve.

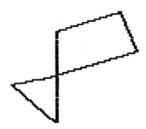

4. b. Not necessarily. If two random directions are the same, two lines will coincide.

EXERCISE SET 10.1: Applications and Skills

1. a. $\dfrac{15}{36}$ **b.** $\dfrac{15}{36}$ **2. b.** $\dfrac{1}{9}$ **3. b.** $\dfrac{4}{9}$

4.

	Tetrahedron	Cube	Octahedron
a. A number less than 3	$\frac{1}{2}$	$\frac{1}{3}$	$\frac{1}{4}$
b. An even number	$\frac{1}{2}$	$\frac{1}{2}$	$\frac{1}{2}$
c. The number 2	$\frac{1}{4}$	$\frac{1}{6}$	$\frac{1}{8}$

5. a. Probability is $\dfrac{1}{13}$. Odds are 1 to 12.   **c.** Probability is $\dfrac{1}{4}$. Odds are 1 to 3.

6. a. $\dfrac{12}{13}$ **c.** $\dfrac{3}{4}$ **7. a.** 7 to 3   **c.** 1 to 499   **8. a.** $\dfrac{4}{7}$

9. a. $\dfrac{9,929,200}{10,000,000} \approx .99$ **c.** $\dfrac{6,800,531}{9,396,358} \approx .72$ **e.** 14   **10. b.** $30

11. a. $1.58   **12. a.** $\dfrac{1}{38}$ **c.** Equal

13. b. Label 3 slips of paper with $1000, $0 and $0 and put them in a hat. Randomly select
one of the slips and repeat this process 8 times. Record the number of $1000 slips. Carry
out this experiment of 8 random selections many times and count the number of times that
exactly four $1000 slips are obtained. The probability is approximately .17.

EXERCISE SET 10.1: Problem Solving

1. a. 23   **d.** 15 sums. Probability $\frac{3}{16}$.

2. Label the faces of one cube with 0, 1, 2, 3, 4 and 5. Label the faces of the second cube with
1, 1, 1, 7, 7 and 7. Find other solutions.

4. No. Label 5 pieces of paper with 5000, 25, 5, 2 and 1 and place them in a hat. Obtain a sequence of 6 numbers by repeatedly selecting from these numbers. Repeat this experiment.

EXERCISE SET 10.1: Computers

2. The probability is approximately .62.

```
5    FOR S = 1 TO 50
10     FOR N = 1 TO 10
20     LET Z =  INT (10000 *  RND (1))
30     IF Z / 2 =  INT (Z / 2) THEN  GOTO 70
40     PRINT "G ";
50     NEXT N
55     PRINT
60     GOTO 90
70     PRINT "B ";
80     GOTO 50
90     NEXT S
100    END
```

3. b. The probability is approximately .17.

EXERCISE SET 10.2: Applications and Skills

1. a. $\left(\frac{18}{38}\right)^2 \approx .224$ **c.** $\left(\frac{18}{38}\right)^3 \approx .106$ **2.** $\frac{1}{2}$

3. a. Independent, $\frac{1}{8}$ **b.** Dependent, $\frac{1}{13} \times \frac{1}{17} \approx .005$

4. a. $\left(\frac{1}{12}\right)^2 = \frac{1}{144}$ **c.** $\left(\frac{11}{12}\right)^3 \approx .77$ **5. b.** $\frac{3}{8}$

6. b. The numbers in the tables are the fourth and fifth rows of Pascal's triangle.

    d. $\frac{5}{32}$

7. a. $\frac{10}{20} \times \frac{9}{19} \approx .24$ **c.** 1   **9. a.** (1) $\frac{1}{10}$ (2) $\frac{1}{5} \times \frac{1}{5} = \frac{1}{25}$

10. a. $1 - \left(\frac{5}{6}\right)^4 \approx .52$ **c.** $1 - \left(\frac{28}{36}\right)^3 \approx .53$ **11. b.** Approximately .59.

12. b. $\left(\frac{1}{365}\right)^2 \approx .0000075$ **13. a.** $\frac{1}{20}$ **c.** $\frac{1}{20} \times \frac{3}{20} \times \frac{1}{20} = .000375$

EXERCISE SET 10.2: Problem Solving

1. a. No.   **d.** Drawer, if there are more than 6 marbles.

2. 11/24

EXERCISE SET 10.2: Computers

2. Theoretical probabilities: $\frac{7}{8}, \frac{3}{4}, \frac{2}{3}, \frac{2}{3}, \frac{7}{8}, \frac{3}{4}, \frac{2}{3}$ and $\frac{2}{3}$

4. **c.** Approximately .49.

EXERCISE SET 10.3: Applications and Skills

1.

Interval	Frequency
7000–7499	1
7500–7999	0
8000–8499	7
8500–8999	8

2. Mode: 10,000 to 10,499
Median: 10,000 to 10,499
The mean income in the United States in 1981 does not fall in the mode or median intervals

4. **a.** Mode **c.** Median

5. **a.** 14 **c.** Mode or median

6. **a.** About 1.57 times longer for a new car; about 4.61 times longer for a sewing machine; 4 times longer for a pound of butter.

7. **a.** National **c.** National

8. **a.** 6 **c.** The test scores for the second class were closer to the mean of 6 and not as spread out as the test scores for the first class.

9. **b.** Great Britain

10. **b.** Approximately 94.5%

12. **a.** 16 to 329 **b.** 4770

13. **a.** Approximately 11.4. Label 5 slips of paper and place them in a container. Compute the average number of selections needed to obtain all 5 colors.

EXERCISE SET 10.3: Problem Solving

1. **a.** Yes **d.** If the guessed mean is too low the average of the differences between the six scores and this mean will increase the mean.

2. No.

EXERCISE SET 10.3: Computers

2.
```
80   LET V = (R / N) ^ .5
90   LET S =  INT (10 * V + .5) / 10
95   PRINT
100  PRINT "THE STANDARD DEVIATION OF THESE "N" NUMBERS IS
     "S"."
110  END
```

4. **b.** The average number of children before a child of each sex can be expected is 3.

```
10  FOR S = 1 TO 20
20  FOR N = 1 TO 9
30  PRINT  INT (9 *  RND (1)) + 1" ";
40  NEXT N
45  PRINT
50  NEXT S
60  END
```

EXERCISE SET 10.4: Applications and Skills

1. **a.** 8908 and 9042

   c. If the coins were tossed in a random manner and the outcomes were not affected by deformed or dirty coins, the conclusion was justified since 9207 is more than 3 standard deviations from the mean.

2. **b.** 23, four times; 66, once; 74, twice; 09, three times

3. **b.** Use groups of five numbers. The first three numbers determine a direction from 0 to 360 degrees. [If the number is greater than 360, subtract 360 (or 720) and use the remainder.] The second two digits determine the distance. (*Note:* There are other ways to answer this question.)

4. **a.** Use pairs of digits and add the two numbers. If the sum is even, the "toss" is heads, and if the sum is odd, the "toss" is tails.

5. **a.** 16% 7. **a.** 34 **c.** 1

8. **a.** Normal **c.** Skewed to the left **e.** Normal 10. **a.** 70% **b.** 95%

11. **b.** When a local percentile score is lower than the national percentile score it means that the local level of achievement is higher than the national level.

12.

Stanine	1	2	3
Lower Percentile	1	4	11
Upper Percentile	3	10	22

EXERCISE SET 10.4: Problem Solving

1. **a.** 1, 4, 6, 4 and 1 for compartments A, B, C, D and E, respectively. **d.** 1260

2. **a.**

A	B	C	D	E	F	G	H	I	J	K	L	M	N	O	P	Q	R	S	T	U	V	W	X	Y	Z
14	3	8	0	3	3	5	12	1	5	4	11	3	0	10	1	9	4	1	0	0	7	8	4	4	18

EXERCISE SET 10.4: Computers

1. **b.**
```
10   DIM A(1000)
20   FOR N = 1 TO 1000
30   LET A(N) =  INT (10 *  RND (1))
35   PRINT A(N)" ";
40   NEXT N
42   PRINT
44   FOR K = 0 TO 9
46   LET S = 0
50   FOR N = 1 TO 1000
60   IF A(N) = K THEN  LET S = S + 1
70   NEXT N
80   PRINT "THE NUMBER OF "K"'S IS "S"."
85   NEXT K
90   END
```

2. **b.** 50, 60, 70 and 80.

d. On the average each digit will occur the same number of times and the mean of the digits from 0 to 9 is 4.5.

3.
```
10   DIM A(500)
20   FOR N = 1 TO 500
30   LET X =  INT (6 *  RND (1)) + 1
40   LET Y =  INT (6 *  RND (1)) + 1
50   LET A(N) = X + Y
60   NEXT N
70   FOR K = 2 TO 12
80   LET S = 0
90   FOR N = 1 TO 500
100  IF A(N) = K THEN  LET S = S + 1
110  NEXT N
120  PRINT "THE NUMBER OF "K"'S IS "S"."
130  NEXT K
140  END
```

Index

segment, 200
English system of measure,
  265–266
ENIAC, electronic calculator, 82
Enlargement, 552–553
Equal additions method of
  subtraction, 121–122, 127
Equality:
  equations, 463
  of fractions, 327–328
  origin of symbol, 110
  of ratios, 425
  of sets, 53
Equations, 463–501
  circle, 485
  ellipse, 492, 500
  exponential, 511–512
  hyperbola, 494, 501
  line, 489, 497–498
  parabola, 491, 499
Equilateral triangle, 205, 215
Equivalent equations, 466
Equivalent sets, 53
Eramus, 103
Eratosthenes, 171
Erman, J.P.A., 150
Error analysis, 112, 125, 144,
  161, 358, 418
Escher, Maurits C., 227, 249,
  536, 537, 578
Euler, Leonard, 240, 579
Euler's formula, 240–241
Even and Odd (game), 5
Even number, 185
Event, 592
Even vertex, 579
Eves, Howard W., 66, 76, 77,
  110, 447
Expanded form:
  decimals, 393
  whole numbers, 70, 106
Exponent, 156–158
  negative, 429
  rule for adding, 157
  rule for subtracting, 157, 429
  scientific notation, 429
Exponential equation, 511–512
Exponential form, 157
Exponential function, 510–512,
  524–525
Expressible, 443

Factor, 168
  common, 182
  computer applications,
    187–188
  greatest common, 182, 184
  models, 168–169, 182–183
  proper, 187

Factor tree, 172, 184, 191
Faded document, 117, 148, 165
Fahrenheit, Gabriel, 270
Fahrenheit scale, 270, 279
Fair game, 596
Fathom, 265
Ferguson, D.F., 447
Fermat, Pierre, 485, 592
Fibonacci, Leonardo, 4–5
Fibonacci numbers, 4–5,
  339–340, 452
Finger counting, 76
Finger multiplication, 139, 145
Finite differences, 513–520,
  529
Finite set, 57
Fixed point, 534–535
Focus:
  ellipse, 492
  hyperbola, 494
  parabola, 490, 499
Folio, 570
Foot, 265
Four-Color Problem, 580, 586
Fourth dimension, 228
Fraction Bars,
  addition, 344–345, 363
  common denominator, 329
  division, 348–349
  equality, 328, 336, 340
  multiplication, 347–348, 350
  subtraction, 345–346, 363
Fractions,
  addition, 344–345
  approximation, 332–333,
    353–354
  Babylonian, 323–324
  calculator, 331, 337, 351–352,
    359
  common denominator, 329
  computer applications,
    333–334, 354–356
  converting to decimals, 331,
    351–352, 395
  denominator, 323–324, 329
  density, 330–331
  division, 348–349
  division of whole numbers,
    324–325
  Egyptian, 323
  equality, 327–329
  error analysis, 358
  Farey sequence, 342
  improper, 331–332
  inequality, 330
  inverses, 351
  lowest terms, 328
  mental calculation, 332–333,
    353–354

mixed number, 331–332,
  345–346
models, 325–327
multiplication, 347–348, 350
music, 360–361
number line, 327, 332,
  344–345, 356
numerator, 323–324
problem solving, 338–341,
  361–364
properties, 349–351
rounding off, 333, 338
subtraction, 345–346
unit, 323
Frame, M.R., 110
Fraser spiral, 14
Frend, William, 366
Frequency:
  distribution, 620–622, 630–631
  polygon, 621
Fujii, John, 586
Functions, 505–512
  domain, 507
  exponential, 510–512
  graphs, 509, 521–526
  linear, 509–510
  many-to-one, 522
  mapping, 532–533
  one-to-one, 522
  range, 507
  sequences, 514
  vertical line test, 509
Fundamental Theorem of
  Arithmetic, 172

Galaxy:
  in Andromeda, 2
  Messier, 56
  Milky Way, 157
Galilei, Galileo, 8, 264, 496, 499
Galton, Sir Francis, 653
Games, 128, 164, 571, 605
Gardner, Martin, 249, 550, 589
Gelosia algorithm, 134, 144
Gematria, 167–167
Generalized distributive
  property, 136
Genus, 578
Geoboard:
  circular, 257
  rectangular, 501–502
Geometric number, 6–7, 11
Geometric sequence, 6, 102
  common ratio, 6
Glide reflection, 536, 548, 550
Goldbach, C., 178
Golden Mean, The, 360
Golden ratio, 452
Golden rule, 426

standard deviation, 625, 642
stanine, 645, 653
Statistics: A Guide to the Unknown, 617
Stevin, Simon, 389
Stories about Sets, 61
Strategies in problem solving, 31–38
Stratified sampling, 627
Subset, 52–53
Subtraction,
  algorithms, 120–123, 407–408
  approximation, 123–124, 353, 413
  closure, 154, 455
  complementary method, 122
  computer applications, 124, 128–129
  decimals, 407–408
  decomposition method, 121
  distributive property, 137, 145
  equal additions method, 121–122
  error analysis, 125, 418
  fractions, 345–346
  integers, 370–371
  mental calculation, 123–124, 353, 413
  models, 120–121, 345–346, 370–371, 407–408
  number line, 126, 345
  problem solving, 128
  regrouping, 120, 408
  signed numbers, 370–371
  whole numbers, 119–129
Supplementary angle, 202, 239
Surface area:
  creating area, 310–311
  cylinder, 306–307
  prism, 306
  pyramid, 316
  scale factors, 558–560, 570
  sphere, 310
Swift, Jonathan, 569
Symbolic Logic and Game of Logic, 15
Symmetry,
  axis of rotation, 250, 535
  beauty, 255
  bilateral, 248–249
  center of rotation, 247, 535
  computer applications, 250–253
  image, 246, 248
  line of symmetry, 245–246
  mirror test, 256
  plane of symmetry, 248–249
  reflectional:
    in a plane, 245–246
    in space, 248–249

rotational:
  in a plane, 246–247
  in space, 250
  vertical symmetry, 249

Tallying, 48–49
Tally sticks, 48–49
Tangent, 491
Tartaglia, 188
Tchebyshev, 178
Temperature:
  Celsius scale, 270
  Fahrenheit scale, 270
  Kelvin scale, 270
Tennyson, Lord, 338
Terminating decimal, 331, 394
Tessellation:
  convex polygons, 217–219
  hexagons, 217
  nonconvex polygons, 219
  quadrilaterals, 219
  squares, 217
  triangles, 217–218
Tetrahedron, 228–229
Thales, 558
Theorem, 472–474
Theoretical probability, 592–593
Theory of Relativity, 228
Topological equivalence:
  computer applications, 580–583
  networks, 579–580
  number of surfaces, 577–578
  simple closed curves, 576
  topological mappings, 576
  torus, 575, 578–579
Topological properties:
  genus, 578
  number of sides, 577–578
Torus, 575
Transformation (*see* Mapping)
Translation, 533
Traversible network, 579–580
Trapezoid, 205
Tree diagram, 609
Triangle,
  altitude, 223
  angles, 215
  area, 286
  congruence, 532
  equilateral, 205, 215
  isosceles, 205
  median, 223
  right, 205
  scalene, 205
  similarity, 557
  tessellation, 217–218
Triangular number, 6
Tricks, 13–14, 45, 478–480
Triskaidekaphobia, 167

Undefined terms, 200, 228
Unexpected Hanging, The, 249, 550, 589
Uniform distribution, 640
Union of sets, 54–55
Unit fraction, 323
Unit of measure:
  ampere, 272
  area (*see* square unit)
  astronomical, 12–13, 423
  candela, 272
  degree Celsius, 270
  early units, 265–266
  gram, 266
  light-year, 435
  liter, 268
  meter, 268
  mole, 272
  second, 272
  volume (*see* cubic unit)
Unit pricing, 436
Universal set, 56
Upper bound, 288
Upper quartile, 644

Valid, 17, 21–23
Van Der Waerden, B.L., 163
Variables:
  algebra, 462–463, 476
  BASIC, 90–92
  dependent, 508
  independent, 508
  Logo, 87–89
Venn diagram, 16–25
Venn, John, 16
Vertex:
  angle, 201
  network, 579–580
  polygon, 204
  polyhedra, 228
  tessellation, 218
Vertical symmetry, 248
Vilenkin, N.Y., 61
Volume (*see also* Cubic unit)
  computer applications, 311–312
  cone, 308–309
  cubic unit, 303–304
  cylinder, 306
  irregular shapes, 310
  liter, 268
  prism, 304–305
  pyramid, 307–308
  sphere, 309–310

Walter, M., 260
Warner, Silvia Townsend, 199
Weight, 266, 269–270
Well-defined set, 51
Wells, H.G., 591

(**Acknowledgements** cont.)

Chapter 3

pg. 104 (top) B.C. by permission of Johnny Hart and Field Enterprises, Inc. **pg. 112** (top) Reprinted by permission from *More About Computers.* © 1974 by International Business Machines Corporation. (bottom right) Drawing by Dana Fradon; © 1976 The New Yorker Magazine, Inc. **pg. 119** (top) B.C. by permission of Johnny Hart and Field Enterprises, Inc. **pg. 130** Courtesy of N.Y. State Office of General Services.
pg. 138 (middle right) Courtesy of Deutsches Museum, Munich. **pg. 141** From the exhibition "Mathematica: a World of Numbers & Beyond", made by the Office of Charles and Ray Eames for IBM Corporation.
pg. 149 Courtesy of TEREX Division, General Motors.
pg. 157 (bottom) Reprinted by courtesy of Hale Observatories. **pg. 167** (top) B.C. by permission of Johnny Hart and Field Enterprises, Inc. **pg. 173** (top right) Reprinted by permission of the United Press International.
pg. 175 Reprinted with permission from Sidney Harris. **pg. 181** Courtesy of U.S. Bureau of Census.
pg. 185 (top) Reprinted by courtesy of Agencia J.B., Rio de Janeiro. **pg. 189** (top) Otis Elevator illustration reprinted from *Architectural Record,* March 1970. © 1970 by McGraw-Hill, Inc. with all rights reserved.
pg. 194 Reprinted from *Aftermath IV,* Dale Seymour et al. Courtesy of Creative Publications.

Chapter 4

pg. 198 Photo by B.M. Shaub. **pg. 199** (top right) Photo by B.M. Shaub. (middle) B.C. by permission of Johnny Hart and Field Enterprises, Inc. **pg. 201** (center) Hirshhorn Museum and Sculpture Garden, Smithsonian Institution. **pg. 202** (bottom right) Reprinted with permission from Sidney Harris. **pg. 207** (bottom) Reproduced from *Handbook of Gem Identification,* by Richard T. Liddicoat, Jr. Reprinted by permission of the Gemological Institute of America. **pg. 208** (top) Photo by B.M. Shaub. **pg. 210** (top right) Copyright © 1974 by Charles F. Linn. Reprinted by permission of Doubleday & Company, Inc. **pg. 214** (top) Courtesy of General Motors Research Laboratories. (bottom right) Photo by B.M. Shaub. **pg. 218** (top) Courtesy of MAS, Barcelona, Spain. **pg. 221** (exercise 1) Reproduced from *Art Forms in Nature,* by Ernst Haeckel (New York: Dover Publications, Inc., 1959). **pg. 225** (middle) Photo by Talbot Lovering. **pg. 227** "Cubic Space Division," by M.C. Escher. Courtesy of the Escher Foundation, Haags Gemeentemuseum, The Hague. **pg. 228** [figures (a) and (c)] Photos by B.M. Shaub. [figure (b)] Courtesy British Museum (Natural History). **pg. 229** (top) Photo by Talbot Lovering. (middle right) Reproduced from *Minerology,* by Ivan Kostov, 1968. Courtesy of the author.
pg. 230 (top) Reproduced from *Polyhedron Models for the Classroom,* by Magnus J. Wenniger. Courtesy of the National Council of Teachers of Mathematics. (middle right) Reproduced from *The Public Buildings of Williamsburg,* by Marcus Whiffen, published by the Colonial Williamsburg Foundation and distributed by Holt, Rinehart and Winston. **pg. 231** (bottom) Photo by B.M. Shaub. **pg. 232** (bottom right) Courtesy of National Aeronautics and Space Administration. **pg. 233** (bottom) Drawings reprinted with permission from *Encyclopaedia Britannica,* 14th edition, © 1972 by Encyclopaedia Britannica, Inc. **pg. 234** Reprinted with permission from *Collier's Encyclopedia.* © 1976 Macmillan Educational Corporation. **pg. 236** (exercise 1a) "Waterfall," by M.C.

Escher. Courtesy of the Escher Foundation, Haags Gemeentemuseum, The Hague. **pg. 239** (exercise 9) Courtesy of Babson College, Wellesley, Massachusetts. **pg. 240** (exercise 11) Courtesy of National Aeronautics and Space Administration. **pg. 244** Courtesy of Government of India Tourist Office, New York. **pg. 245** (bottom) Courtesy University of New Hampshire Media Services. **pg. 247** (middle) B.C. by permission of Johnny Hart and Field Enterprises, Inc. (bottom left and right) Reproduced from *Snow Crystals,* by W.A. Bentley and W.J. Humphreys. Reprinted by courtesy of McGraw-Hill Book Company, Inc. **pg. 248** (top right) Photograph reprinted by courtesy of The Magazine *Antiques.* (bottom left and right) Courtesy of Entomology Department, University of New Hampshire. **pg. 249** (top left and right) Courtesy of Lothrop's and Ethan Allen, Dover, New Hampshire. Photo by Ron Bergeron. (top middle) Courtesy of Jamaica Lamp Company, Queensville, New York. (bottom right) "Sphere with Fish," by M.C. Escher. Courtesy of the Escher Foundation, Haags Gemeentemuseum, The Hague. **pg. 250** (top right) Courtesy of Lothrop's Ethan Allen, Dover, New Hampshire. **pg. 254** Courtesy of Spanish National Tourist Office, New York. **pg. 255** (exercise 2) Courtesy of The American Numismatic Society, New York. (exercises 3a, b, c) Reproduced from *Art Forms in Nature,* by Ernst Haeckel (New York: Dover Publications, Inc., 1959). **pg. 256** (top right) photo by Ron Bergeron. **pg. 258** (exercise 12, left) Courtesy of Erie Glass Company, Parkridge, Illinois. (exercise 12, middle and right) Courtesy of Lothrop's Ethan Allen, Dover, New Hampshire. **pg. 258** (exercises 14a, b, c) Reprinted from *Early American Design Motifs,* by Suzanne E. Chapman (New York: Dover Publications, Inc. 1974) **pg. 259** (exercises 15a, b, c, d) Reprinted from *Symbols, Signs and Signets,* by Ernst Lehner (New York: Dover Publications, Inc. 1950). **pg. 260** (top) Photo by B.M. Shaub.

Chapter 5

pg. 264 (top) Courtesy of British Tourist Authority, New York. **pg. 265** (bottom right) From *The Book of Knowledge,* © 1960, by permission of Grolier Incorporated. **pg. 269** Reprinted from *Popular Science* with permission. © 1975 Times Mirror Magazines, Inc.
pg. 271 (top right) Courtesy of O.L. Miller and Barbara A. Hamkalo, Oak Ridge National Laboratory.
pg. 273 General Electric Research and Development. **pg. 274** (exercise 1 cartoon) Reprinted with permission from Sidney Harris. **pg. 276** (exercise 4) Photo by Ron Bergeron. **pg. 278** (exercise 11) Courtesy of Science Museum, London. **pg. 281** (exercise 3) Reprinted with permission from The Associated Press.
pg. 282 Courtesy of the Federal Reserve Bank of Minneapolis. **pg. 285** (top) Photos by Ron Bergeron.
pg. 286 (top) Photo by Talbot Lovering. **pg. 287** (bottom) Based on drawing from *Mathematics and Living Things,* Student Text, School Mathematics Study Group, 1965. Reprinted by permission of Leland Stanford Junior University. **pg. 288** (bottom) Reproduced from Encyclopaedia Britannica (1974), Volume VIII of the Micropaedia. Reprinted by permission of Encyclopaedia Britannica, Inc. **pg. 289** (bottom) Courtesy of Gemological Institute of America. **pg. 290** Photos by Ron Bergeron.
pg. 294 (exercise 1) Courtesy of Gunnar Birkerts and Associates, Architects. **pg. 296** (exercise 7) Reprinted by permission of Peachtree Plaza. **pg. 297** (bottom right)

Photo by Ron Bergeron. **pg. 299** (exercise 12) Courtesy of Dr. William Webber, University of New Hampshire. (exercise 13) Photo by Talbot Lovering. **pg. 301** (exercise 5) By John A. Ruge, reprinted from the *Saturday Review*. **pg. 303** Courtesy of General Dynamics, Quincy Shipbuilding Division. **pg. 306** (top right) Courtesy of Renaissance Center. **pg. 307** Reprinted by permission of Transamerica Corporation. **pg. 308** (middle right) Courtesy of Leslie Salt Co., Newark, California. **pg. 309** Courtesy of National Aeronautics and Space Administration. **pg. 310** (bottom left and right) Photos by Talbot Lovering. **pg. 313** (exercise 1) Courtesy of John P. Adams, University of New Hampshire. **pg. 314** (top) B.C. by permission of Johnny Hart and Field Enterprises, Inc. **pg. 315** (exercise 8) Collection: Mrs. Harry Lynde Bradley, Andre Emmerich Gallery. **pg. 319** (top) Photo by Ron Bergeron. (exercise 2) Courtesy of General Dynamics, Quincy Shipbuilding Division. **pg. 320** Photo by Herb Moyer, Exeter, New Hampshire. Reprinted by permission from Rodney Sanderson.

Chapter 6

pg. 322 Courtesy Edward C. Topple, N.Y.S.E. Photographer. **pg. 323** Reproduced by courtesy of the Trustees of the British Museum. **pg. 325** (bottom left) From *The Book of Knowledge,* © 1960, by permission of Grolier Incorporated. (bottom right) Official U.S. Navy photograph. **pg. 326** (lower rt.) Redrawn from The Encyclopedia Americana, © 1978 The Americana Corporation. **pg. 335** (exercise 1) Courtesy of Minolta Corporation. **pg. 340** Reproduced from *Patterns In Nature* by Peter S. Stevens. Atlantic Monthly/Little Brown, 1974. **pg. 343** (top, and bottom right) Courtesy of Rockwell International. **pg. 344** (top left, middle, and right) Illustrations from *Webster's New International Dictionary,* Second Edition © 1959 used by permission of G. & C. Merriam Co., publishers of the Merriam-Webster Dictionaries. **pg. 354** Courtesy of American Stock Exchange. **pg. 358** (exercise 7) Illustration from *Webster's New International Dictionary,* Second Edition © 1959 used by permission of G. & C. Merriam Co., Publishers of the Merriam-Webster Dictionaries. **pg. 360** (exercise 12) Drawing by Dana Fradon; © 1973 The New Yorker Magazine, Inc. **pg. 363** (exercise 3) Photo by courtesy of the Trustees of the British Museum. **pg. 365** B.C. by permission of Johnny Hart and Field Enterprises, Inc. **pg. 366** Redrawn from *Sports Illustrated.* © 1976 Time Inc. **pg. 367** (bottom) Courtesy of National Aeronautics and Space Administration. **pg. 368** Adapted from *The Book of Popular Science,* Courtesy of Grolier Incorporated. **pg. 379** (exercise 1) Official U.S. Navy Photograph. **pg. 383** (exercise 11) Redrawn from a photo by courtesy of National Aeronautics and Space Administration.

Chapter 7

pg. 388 Courtesy of Professor Erwin W. Mueller, The Pennsylvania State University. **pg. 400** (exercise 1) Courtesy of the National Bureau of Standards. **pg. 402** (exercise 5) Redrawn by permission from D. Reidel Publishing Company. **pg. 406** (top) Courtesy of Bulova Watch Co. Inc. **pg. 423** Courtesy of National Aeronautics and Space Administration. **pg. 424** (bottom) Photo by Herb Moyer, Exeter, New Hampshire. **pg. 427** Courtesy of the Stanford Research Institute, J. Grippo (Project Manager). **pg. 429** (top) Redrawn from data by courtesy of the Population Reference Bureau, Inc. **pg. 434** (exercise 2) Courtesy U.S. Department of Transportation and The Advertising Council. **pg. 436** (exercise 8) Reprinted by permission of the Pillsbury Company. **pg. 437** (exercise 9) Courtesy of National Aeronautics and Space Administration. **pg. 440** Reprinted from *Leonardo da Vinci* by permission of International Business Machines corporation. **pg. 442** Reprinted with permission from Sidney Harris. **pg. 444** (top right) Courtesy of Yale University. **pg. 445** (middle) Courtesy of Lehnert and Landrock, Cairo.

Chapter 8

pg. 426, 466 Reprinted by special permission of Sidney Harris. **pg. 475** Reprinted by permission of the publisher from *Math Puzzles* by Sam Loyd, editor Martin Gardner (New York: Dover Publications. Inc., 1959). **pg. 480** Reprinted from the *Arithmetic Teacher,* (October 1972) copyright 1972 by the National Council of Teachers of Mathematics, Inc. Used by permission. **pg. 482** Photo courtesy of American Institute of Steel Construction. **pg. 483** (top left) Courtesy of N.Y. State Office of General Services. (top right) Courtesy of Calvin Campbell, M.I.T. (middle) Drawing by Chas. Addams; © 1974 The New Yorker Magazine, Inc. **pg. 487** (top right) Courtesy of Daytona International Speedway. **pg. 490** (top right) Fountain at Swirbul Library, Adelphi University. Photo courtesy of Madeleine Lane, sculptor. **pg. 496** Reproduced from *The Earth and Its Satellite,* John E. Guest, Editor. Reprinted by courtesy of David McKay Company, New York. **pg. 499** (top) (exercise 8) Courtesy of General Dynamics. **pg. 503** (bottom) Reprinted from *Mathematics and Humor.* Edited by Aggie Vinik, Linda Silvey and Barnabas Hughes, © 1978 by the National Council of Teachers of Mathematics. Used by permission. **pg. 505** Courtesy U.S. Parachute Association. **pg. 521** Courtesy of Jim and Lisa Aschbacher. **pg. 522** (exercise 2c) Courtesy of Dr. Richard A. Petrie. **pg. 511** (top right, middle left and right) **and 525** Redrawn from *Mathematics and Living Things,* Teacher's Commentary, School Mathematics Group, © 1965, by permission of Leland Stnaford Junior University. **pg. 526** (bottom) Adapted from The Limits to Growth: A Report for The Club of Rome's Project on the Predicament of Mankind, by Donella H. Meadows, Dennis L. Meadows, Jørgen Randers, William W. Behrens III. A Potomac Associates book published by Universe Books, New York, 1972. Graphics by Potomac Associates. Permission from Universe Books.

Chapter 9

pg. 532 (top) Reprinted by permission Saturday Review, © 1976 & V. Gene Myers. **pg. 533** (bottom) right) Courtesy of U.S. Geological Survey. **pg. 534** (bottom) Courtesy of U.S. Department of State. **pg. 535** (bottom right) Courtesy of Travel Marketing, Inc., Seattle. **pg. 536** (top right) "Swans," by M.C. Escher. Courtesy of the Escher Foundation, Haags Gemeentemuseum, The Hague. **pg. 537** (top) Alhambra drawings by M.C. Escher. Escher Foundation, Haags Gemeentemuseum, The Hague. **pg. 538** (bottom right) Courtesy of

Rival Manufacturing Company. **pg. 539** (top right and bottom) Reproduced from the book, *Let's Play Math,* by Michael Holt and Zoltan Dienes. Copyright © 1973 by Michael Holt and Zoltan Dienes used by permission of publisher, Walker and Company. **pg. 544** (top) Courtesy of John P. Adams, University of New Hampshire. **pg. 545** (exercise 3) Courtesy University of New Hampshire Media Services. **pg. 550** (top) Reproduced by permission of the publisher from *Aesthetic Measure* by George D. Birkhoff (Cambridge, MA: Harvard University Press). © 1933 by the President and Fellows of Harvard College; 1961 by Garrett Birkhoff. **pg. 552** (top) Courtesy of J.E. Cermak, Fluid Dynamics and Diffusion Laboratory, Colorado State University. (bottom) Courtesy of Rockwell International Corporation. **pg. 556** Reproduced by permission of the National Ocean Survey (NOAA). U.S. Department of Commerce. **pg. 559** (top) Photo by Wayne Goddard, Eugene, Oregon. Description of knife reprinted by permission of the Register Guard, Eugene, Oregon. **pg. 560** Courtesy of Sally Ann Sweeney. **pg. 564** (exercise 1) Courtesy of Boeing. **pg. 568** (exercise 10) Courtesy of Sidney Rogers Chair Company, Georgetown, Massachusetts. **pg. 571** (exercise 4) Courtesy of Sally Ann Sweeney. **pg. 574** (top) Photo. Science Museum, London. (bottom right) British Crown Copyright. Science Museum, London. **pg. 575** (top right) and middle) Photos by Talbot Lovering. **pg. 576** (middle) Photos by Ron Bergeron. **pg. 578** (right)"Moebius Strip II," by M.C. Escher. Courtesy of the Escher Foundation, Haags Gemeentemuseum, The Hague. (left) Drawing from patent 3267406, filed by R.L. Davis. Reprinted by courtesy of Sandia Laboratories, Albuquerque, New Mexico. **pg. 586** (exercise 9) Redrawn from *Puzzles and Graphs,* John Fujii. Published by NCTM.

Chapter 10

pg. 595 (top) Courtesy of Joseph Zeis, cartoonist. **pg. 596** (top right) Photo by Ron Bergeron (bottom) Reprinted by permission of New Hampshire Sweepstakes Commission. **pg. 600** (exercise 2) B.C. by permission of Johnny Hart and Field Enterprises, Inc. **pg. 602 and 603** Mortality Table from *Principles of Insurance,* by Robert Mehr and Emerson Cammack, © 1966. Reprinted by courtesy of R.D. Irwin, Inc. **pg. 605** (top and bottom) Reprinted by permission of New Hampshire Sweepstakes Commission. **pg. 606** (top) Courtesy of National Aeronautics and Space Administration. **pg. 613** (exercise 4) Reprinted from *Ladies Home Journal,* February 1976, by courtesy of Henry R. Martin, cartoonist. **pg. 615** (exercise 12) Photo courtesy of The Edward Petritz Family, Butte, Montana. **pg. 619** (top) Drawing by Ziegler; © 1974 The New Yorker Magazine, Inc. **pg. 620** Reproduced from "Bills of Mortality" (p. 83) in *Devils, Drugs, and Doctors,* by Howard M. Haggard, M.D. Reprinted by permission of Harper & Row, Publishers, Inc. **pg. 622** (gestation table) Reprinted by permission of *The World Almanac and Book of Facts 1984.* Copyright © 1976 by Newspaper Enterprise Association, Inc., New York. **pg. 623** (earthquake table) Reprinted by permission of *The World Almanac and Book of Facts 1984.* Copyright © 1976 by Newspaper Enterprise Association, Inc., New York. **pg. 627** Drawing by Webber; © 1975 The New Yorker Magazine, Inc. **pg. 633** (exercise 6) From *The Peoples Almanac,* 1975 Edition by David Wallechinsky and Irving Wallace. Copyright © 1975 by David Wallechinsky and Irving Wallace. Reprinted by permission of Doubleday & Company, Inc. (exercise 7, home run table) Reprinted by permission of *The World Almanac and Book of Facts 1984.* Copyright © 1976 by Newspaper Enterprise Association, Inc., New York. **pg. 639** Reproduced from the collection of the Library of Congress. **pg. 640** Reproduced from *A Guide to the Unknown,* by Frederick Mosteller et al., 1972. Reprinted by courtesy of Holden-Day, Inc. **pg. 644** (bottom) Redrawn from the Differential Aptitude Tests. Copyright © 1972, 1973 by The Psychological Corporation. Reproduced by special permission of the publisher. **pg. 648** (exercise 1) Reprinted by courtesy of *Washington Post.* **pg. 652** (exercise 11) Reproduced by permission from the Stanford Achievement Test, 7th edition. Copyright © 1982 by Harcourt Brace Jovanovich, Inc. All rights reserved. **pg. 653** (exercise 1) From the exhibition "Mathematica: a World of Numbers & Beyond" made by the Office of Charles and Ray Eames for IBM Corporation.